REVISED
NUFFIELD CHEMISTRY TEACHERS' GUIDE II

Published for the Nuffield Foundation
by Longman Group Limited

Longman Group Limited London
Associated companies, branches, and representatives throughout the world

First published 1966
Revised edition first published 1978, reprinted 1979, 1980
Copyright © The Nuffield Foundation, 1966, 1978

Design and art direction by Ivan and Robin Dodd

Filmset in Monophoto Times New Roman 327
by Photoprint Plates Limited, Rayleigh, Essex
and made and printed in Great Britain
by Butler & Tanner Ltd, Frome and London

REVISED
Nuffield Chemistry
TEACHERS' GUIDE II

ISBN 0 582 04636 X

General editor,
Revised Nuffield Chemistry:
Richard Ingle

Editor of this volume:
B. E. Dawson

Associate editor:
B. J. Stokes

Contributors:
E. H. Coulson
B. E. Dawson
A. M. Dempsey
H. F. Halliwell
Richard Ingle
M. J. W. Rogers
B. J. Stokes
G. Van Praagh
M. D. W. Vokins

Contents

Foreword

It is now more than ten years since the Nuffield Foundation undertook to sponsor curriculum development in science. The subsequent projects can now be seen in retrospect as forerunners in a decade unparalleled for interest in teaching and learning not only in, but far beyond, the sciences. Their success is not to be measured simply by sales but by their undoubted influence and stimulus to discussion among teachers – both convinced and not-so-convinced. The examinations accompanying the schemes of study which have been developed with the ready cooperation of School Certificate Examination Boards have provoked change and have enabled teachers to realize more fully their objectives in the classroom and laboratory. But curriculum development must itself be continuously renewed if it is to encourage innovation and not be guilty of the very sins it sets out to avoid. The opportunities for local curriculum study have seldom been greater and the creation of Schools Council and Teachers' Centres have done much to contribute to discussion and participation of teachers in this work. It is these discussions which have enabled the Nuffield Foundation to take note of changing views. correct or change emphasis in the curriculum in science, and pay attention to current attitudes to school organization. As always, we have leaned on many, particularly those in the Association for Science Education who, through their writings, conversations, and contributions in other varied ways, have brought to our attention the needs of the practising teacher and the pupil in schools. This new edition of the Nuffield Chemistry materials draws heavily on the work of the editors and authors of the first edition to whom an immense debt is owed. The work leading to the first edition, published in 1966, was directed by Professor H. F. Halliwell, organizer of the Chemistry Project which carried out the trials in schools of the original draft materials. The editors of the first publications were:

H. F. Halliwell
Introduction and Guide
Book of Data

M. J. W. Rogers
The Sample Scheme Stages I and II: The Basic Course

G. Van Praagh
The Sample Scheme Stage III: A Course of Options
Laboratory Investigations

B. J. Stokes
Collected Experiments

E. H. Coulson
Handbook for Teachers

H. P. H. Oliver
Background Books

The new edition contains a preponderant part of these authors' material, either in its original form or in edited versions. They have all acted as consultants on the course the revision should take. They are credited among the authors of the new edition but their wider contribution in providing a firm basis for further developments must be gratefully acknowledged here.

I particularly wish to record our gratitude to Dr Richard Ingle, the General Editor of this new series. It has been his responsibility, together with Professor E. H. Coulson as consultant, to organize and coordinate this revision and it is largely through their efforts that we have been able to ensure the fullest cooperation between teachers and the authors. Thanks are due also to the Consultative Committee under the chairmanship, first, of the late Professor Sir Ronald Nyholm and, since his death, of Professor J. Millen, and especially to the group of practising teachers which met regularly to hear reports on the progress of the revision and to give their advice. This group consisted of:

H. S. Finlay
A. D. Gazard
J. A. Hunt
David J. Keeble
Michael Shayer
B. J. Stokes

As always I should like to acknowledge the work of William Anderson, our publications manager, his colleagues, and, of course, our publishers the Longman Group Ltd for their continued assistance in the publication of these books. The editorial and publishing contribution to the work of the projects is not only most valued but central to effective curriculum development.

K. W. Keohane
Coordinator of the Nuffield Foundation Science Teaching Project

Acknowledgements

The Editor of this volume of the *Teachers' guide* wishes to record his thanks to all those teachers who responded to various enquiries about the use of Nuffield Chemistry materials in secondary schools, to other teachers who contributed specialist studies related to this revision during a period of study at one of a number of centres, to all who helped in any way with small scale trials of materials associated with this revision, and to others who contributed their expertise in devising and testing new apparatus and experiments. A list of names appears below.

The continuous development of public examinations associated with Nuffield Chemistry has played no small role in guiding the interpretation of the proposals since 1965, through the arrangements made with the University of London School Examinations Department in close association with other GCE examination boards, and acknowledgement is made to those most closely concerned with such work.

Grateful acknowledgement is also made to Mr W. Anderson, Mrs H. Ellis, and Mrs D. Williams, and their colleagues, of the Publications Department of the Nuffield Foundation Science Teaching Project for their advice and expert assistance at all stages of the production of this book.

Dr A. W. B. Aylmer-Kelly
Dr D. Barlex
Mrs A. M. Black
W. Bleyberg
P. Borrows
D. Bradford
Dr H. D. Campbell-Ferguson
R. B. Chenley
Miss B. A. Clarke
Professor E. H. Coulson
P. Craggs
Dr A. M. Dempsey
Mrs S. Dempsey
Mrs J. Elson
H. Finlay
Dr A. D. Gazard
A. P. Giddings
Miss C. Groves
Professor H. F. Halliwell

J. Head
D. C. Hobson
Dr M. Hudson
P. G. Hudson
J. A. Hunt
Dr G. Huse
Dr Richard Ingle
E. W. Jenkins
Mrs J. R. Kalicki
D. J. Keeble
A. J. Malpas
Dr A. L. Mansell
D. H. Mansfield
Professor D. J. Millen
Miss B. J. Myers
The late Sir Ronald Nyholm
H. P. H. Oliver
C. Othen

C. V. Platts
J. R. Pounds
Dr D. Redshaw
G. P. Rendle
I. F. Roberts
M. J. W. Rogers
M. Shayer
D. A. Stephens
B. J. Stokes
D. Tee
R. W. Thomas
P. Thompson
C. W. Thorpe
M. T. Thyne
Dr R. Tremlett
Dr G. Van Praagh
M. D. W. Vokins
Dr D. Wharry
W. J. U. Woolcock

Editor's Preface

This book has been made possible through the combined suggestions of many teachers obtained either as a result of enquiries through questionnaires and interviews, or from detailed studies and trials of materials for use in Stage II Nuffield Chemistry. The presence of alternative teaching approaches may convey an impression that the Basic course has been extended. This is not the case. Several variations relate to areas in which teachers have reported that their pupils experience difficulty – such as the mole (gram-atom) concept, the writing of chemical equations, and energy considerations. The text also contains commentaries at various points on the use of mathematics in the presentation of chemical concepts.

The Nuffield Chemistry proposals described in this *Teachers' guide* were first published in 1966 as a contribution to the reappraisal of the place of science in the curriculum that was then taking place. The teaching schemes proposed were intended for children of average or above average ability in the 11–16 year age group who were expected to take an examination equivalent in standard to G.C.E. O-level, and were based on the Policy Statement of the Association for Science Education (then the Science Masters' Association and the Association of Women Science Teachers) and the work of the Chemistry Panel of that association (1961). Since the proposals were first published, they have been adopted by many teachers, and in this second edition a substantial revision has been undertaken in the light of the experience that has been gained.

The principal points of revision are:
1. The range of publications for both teachers and pupils has been restructured: the new collection of books and other materials is described in Chapter 4 of this volume.
2. Although the main outlines of the teaching schemes have been retained, many amendments have been made to points of detail and two schemes are offered for Stage II (see Chapter 3).

This volume is the second of three self-contained volumes of the new *Teachers' guide*. It is itself in three parts. In Part 1, the approach to the teaching of chemistry is discussed briefly together with other related themes, such as the aims and objectives for Stage II, and an outline of the content of the two schemes.

The second part of the book describes in detail two alternative schemes which may be followed during the third, fourth, and part of the fifth year of a five-year course. Alternative IIA uses the same general order as that adopted for Stage II of the original (1966) Sample Scheme, together with such modifications as experience has shown to be necessary (e.g. a delayed introduction of conceptually demanding material). Alternative IIB also observes these modifications but is more radical in its approach. In addition, this second scheme provides for an additional possible terminal point for the course at the end of the third year, thereby meeting the needs of schools who do not require all pupils to study chemistry for five years. The content of both schemes is approximately the same, thereby providing another illustration of the flexibility of the original proposals and of the scheme of public examinations. Both alternative schemes follow the general pattern adopted for the first Sample Scheme in that they offer teachers detailed suggestions on presentation and exact detail for suggested experiments. As in the first edition, these schemes probably contain more detail and information than any one teacher may require.

It is necessary to emphasize that any one of the schemes must be regarded as an interpretation, a sample scheme, written for the guidance of those teachers who want them. In no way are they intended to limit those other teachers who like to teach chemistry using the same principles but with the content designed differently and yet drawing from the suggestions contained in this guide.

In preparing this three-volume guide, three main groups of teachers have been in mind. There are those who will read Part 1 of each book and decide to produce their own scheme based on the principles outlined. Others may produce their own scheme using the Topics described in Part 2 but in a different order and incorporating those sections that they find most useful. Yet others may prefer to accept one of the schemes as it stands, developing their own variations and amendments as they proceed.

Part 3 of this book consists of a series of appendices which relate to Stage II of Nuffield Chemistry.

The first volume of this *Teachers' guide* describes the teaching schemes proposed for Stage I and is also divided into three parts, as described above. *Teachers' guide III* covers the Options for Stage III. Each volume is complete in itself.

Part 1 Introduction

Chapter 1

A modern approach to the teaching of chemistry

The nature and extent of the Nuffield Chemistry proposals are discussed at length in volume I of the *Teachers' guide*. However, not all teachers who seek to use Stage II materials will have used Stage I materials with pupils, and their pupils may have received a different and somewhat broader introduction to science. Accordingly, this chapter relates primarily to the content and approach adopted in Stage II. This stage follows immediately from some introductory study on the exploration of materials and extends over a period of rather more than two years.

Teaching chemistry at this level requires a great deal of skill. The tasks involved include the formation of pupils' attitudes, conveying knowledge about chemical aspects of the material world, showing others how to use ideas about atoms and particles, and involving young people in the process of enquiry. In addition, such tasks must necessarily require the teacher to indicate the relevance of the work done to daily life and to other studies undertaken by pupils elsewhere in the school, and to discuss with pupils social and moral issues, as and when the occasion demands. In so many ways, the teaching of chemistry differs from the practice of chemistry in the professional sense and many would claim that it is equally demanding.

In almost every lesson, pupils need to be given opportunities to learn to distinguish between observed fact and explanation, to learn to appreciate how chemistry develops through an interplay of observed fact and the ideas which may be expressed in the form of hypotheses or laws. The pupil learns something of the contribution of chemical science to our society. In undertaking such studies, pupils should find that imagination plays an important role in the formulation of scientific explanation. So, we hope that teachers will encourage their pupils to participate in discussions about the variety of situations presented in the course materials and to suggest possible explanations and additional experiments. Clearly this type of activity requires practice before precision can be acquired in using the ideas the chemist has about atoms and particles – the essence of Nuffield Chemistry at Stage II.

In broad outline, Stage II sets about establishing an interpretation of chemistry as an activity which depends on observable changes in materials which, in turn, can be interpreted simply and generally in terms of a model. Towards the end of Stage II, this level of interpretation is extended to include the energy changes which accompany changes in materials. The relationships that exist between these three types of activity can be summarized as shown in figure 1.

Figure 1

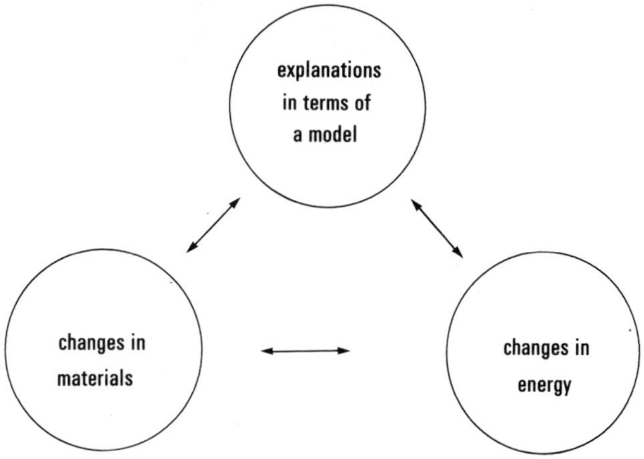

When attempting to convey this interpretation of chemistry to pupils, it is all too easy to become over-precise and to lose personal enthusiasms for specific parts of the subject matter in an elementary course. Also, the literature contains many reports of teachers who find it difficult to interpret such an approach. The literature also contains other reports commenting on the value of this approach to teaching. That arch-advocate of heurism, the late Professor H. E. Armstrong, had much to say on this matter and anticipated a great deal of the substance of current correspondence. For example, he wrote:

> The greatest nonsense is talked about the impossibility of making the student discover everything for himself: no one asks that he should or believes that he can, only that he should learn how discoveries are made. . . .

To-day we might hope that this is accepted and applied in a variety of contexts in secondary education! It follows that the presentation of chemistry at this level will require the teacher to adopt a variety of roles to suit specific occasions: he will lead some enquiries; he will demonstrate on other occasions; he will question and will listen to comments and observations made by pupils; and, he will guide his pupils' reading about science and the activities of scientists. Indeed, we could go even further: the teacher must be true to himself *if* he is to convey conviction, sincerity of purpose, and enthusiasm for his subject.

Time allocation In planning the amount of work to be done, the allocation of teaching time adopted has been that recommended in the Policy Statement of the Association for Science Education, published in 1961: for a five-year course leading to a public examination at 16 +,

Years 1 and 2 (which correspond to Stage I)	1 double period in a laboratory, and 1 homework period per week
Years 3, 4, and 5 (which correspond to Stages II and III)	1 double period and 1 single period in a laboratory and 1 homework period per week

A school teaching year has been taken to consist of about 30 weeks.

Chapter 2

The aims and objectives of Stage II

General aims

The Nuffield proposals include a set of general aims to be achieved in the teaching of chemistry. These aims are not equally appropriate for all stages of a five-year course. Emphasis must change with time to meet specific needs as pupils develop and acquire knowledge and attitudes. In *Teachers' guide I*, emphasis was given to the development of materials to illustrate those aims considered to be most appropriate for Stage I of the course. The requirements of Stage II are more extensive but still dependent upon the foundations provided by Stage I or by some quite different introduction, such as that given by the Nuffield Combined Science course. A formulation of general aims for Stage II can be helpful for the teacher and assist him to avoid gaps in an overall provision for his pupils. Thus, during Stage II, it is proposed that the pupils should:

1. Acquire basic knowledge about the behaviour of substances.
2. Seek patterns in the behaviour of substances.
3. Develop concepts related to the classification of substances and their observable behaviour.
4. Seek interpretations of all the experience acquired in terms of an appropriate model.
5. Test interpretations of phenomena by further exploration which may lead to an adjustment of the original interpretation.
6. Relate interpretations and observations to associated changes in energy for a given chemical system.
7. Develop manipulative skills in laboratory procedures using simple apparatus.
8. Gain confidence in learning by discovery and in applying knowledge.
9. Record their work in a personal style suited to their maturity and in an accurate manner.
10. Find out about the sources and uses of the substances they meet during the course, and any hazards associated with them.
11. Relate their personal experience of chemical phenomena to the needs of society.

Objectives and assessment

The general aims stated in the previous section are broad statements of intent. They may be distinguished from *objectives* which provide a clearer picture of what a pupil should be able to do, or understand, after doing a piece of work. In this book, we shall distinguish between *general objectives*, which are mentioned in this section, and the *objectives for pupils*, which are given at the start of each section of Part 2. For example, the general aim quoted as number 1 above, 'Acquire basic knowledge about the behaviour of substances', is related to the general objective for the course, 'Facility in recalling information and experience'. In addition, there are a variety of specific objectives from a large number of sections of course materials in Part 2 which require pupils to acquire knowledge of a range of chemical phenomena and processes. If the pupil has achieved these objectives, he should be able to remember that knowledge (a general objective), and for any particular section of the course recall specific information about a chemical or a process (a specific objective). Since the publication of the original *Sample Scheme*, many teachers have become aware of the value and range of educational objectives which can be applied to the interpretation of a course. One such taxonomy of objectives covers needs in the cognitive and affective domains and some readers may wish to refer to this collection* in order to extend their knowledge of educational objectives, and learn something about their formulation and systematic presentation.

A general statement of objectives of the Nuffield proposals now follows. As stated in *Teachers' guide I*, the first five are the most important at Stage I level or its equivalent. The remaining three are of lesser importance at Stage I but should not be overlooked. At Stages II and III, all eight general objectives are relevant and, as will be seen from Appendix 3, all play an important part when examination papers are devised to assess the progress of pupils.

1. Facility in recalling information and experience.
2. Skill in handling materials, manipulating apparatus, carrying out instructions for experiments, and making accurate observations.
3. Skill in devising an appropriate scheme and apparatus for solving a practical problem.
4. Skill in handling and classifying given information (including, in Stage II, graphical representations and quantitative results).
5. Ability to interpret information with evidence of judgment and assessment.
6. Ability to apply previous understanding to new situations and to show creative thought.
7. Competence in reporting on, commenting on, and discussing matters of simple chemical interest.
8. Awareness of the place of chemistry among other school subjects and in the world at large.

The more detailed and specific *objectives for pupils*, given at the beginning of each section of Part 2 of this book, are offered in the hope that they will assist teachers to focus attention on what is required in that particular section. Like other aspects of the Nuffield proposals, such indications must not be regarded as invariable: lessons may develop in ways other than those described and achieve equally worthwhile objectives. Indeed, objectives should be varied by teachers who wish to teach according to the Nuffield proposals but develop their own schemes of work to achieve the same general aims.

*Association of College and University Examiners (ed.) Bloom, B. S. (1965)
A taxonomy of educational objectives: Handbook I—Cognitive domain. Longman.
Association of College and University Examiners (ed.) Bloom, B. S., Krathwohl, D. R., and Masea, B. B. (1965) *A taxonomy of educational objectives: Handbook II—Affective domain.* Longman.

It was suggested in Chapter 1 that much of the intent of Nuffield Chemistry depends on the enthusiasms and abilities of teachers to encourage their pupils. Analyses of the kind mentioned in this chapter can do much to direct the attention of teachers to basic requirements of a course but do little to convey the total requirement and design of a course. Educational objectives do not provide a basis for a complete analysis of a curriculum: they are no more than aids to an understanding of specific needs. They are also useful guides to assessment procedures but even here they do not provide a complete prescription.

Perhaps the chief danger of accepting an objectives model for a course of study is that it may trivialize the education that emerges from it. Again if we attach too much importance to evaluating what we can specifically predict, there is considerable danger that we will tend to teach more and more predictable specifics and fewer of those aspects of chemistry that are really important but more difficult to measure. Accordingly, it is suggested that teachers use both the general aims and objectives mentioned in this chapter, and the specific objectives for pupils in Part 2, as *indicators* rather than complete descriptions of intent for Stage II.

Perhaps the two opposing views of the role of aims and objectives in a programme such as Nuffield Chemistry might be summed up as follows. *Either* one may regard the teacher as a provider of stimuli to which the pupil must respond (i.e. a behaviourist view of learning) *or* one may regard education as not being totally concerned with behaviours but rather with making a change in behaviour a possibility. This second view provides for the reality that no two pupils will be sufficiently alike to warrant their being treated in an identical manner. To provide a detailed guide for teachers, using the ideas of Nuffield Chemistry so as to meet such situations, would be difficult if not impossible. However, it is hoped that teachers will adapt the suggestions to suit the conditions and situations that obtain in schools rather than use only those ideas and details that can be conveyed by the printed word.

Chapter 3

The content of Stage II

This stage of *Revised Nuffield Chemistry* is intended to be a basis for courses with pupils in the 13 to 16 year age group. The school teaching year has been assumed to consist of 30 weeks, and Stage II requires about 7 school terms, given a lesson allocation of one double and one single period each week. Stage II is designed to follow *either* Stage IA or IB of Nuffield Chemistry, *or* Nuffield Combined Science, *or* other comparable course materials of an introductory and exploratory nature for 11 to 13 year old pupils of average or above average ability.

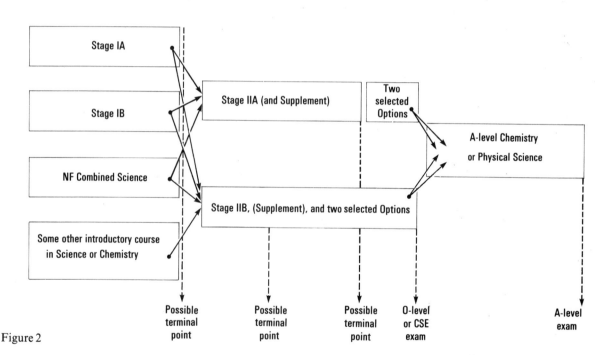

Figure 2

Stage II is concerned with the ideas that chemists use, including atomic theory, structure, and energy. It provides an introduction to the skills required to use chemicals and apparatus. In addition, it is a basis for optional studies, such as the *Options* of Stage III of the course. The Revised Nuffield Chemistry Scheme can be either a terminal course or a preparation for more advanced studies in chemistry, as indicated in figure 2.

Alternative IIA

Alternative IIA uses the same topic order in Stage II as the first edition of this guide, *The Sample Scheme*, but it contains some new experiments and a reorganization of the material relating to difficult concepts. These changes are based on an extensive analysis of comments and suggestions by teachers who have used the original materials.

Alternative IIA opens with a short discussion about the particulate nature of matter. In the main route through Topic A11, the mole (or gram-atom) is introduced and used to determine the formulae of some binary compounds. An alternative route is also provided, in which evidence for the particulate nature of matter is used as a rationale for Daltonian-style simplification rules which are in turn used for the determination of formulae. The 'salt-gas' problem follows (A12) and is used to extend pupils' knowledge of the composition of simple compounds. It also serves as a means for extending pupils' appreciation of the terms 'element' and 'compound'. In this way, the first two Topics provide a preparation for a series of lessons on the Periodic Table in Topic A13, in which pupils explore the properties of various groups of elements. If the alternative route were followed in A11, the mole (gram-atom) should be introduced in section A14.1. The two routes converge in section A14.2, where the different volumes of moles (gram-atoms) of solid elements lead into a discussion of packing, and hence of structure. The structures of some simple compounds are also studied in A14. Topic A15 is about the three states of matter and the energy transfers required for changes of state.

An opportunity is provided for pupils who have not studied electricity to become acquainted with some simple ideas about the subject, including electrical circuits, at the beginning of Topic A16. The restructured Topic also includes some discussion on the structure of the atom and on chemical bonding which are seen to be dependent on the main theme of the Topic, the behaviour of electrolytes. The problem of finding the relative number of particles involved in reactions forms the basis of Topic A17. Other matters which receive attention include ways of expressing the concentrations of solutions and a formal derivation of chemical equations, including ionic equations, from experimental evidence.

With this background knowledge of chemistry, more detailed studies of chemical systems now follow and include an appreciation of reaction kinetics in Topic A18; chemical equilibrium in Topic A19, and acidity and basicity in Topic A20. Topic A21 provides an introduction to the study of macro-molecules and some simple molecular carbon compounds. The chemistry of ammonia and its importance for solving the world food problem follow in Topic A22.

The final Topics are A23, on energy changes in chemical systems, and A24, on radiochemistry. These Topics are not developed in isolation from the others in Alternative IIA but serve as a means for reviewing and extending aspects of chemistry introduced during the course. The treatment of these topics differs from that given in the *Sample Scheme*.

This general outline of Alternative IIA is now reviewed in greater detail.

An outline of Alternative IIA

Topic A11   *Atoms in Chemistry*
A11.1 What evidence have we that matter is made up of particles?
A11.2 How big are the particles?

A11.3 Comparing numbers of atoms by mass measurement
A11.4 Determination of the formulae of simple compounds

Alternative route: omitting A11.3
A11.4 Determination of the formulae of simple compounds without using the mole concept

Two approaches are offered for this Topic: the main approach introduces and uses the mole (gram-atom) concept and in broad outline follows the original Sample Scheme. The alternative route uses another approach to the derivation of the formulae of some binary compounds by requiring simplification rules not unlike those put forward by Dalton. If the alternative route is adopted, the term mole (gram-atom) can be deferred until Topic A14. *Suggested time allocation: 4–5 weeks.*

Topic A12   *Investigation of 'salt gas'*
A12.1 What are the properties of 'salt gas'?
A12.2 What elements are present in 'salt gas'?
A12.3 Is 'salt gas' a compound of hydrogen and chlorine only?
A12.4 What is the formula of hydrogen chloride?

In this revision, the Topic is presented as a 'bridge' between A11 and A13. Also, the terms analysis, synthesis, element, and compound are revised and extended. A new experiment requiring the analysis of hydrogen chloride gas avoids the need to synthesize this gas from its elements in the presence of a platinum catalyst, as suggested in the original Sample Scheme. *Suggested time allocation: 2 weeks.*

Topic A13   *Looking at the elements in the light of the Periodic Table*
A13.1 The story of the Periodic Table
A13.2 The alkali metals
A13.3 The halogens (and inert gases)
A13.4 The transition metals
A13.5 Carbon: an element found in living things
A13.6 Silicon: an element found in many rocks

A historical development of the ideas which led to the Periodic Table is used to introduce the Topic. The properties of elements in vertical groups are used to indicate the value of this classification.

The properties of some transition elements are compared and the possibilities of 'horizontal' relationships explored simply. Some studies on the chemistry of carbon and silicon complete the Topic. *Suggested time allocation: 6 weeks.*

Topic A14   *Finding out how atoms are arranged in elements and other substances*

> *Alternative route*
> A14.1 Comparing numbers of atoms by mass measurements
> *(relates only to the alternative route for Topic A11)*

A14.2 Looking at the volumes occupied by 1 mole of atoms (gram-atom) of several solid elements

A14.3 Watching crystals grow

A14.4 The problem of trying to investigate things too small to be seen

A14.5 How are atoms arranged in metals?

A14.6 The importance of the structure of metals to engineers

A14.7 The effect of packing sulphur atoms in different ways

A14.8 Another element which has different forms: carbon

A14.9 What can we find out about the structure of gases?

A14.10 Differences between giant and molecular structures

A14.11 The arrangement of atoms in compounds, such as magnesium oxide and sodium chloride

A14.12 Starting to write chemical equations

If the alternative route was followed in A11, this Topic opens with the presentation of the mole (gram-atom) concept. Both sequences continue with the packing of atoms in solid elements and in compounds, using various analogues. The structure of metals and the implications of this work for the engineer are considered next. An investigation of the structure of gases leads to an understanding of the term molecule, and some giant and molecular structures are compared. A study of the structures of simple compounds is used to introduce chemical equations as summaries of chemical reactions. Various appendices on teaching aids are provided. *Suggested time allocation: 7–8 weeks.*

Topic A15   *Solids, liquids, and gases*

A15.1 The kinetic theory of matter

A15.2 Finding out how much energy is needed to melt solids and to vaporize liquids

A15.3 Using data on heats of vaporization to provide information about structure.

Some evidence to support a kinetic theory of matter is given and the kinetic theory is used to account for differences between the three states of matter. Is energy needed to melt a solid and to vaporize a liquid? Subsequent discussion reveals a relationship between the heat of vaporization and the structure of a substance (i.e. whether it has a giant structure or a molecular structure). *Suggested time allocation: 4 weeks.*

Topic A16   *Explaining the behaviour of electrolytes*
        A16.1 Elementary electricity
        A16.2 An investigation of the electrolysis of lead bromide
        A16.3 Looking at the migration of ions
        A16.4 How much electricity is needed to deposit 1 mole of atoms
        (gram-atom) of lead, copper, and silver?
        A16.5 Investigating the electrolysis of solutions of electrolytes
        A16.6 Is the same quantity of electricity always needed to liberate
        1 mole of atoms (gram-atom) of copper?
        A16.7 Implications of the ionic theory for the structure of the atom
        A16.8 Molecules, giant structures, and bonding

        Provision is made in section A16.1 for those pupils who have not
        studied electricity before doing this work and it includes a descrip-
        tion of conductors and non-conductors and the concept of charge.
        An investigation of the electrolysis of fused lead bromide leads to
        a study of the migration of ions and to the quantitative investigation
        of electrolysis. The evidence obtained is used to introduce a *simple*
        interpretation of atomic structure and the properties of elements
        and compounds in terms of their structures. *Suggested time alloca-
        tion: 6 weeks.*

Topic A17   *Finding the relative number of particles involved in reactions*
        A17.1 A review of terminology: the mole as a unit
        A17.2 Reactions involving the formation of a precipitate
        A17.3 Reactions involving gases
        A17.4 A replacement reaction between a metal and a salt
        A17.5 Some uses of chemical equations

        By this point in the course (which approximates to the beginning of
        the fourth year of a 5-year course), it is desirable that pupils should
        find that they need to review the various usages adopted for the
        mole concept and it is convenient to include a reference to work
        which requires the use of solutions. A variety of types of reaction
        are then investigated with the view to seeking evidence to support
        the use of chemical equations as summaries of reaction. *Suggested
        time allocation: 4–5 weeks.*

Topic A18   *How fast? Rates and catalysts*
        A18.1 Measuring the rate of a reaction: how does concentration
        of a reactant affect the rate of a reaction?
        A18.2 How do changes in temperature affect the rate of reaction?
        A18.3 The effect of particle size on the rate of a reaction
        A18.4 How do catalysts affect the rates of reactions?

        This treatment of rates of reactions is intended to be a simple semi-
        quantitative study of the factors mentioned in the titles of the
        various sections rather than a detailed quantitative and mathe-
        matically orientated survey. A simple qualitative interpretation of
        chemical reactions as collision processes can emerge from the
        suggestions made. *Suggested time allocation: 4 weeks.*

Topic A19   *How far? The idea of dynamic equilibrium*
A19.1 Looking at reactions which can go both ways
A19.2 Looking more closely at reversible reactions: the theory of
dynamic equilibrium
A19.3 Evidence for the dynamic nature of equilibrium

The distinction is made between reactions which can be made to go
in 'both directions', by altering the conditions, and the establishment of a condition of equilibrium. As in all the Topics in the
scheme, evidence is obtained through experimental investigation.
Essentially the approach uses qualitative evidence, models, and a
radioactive tracer technique to provide evidence for the dynamic
equilibrium between lead chloride and its saturated solution.
Suggested time allocation: 3 weeks.

Topic A20   *Acids and bases*
A20.1 Investigating the acidic properties of some compounds
A20.2 Concerning acids and bases
A20.3 Assessing the role of water
A20.4 Investigating the reaction of an acidic solution and an
alkaline solution
A20.5 Preparing salts

This Topic illustrates the use of evidence which enables scientists
to move from simple descriptions (such as the identification of a
substance by what it is seen to do) to conceptual considerations
(such as the theoretical implications of what is observed). A revision
of acidity-alkalinity from Stage I leads to a greater understanding
of the term acidity and to operational definitions of the terms acid
and base. An investigation into the role of water in acid-base
systems is used to introduce the Brønsted-Lowry theory of acids
and bases. This is then used to interpret the reaction between an
acidic solution and an alkaline solution. The Topic is rounded off
by a series of related experiments on the preparation of salts.
Suggested time allocation: 6 weeks.

Topic A21   *Breaking down and building up large molecules*
A21.1 Molecular materials; carbon chemistry
A21.2 The breakdown of starch
A21.3 Can glucose be broken down further?
A21.4 Can ethanol also be broken down further?
A21.5 Cracking petroleum to obtain new products
A21.6 Preparing soaps and soapless detergents from castor oil
A21.7 Breaking down and building up perspex
A21.8 Making nylon

The Topic is primarily concerned with large molecules, both natural
and synthetic, and their importance in biochemistry and the chemical industry. The Topic also includes some introductory work on
polymers. *Suggested time allocation: 6 weeks.*

Topic A22   *Ammonia, fertilizers, and food production*
        A22.1 Getting ammonia from natural substances
        A22.2 What elements are present in ammonia?
        A22.3 What is the formula of ammonia?
        A22.4 Making ammonia from nitrogen and hydrogen
        A22.5 Ammonia as an alkali
        A22.6 Fertilizers and food production

This Topic provides further instances of the usefulness of chemistry. *Suggested time allocation: 6 weeks.*

Topic A23   *Energy changes in chemical systems*
        A23.1 Energy changes in chemical systems
        A23.2 What is the source of the energy liberated in a chemical change?
        A23.3 Chemicals as fuels
        A23.4 Energy transformation and related topics
        A23.5 Ways of obtaining energy for use in industry

The Topic is intended to provide a simple review of some aspects of energy changes in chemical systems. Energy-level diagrams are introduced as a means of representing the energy changes associated with specific chemical reactions. Calorimetric procedures are used in quantitative determinations of a number of heats of reaction, the principle of conservation of energy being assumed. Attention is then focused on combustion processes and the use of chemicals as fuels. A method for measuring the heat evolved when different fuels are burnt successively in air is worked out and the results are expressed so that useful comparisons may be made. Finally, energy transformations are considered and fuel cells are introduced as potentially more efficient sources of power. Ways of obtaining energy for use in industry receive attention in the last section of the topic. *Suggested time allocation: 5 weeks.*

Topic A24   *Radiochemistry*
        A24.1 An introduction to radioactivity
        A24.2 The study of the decay and growth of some radioactive chemicals
        A24.3 The uses of radioactive isotopes as 'detectives'

The historical development of this branch of chemistry is illustrated by a series of demonstrations in which α, β, and γ-radiations are identified. The idea of a radioactive element undergoing transformation to another element is introduced and the concept of isotopes is treated simply. The Topic is concluded by reviewing some of the uses for radioactive isotopes. *Suggested time allocation: 3 weeks.*

Alternative IIB

This alternative seeks to serve the same objectives as Alternative IIA, to cover more or less the same content, and to lead up to the same public examination. However, Alternative IIB has a radically different sequence and approach which has been determined after detailed consultations with many teachers who have used the original course materials and who have also contributed in various ways to the preparation of the scheme, including participation in limited trials of sections of the materials.

This alternative to the *Sample Scheme, Stage II*, meets several needs:
1. the incorporation of social aspects of chemistry in earlier sections of Stage II;
2. the avoidance of an initial high-level conceptual demand which characterized Stage II of the original Sample Scheme;
3. the provision of a balanced course for those pupils who terminated their study of chemistry at the end of their third year in a secondary school of the 11–16 or 11–18 type;
4. the provision of opportunity to introduce Stage III Options during the fourth-year work rather than at the end of the course, now that the conceptual levels of treatment at Stages II and III are similar.

Alternative IIB is itself in three parts. *Part 1* begins with a simple recapitulation of ideas and facts established either during Alternative IA or IB or through some other introductory study of science, such as Nuffield Combined Science. Topic B11 is concerned with the characteristics of chemical elements, the properties of oxides, and related matters. The 'salt-gas' problem also occurs in this first Topic, and provides a basis for further investigations using acids, bases, and salts in Topic B12. At this point in the scheme, only operational definitions of these terms are needed (i.e. those which enable the indentification of these classes of compounds to be effected). Word equations are used as reaction summaries and pupils are introduced to the symbols for various elements. The formulae of a number of simple compounds may be used (without formal proof) to establish an awareness of the use of chemical equations as reaction summaries. By using a knowledge of the properties of acids, an introductory study of reaction rates follows in Topic B13. This treatment of reaction rates is introductory and differs from that adopted in Alternative IIA. It is concerned with establishing the time required for a reaction to finish rather than with establishing the experimental characteristics of a given rate of reaction. The effects of the concentration of reactants, the presence of catalysts, and temperature are also investigated simply.

Part 2 covers Topics B14 to B17. The introductory theme of acidity and basicity is developed a little further in 'Chemistry and the world food problem' to provide a different approach to that used for Topic A22. This opportunity is used to demonstrate the relevance of chemistry to mankind, a theme which recurs throughout Alternative IIB.

Teachers are encouraged to use models to represent specific species in reaction summaries, thereby complementing the more conventional presentation of chemical equations.

The production of large molecules from small ones by plants provides an appropriate linking theme between the 'world food problem' and 'everyday materials'. The activity of the chemist is presented as making new materials from those around us. The properties and uses of synthetic materials, such as plastics and fibres, contrast sharply with those of metals. Clearly, such differences must depend in some way on the presence of different 'internal' structures.

So far in the development of Alternative IIB, no formal comment has been made about the particulate nature of matter, or its relevance for the analysis of the structure of chemicals. Accordingly, two important topics now follow: 'Atoms and the Periodic Table' and 'The arrangement of atoms in elements and compounds'.

Part 2: the alternative route covers the same material as the main route, but in a different order. Topic B17 (an introduction to structure) can follow immediately after Topic B13 to allow pupils to adopt an enquiring approach to the problem before being called upon to accept the findings of structure investigations. (This must clearly depend upon the teacher adapting these detailed suggestions to suit the circumstances.)

The application of these ideas on structure to everyday materials follows (B15.5 to B15.8) and due consideration is given to large molecules and to metals. An aspect of the social significance of chemistry comes next: 'Chemistry and the world food problem' and, in this context, the residual material of Topic B15 (i.e. B15.1 to B15.4) and Topic B14 can follow. The unifying theme of structure can be extended now to Topic B16 in which the relevance of the units of structure, atoms, are related to that other unifying theme in chemistry, the Periodic Table.

This second route through Topics B14 to B17 provides pupils with some understanding of unifying themes in chemistry at a relatively early age. It also provides pupils with an opportunity to use 'simple' molecules before 'large and complex' ones. Both routes yield a convenient theme on which to end a third-form course, 'periodicity' (in Topic B16) in one case and 'structure' in the other.

Part 3: Topics which are more dependent on the application of theoretical ideas now follow and can be integrated with the optional material of Stage III. Topics B16 and B17 provide adequate preparation and context for another unifying concept in Topic B18, the mole. This enables pupils to determine the formulae of simple compounds within a framework of modern chemical theory.

A study of electrolysis follows and leads naturally to investigations of the properties of ions in solution – a general theme which enters into much inorganic chemistry. Other themes include a simple treatment of the Brønsted-Lowry theory of acids and bases, and a study of several chemical systems which can 'go in both directions'. Chemical equilibrium is considered separately (Topic B21).

The general theme of rate processes in chemistry (Topic B22) comes next and is viewed in two contexts: the phenomenon of radioactive decay and growth and the study of reaction rates. This process of unification is not without its advantages but it is important for pupils to distinguish between radioactive decay (involving the nuclei of atoms which are unaffected by chemical environment) and chemical change (involving only part of the structure of atoms—the outer electrons—which can be affected by the chemical environment.)

The energy changes which are associated with chemical reactions form the central theme of Topic B23. The final Topic, B24, offers an opportunity for pupils to revise their ideas on periodicity in a much wider context. The Topic is concerned mainly with the chemistry of typical transition elements and group IV elements, and so the content of the Topic could be merged with that of Stage III Option on Periodicity.

This general outline of Alternative IIB is now reviewed in greater detail.

Part 1

Topic B11 *Elements and compounds*
B11.1 Elements, compounds, and mixtures
B11.2 Properties of oxides
B11.3 The 'salt gas' problem
B11.4 The composition of 'salt gas'

This Topic can be used to revise earlier work in Chemistry, or serve as the starting point. If the second course is followed, B11.1 will need to be expanded to enable pupils to gain experience of materials and ideas. After B11.2, pupils investigate the properties and composition of 'salt gas' (hydrogen chloride) in the remaining sections of the topic. *Suggested time allocation: 3–4 weeks.*

Topic B12 *Acids, bases, and salts*
B12.1 Investigating acids
B12.2 Concerning bases and alkalis
B12.3 Salts and their occurrence as minerals

Some recapitulation of ideas is provided for those pupils who have studied Stage I or comparable materials, together with an opportunity to extend their knowledge and experience of acids, bases, and salts. If the content of this Topic is entirely new to pupils, then additional time will be required to enable pupils to acquire some experience of the use of materials. *Suggested time allocation: 4 weeks.*

Topic B13 *How fast? A study of reaction rates*
B13.1 Measuring the rate of a reaction; how does the concentration of a reactant affect the rate of a reaction?
B13.2 Investigating the influence of temperature on the rate of a reaction
B13.3 What is a catalyst?

The Topic uses the knowledge gained in Topic B12 together with everyday experience of rates of reaction to introduce simple rate studies. It starts with the effect of the concentration of acid on the time required for an acid-metal system to finish reacting. Other experimental work is used to quantify the effect of concentration of reactants and of temperature on the rate of reaction. A simple interpretation in terms of collision processes is possible from the results obtained. A series of separate studies on catalysis leads to a pupil-directed study on catalysts. *Suggested time allocation: 3 weeks.*

Part 2: main route

Topic B14   *Chemistry and the world food problem*
B14.1 Identifying the problem and the ways in which the chemist can help
B14.2 Analysis of ammonia
B14.3 Synthesis of ammonia
B14.4 Making fertilizers
B14.5 The balance of nature

The first and last sections contain material which some schools may cover in other subjects. It calls for some coordination between teachers from different departments to prevent needless repetition. Emphasis is placed on the role of science in society. *Suggested time allocation: 5 weeks.*

Topic B15   *Everyday materials: large molecules and metals*
B15.1 Food
B15.2 Starch – an example of a carbohydrate
B15.3 Breaking down glucose
B15.4 Breaking down fats and oils
B15.5 What compounds are there in crude oil?
B15.6 Breaking down big molecules in crude oil
B15.7 Making and using plastics
B15.8 Metals and their importance to engineers

Section B15.1 serves as a bridge theme between Topic B14 and the main theme for Topic B15, 'everyday materials'. The compositions of some everyday foods are tested and molecular models are used to distinguish between typical carbohydrates, fats, and proteins. Starch is then shown to contain carbon and hydrogen and some of its reactions are investigated. The breaking down of starch into smaller molecules is studied, leading to the study of glucose and fermented liquors. Subsequent investigations relate to fats and oils. In section B15.6, the fractionation of crude oil is revised and interpreted simply. The cracking of oil is considered in some detail. The uses of the various fractions are described, and some time is spent on making and investigating plastics. The final section offers some investigations to illustrate the importance of metals to engineers. *Suggested time allocation: 7 weeks.*

Topic B16   *Atoms and the Periodic Table*
B16.1 Getting some idea of the size of atoms
B16.2 Getting some idea of the mass of atoms
B16.3 How can the elements be classified?
B16.4 The alkali metals
B16.5 The inert gases and the halogens
B16.6 Carbon and silicon

This Topic provides a novel approach to the size of atoms and to relative atomic mass. The approach to periodicity is again from a historical point of view but on this occasion a simulation exercise using a 'card game' is presented. *Suggested time allocation: 5 weeks.*

Topic B17   *The arrangement of atoms in elements and compounds: an introduction to structure*
B17.1 The arrangement of atoms in solids, liquids, and gases
B17.2 How are atoms arranged in salts?
B17.3 The idea of the molecule
B17.4 Carbon and sulphur: some structural studies
B17.5 Chemical equations

After a brief consideration of the differences between solids, liquids, and gases, pupils study the shapes of some crystals and are led to appreciate that such regularity provides strong evidence for the orderly way in which the constituent atoms are arranged. The term giant structure is introduced and the structure of sodium chloride is discussed. The evidence for this interpretation is reviewed briefly and analogues of X-ray diffraction patterns may be introduced as part of the Topic. The term molecule is revised and it is emphasized that atoms are held together more strongly in giant structures than molecules are in molecular structures. Chemical equations are then used to relate these ideas to earlier views on the nature of chemical change. *Suggested time allocation: 4 weeks.*

Part 2: Alternative route

Topic B17   *The arrangement of atoms in elements and compounds: an introduction to structure*

B17.1 The arrangement of atoms in solids, liquids, and gases
B17.2 How are atoms arranged in salts?
B17.3 The idea of the molecule
B17.4 Carbon and sulphur: some structural studies
B17.5 Chemical equations

Topic B15 (modified)   *Everyday materials: large molecules and metals*

B15.5 What compounds are there in crude oil?
B15.6 Breaking down big molecules in crude oil
B15.7 Making and using plastics
B15.8 Metals and their importance to engineers
B15.1 Food
B15.2 Starch—an example of a carbohydrate
B15.3 Breaking down glucose
B15.4 Breaking down fats and oils

Topic B14   *Chemistry and the world food problem*

B14.1 Identifying the problem and the ways in which the chemist can help
B14.2 Analysis of ammonia
B14.3 Synthesis of ammonia
B14.4 Making fertilizers
B14.5 The balance of nature

Topic B16   *Atoms and the Periodic Table*

B16.1 Getting some idea of the size of atoms
B16.2 Getting some idea of the mass of atoms
B16.3 How can the elements be classified?
B16.4 The alkali metals
B16.5 The inert gases and the halogens
B16.6 Carbon and silicon

This presentation requires the same time allowance and covers the same ground as the main route through part 2. Its advantage is that rates of reaction lead directly to an introduction to structure and thence to the results of structural investigations (Topics B15 and B16). The approach also has the advantage of grouping all the food chemistry together.

Part 3   *Topic B18*   *The mole and its use in the determination of the formulae of compounds*

B18.1 Examining a ballasted model of magnesium oxide
B18.2 Finding the formula of magnesium oxide by experiment
B18.3 The idea of a mole of atoms
B18.4 Using the 'mole of atoms' to find the formulae of several compounds
B18.5 The mole of molecules
B18.6 What volumes do moles of gases occupy?
B18.7 Using the 'mole of molecules' to determine the formulae of gases

The approach used is quite different to that adopted in Alternative IIA. It is based on the use of a ballasted model of the structure of a binary compound. Arguments are used which are based on a Daltonian approach to formulae and can be tested experimentally by pupils. A selection of studies is undertaken to illustrate the ideas discussed. The work with gases requires the use of Avogadro's law and of Gay-Lussac's law of combining volumes. *Suggested time allocation: 5 weeks.*

Topic B19   *Electrolysis*
B19.1 What happens when an electric current is passed through various solutions and molten substances?
B19.2 Elementary electricity
B19.3 Observing the migration of ions
B19.4 The structure of the atom: some implications
B19.5 Some studies in electrolysis

The Topic opens with a circus of activities to remind pupils of work done in this area of chemistry during Stage I. A study of the migration of ions is followed by a simple explanation of electrolysis. The structure of the atom is considered. The amount of charge on various ions is measured, and other themes are considered in the last section of the Topic. *Suggested time allocation: 5–6 weeks.*

Topic B20   *Ions in solution and related chemical systems*
B20.1 The mole as a unit
B20.2 Formation of precipitates
B20.3 Reactions involving gases
B20.4 Properties of ions in solution
B20.5 Reversible reactions
B20.6 Looking at the role water plays

The opening section is used to review the use of the mole so far and to introduce its use when referring to the concentration of a solute in a solution. The systems studied include precipitation reactions, reactions in which gases are formed or in which gases react together, and reactions in which reactants and products are ions. Reversible reactions receive a separate consideration, as does the role that the solvent plays in studies of reactions using solutions of solutes. This last section is extended to provide a simple treatment of the Brønsted-Lowry theory of acids and bases. *Suggested time allocation: 6 weeks.*

Topic B21   *Chemical equilibria*
B21.1 Characteristics of chemical reaction
B21.2 A study of systems in equilibrium

The Topic opens with a survey of the characteristics of a chemical reaction, with special reference to reversible reactions mentioned in the previous Topic. Some evidence is provided to suggest that not all reactions proceed to completion and the idea of dynamic equilibrium is introduced. The need for radioactive tracer techniques to support the theory is mentioned. *Suggested time allocation: 2 weeks.*

Topic B22   *Rate processes in chemistry*
             B22.1 Introduction to radiochemistry
             B22.2 The study of a radioactive element
             B22.3 Relating products and radiation to the source of activity
             B22.4 Uses of radioactive materials
             B22.5 Studies of rates of reaction

Arising from the previous Topic, there is a need to assess a new technique to solve a problem. The Topic presents the evidence for radiochemistry briefly and, in due course, the possibility of using a tracer technique to investigate a system in chemical equilibrium is reached.

Other forms of rate process are then reviewed. Topic B12 is revised and other systems are introduced to extend the pupils' knowledge. Pupils should appreciate the distinction between these two types of rate process: radioactivity which is immune to the chemical environment and is dependent on a nuclear process; chemical reactions which can be influenced by changing the chemical environment and are dependent on the electronic structure of the species taking part. *Suggested time allocation: 4 weeks.*

Topic B23   *Chemicals and energy*
             B23.1 Energy changes and changes of state
             B23.2 Energy changes and chemical reactions
             B23.3 Chemicals and fuels
             B23.4 Energy transformations and related subjects

The overall content of this Topic is similar to A23 although there are differences on points of detail. *Suggested time allocation: 5 weeks.*

Topic B24   *A second look at the Periodic Table*
             B24.1 The Group IV elements
             B24.2 Transition metals
             B24.3 The position of an element in the Periodic Table

This Topic provides an opportunity to apply some of the concepts developed earlier in the course as well as a context for investigating Group IV elements and some transition elements. The Topic also provides a link to Stage III, Option 7. *Suggested time allocation: 4–5 weeks.*

A comparison of the Alternative Schemes for Stage II

The two new alternatives for Stage II differ a great deal in flavour but not in content. No simple comparison of the type used in *Teachers' guide I* for Stage I alternatives is feasible. However, teachers who have used the 1966 Sample Scheme may be interested to know how these new proposals relate to the original scheme at Stage II. This is summarized in the Table below. (All references are based on the original *Sample Scheme*.)

The Sample Scheme (1966) *Stage II: order of sections*		**Stage IIA** Approximate equivalent section	**Stage IIB** Approximate equivalent section
Topic 11	*Atoms in chemistry*		
11.1	What evidence have we that matter is made up of particles?	A11.1	various, e.g. B16.1, B17.1
11.2	How big are the particles?	A11.2	B16.1
11.3	Comparing numbers of atoms by weighing	A11.3 or A14.1, A14.2	B16.2, B18.3
11.4	Using the concept of gram-atoms to find formula	A11.4	B18.2, B18.4
Topic 12	*Investigation of salt and 'salt gas'*		
12.1	What elements are present in 'salt gas'?	A12.1, A12.2	B11.3
12.2	Is 'salt gas' a compound only of hydrogen and chlorine?	A12.3	B11.4
12.3	What is the formula of hydrogen chloride?	A12.4	B18.7
Topic 13	*Looking at the elements in the light of the Periodic Table*		
13.1	The story of the Periodic Table	A13.1	B16.3, B24.1
13.2	The alkali metals	A13.2	B16.4
13.3	The halogens	A13.3	B16.5
13.4	The heavy metals	A13.4	B24.2
13.5	Carbon: an element common in the living world	A13.5	B16.6, B24.1
13.6	Another way of oxidizing carbon compounds	A13.5	B16.6
13.7	Silicon: an element common in rocks	A13.6	B16.6, B24.1
Topic 14	*Finding out how atoms are arranged in elements*		
14.1	Watching crystals grow	A14.3	B17.2
14.2	The problem of trying to investigate things too small to be seen	A14.4	
14.3	How are atoms arranged in metals?	A14.5	B15.8
14.4	The importance of the structure of metals to engineers	A14.6	B15.8
14.5	The effect of packing sulphur atoms in different ways	A14.7	B17.4
14.6	Another element which has different forms: carbon	A14.8	B17.4
14.7	What do we know about the structure of gases?	A14.9	B17.3, B18.5
14.8	How are atoms arranged in compounds?	A14.10, A14.11	B18.1
14.9	Gram-formulae and gram-molecules	A14.11, A14.12	B17.5, B18.7

Supplementary material Several teachers who participated in one of the surveys leading to *Revised Nuffield Chemistry* commented on the need for detailed guidance on atomic structure and chemical bonding. The provision of *non-examinable* material on these subjects could be used with advantage at Stage II, *either* to relate Stages II and III more closely *or* to assist those pupils who aim to by-pass a public examination at 16 + in chemistry.

Teachers who are familiar with the first edition of this *Teachers' guide* will note that energy receives a less detailed treatment in this edition. The *Handbook for Teachers* provides much information on this theme, thereby supporting a more extensive coverage than has been advocated in this present edition, and so no additional supplement is made here.

An outline of the supplementary units for Stage II is as follows.

Supplement S.1 *The structure of the atom*
S1.1 Introduction: Dalton's atomic theory
S1.2 Can atoms be divided?
S1.3 What does the nucleus consist of?
S1.4 Isotopes and relative atomic mass

Supplement S.2 *How are the electrons arranged in atoms?*
S2.1 First ionization energies of the elements and the concept of 'shells'
S2.2 Successive ionization energies of sodium

Supplement S.3 *Chemical bonding*
S3.1 Ionic bonding
S3.2 Covalent bonding

If these supplements are used, the total time required is about 15 periods (S1, 7 periods; S2, 3 periods; S3, 5 periods). In any one of the schemes referred to, it would be advantageous if the pupils had developed a concept of the 'atom' beyond that of the simple 'Daltonian atom', as used in Topic A11, and had clear ideas of the meaning of ionic and covalent bonding.

Suggested patterns for the use of 'supplementary units'
In making suggestions for the placing of these units into a sample scheme, due care is necessary if the teacher wishes to avoid changing the nature of the scheme. Thus, with Alternative IIA, it is feasible to use the suggestions made in S1.1 and S1.2 after Topic A15, and then to use S1.3 and S1.4 after Topic A18. Units S2 and S3 could follow after Topic A20. Other relationships between these supplementary units and Alternative IIA are set out in Part 2 of this book. In Alternative IIB, Units S1 and S2 could be adapted for use after Topic B19. Unit S3 could be adapted to extend Topic B20. Alternatively, all three units could be used sequentially to prepare for Topic B24.

Chapter 4

Publications and teaching aids

A brief description of the work which led to the revision of Nuffield Chemistry appears in *Teachers' guide I*. The basic aim of the project remains the same but in this second edition many alterations have been made to the range of publications now familiar to many teachers.

Books for the teacher
Five books are now available which offer advice to teachers on planning their teaching schemes, and on the teaching resources available to them. These books, to which many teachers have contributed, have two main purposes:
1. To enlarge upon the general aims of the scheme with suggestions about topics of study and how to handle them and about the scope and content of possible courses arising out of the scheme.
2. To help teachers to adapt the scheme to the particular needs of their pupils, through suggestions about teaching approaches and the organization of lessons, and about the availability of teaching aids.

The titles and details of these books are as follows:

Teachers' guide I
Teachers' guide II
These two volumes are self-contained books. Volume I deals with Stage I (years 1 and 2) and Volume II deals with Stage II (years 3, 4 and part of 5) of the five-year course leading to the first public examination at 16+ (G.C.E. O-level). These books take the place of the *Introduction and Guide* and *The Sample Scheme Stages I and II: The Basic Course*.

Teachers' guide III
This book contains the revised proposals for the options in the final year of the five-year course and is in parallel with the publications for pupils at Stage III.

Handbook for teachers
This book has *not* been revised. It contains essays covering the intellectual and theoretical grounds on which the scheme is based together with much information and advice. Its main objects are:
1. To show how the major ideas of present-day chemistry, those concerned with structure, rates, equilibria, and energy changes, are integrated in the Sample Scheme.
2. To examine the possible consequences of introducing these topics to work in the sixth form. Its contents therefore in some of the Topics discussed deliberately go beyond that needed for the 11-16 age group. It is not intended solely as a guide to the Sample Scheme; those wishing to devise other teaching programmes should find

much that is helpful in its pages. No attempt is made to avoid controversial issues, but these are discussed in a constructive manner in the hope that a contribution can be made to future progress in the teaching of chemistry.

Collected experiments
This book has *not* been revised. It is for the use of teachers and has two purposes:
1. To provide alternative experiments in case those given in the *Teachers' guides* are not suited to the particular conditions under which the teacher is working.
2. To provide some experimental materials for teachers who want to use the Nuffield approach, but who want to teach subject matter other than that given in the *Teachers' guides*.

The experiments are arranged in this book by themes.

Books for the pupil

As a result of the inquiries on which this revision is based, the books for pupils have undergone major changes in content and format. They are:

Stage I
- *Experiment sheets I*
- *Study sheets*
- *Growing crystals* (Background book)

Stage II
- *Experiment sheets II*
- *Handbook for pupils*
- *Chemists in the world*

Stage III *Option booklets*

Experiment sheets I and II
These titles replace the *Laboratory Investigations I and II* of the first edition. The format of these titles has been changed from that of the first edition. They are available as pamphlets in A5 format, with each page punched.

Experiment sheets I contains the sheets for both IA and IB to enable those teachers who use materials from both alternatives to continue to do so. *Experiment sheets II* contains the instructions relating to both IIA and the new IIB. Both books have been written in the light of comment received from schools. As a further help, each experiment sheet is reproduced for the teacher in the appropriate volume of the *Teachers' guide*.

Study sheets
The twelve *Study sheets* replace the six *Background Books* originally designed for Stage I. They provide a more balanced coverage of background reading for this stage and suggestions for when they could be used are given in Part 2 of *Teachers' Guide I*. The titles of the *Study Sheets* are:
Analysis
Where chemicals come from
Heating things
Burning and Lavoisier
Fresh air?

The chemical elements
Competitions
Water
Chemistry and electricity
Chemicals and rocks
The halogens
The words chemists use.

Handbook for pupils
This book (to which many practising teachers have contributed) for use at Stage II has been written in response to requests from schools. The first part of the book contains chapters on topics of chemistry, ranging from periodicity to structure, to which the teacher can refer a pupil needing an exposition of a particular chemical idea. Part 2 contains articles on the chemical industry, the world food problem, and the social implications of chemistry. Part 3 contains the basic data necessary for the course, a glossary of terms, and answers to numerical questions. The chapter titles are:
  1 Periodicity
  2 Finding out about matter
  3 Atomic structure and bonding
  4 Formulae and equations
  5 Studying chemical reactions
  6 Energy changes and material changes
  7 Electricity and matter
  8 Radioactivity
  9 The structures of elements and compounds
  10 The chemical industry
  11 World food problems
  12 Man, chemistry, and society
     Reference tables

Chemists in the world
This book draws upon much of the material from the *Background Books* for Stage II not incorporated in the *Handbook for pupils*. The material has been revised and up-dated. The first part of the book covers the history of chemistry from the development of the atomic theory by Dalton to modern times. The second part deals with the industrial applications of chemical discoveries. The chapter titles are:

1 Finding out about the atom
2 Using ideas about atoms.
3 The way of discovery
4 Davy and Faraday
5 Energy and chemicals from coal, gas, and oil
6 Polymers from petroleum
7 Fertilizers
8 Ceramics and glasses

Pupils' option booklets
Ten of the eleven options proposed in this revision are provided with a booklet for use by pupils. Each contains background material and experiments. The titles of these booklets are:
1 Water
2 Colloids
3 Drugs and medicines
4 Metals and alloys
5 Plastics
6 Change and decay
7 Periodicity, atomic structure and bonding
8 The chemical industry
9 Analysis with a purpose
10 Historical topics

Visual aids

Film loops
Stage I:
1–1 Salt production
1–2 Chlorophyll extraction
1–3 Whisky distillation
1–4 Oil prospecting
1–5 Petroleum fractionation
1–6 Liquid air fractionation
1–7 Gold mining
1–8 Iron extraction
1–9 Copper refining
1–10 Limestone
1–11 Fluorine manufacture
1–12 Uses of fluorine compounds
1–13 Chlorine manufacture
1–14 Chlorine – uses
1–15 Bromine manufacture
1–16 Bromine – uses
1–17 Iodine manufacture
1–18 Iodine – uses

Stage II:
2–1 Measuring the very small
2–2 Gram-atoms (The mole of atoms)
2–3 Sulphur crystals
2–4 Heating water
2–5 Liquid-gas equilibrium
2–6 Solid-liquid equilibrium
2–7 Movement of molecules
2–8 Electrolysis of lead bromide
2–9 Cracking hydrocarbons
2–10 Plastics
2–11 Ammonia manufacture
2–12 Ammonia – uses
2–13 Catalysis in industry
2–14 Energy changes in HCl formation
2–15 Radioactive materials – uses

New loops for Stage II:
2–16 The formula of hydrogen chloride
2–17 Solids, liquids, and gases
2–18 The electrolysis of potassium iodide solution
2–19 Dynamic equilibrium

Stage III:
3–1 Growth of crystals
3–2 Metallurgical techniques
3–3 Metals: mechanical properties
3–4 Giant molecules – proteins

Overhead projection originals
These relate to Stages II and III and consist of printed originals which can be used to make transparencies for use on an overhead projector. A list of overhead projection originals for a particular Topic is given at the beginning of that Topic, and they are reproduced in the text.

Slide series
For the convenience of teachers, these will be made available in the form of film strips. (Details appear in the publisher's brochure.) They are listed at the beginning of Topics in the same way as the overhead projection originals.

Part 2 Alternative A

Safety in science laboratories

Teachers are reminded of the need to observe those safety regulations which apply to schools in their area. General advice on safety in school science laboratories in England and Wales is conveyed through Department of Education and Science publications (see page 650 for the list relating to the period immediately before this guide was prepared for publication). Teachers are advised that it *may* be necessary for them to adapt some of the suggestions in this guide in order to conform to their local safety regulations.

Topic A11

Atoms in Chemistry

Purposes of the Topic

1. To provide evidence to support the atomic theory and to point out that the 'proof' of this theory rests principally in its usefulness.
2. To show that the particles of which matter is composed are very small indeed.
3. To introduce and use the terms atom and relative atomic mass.
4. To introduce and use the symbols for the chemical elements.
5. To introduce the mole of atoms (gram-atom).
6. To illustrate the processes of analysis and synthesis.

Contents

Note. Two routes are proposed for this Topic, which is linked with Topic A14.
A11.1 What evidence have we that matter is made up of particles?
A11.2 How big are the particles?

A11.3 Comparing numbers of atoms by mass measurement A11.4 Determination of the formulae of simple compounds	*Alternative route: omitting A11.3* A11.4 Determination of the formulae of simple compounds without using the mole concept

Timing

In this book, the time suggested for each Topic is expressed in weeks. It is assumed that a week contains 1 double period and 1 single period. A11.1 will probably need a double period and a single period and A11.2 at least a double period. A total allowance of between 4 and 5 weeks is suggested for the whole Topic.

Introduction to the Topic

The idea that matter is made up of particles is now so well known that we can probably assume that our pupils are familiar with the term atom. Few, however, will have considered what evidence there is to support their knowledge of atomic theory and the structure of matter. We therefore suggest that pupils spend the first lesson considering the question 'What evidence do we have that matter is composed of particles?' The evidence discussed at this point in the course is based on diffusion phenomena and is not intended to be conclusive.

'If there are particles, how big are they?' Our approach here is to measure the thickness of the thinnest thing we can easily make. For practical purposes this turns out to be an oil film, and this experiment tells us that the particles are about 1 nanometre (10^{-9} m) in diameter. The film strip 2–1 'Measuring the very small' is designed to help pupils to grasp the significance of the size of such units as the nanometre.*

*The earlier version of this visual aid, film loop 2–1, refers to obsolescent units.

The third question is about the masses of these particles. 'If they have volume, they must surely have mass – but how can we find the mass of particles which are so small?'

The determination of the relative masses of atoms, an idea first put forward by Dalton, is a long process which is much more suitable for discussion in detail in the sixth form than at this stage. We therefore introduce pupils to relative atomic masses (a term which is to be used in preference to the now obsolescent term atomic weight), explaining that they will consider the methods by which these values were originally obtained at a later stage.

Two alternative routes   At this point, we offer two routes to the final problem in the Topic, which is the derivation of the formulae of simple chemical compounds: the main route is based on the use of the mole of atoms (gram-atom), and the alternative route is dependent upon the use of 'common-sense' (Daltonian) simplification rules. The alternative route is printed inside a box of dashed lines to enable the reader to distinguish clearly between the two routes.

The main route. This is concerned not so much with the actual mass of an atom as with the relative mass of a fixed number of atoms. Thus, we show that we can 'count out' the number of atoms by weighing. To take an example: one atom of carbon is twelve times as heavy as one atom of hydrogen; so, if we have 12 g of carbon and 1 g of hydrogen, we know that we must have the same number of atoms in each sample.

Through examples of this kind, we can introduce pupils to the notion of the mole of atoms (or gram-atom – if this term is preferred at this stage). One mole of atoms is a certain number (about 6×10^{23}) of atoms of an element. The value of the mole emerges as the course develops. Teachers will find it helpful if bottles each containing a mole of atoms (gram-atom) of various elements are available in the laboratory from this time onwards.

The use of the mole of atoms (gram-atom) concept to determine the formulae of a number of simple compounds follows. Formulae are presented as an expression of the proportion by moles of atoms (gram-atoms) of the elements present in a compound, and not as a shorthand notation to be learned.

> *The alternative route.*
> The introduction of the mole of atoms (gram-atom) is left until Topic A14, and 'simplification rules' are used to obtain formulae from experimental data. It requires the use of relative atomic masses.

Whichever route is used through the Topic, the naming of particles can represent some difficulties. It is suggested that whenever possible the words 'atom' and 'molecule' should be avoided and the word 'particle' used instead. Atoms are bound to be mentioned when Dalton's theory is mentioned in section A11.1. Ideally, they should be referred to simply as particles of elements and such a treatment will lead naturally to the need for specialist nomenclature. In section A11.2, the particles whose sizes are measured are, of course, molecules, but it would be wise not to refer to them as molecules here since this term is not properly introduced until later. It is more appropriate to use such terms as 'particles' or 'groups of atoms' at this point.

Again, whichever route is used, certain essential points must be established. The pupils need to know that an element consists of particles of the same kind (atoms); and that when elements combine to form a new substance – known as a compound – each of the small particles present in that compound will be made from particles of each of the elements taking part in the reaction. Indeed, as was established during Stage I, compounds are very different from the elements from which they are made and this point can be revised by referring to examples. A number of experiments should be performed to establish the procedure for finding the formulae of simple compounds. Once this has been done, pupils may be given other examples of compounds and their formulae, provided it is emphasized that before the formula for a compound can be written, someone has to do an experiment to find out just what it is. The experimental work offered in this Topic provides further opportunities to use the terms analysis and synthesis, to use symbols, and construct formulae.

Subsequent development

The dominant theme of Topic A14 is the structure of the elements. The mole of atoms (gram-atom) is included in the main route of this Topic (A11) and is used in a study of the structure of simple compounds. However, if the alternative route is followed the mole of atoms (gram-atom) must first be introduced in A14.1. (Other aspects of the mole concept appear in Topics A16 and A17.)

Alternative approach

Alternative IIB offers quite a different presentation of the mole concept. It depends on a much delayed introduction and on an appreciation of the nature of scientific models. The basic idea is introduced in Topic B18 and other aspects occur in Topics B19 and B20.

Background knowledge

The Topic follows from much of the work in Stage I, or in Nuffield Combined Science. In particular it requires a knowledge of the term element (see Topics A5 or B7) and of the formation of simple compounds, such as copper oxide and magnesium oxide.

Pupils will need to be able to multiply and divide decimal fractions, to know how to use standard form, and be able to work with ratios.

Standard form (sometimes known as scientific notation) means the expression of any number as a number lying between 1 and 10 multiplied by some power of 10. Thus, we may express 0.000 000 2 as 2×10^{-7}, and 201 as 2.01×10^2.

For pupils to use moles of atoms (gram-atoms) successfully, it is necessary for them to handle both a ratio (as in the conversion of grams to moles of atoms or gram-atoms) and a ratio between ratios (in the subsequent determination of formulae). The mathematics necessary for this, and the conceptual demands made on the pupil, are discussed as necessary in this Topic.

Note. A number of simplifications can be made if a graphical approach is adopted, as shown on page 71 and also in Topic B18. The alternative route also enables pupils to determine the formulae of simple compounds without using the mole concept.

<table>
<tr><td>

Further references

for the teacher

</td><td>

Handbook for teachers, Chapters 4 and 5, gives background information on teaching the Topic.
Collected experiments, Chapter 12, gives additional experiments.

Nuffield physics
Teachers whose pupils are following the Nuffield Physics course should note carefully the approach to particles in Year I of that course (see Nuffield Physics *Teachers' guide I*). In particular pupils should be aware that in chemistry the word particle is used as a general term and can refer to an atom, molecule, ion, etc. The term particle is used until the meaning of these more specific terms can be properly appreciated. In the Physics course, the words atom and molecule are used interchangeably without explanation in Year I, while the word particle refers to, for example, the particles of smoke seen in the observation of Brownian motion in a smoke cell. Pupils acquire the precise meaning of the terms atom and molecule later in the course. In addition, one form of the oil-film experiment appears in Year I of the Physics course.

</td></tr>
<tr><td>

Supplementary materials

</td><td>

Film strips and loops
2–1 'Measuring the very small'
2–2 'Gram-atoms (The mole of atoms)'

Film
'Gram-atoms, formulae, and equations' ESSO. Films for Science Teachers No. 17

Overhead projection originals
1 Conversion scale for magnesium *(Figure A11.7)*
2 Conversion scale for oxygen *(Figure A11.8)*
3 Conversion scale for copper *(Figure A11.11)*
4 Conversion scale for oxygen *(Figure A11.12)*
5 Conversion scale for mercury *(Figure A11.16)*
6 Conversion scale for chlorine *(Figure A11.17)*
7 Reaction of magnesium with oxygen *(Figure A11.18)*
8 Reaction of magnesium with oxygen, with lines corresponding to the compositions, Mg_2O, MgO, MgO_2 *(Figure A11.21)*
9 Reaction of copper with oxygen, with lines corresponding to the compositions Cu_2O, CuO, CuO_2 *(Figure A11.22)*

</td></tr>
<tr><td>

Reading material

for the pupil

</td><td>

Chemists in the world, Chapters 1 'Finding out about the atom' and 2 'Using ideas about atoms.
Handbook for pupils, Chapter 2 'Finding out about matter' and 4 'Formulae and equations'.

</td></tr>
</table>

Topic A11 Atoms in Chemistry **35**

A11.1
What evidence have we that matter is made up of particles?

In this section pupils discuss the theory that matter is made up of particles and find some experimental evidence to support the theory.

A suggested approach

Objectives for pupils

1. Awareness of evidence for and against the particulate theory of matter (as opposed to the 'continuous' theory of matter)
2. Knowledge of the process of diffusion, and its use as evidence that matter is capable of division into minute particles which are in continuous motion

From their work in Stage I pupils should be familiar with chemical reaction. They should also know that all matter is composed of substances which can be broken down into elements. But they have yet to answer the questions: 'Of what are these elements themselves composed?' and 'What is happening when a chemical reaction takes place?' The first person to give a satisfactory answer to these questions was John Dalton when he postulated that chemical reactions could be described in terms of the fundamental particles, or atoms, of which matter is composed.

We might begin the section with a discussion about the nature of matter. The choice between a particulate and a continuous theory could be presented as follows. If a piece of, say, copper is divided into halves, one half is then cut into two pieces, and the process repeated many, many times, will there come a time when the smallest possible particle having the properties of copper is obtained, or will it, in theory, be possible to go on cutting indefinitely?

Samples of Dutch metal or gold leaf can be used to stress the point to be made. Is it possible to make samples of metals even thinner than sheets of Dutch metal or gold leaf?

A discussion using these ideas is almost bound to involve the word 'atom', and it may be interesting to hear from pupils how much they know (or think they know!) about atoms. A brief account of the history of the atomic theory up to (*but not including*) Dalton's time will lead to a number of experiments, each of which may be discussed in the light of both the particulate and continuous theories of matter. Pupils may be surprised to find how difficult it is to 'prove' the particulate theory and to find any really conclusive evidence in its favour. The suggested experimental work includes examples of the diffusion of particles in gases and in liquids, and of diluting solutions of coloured materials.

In these experiments and in the discussion which may follow, it is suggested that the words 'atom' and 'molecule' be avoided, and only the word 'particle' be used. (To avoid confusion, the teacher may need to point out to his pupils that this is being done.) The term atom is introduced in A11.3, as meaning the smallest particle of an element, and the term 'molecule' appears somewhat later in the course.

Diffusion in solutions
This experiment should be done by the teacher.

Apparatus

The teacher will need:

Small wide-mouthed bottle with screw cap

Large beaker

Potassium permanganate (manganate (VII)) solution

Ammonium dichromate crystals

Procedure

1. Fill the small bottle with a strong solution of potassium permanganate, screw on its cap, and put it in the middle of a large glass beaker. Run water into the beaker until it is nearly full. Carefully unscrew and remove the cap, and leave the apparatus undisturbed until the next week, when the colour of the permanganate will have diffused throughout the water.

2. Drop a small crystal of ammonium dichromate into a beaker of water. Within ten to fifteen minutes the colour will spread throughout the water.

Pupils will see that the solids mix completely with water, and it is evident that in this process the solids must be divided up into extremely small particles, since they do not settle out on standing (nor can they be removed by filtering). It should be pointed out that this mixing can only be explained if you suppose that both the solid and the water can be divided into minute particles, so that these particles can get in between each other. However, the teacher should be clear in his own mind that this says nothing about whether they are particles in the sense of atoms and molecules, incapable of further subdivision without change of properties, or whether matter is infinitely divisible. Evidence for this must lie elsewhere and be considered later. Attention is now directed to gases.

Diffusion of gases
Parts 1, 2, and 4 of this experiment should be done by the teacher.

Apparatus

The teacher will need:

1 piece of glass tubing, approximately 2 cm diameter and between 0.5 and 1.0 m long

2 stands, bosses, and clamps

Cotton wool

Tongs or forceps

2 corks to fit glass tube

0.880 ammonia solution

Concentrated hydrochloric acid

1. *The diffusion of ammonia and hydrogen chloride in air*
Procedure

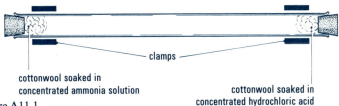

clamps

cottonwool soaked in concentrated ammonia solution

cottonwool soaked in concentrated hydrochloric acid

Figure A11.1

Support the piece of glass tubing horizontally by means of two stands. Have ready two small plugs of cotton wool and two corks to fit the tube. Dip one piece of cotton wool in concentrated hydrochloric acid and the other in 0.880 ammonia solution. Allow the surplus liquid to drain off. Place the cotton wool soaked in concentrated hydrochloric acid in one end of the tube, and the other prepared pad in the other end. Close the ends of the tube with the corks to prevent unnecessary fumes in the room. After a short time a white cloud of ammonium chloride will be seen at the point at which the two gases have met after diffusing through the air in the tube towards each other. This is best viewed against a dark background.

2. *The diffusion of bromine in air*
This experiment should be done by the teacher.

Apparatus

The teacher will need:

2 gas jars and covers

Vaseline

Teat pipette

White cardboard (or other material) for background

Bromine

Procedure
Use a long enough teat pipette to place a few drops of liquid bromine in the bottom of a gas jar. Invert a second jar over the first, making sure that the joint between them is gas-tight by lightly smearing the top of one of the jars with Vaseline. Notice the absence of colour in the upper jar.

Leave the jars to stand for a time. Notice the eventual appearance of the colour of bromine in the upper jar, and how long it takes to appear.

The apparatus is best viewed against a white background.

3. *Are the particles of a gas in motion?*

Apparatus

Each pupil (or pair) will need:

Experiment sheet 41

2 test-tubes, 125 × 16 mm, with corks

Limewater

Supply of carbon dioxide

Procedure
Details of the procedure are given in *Experiment sheet* 41 which is reproduced below.

Experiment sheet 41
When doing this experiment you will need to know a test for carbon dioxide. If you have forgotten this look it up in your notes of past work, or in a chemistry text book. Describe it here.

Carbon dioxide gas is denser than air. What would you expect to happen if you set up the arrangement shown in the figure below, with carbon dioxide in tube A and air in tube B?

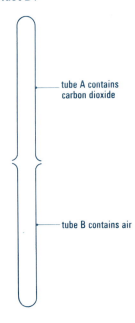

tube A contains carbon dioxide

tube B contains air

Figure A11.2

Test your answer by an experiment.

Fill a test-tube with carbon dioxide (your teacher will provide a suitable source of the gas) by holding the tube mouth *upwards* and passing the gas into it through a glass delivery tube which reaches nearly to the bottom of the test-tube. Pass the gas for one minute, remove the delivery tube, and cork the test-tube. Hold the test-tube upside down over a similar tube containing air.

Remove the cork and place the tubes mouth to mouth. After five minutes cork both tubes and test the contents for carbon dioxide. What happens:
in tube A?
in tube B?

What would you expect to happen if you put the carbon dioxide in tube B and the air in tube A?

Test your answer by another experiment. What happens?

Do the results of these experiments support the idea that the particles of a gas are in motion? Give your reasons.

Apparatus
The teacher will need:
2 gas jars and covers
Bunsen burner
Splints
Supply of hydrogen

4. *The diffusion of hydrogen in air*
This experiment should be done by the teacher.

Procedure
Fill a gas jar with hydrogen and cover it with a cover slip (gas-jar lid). Invert the filled jar over a gas jar of air and then remove the cover slip. After a few moments separate the jars and apply a flame to the upper one. There will be a loud bang, as the hydrogen ignites. This demonstrates that hydrogen is a light gas and will rise into a jar containing air. Now repeat the experiment but this time have the hydrogen in the upper jar inverted over a jar of air. If the jars are left for about five minutes some hydrogen will be carried into the lower jar by diffusion and its presence can be detected by ignition.

After discussing the results of Experiments A11.1a and A11.1b, make the point that solids, liquids, and gases can be divided into minute particles. If this were not the case, they would not mix so easily, for one *indivisible* substance could not penetrate, and mix completely, with another. This is not, however, proof of a particulate theory, nor is it evidence of whether matter consists of particles incapable of further subdivision without change of properties, or whether matter is *capable* of being divided indefinitely. But the experiments do suggest one further thing, namely that the particles of matter must be in motion, for the colour of the permanganate is seen to spread through the water and the gases are seen to spread through the air in the containers in which they are placed. Teachers may like to illustrate this motion of particles more clearly by a demonstration of Brownian motion using a smoke cell but should note that pupils may have seen this experiment in Years 1 or 3 of Nuffield Physics.

Demonstration of Brownian motion by smoke particles

Apparatus

Each pupil (or pair) will need:

Microscope

Whitley Bay smoke cell
(Nuffield Physics item 29)

12 V power supply

Microscope cover glass

Waxed paper drinking straw
(plastic straws are unsuitable)
or short (10 cm) length of sash
cord

2 lengths of connecting wire

Procedure
Of all the available forms of smoke cell for looking at Brownian motion under a microscope, the Whitley Bay form is the simplest.

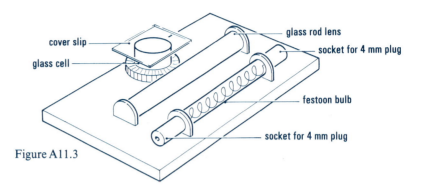

Figure A11.3

The diagram shows the general arrangement. The smoke is put in a small vertical glass tube, closed at the bottom, with the bottom blackened, and the cell is covered with a microscope cover glass. Light is concentrated into a narrow line across the tube by a cylindrical lens. The light source is a horizontal festoon lamp and the lens is a horizontal glass rod arranged to produce a line image of the filament in the middle of the glass tube of smoke.*

The smoke may be provided by a smouldering drinking straw, or a smouldering piece of sash cord. The drinking straw is held almost vertical and lit at the *top* end, so that smoke pours down through the straw into the cell at the lower end. Alternatively, a piece of sash cord is lit and then held just over the mouth of the cell; but pupils must be warned not to leave it smouldering. (More elaborate schemes in which smoke is picked up in a small teat pipette and squirted into the cell, are an unnecessary complication, and do not work so well after some use.) Particles of smoke will be seen as individual points of light in continuous, random, motion.

1. Try to explain the diffusion of potassium permanganate in water by supposing:
a. both substances are made up of particles.
b. they are both capable of infinite subdivision.
2. Describe the smoke cell experiment and explain in your own words the behaviour of the particles of smoke as viewed through a microscope.

*In order to minimize convection, the lamp is placed *below* the level of the glass rod so that the image in the tube is *very nearly* at the top. For proper imaging (necessary to make the cell work well), the lamp, rod, and cell must be spaced as follows: from filament to surface of rod, one rod-diameter; from surface of rod to centre of cell, one rod-diameter. If teachers make their own smoke cells they should follow this geometry carefully.

Summary

During this section, pupils should have considered the form that matter might take, and have met alternative interpretations: either that matter might be made up of particles, or that matter is capable of infinite subdivision. They should have seen experiments on diffusion, which show that two types of matter can mix spontaneously, and should realize that such observations lead us to suppose (1) that matter can be divided into very small particles; and (2) that if particles exist they must be in continuous motion.

A11.2
How big are the particles?

In this section the pupils are first asked to assume that matter *does* consist of particles. They then perform experiments which give some idea of the maximum size that such particles can be.

A suggested approach

Objective for pupils

Awareness of the order of magnitude of the size of the particles of matter, and of its relation to the size of more familiar objects

The teacher will need

Film strip 2–1 'Measuring the very small'
Electron micrograph of a virus crystal

Ask pupils to assume that the particulate theory is correct. Then ask them to try to devise an experiment which will give some idea of the size that the particles might be.

If they suggest looking under a powerful microscope, explain that the most powerful optical microscope does not show the particulate nature of matter. The electron microscope does not show the particulate nature of matter. But the electron microscope has shown that certain substances do appear to be composed of regular particles, and an electron micrograph of a virus crystal, if available, could be shown to the class at this point. An idea of the order of size of a particle can be gained by taking a very small crystal of potassium permanganate, dissolving it in water, and diluting a number of times. Alternatively a little ether (ethoxyethane) (say 5 cm^3) can be dropped on to a watch-glass by the teacher. Eventually everyone in the class will smell it, showing that at least one particle of ether must be present in each lung-full of air in the room. These experiments are described below.

Experiment A11.2a

Apparatus

Each pair of pupils will need:

Experiment sheet 42

Test-tube, $150 \times 16 \text{ mm}$, with cork to fit

Beaker, 250 cm^3, with graduation marks

White paper as background

Crystal of potassium permanganate (manganate (VII))

Into roughly how many particles might a crystal of potassium permanganate be divided?

Procedure
This is described in *Experiment sheet* 42 which follows. Potassium permanganate is chosen for this experiment because of its intense colour, which makes it visible even in very dilute solutions. The calculation associated with the experiment is minimal, and even without the arithmetic it can be seen that the particles, if they exist, must be very small.

Experiment sheet 42

Potassium permanganate dissolves in water to form a deep purple solution. You are going to make a solution from one crystal of potassium permanganate and find out how much you can dilute the solution without losing the colour completely.

You will need some means of measuring out $20\,cm^3$ of solution and $200\,cm^3$ of solution. A 150×16 mm test-tube holds about $20\,cm^3$ and a $250\,cm^3$ graduated beaker can be used to measure $200\,cm^3$.

Select a potassium permanganate crystal about 3 mm long and proceed as follows:
1. Put the crystal into a 150×16 mm test-tube.
2. Half fill the tube with water.
3. Warm the test-tube, shaking it gently from side to side, for 3 minutes. The crystal should then have dissolved.
4. Put about $50\,cm^3$ water into the graduated beaker.
5. Pour in the potassium permanganate solution from the test-tube.
6. Add more water to the test-tube and pour it into the beaker. Go on doing this until all the potassium permanganate solution has been transferred to the beaker. How do you know when to stop?
7. Pour more water into the beaker until you have filled it up to the $200\,cm^3$ mark.
8. Stir the contents of the beaker with a glass rod until the colour is even throughout.
9. Fill the test-tube (which holds about $20\,cm^3$) with solution from the beaker. What fraction of the original crystal is now dissolved in this test-tube full of solution?
10. Pour the rest of the solution into the sink. Wash out the beaker.
11. Stand the beaker on a sheet of white paper. Pour the contents of the test-tube into it. Wash the test-tube several times as before, adding the washings to the beaker.
12. Fill the beaker with water up to the $200\,cm^3$ mark. Stir well. Can you still see the colour of potassium permanganate in the beaker?

Repeat operations 9–12 until you can no longer see the permanganate colour in 12. How many times have you now diluted the original solution?

What fraction of the original crystal was dissolved in the test-tube which contained the last solution to have a colour that you could see?

The test-tube holds $20\,cm^3$ solution and you can take it that $1\,cm^3$ of solution contains 20 drops. How many drops of solution can the test-tube hold?

What fraction of the original crystal is contained in one drop of the final solution?

If you assume that each drop of the final solution contains at least one particle of potassium permanganate, what is the smallest number of particles into which the original crystal can be divided?

How small are the particles of matter?
This experiment should be done by the teacher.

Apparatus
The teacher will need:
Measuring cylinder ($10\,cm^3$)
Watch glass, 150 mm diameter
Ether (ethoxyethane)

Procedure
Arrange your pupils all around the laboratory, and then pour out $5\,cm^3$ of ether (ethoxyethane) onto a large watch glass on the front bench. Ask the pupils to put up their hands when they smell the ether. Soon, all will be able to detect the smell: it has 'filled' the room.

When this has happened it must be evident that each 'lung-full' of air must have at least one ether particle in it. Suppose a lung-full is $100\,cm^3$: to find how many 'lung-fulls' in the laboratory it will be necessary to find its volume.

To avoid possible difficulties in converting units, measure the laboratory length, breadth, and height in centimetres. Multiply them together to get the volume in cm^3. Then:

$$\text{number of ether particles} \geqslant \frac{\text{volume of laboratory in } cm^3}{100\,cm^3}$$

where $100\,cm^3$ is the volume assumed for one 'lung-full'.

As $5\,cm^3$ of ether was taken to start with, each particle can be no bigger than:

$$\frac{5}{\text{number of particles}}\,cm^3$$

where $5\,cm^3$ is the total volume of all the ether particles.

In both these experiments, stress that the results obtained are *not* 'the size of the particles' but merely a rough value, an indication of the biggest size such particles could have, assuming that they exist. Point out that on this evidence, the particles could be a great deal smaller.

Experiments A11.2a and b show that particles of matter must be very small indeed. An experiment which gives a better idea of the size of the particles may be attempted. It involves making an estimate of the thickness of a film of oil spread out over water and gives, once again, a maximum possible size for the particles.*

Experiment A11.2c

How small are the particles of oil?

Apparatus

Each pupil (or pair) will need:

Experiment sheet 43

Flat dish or tray, made of plastic (polythene) or a painted 'tin' tray, at least 25 cm square, preferably larger

Glass rod or tube drawn out to a solid point

Pipette, graduated $1\,cm^3$, and pipette filler

Procedure

The volume of a small drop of oil can be measured either as described in *Experiment sheet* 43 or by holding the drop against a millimetre scale (as described in Nuffield Physics *Teachers' Guide 1*), using a magnifying glass.

The oil film may be a monomolecular layer, but this experiment does not give any evidence one way or the other; it is *not* necessary to assume that it is monomolecular, and the conclusion does not depend upon it. Pupils need not be told about this unless they ask.

It is assumed in this experiment that the pupils understand that the volume of the oil remains unchanged, although its shape is changed

(Continued)

---

*If the pupils have been following the Nuffield Physics Course they may have done a similar experiment in their first year (see Nuffield Physics *Teachers' Guide 1*); in that case it is suggested that Experiment A11.2c be omitted and that pupils be reminded of the results of the Physics experiment.

Powdered talc (baby powder), or lycopodium powder, placed in a specimen tube with a piece of muslin stretched over its mouth and held in place with a rubber band, or a chalky blackboard rubber

Camphorated oil

from a spherical drop to a thin layer. Some pupils may be a little uncertain about this, and the teacher should make the point clear, perhaps illustrating it with a diagram such as figure A11.4.

When this drop of oil spreads out   to make a pool of oil on top
                                     of the water, the volume
Figure A11.4   of the oil is still the same

An alternative experiment using a solution of a carboxylic acid in a volatile solvent, included in the first edition, is not given in this revision. The calculation required is difficult for many pupils at this age, and an approximate idea of the size can be more easily obtained using the simple oil drop experiment. (The experiment using a solution of a carboxylic acid appears in Nuffield Advanced Chemistry (Experiment 3.6) where its full value can be explored in a determination of the Avogadro constant.)

Experiment sheet 43
In earlier experiments, you will have seen that a small amount of matter can be divided into a large number of separate parts. In this experiment you are going to try to get a very rough idea of the size of a particle of matter.

If a drop of oil is placed on the surface of water it spreads out to form a very thin film, the area covered by the film being very large compared with the size of the drop. The volume of the oil does not change, however, so the volume of the film is the same as that of the drop of oil from which it is formed. The oil film must be *at least* one particle thick, so if we can calculate its thickness we can get some idea of the upper limit to the size of an oil particle.

We have three problems to solve: how to find the volume of the oil drop used, how to find the area of the film that it produces, and how to find the thickness of this film.

A very small volume of oil will be used, just enough to cover the tip of a pointed glass rod.

To find the volume of oil that you will use
Put some camphorated oil in a dry burette, open the tap gently until the jet under the tap is filled with oil (put a dry test-tube under the tap to catch any oil which flows out). Adjust the oil level until it is on one of the cm^3 graduations. Turn the tap on so that the oil escapes a drop at a time and count the number of drops formed when 1 cm^3 of oil flows out.

Number of drops formed by 1 cm^3 of oil is

Take a small, clean watch glass and let one drop of oil fall on to it. Dip the point of your glass rod into the oil drop, withdraw it, and wipe it clean on the edge of a piece of filter paper. Dip the rod into the oil again and repeat the process, using a new, clean spot on the edge of the filter paper to clean it. Keep doing this until no more oily marks appear on the filter paper and find whether the number of 'points full' of oil that you remove from one drop is nearest to 10, 20, 50, 100, 200, or 500.

Number of oil portions obtainable from one drop is roughly

Number of oil portions obtainable from 1 cm^3 oil is roughly

Volume of one oil portion is about...................................cm^3.

To find the area of the oil film formed from one portion of oil
Put water into a clean, flat tray to a depth of about 1 cm. Sprinkle the powder provided over the surface of the water until it is evenly covered.

Obtain another drop of camphorated oil from the burette in your watch glass. Have a ruler graduated in centimetres handy.

Dip the clean point of the glass rod into the oil in the watch glass and touch the centre of the dusted surface in the tray with it. Measure the diameter of the clear area on the surface to the nearest centimetre.

The diameter is..................................cm.

Describe what happened when the point of the rod touched the water surface.

Now that you know the diameter of the oil film, there are two possible ways of estimating its area.
1. Use the formula, area of a circle = $3.14 \times (radius)^2$.
2. Find the area of the square whose side is equal to the diameter of the film and subtract one-fifth of this area to allow for the 'corners'.

Figure A11.5

(For example, if $d = 16$ cm, $d^2 = 256$ cm^2; subtract 51 to give 205 cm^2 as the area required.)

The area of the oil film that you obtained is..................................cm^2.

To find the thickness of the oil film
Divided the volume of oil used by the area of the oil film.

Thickness of oil film $= \dfrac{\text{volume}}{\text{area}} = \dfrac{\text{..................................}}{}$cm

$= $..................................cm

This is the *largest* length that an 'oil particle' can have.

If an oil particle has the shape of a cube, can you calculate how many particles there would be in 1 cm^3 of oil?

After the experiment collect the results and put them on the board for all to see. Results will vary somewhat, but thickness in the region of 2 to 5 nanometres ($2-5 \times 10^{-9}$ metres) may be expected. The experiment has therefore shown that *if* there are particles, *then* they cannot be bigger than, say, 2 nanometres in diameter. It may be helpful to express this result another way: thus, expressing the thickness as 1/2 000 000 cm *or* 2 000 000 layers thick $\equiv$ 1 cm may have a greater effect.

The teacher now faces two main problems.

1. What should be done about the use of the words atom and molecule, in place of particle (particularly if these words are familiar to pupils)?
2. How can pupils get any idea of what is meant by the extremely small distances being referred to; for example, how can they understand what a nanometre is?

With regard to the first question, teachers should use their discretion about how to introduce these words with a particular class. Two possible ways are now suggested and another is given in Alternative IIB.

One suggestion is that the meanings of these words are mentioned at this point. Pupils already know about elements and compounds from Stage I, and a discussion led by the teacher at this point might therefore go along the following lines. In the debate about particles, it was suggested that the smallest possible piece of an *element* is known as an *atom*; and that the smallest possible piece of certain types of *compound* (but not all) is known as a *molecule*. Since compounds consist of more than one element joined together, molecules of compounds must be made up of more than one atom. The oil used in the previous experiment is a compound, and so its particles must be made up of more than one atom. Atoms must therefore be even smaller than the calculated thickness of the oil film.

The second suggestion is to defer the matter and let the ideas arise as they are met in later Topics. If this step is taken, then it may be necessary to point out to pupils that a precise discussion of the meaning of such terms as atom and molecule is being deferred.

Pupils may be helped towards an understanding of the smallness of the sizes being discussed by a teacher demonstration, described in Alternative IIB, B16.1. This demonstration uses pictures of objects in a sequence of successive steps of ten-fold reduction in size, and it could be used here as well.

The (revised) film strip 2–1 'Measuring the very small'* may also be shown at this point.

Suggestion for homework

Write a 'science fiction' story in which you shrink first to a height of one millimetre and then to a height of one nanometre!

Summary

By the end of this section, pupils should know how to find experimentally the maximum size which supposed particles of matter are likely to have. They should have some idea of what this size is, and of the relation of this size to more familiar objects.

---

*The earlier version of 'Measuring the very small' (a film loop and two charts) refers to the non-SI unit, the Ångstrom and to the gram-atom.

Section A11.3 introducing the mole of atoms (gram-atom) follows. Some teachers, however, may prefer to defer this until Topic A14. They should turn to page 66 where there is an alternative route to finding formulae. This alternative route is printed inside a box of dashed lines. Then, when they reach Topic A14, they will need to begin with section A14.1 which introduces the mole of atoms.

A11.3
Comparing numbers of atoms by mass measurement

Pupils are told that elements *are* considered to be capable of subdivision into particles, and that these particles are known as *atoms*. They consider the relative masses of different elements and are led to see that if samples of different elements have masses which are in proportion to their relative atomic masses, then each one of these samples must contain the same number of atoms. The mole of atoms (gram-atom), treated as an amount of an element, arises out of this presentation.

A suggested approach

Objectives for pupils

1. Understanding the meaning of the term relative atomic mass of an element
2. Understanding the meaning and use of the terms mole of atoms (gram-atom) and the Avogadro constant
3. Knowledge of the symbols of some common elements
4. Ability to use data tables to look up relative atomic masses

The teacher will need:

Bottles of clear glass containing, respectively: 12 g carbon, 24 g magnesium, 32 g sulphur, 64 g copper

'Half-cube' box of capacity 12 dm^3 (see text)

Supply of paper bags

12 each of three different sizes of ball, say polystyrene spheres, wooden balls, and golf balls, or ballasted polystyrene spheres (see B16.2)

Film-loop projector

Film loop 2–2 'Gram-atoms (The mole of atoms)' (with notes)

(Continued)

Pupils know that matter may well consist of particles, and that the particles must be extremely small. They should be told that *if* we assume that matter is made up of particles, *then* a great deal of its behaviour becomes easy to explain and understand.

For a start we shall consider elements to be substances which are capable of subdivision into similar particles called *atoms*. We use the word atom to mean the smallest possible piece of an element. Pupils could be told that this idea was introduced into chemistry by John Dalton in 1803, and that he further assumed that atoms of the same element are all alike and that atoms of different elements differ from each other, in mass and other properties. In fact Dalton proposed a method for comparing these masses, and, subsequently, other methods were proposed by other chemists. The actual experiments and calculations by which chemists determined these relative atomic masses* are too difficult for pupils of this age, and it is therefore suggested that they should be told that we will make use of the results of these discoveries without explaining how the work was done.

Some teachers may like to discuss the matter by using an analogue of a mass spectrometer at this point (for example, the model described in Nuffield Advanced Chemistry, *Teachers' guide II*, Appendix 2, page 262). *However, this possibility should not be allowed to detract from the main thread of thought of this section.*

To familiarize pupils with relative atomic masses it is a good idea to let them look up some values in the Reference section of the *Handbook for pupils* (Table 1) and answer some simple questions such as 'How much heavier is the sulphur atom than the hydrogen atom?'

*Relative atomic masses were originally referred to as 'atomic weights'.

Optional: model to illustrate
the principle of the mass
spectrometer.

and 'How much heavier is the magnesium atom than the carbon atom?' The time spent on this sort of exercise will, of course, vary with the ability of the class.

Now that pupils know something about relative atomic masses of the elements, what use can we make of them? We must first try to take the idea away from the realms of abstract thought and make it into something concrete and visible to the pupils. For this reason the mole of atoms (gram-atom) is now introduced.

Tell the pupils that chemists find it helpful to use amounts of elements containing the numerical value of their relative atomic masses in grams; and that they refer to these amounts as moles of atoms (gram-atoms) of the elements. Show the pupils bottles containing one mole of atoms (gram-atom) each of as many solid elements as you can – for example 12 g of carbon, 24 g of magnesium, 32 g of sulphur, and 64 g of copper. In addition, have a 'half-cube' box of 12 dm³ capacity to represent 16 g of oxygen* (see also figure A11.6).

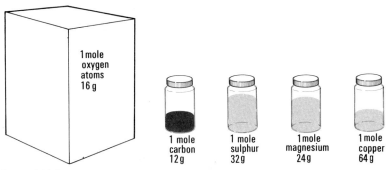

Figure A11.6

Ask the pupils if they can see any relationship involving the numbers of atoms in each bottle. Try to lead the pupils to the idea that *each bottle contains the same number of atoms*, and that the whole point of introducing the idea of the mole of atoms (gram-atom) is to be able to use equal numbers of atoms of different elements. If this fact is not evident, then it must be stated and made quite clear before proceeding; once it has been grasped, it makes many chemical phenomena easier to understand.

One way which may help to make the idea clear is to show pupils a series of articles – say polystyrene balls, wooden balls, and golf balls, or even a series of ballasted polystyrene spheres, each of known relative mass.

Now take three paper bags and put ten balls of relative mass one in one bag, ten balls of relative mass two in another bag, and so on.

*1 mole of gas occupies 24 000 cm³ at room temperature and this corresponds to a cube of side 28.8 cm. A box of dimensions 28.8 × 28.8 × 14.4 cm may therefore be used to represent 1 mole of oxygen atoms.

What can the pupils say about the relative masses of the three bags? Then, without showing the pupils, put simple proportionate numbers of one type of ball, say ten, twenty, and five, in separate bags and find their masses. Ask pupils what they can say about the relative numbers of balls in each bag, knowing only the relative masses of the balls and not the absolute masses. This exercise can be varied in a number of ways to make the point. Explain that it is a process of 'counting out by finding mass'. We are finding the number of balls by finding the mass of the bags. Similarly, we can find the number of atoms in a known quantity of an element by finding its mass. (Banks do the same when they weigh out £5-worth of 10p pieces.)

Another way to make the idea clear is to show film loop 2–2 'Gram-atoms (The mole of atoms)'. This has been made especially for this purpose. In the loop three different objects are used, having masses of 1, 2, and 3 units. 10 g of the small, 20 g of the medium, and 30 g of the large objects are then produced and each amount is shown to contain the same number of objects. The idea is then repeated using atoms of carbon (12 units), sulphur (32 units), and copper (64 units). 12 g of carbon, 32 g of sulphur, and 64 g of copper are shown to have the same number of atoms.

It is important to read the notes supplied with the film loop carefully before showing the loop to pupils.

Pupils may be curious to know the actual number of atoms in a mole of atoms (gram-atom) of an element. Although it is not important to the argument at this point many teachers have found that this knowledge helps the pupils' understanding. Pupils may therefore be told that it is a very large number, about 6×10^{23}; that it can be found experimentally by a number of independent methods too difficult to explain at this point; and that it is known as the Avogadro constant after a famous nineteenth-century Italian scientist.

As we need to use moles of atoms (gram-atoms) of elements frequently, it is convenient to have some shorthand way of referring to them. We suggest the introduction of the symbol for each element for this purpose.

Write some symbols on the board and let the pupils look up others in the Reference section of the *Handbook for pupils*. (Note that the symbol will later be used to represent one atom of an element.) At this point of the course, Zn represents one mole of atoms (gram-atom) of zinc and S one mole of atoms (gram-atom) of sulphur.

Finally, in order to bring home the physical reality of moles of atoms (gram-atoms) of elements, get the pupils to measure out different fractions of a mole of atoms (gram-atom) of several elements. They will become more familiar with the elements themselves, and will have some time to get used to thinking in terms of moles of atoms (gram-atoms) through using the idea. The following are some suggestions for such an exercise.

Exercises in obtaining moles of atoms (gram-atoms) of various elements

Apparatus

The following apparatus and chemicals will enable all the suggested exercises to be carried out by pupils:

Copies of the *Handbook for pupils*

Access to a rough balance, measuring mass to the nearest gram

Supply of paper squares to place on balance pans

Supply of 100 cm³ beakers in which to place known masses of elements

Carbon (graphite)

Copper, metal turnings

Sulphur, powdered roll

Zinc, metal granulated

Procedure

Pupils should be asked to use Table 1 in the Reference section of the *Handbook for pupils* as a source of the relative atomic masses. Ask them to obtain the quantities suggested below, on paper using a rough balance. Then place these amounts in beakers (suitably labelled) so that they may be kept until the end of the lesson and compared.

1. Obtain one mole of atoms (gram-atom) of sulphur
2. Obtain half a mole of atoms (gram-atom) of copper
3. Obtain one-tenth of a mole of atoms (gram-atom) of zinc
4. Obtain as many atoms of copper as there are in 16 g of sulphur
5. Obtain twice as many atoms of carbon as there are in 64 g of copper

In doing these exercises the pupils will begin to see the relation between mass in grams and moles of atoms (gram-atoms), namely,

$$\text{number of moles of atoms (gram-atoms) of an element} = \frac{\text{number of grams of the element}}{\text{relative atomic mass of the element}}$$

or in other words:

$$\text{the amount of an element in moles of atoms (gram-atoms)} = \frac{\text{mass of the element in grams}}{\text{relative atomic mass of the element}}$$

The numbers involved in these exercises have deliberately been kept simple; in the experiments of the next section the numbers will not be simple. Because of this special conversion scales are given in figures A11.7, A11.8, A11.11, and A11.12 (also available as Overhead Projection originals) for use with the *Experiment sheets*. These scales should also help pupils in understanding the relationship.

1. What is the mass of one-tenth of a mole of atoms (gram-atom) of:
(*a*) sodium   (*b*) hydrogen   (*c*) magnesium   (*d*) chlorine
(*e*) mercury   (*f*) silver   (*g*) potassium   (*h*) manganese   (*i*) lead
(*j*) copper.

2. Write down the number of moles of atoms (gram-atoms) present in:
(*a*) 54 g aluminium   (*b*) 44 g boron   (*c*) 48 g helium
(*d*) 11 g manganese   (*e*) 1.6 g sulphur   (*f*) 2 g oxygen
(*g*) 30 g calcium   (*h*) 42 g iron   (*i*) 16 g copper   (*j*) 2.07 g lead.

3. Write down the masses of each element in the following:
(*a*) $\frac{1}{4}$ mole of atoms (gram-atom) magnesium
(*b*) 5 moles of atoms (gram-atoms) bromine
(*c*) 0.5 mole of atoms (gram-atom) chromium
(*d*) $\frac{1}{6}$ mole of atoms (gram-atom) silver
(*e*) $\frac{3}{4}$ mole of atoms (gram-atom) iron

(f) $\frac{1}{3}$ mole of atoms (gram-atom) carbon
(g) $\frac{1}{7}$ mole of atoms (gram-atom) nitrogen
(h) 0.125 mole of atoms (gram-atom) copper
(i) $\frac{1}{6}$ mole of atoms (gram-atom) barium
(j) 0.2 mole of atoms (gram-atom) chlorine

4. Write down the mass of:
(a) Calcium having the same number of atoms as 11 g manganese
(b) Zinc having the same number of atoms as 0.1 mole of atoms (gram-atom) mercury.
(c) Potassium having the same number of atoms as 8 g magnesium
(d) Oxygen having the same number of atoms as 10 g bromine
(e) Copper having the same number of atoms as 3 g aluminium.
(f) Silicon having the same number of atoms as 10 moles of atoms (gram-atoms) sulphur.
(g) Zinc having the same number of atoms as 0.25 moles of atoms (gram-atoms) chlorine.
(h) Sodium having five times as many atoms as there are in 39 g potassium.
(i) Hydrogen having the same number of atoms as in 197 g gold.
(j) Carbon having four times as many atoms as there are in 8 g oxygen.

5. Write down the number of moles of atoms (gram-atoms) of the first named element having the same number of atoms as the second named element.

(a) nitrogen – 36 g carbon   (b) sulphur – 46 g sodium
(c) phosphorus – 80 g oxygen   (d) mercury – 3.55 g chlorine
(e) magnesium – 28 g nitrogen   (f) oxygen – 0.125 g hydrogen
(g) sulphur – 3.2 g oxygen   (h) lead – 0.64 g oxygen
(i) sulphur – 7.8 g potassium   (j) chromium – 71 g chlorine

Summary

By the end of this section pupils should know:
1. That elements are considered to be made up of atoms; all atoms of one element are alike, and differ from atoms of every other element.
2. It is possible to compare the masses of atoms of different elements, and thus build up a list of relative atomic masses.
3. The amount of an element equal to the relative atomic mass in grams is known as the mole of atoms (gram-atom) of the element.
4. One mole of atoms (gram-atom) of all elements contains the same number of atoms. This number, 6×10^{23}, is known as the Avogadro constant.
5. Symbols for the elements, such as Zn for zinc, stand for one mole of atoms (gram-atom) of the element.

Pupils should also be able to use data tables to look up relative atomic masses, and should understand the relation between grams and moles of atoms (gram-atoms).

A11.4
Determination of the formulae of simple compounds

In this section pupils use their knowledge of relative atomic masses, and of the mole of atoms (gram-atom), to find the formulae of some simple compounds experimentally. Formulae are thus presented as an expression of the proportions, by moles of atoms (gram-atoms), of the elements present in a compound, and not as a short-hand notation to be learned.

Note. Gram-formulae may be introduced as a natural extension of gram-atoms, and a formula such as CuO may then be seen to refer to an actual amount of the substance which it represents.

Objectives for pupils

1. Awareness of the experimental basis of the formulae of compounds
2. Ability to use the mole of atoms (gram-atom) to find a formula from the reacting masses of the elements concerned

The pupils should know from their work in Stage I that compounds like water, magnesium oxide, and copper oxide are made up from the elements hydrogen and oxygen, magnesium and oxygen, and copper and oxygen, respectively, but they do not know in what proportion the atoms of each element combine together to form such compounds.

The question which we now try to answer is 'How can these "combining numbers" be found?' This is where we make use of the knowledge of the mole of atoms (gram-atom) which has been acquired in the previous section. By measuring the masses of two elements which react, we can work out the relative numbers of moles of atoms (gram-atoms) present, and hence the ratio of the numbers of each atom, in the compound.

The experiments in this section are to find the chemical formulae of some compounds. It should be understood that the symbols such as Mg, Cu, and O have been introduced as standing for 1 mole of atoms (gram-atom) of the element; so a formula such as MgO or CuO must be an expression of the relative numbers of moles of atoms (gram-atoms) of each element present in the compound. Only when this is clearly understood should it be pointed out to pupils that because moles of atoms (gram-atoms) of different elements each contain the same numbers of atoms, a formula is also an expression of the relative numbers of *atoms* present in the compound.

The first two experiments below are intended for the pupils, and the third one for the teacher. Many teachers consider that good results are essential in quantitative work if pupils are to be convinced of its value; these experiments will all give such results *provided* attention is given to detail, as the specimen class results given after the *Experiment sheets* show. The order in which these experiments are attempted is not important. However, the synthesis of magnesium oxide is conceptually simpler than the analysis of copper oxide and so appears first.

To find the formula of magnesium oxide

Apparatus

Each pupil (or pair) will need :

Experiment sheet 44

Crucible and lid

Tongs

Pipeclay triangle

Bunsen burner and heat resistant mat

Tripod

Access to balance

Access to desiccator

Dry asbestos (or equivalent material) discs (to fit crucibles)

Magnesium ribbon, 15 cm length (previously cleaned so that it is free from oxide)

Procedure

Before the lesson, crucibles should be fitted with a disc of dried asbestos wool* to protect the bottom of the crucible from burning magnesium. Each prepared crucible and lid should be heated to constant mass, using the conditions to be employed for the determination, and then stored for use in a desiccator. Teachers will appreciate that the disc must be treated as part of the 'crucible and lid' in the experiment. Pupils should use the same mass of magnesium ribbon (about 0.16 g). *Experiment sheet* 44 is reproduced below.

Experiment sheet 44

There are two ways of finding the masses of metal and oxygen present in a sample of a metal oxide. One is to start with a known mass of metal oxide and to remove the combined oxygen by chemical means, and then to determine the mass of the residual metal. The other is to start with a known mass of metal, to convert it to the oxide and to find the mass of metal oxide formed, and then to calculate the mass of oxygen used. In this experiment you will use the second method. You will convert a known mass of magnesium to magnesium oxide by heating it in air.

Find the mass of a clean, dry crucible fitted with a dry asbestos disc and lid, and prepared ready for use. Enter the mass of the crucible, disc, and lid, in the table of results below.

You will be given a length of magnesium ribbon. Clean it with emery paper (or a blunt penknife).

Coil a 15 cm length of ribbon loosely, place it in the crucible on the dry asbestos disc, and find the total mass of magnesium + crucible, disc, and lid.

Support the crucible on a pipeclay triangle placed on a tripod stand.

Light a Bunsen burner and adjust the flame so that it is non-luminous and just touches the bottom of the crucible. Heat the crucible for 2–3 minutes and then use crucible tongs to lift the lid cautiously so as to admit a little air. (Why is this necessary?)

Continue heating and raising the lid at intervals until the magnesium catches fire. (You may need to increase the size of the flame.) When this happens, remove the burner and keep the magnesium burning by lifting the lid every minute or two. Try not to allow too much white smoke to escape. (Why?)

When the magnesium no longer burns, remove the lid and place it on a heat-proof mat. Replace the Bunsen burner and heat the crucible strongly for five minutes. Replace the crucible lid and allow the crucible to cool in a desiccator until you can hold it in your hand (CARE!). Find the mass of crucible, disc, and lid + magnesium oxide. Complete the table.

Mass of crucible, disc, and lid	....g
Mass of crucible, disc, and lid + magnesium	....g
Mass of crucible, disc, and lid + magnesium oxide	....g
Mass of magnesium used	....g
Mass of oxygen which combines with magnesium	....g

Complete the calculation after discussing the method with your teacher.

---

*In areas where safety regulations relating to asbestos are in force, an alternative material should be used. Asbestos paper is *not* an effective substitute for an asbestos wool disc.

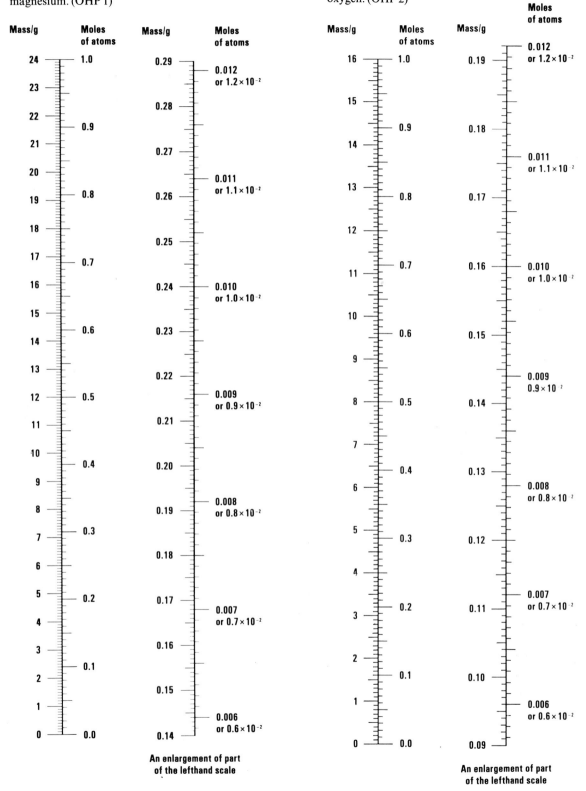

Figure A11.7
Conversion scale for
magnesium. (OHP 1)

Figure A11.8
Conversion scale for
oxygen. (OHP 2)

An enlargement of part
of the lefthand scale

An enlargement of part
of the lefthand scale

After the experiment discuss the results. Pupils are told in the *Experiment sheet* to complete their calculations after discussion with the teacher.

First, a quick check as to the accuracy of their results can be made using figure A11.9. Of course, if all pupils have used the same mass of magnesium, the check is even easier, and a bar chart can then be used to summarize the results. Any obviously out-of-line results can be discussed and discarded. If there is no agreement, the experiment will have to be done again. But here is a valuable lesson – provided, of course, that more accurate results are obtained next time.

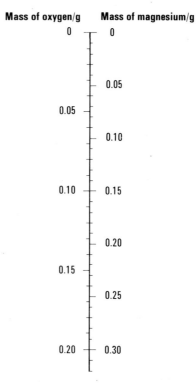

Figure A11.9

Results may be low in oxygen; three possible reasons for this are:
1. Loss of magnesium as smoke
2. Incomplete oxidation of magnesium
3. Formation of magnesium nitride.

If the experiment is carefully carried out loss of magnesium oxide as smoke is very small indeed. It is, however, difficult to ensure complete oxidation of the magnesium, as the oxide forms a protective coating on the outside, especially when a limited air supply is used. The magnesium should be given as much air as possible during the experiment provided that this condition does not lead to an excessive loss of magnesium oxide as smoke. If the class results are low, ask the pupils if they can suggest why this is so. If they suggest incomplete oxidation, get them to test their theory by adding dilute hydrochloric acid to the contents of the cold crucibles to see if any gas (hydrogen) is evolved.

The addition of a few drops of dilute nitric acid to the cooled crucible's contents, followed by a further period of heating to constant mass, will ensure the removal of both the protective coating of oxide and any magnesium nitride from the products of reaction.' (Not everyone will wish to observe this precaution at this stage of the course and so it has not been included in the general procedure for the determination.)

If after taking all reasonable precautions, the class results are *very* low in oxygen, it is suggested that rather than trying to pretend that the results 'prove' the formula is MgO it would be better to tell pupils that chemists have carried out other experiments and have strong reasons for believing that the correct formula is MgO.

The use of cleaned magnesium and pure oxygen as a demonstration experiment with special apparatus yields good results and some teachers prefer this approach to the determination.

Mathematical considerations

Although the principles underlying the determination of formulae and the relevant calculations appear simple to the teacher, it must be remembered that this is not necessarily so for the 13 to 14-year-old pupil. Pupils of this age often experience difficulty when dealing with proportion and on occasion with quite simple arithmetical operations.* In some mathematics courses, although pupils do carry out basic arithmetical processes successfully, they generally solve problems which give integers as answers. In chemistry, we use experimentally-determined data and the arithmetic seldom works out quite so easily. Some consideration ought to be given to this aspect. For example, it is suggested that relative atomic masses should be used to the nearest whole number: for this investigation 24 for magnesium and 16 for oxygen.

In addition, the rounding off of results to show the ratio $1:1$ in the final calculation may appear arbitrary and confusing to pupils. It may need to be justified.

From the bar chart of results discussed earlier the average value of magnesium oxide formed from a definite quantity of magnesium can now be found.

We now convert these masses into numbers of moles of atoms (gram-atoms). (Why? So that we can see if moles of atoms combine in simple ratios.) Pupils have had some experience of this in the previous section; but in the examples used there, the numbers were kept simple. In this experiment, the numbers are likely to be awkward, and conversion scales (figures A11.7 and A11.8, also available as Overhead Projection originals) may be used. For example, from figure A11.7, pupils can see that 12 g of magnesium is the 'same

---

*For example, there are several different methods in common use for subtraction. Pupils can be confused by such differences in procedure for these sums (often for understandable reasons). It is important for the teacher to be aware of the procedures for all arithmetical processes which are familiar to the pupils.

thing' as 0.5 mole of atoms (gram-atom) of magnesium. By using these conversion scales, pupils can read off the number of moles of atoms (gram-atoms) corresponding to the masses of magnesium and oxygen which react together. (*Note.* Care is needed to ensure the correct scale is used for each element. Pupils should be encouraged to abandon their reliance on the scales as quickly as possible.)

If pupils have been allowed to use different initial masses of magnesium in their experiments, a slightly different approach will be needed. The approach used in the Alternative route (pages 68–71) might be used, or each group could be asked to calculate their result and a bar chart made of the results of their calculations. Out-of-line results could then be discussed and discarded as before.

A specimen set of class results is shown in Table A11.1 and is followed by a specimen calculation.

Mass of magnesium oxide formed /g	
A	0.242
B	0.248
C	0.246
D	0.244
E	0.246
[F	0.220] ← ——discard
G	0.246
H	0.248
Average value	0.246

	Magnesium	Oxygen
Reacting mass	0.152 g	0.094 g
Divide reacting mass by relative atomic mass of element	$\frac{0.152}{24} = 0.0063(3)$	$\frac{0.094}{16} = 0.0058(8)$
Divide by smallest ratio	$\frac{0.0063(8)}{0.0058(8)} = 1.07$ $= 1$ (to nearest whole number)	$\frac{0.0058(8)}{0.0058(8)} = 1$

Table A11.1 Specimen results – Experiment A11.4a
Mass of magnesium used for each determination 0.152 g

And so the formula of magnesium oxide is most probably MgO.

Finally it is suggested that the teacher shows a bottle containing 24 g magnesium, a 12 dm^3 half-cube to represent 16 g oxygen, and a bottle containing 40 g magnesium oxide as a summary for this study. No proof need be offered for assuming 16 g oxygen occupies 12 dm^3. The summary may be presented by stating '*This* amount of magnesium and *this* amount of oxygen combine to produce *this* amount of magnesium oxide'.

To find the formula of black copper oxide

Apparatus

Each pupil (or pair) will need:

Experiment sheet 45

Hard-glass test-tube, 125 × 16 mm, with small hole near the closed end

Length of rubber tubing fitted with cork and glass tube

Bunsen burner and heat resistant mat

Access to two gas taps (see *Note 1*) and balance

Stand, boss, and clamp

Pure dry black copper(II) oxide (exactly 2 g), analytical grade (see *Note 2*)

Desiccator in which to store copper(II) oxide

Procedure
Experiment sheet 45 is reproduced opposite.

In this experiment, natural gas (see *Note 1*) is used for the reduction of black copper(II) oxide. Each pupil (or pair of pupils) should use exactly 2.00 g of good quality dry copper(II) oxide (see *Note 2*) and this could be given to them in weighed test-tubes. The tube is then set up as shown in figure A11.10 in the *Experiment sheet*.

When reduction is complete and the burner is removed, it is important to keep a small stream of gas passing through the apparatus until it is quite cold, otherwise air will enter and oxidize the copper.

Note 1. The experiment may be carried out with town gas in a similar way but less heat is required; a 'colourless' Bunsen burner flame should be used instead of a roaring one.

Note 2. It is *essential* to use *dried analytical grade* copper(II) oxide to obtain good results. Heat the copper(II) oxide before the lesson in an open dish to 300–400 °C for ten minutes to drive off water vapour, and allow it to cool in a desiccator.

Concerning natural gas
The composition of natural gas varies slightly depending on its original source, but it is mainly methane. It will reduce dry analytical grade copper(II) oxide satisfactorily – PROVIDED STRONG HEAT is employed.

Town gas and pure hydrogen (the latter *must* be restricted to teacher demonstrations) reduce copper oxide at lower temperatures than natural gas. If hydrogen is used, it should be taken from a cylinder of gas and *not* from a laboratory preparation, since it is difficult to ensure *complete* removal of air from the apparatus used to prepare the gas, and explosions may result when the surplus gas is ignited. Table A11.2 shows typical gas analyses (South Eastern Gas Board area) and enables a comparison to be made between natural gas and manufactured gas.

Constituent	Natural gas (%v/v)	Manufactured gas (%v/v)
Methane	94.7	29.5
Ethane	3.0	1.0
Propane	0.6	0.2
Butane	0.2	0.1
Carbon monoxide	0.0	2.0
Carbon dioxide	0.0	12.0
Hydrogen	0.0	54.7
Nitrogen	1.5	0.5

Table A11.2 A comparison between natural gas and manufactured gas

Experiment sheet 45

In this experiment, you will convert a known mass of black copper oxide to copper by heating it in a stream of natural gas (or town gas or hydrogen).

The apparatus to use is shown in the diagram.

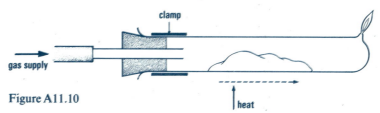

Figure A11.10

The test-tube must be dry. Find its mass (without the cork and glass tube) and enter it in the table below. Add some pure, dry black copper oxide and find the mass of tube + oxide. Your teacher will tell you the exact mass of copper oxide to use. Fix the tube horizontally in a clamp (be careful not to spill any oxide) and insert the cork and glass tube. Connect the glass tube to the bench gas supply.

You need a gentle stream of gas through the special test-tube; if the gas flows too quickly it will blow the oxide out of the hole at the end of the tube. Place the palm of one hand just above the hole at the end of the tube and use the other hand to slowly turn the gas tap. You will feel a coolness on your palm when gas is flowing through the tube. When this happens leave the gas tap at this position, wait for ten seconds, and light the gas when it leaves the tube. Adjust the flame so that it is about 3 cm high.

Light a Bunsen burner, turn the gas full on, and adjust the air hole to give a roaring flame (this is for natural gas, your teacher will tell you how to adjust the burner if you use town gas). Heat the copper oxide at the point indicated by the arrow in the diagram, until a red glow appears in it. By slowly moving the flame to the right (dotted arrow) lead this glow through the copper oxide. Finally heat the whole of the solid residue in the tube by moving the flame to and fro along it for 3–5 minutes. The copper oxide should then have been changed to copper, a pink powder.

Turn out the Bunsen flame and allow the test-tube to cool with gas still passing through it. (Why?)

When the test-tube is cool enough for you to touch the under surface comfortably with your hand (CARE!), turn out the gas flame, remove the cork and delivery tube, and find the mass of the test-tube + copper.

Complete the table.

Mass of test-tube	.... g
Mass of test-tube + copper oxide	.... g
Mass of test-tube + copper	.... g
Mass of copper remaining	.... g
Mass of oxygen removed	.... g

Complete the calculation after discussing the method with your teacher.

After the experiment discuss the results. The *Experiment sheet* does not give instructions for the calculation, although pupils who worked through the magnesium oxide experiment successfully should be able to do it, given the conversion scales shown in figures A11.11 and A11.12 (also available as Overhead Projection originals). However, some pupils may need more help.

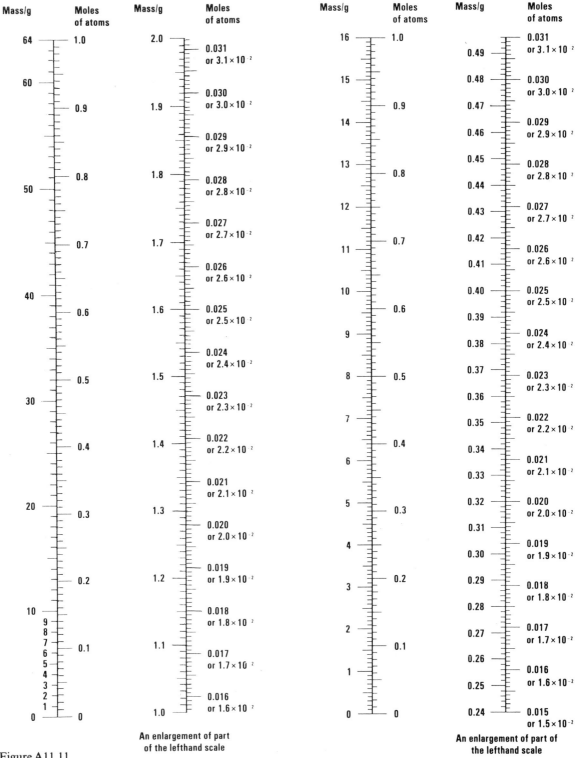

| Mass/g | Moles of atoms | Mass/g | Moles of atoms | Mass/g | Moles of atoms | Mass/g | Moles of atoms |

Figure A11.11
Conversion scale for copper. (OHP 3)

First, discuss the actual values obtained for quantities of copper and oxygen. (All pupils should have started with the same amount of copper oxide.) As in A11.4a, they can be plotted on bar charts (see figure A11.13) and obviously poor results discarded. In order to make a reasonable inference that the ratio of copper to oxygen is

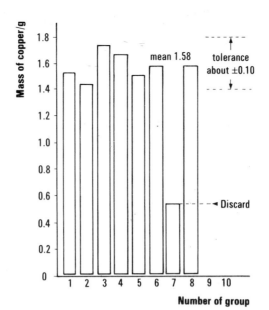

Figure A11.13
A bar chart like this
provides a good way of
comparing the experimental
results of various groups in
the class and of obtaining a
mean value. Some idea of
the tolerance may also be
obtained. (The value
obtained by group 7 would,
of course, be neglected in
working out the class
average.)

1 : 1, the value for copper must lie between 1.550 g and 1.645 g and the value for oxygen between 0.450 g and 0.355 g if the initial mass of copper oxide was 2.000 g. For this reason it is probably better to work with the class average than with individual results.*

In Table A11.3 the results from figure A11.13 (barring those for group 7) were used to work out the ratio of copper to oxygen, for individual groups.

Group	Mass of tube /g	Mass of tube + copper oxide /g	Mass of tube + copper /g	Moles Cu	Moles O	Ratio Cu : O
1	40.641	42.641	42.233	0.0249	0.0255	1/1.02
2	27.479	29.479	29.001	0.0238	0.0293	1/1.22
3	17.629	19.629	19.219	0.0248	0.0256	1/1.03
4	40.537	42.537	42.123	0.0248	0.0259	1/1.04
5	28.849	30.849	30.446	0.0250	0.0252	1/1.01
6	18.111	20.111	19.706	0.0249	0.0253	1/1.02
8	18.117	20.117	19.728	0.0252	0.0243	1/0.96

Table A11.3 Specimen results – Experiment A11.4b

Specimen calculation (using the class average value)
In one class, the class average for the mass of copper in 2.00 g samples of copper oxide was found to be 1.58 g. The difference, 0.42 g, is therefore the average mass of oxygen in these samples.

Figure A11.12
Conversion scale for
oxygen. (OHP 4)

These reacting masses now need to be converted into numbers of moles of atoms (gram-atoms). To do this we divide the masses in grams by the relative atomic mass of the appropriate element, thus moles of atoms of copper $\frac{1.58}{64} = 0.024(7)$ and moles of atoms of oxygen $\frac{0.42}{16} = 0.026(3)$.

*For a fuller discussion see Malpas, A. J. (1973) 'Combining masses, ratios, and chemical formulae' *School Science Review*, **54**, No. 188, 542.

Alternatively the conversion scales (figures A11.11 and A11.12) can be used.

The calculation could be set out as follows:

	Copper	Oxygen
Reacting mass	1.58 g	0.42 g
Divide reacting mass by relative atomic mass of element	$\frac{1.58}{64} = 0.0247$	$\frac{0.42}{16} = 0.0263$
Divide by smallest ratio	$\frac{0.0247}{0.0247}$ $= 1$	$\frac{0.0263}{0.0247}$ $= 1.07$ $= 1$ (to nearest whole number)

And so the formula of the black copper oxide is most probably Cu_1O_1 or more simply CuO.

When this conclusion has been reached, produce a bottle containing 64 g copper, a 12 dm³ half-cube to represent 16 g of oxygen, and a bottle containing 80 g copper(II) oxide, with a remark such as 'This amount of copper and this amount of oxygen combine to produce this amount of copper oxide' (see figure A11.14).

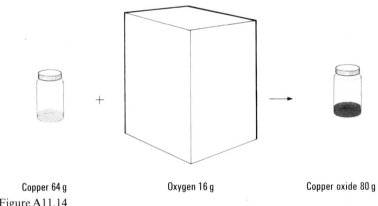

Copper 64 g   Oxygen 16 g   Copper oxide 80 g

Figure A11.14

Some teachers find it helpful to introduce the term *gram-formula* during this section. It could be used, for example, to describe the 80 g of copper oxide. Statements like '1 mole of atoms (gram-atom) of copper combines with 1 mole of atoms (gram-atom) of oxygen to produce 1 gram-formula of copper oxide' should be used. There is no reason why we should not replace the words:

| 1 mole of atoms (gram-atom) of copper | combines with | 1 mole of atoms (gram-atom) of oxygen | to give | 1 gram-formula of copper oxide |

by the following symbols:

$$Cu \quad + \quad O \quad \longrightarrow \quad CuO$$

providing we remember that the pupils at this stage associate these symbols with moles of atoms (gram-atoms) *and not with individual atoms, etc.*

We should at this stage refrain from writing the usual balanced equation:

$$2Cu \quad + \quad O_2 \quad \longrightarrow \quad 2CuO$$

because we have not yet introduced molecules; this will be done later in the scheme during Topic A14.

It is, however, desirable to get pupils used to including the state symbols, s for solid, l for liquid, and g for gas, right from the beginning. It is suggested that the equation which pupils should write at this stage is:

$$Cu(s) \quad + \quad O(g) \quad \longrightarrow \quad CuO(s)$$

At the same time draw the pupils' attention to what is happening in the 'world of atoms' by making statements like those shown in figure A11.15.

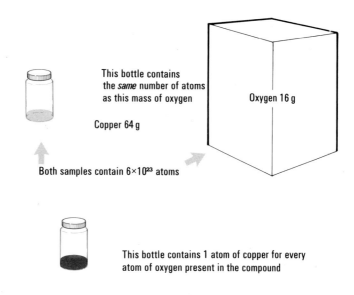

This bottle contains the *same* number of atoms as this mass of oxygen

Oxygen 16 g

Copper 64 g

Both samples contain 6×10^{23} atoms

This bottle contains 1 atom of copper for every atom of oxygen present in the compound

Copper oxide 80 g

Figure A11.15

Experiment 11.4c

What is the formula of mercury chloride?
This experiment must be done by the teacher.

Apparatus
The teacher will need:
$100 \, cm^3$ beaker
Water-bath or large beaker
Filter paper
Access to balance and oven
(Continued)

It gives convincing results, and is important as it gives an example of a compound having a ratio of moles of atoms (gram-atoms) of 1:2. For these reasons it should not be omitted.

Procedure
Transfer a known mass (about 5 g) of mercury(II) chloride into a $100 \, cm^3$ beaker of known mass. Add $30 \, cm^3$ of distilled water and heat on a water bath. Add about $10 \, cm^3$ of hypophosphorous acid solution. The mercury(II) chloride is soon reduced to the metal

Bunsen burner, tripod, gauze, and heat resistant mat

Measuring cylinder, 25 cm³

Glass rod

Acetone (propanone)

Mercury(II) chloride

Hypophosphorous acid (phosphinic acid) solution (about 50%)

which, on further heating and stirring, collects into one or more globules. When this stage has been reached, the metal is washed by decantation with water and acetone (propanone) in succession, the last few drops being removed by filter paper. Allow to stand for two or three minutes and then find the total mass of the beaker and mercury.

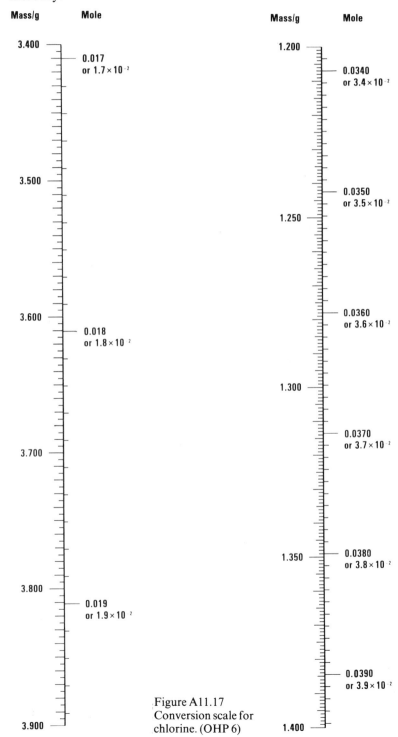

Figure A11.16
Conversion scale for mercury. (OHP 5)

Figure A11.17
Conversion scale for chlorine. (OHP 6)

Topic A11 Atoms in Chemistry

Treatment of results

Mass of beaker empty	m_1 g
Mass of beaker and mercury(II) chloride	m_2 g
Mass of beaker and mercury	m_3 g
Mass of mercury(II) chloride taken	$(m_2 - m_1)$ g
Mass of mercury formed	$(m_3 - m_1)$ g

The calculation may be carried out in the same way as in the previous experiment (A11.4b). Conversion scales for use with this experiment are shown in figures A11.16 and A11.17, and are also available as overhead projection originals. One mole of atoms (gram-atom) of mercury is found to combine with two moles of atoms (gram-atoms) of chlorine. Hence the formula of mercury chloride is $HgCl_2$.

Once again it may be useful to produce a bottle containing 201 g of mercury, two half-cubes* to represent two moles of atoms (gram-atoms) of chlorine, and a bottle containing 272 g (1 mole or gram-formula) of mercury(II) chloride.

Teachers who are familiar with the first edition of the Sample Scheme may note that the formula of water experiment has been omitted. Those who wish to perform the experiment will find full details in Topic B18, Experiment B18.4d. Any teachers not already familiar with the experiment and who wish to carry it out are advised that it requires practice if accurate results are to be obtained. Since hydrogen is used, *full safety precautions MUST be taken.*

Now turn to page 73 to complete the Topic.

---

*The half-cubes are used to represent visually two moles of atoms (gram-atoms) of chlorine, i.e. 71 g of chlorine. At the same time they show the volume which this mass of chlorine occupies under room conditions – *but it is the mass of the chlorine which is relevant at the moment.* If the pupils should observe that you have represented one mole of atoms (gram-atom) of oxygen and one mole of atoms (gram-atom) of chlorine by a half-cube of the same volume, one might reply that they have made an interesting and important observation, which will have important repercussions later in the course (A14.7).

An alternative treatment leading to the derivation of the
formulae of simple compounds without reference to the mole
of atoms (gram-atom), that is, by omitting section A11.3.

*This replaces sections A11.3 and A11.4 and requires the mole of
atoms (gram-atom) to be introduced in section A14.1 (see
page 110).*

A11.4
Alternative route: Determination of the formulae
of simple compounds without using the mole concept

In this section, pupils are introduced to a simplified form of
Dalton's theory and use relative atomic masses to determine
the formulae of some compounds.

A suggested approach

Objectives for pupils

1. Knowledge of the symbols
and relative atomic masses of
some common elements
2. Awareness of the
experimental basis of the
formulae of compounds
3. Ability to find the formula
of a simple binary compound
from the reacting masses of the
elements concerned
4. Knowledge of the constant
composition of chemical
compounds

Earlier sections of this Topic (A11.1 and A11.2) reviewed
some evidence to support a particulate theory of matter. Tell
pupils that the smallest particles of elements are called atoms
and we can represent these atoms by symbols: we could use
circles or triangles to represent atoms taking part in chemical
reactions, for example. Thus, the synthesis of magnesium
oxide from magnesium and oxygen could be represented by
coloured spheres or by drawing something like figure A11.18.
(Pupils know from their work in Stage I that compounds like
water, magnesium oxide, and copper oxide are made up from
the elements hydrogen and oxygen, magnesium and oxygen,
and copper and oxygen, respectively – but they do not know
the relative number of atoms of each element in these
compounds.)

John Dalton used the same sort of approach as in figure
A11.18 to understand chemical reactions as long ago as 1803.
His symbols were very like those in figure A11.18 but were
more varied so that he could represent atoms of all the
elements known to him. We still use symbols to represent the
atoms of elements but these are now related to the names of
elements: H represents one atom of hydrogen; O represents
one atom of oxygen, I one atom of iodine; etc. Many
discoveries of elements have been made since this system was
introduced and several elements have names beginning with
the same letter. Our system of symbols allows for this as with
Carbon C; Calcium Ca; Chlorine Cl; Cobalt Co. Other
elements have symbols related to their Latin names, as with
lead Pb (*plumbum*); iron Fe (*ferrum*); and copper Cu (*cuprum*).

Chemists have found that elements possess distinctive
properties: thus during Stage I pupils found that magnesium
was very different from oxygen or chlorine. Dalton was the
first to suggest that the smallest particles of elements (atoms)

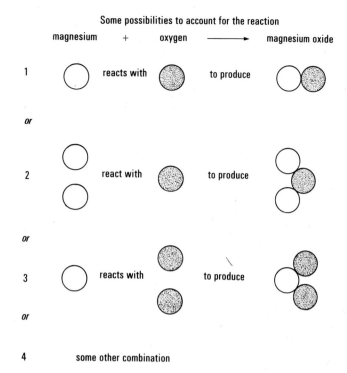

Figure A11.18
Reaction of magnesium with oxygen. (OHP 7)

also had the distinctive properties associated with these elements. Indeed, we now know that not only is this so but that one atom of hydrogen possesses only $\frac{1}{12}$th the quantity of matter present in an atom of carbon or only $\frac{1}{16}$th of that present in an atom of oxygen. Chemists refer to this property of atoms as the relative atomic mass of an element. A list of relative atomic masses and symbols occurs in the Reference section of the *Handbook for pupils* (Table 1).

After discussing the table, tell pupils that this information can be used to determine the formulae of chemical compounds. We could use it, for example, to find out the composition of magnesium oxide, distinguishing whether it could be represented by the formula MgO or Mg_2O or MgO_2 or some other combination – as suggested by figure A11.18.

This introduction must not be hurried in any way and questions raised by the pupils may require carefully worded answers to avoid giving answers to the investigations in this section. One problem arises: does a compound always have the same composition? The atoms from which compounds are made are extremely small. They are too small to see but we need to know whether atoms always react together in the same way. Does magnesium always react with oxygen in the same way? If it does, how could we determine the composition of magnesium oxide? How could we set about solving this puzzle?

The listing of pupils' suggestions and the discussion and subsequent outlining of the procedure for the next experiment should not be hurried. The experimental determination will require a double period and should *not* be considered in the same lesson as this introductory material.

During the investigation, make the point that pupils need to establish whether magnesium will always react with the same proportion of oxygen. Each group of pupils must begin with a known quantity of clean magnesium ribbon in the range 0.2 grams to 1.5 grams. During the preliminary discussion, the teacher can set the scene for a discussion of experimental results by collecting results from the groups either on the blackboard or by using an acetate sheet for use with an overhead projector.

Alternative route
Finding the formula of magnesium oxide by experiment

Procedure
Experimental details are given in *Experiment sheet* 117 which is reproduced below. It is suggested that each group uses a known mass of magnesium in the range 0.2 to 1.5 g.

Before the lesson each crucible should be fitted with a disc of dried asbestos*to protect the bottom of the crucible from the burning magnesium. Each prepared crucible and lid should be heated to constant mass using the conditions to be employed for the determination, and then stored in a desiccator. Point out that the disc must be considered to be part of the 'crucible and lid' in the experiment.

Experiment A11.4a

Apparatus

Each pupil (or pair) will need:

Experiment sheet 117

Crucible and lid (*Note*. Some teachers may prefer to prepare this apparatus before the lesson)

Dry asbestos* (or equivalent material) discs (to fit crucibles)

Tongs

Pipeclay triangle

Bunsen burner and heat resistant mat

Tripod

Emery paper

Access to desiccator

Access to balance

Magnesium ribbon (free from oxide coating) (quantity between 0.2–1.5 g)

2M nitric acid

Experiment sheet 117
You are going to measure the masses of magnesium and oxygen which combine together to form magnesium oxide.

Weigh the crucible, containing an asbestos disc, and a crucible lid. Place the apparatus on a pipeclay triangle on a tripod, underneath which is an asbestos square to protect the bench from reflected heat. Heat the crucible gently by means of a Bunsen burner flame and then more strongly for several minutes. After allowing the crucible to cool, the crucible, lid, and asbestos inside the crucible are reweighed as a single unit of apparatus. This exercise is repeated until the three together are found to have a constant mass. Any one of these items may contain some water.

Why do you think we take this trouble to heat the apparatus to a constant mass?

The purpose of the asbestos disc is to protect the surface of the crucible from the action of burning magnesium.

Use a known length of magnesium ribbon for your experiment. Discuss the length you are to use for your experiment with your teacher. It will be between 10 cm and 30 cm long. Scrape the surface of the magnesium free from magnesium oxide by using some emery paper.

*In areas where safety regulations relating to asbestos are in force, an alternative material should be used. Asbestos paper is *not* an effective substitute for an asbestos wool disc.

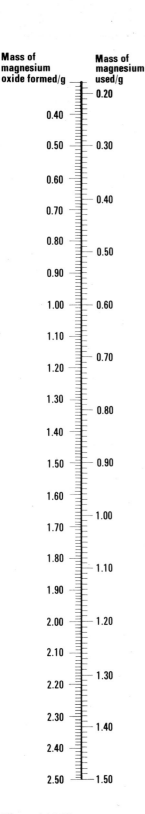

Mass of magnesium oxide formed/g	Mass of magnesium used/g
	0.20
0.40	
0.50	0.30
0.60	
0.70	0.40
0.80	
	0.50
0.90	
1.00	0.60
1.10	
1.20	0.70
1.30	
	0.80
1.40	
1.50	0.90
1.60	
1.70	1.00
1.80	
	1.10
1.90	
2.00	1.20
2.10	
2.20	1.30
2.30	
	1.40
2.40	
2.50	1.50

Figure A11.19

Why should you clean the surface of your piece of magnesium?

Coil the magnesium tightly and place it on the dried asbestos wool in your crucible. Replace the lid of the crucible and weigh the crucible and its contents as a single item of equipment.

Heat the crucible – just as you did first of all. When the magnesium ignites, heat it more strongly. Your Bunsen burner flame should never be too big otherwise the loss of magnesium oxide in the experiment is excessive. You may lift the lid CAREFULLY during the experiment to allow a little air (oxygen) into the crucible when the magnesium has ignited. To do this, the Bunsen burner should be moved well away from the crucible and the lid lifted slightly but kept level just over the rim of the crucible.

How would the loss of magnesium oxide alter your experimental result?

As soon as the magnesium ceases to flare up when the crucible lid is raised and appears to have finished burning, heat the crucible strongly but lift the crucible lid carefully a few times as in the previous part of the experiment. Allow the crucible and its contents to cool and reweigh them.

Add six drops of dilute nitric acid to the contents of the crucible. Replace the crucible lid and heat the crucible gently, carefully raising and lowering the lid of the crucible as in the previous part of the experiment. When the contents of the crucible are dry, heat the crucible more strongly. Allow the crucible and its contents to cool and then reweigh until a constant mass is obtained.

Why is it necessary to repeat the heating and reweighing to obtain a constant mass?

Results

Mass of crucible (asbestos and lid) + magnesium .... g

Mass of crucible (asbestos and lid) .... g

Mass of magnesium .... g

Mass of crucible (asbestos and lid) + magnesium oxide .... g

Mass of crucible (asbestos and lid) + magnesium .... g

Mass of oxygen .... g

Use the mass of magnesium and the mass of oxygen that you have determined by your experiment to obtain the formula of magnesium oxide. The relative atomic mass of magnesium is 24 and that of oxygen is 16.

Figure A11.19 offers a simple way for the teacher to check experimental results. Even so, it will be necessary for the teacher to discuss each set of results and to show whether unreacted magnesium is present in at least one instance. Class results should indicate a linear relationship between the reacting masses of magnesium and oxygen (see figure A11.20) and so we can deduce that magnesium oxide possesses constant composition.

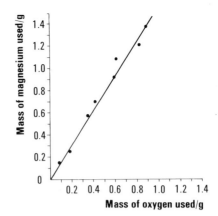

Figure A11.20

If inadequate care has been taken with the experimental work (as indicated by the use of figure A11.19 by the teacher), it may be more appropriate to base an initial discussion on figure A11.21 rather than on figure A11.20.

The slope of the graph shown in figure A11.20 provides additional information about the composition of magnesium oxide using Daltonian-style simplification rules of the type shown in figure A11.18. Thus, *if* we believe that – in the ultimate sense – one atom of magnesium combines with one atom of oxygen to form magnesium oxide, then the ratio of the relative atomic masses of magnesium and oxygen should also agree with the experimentally determined value. It would then follow that magnesium oxide could be represented by the formula MgO. On the other hand, two atoms of magnesium might combine with one atom of oxygen to form magnesium oxide. This would require a different relationship between the reacting masses of the elements and the formula of magnesium oxide would then be Mg_2O. Other possibilities exist and may be discussed.

Figure A11.20 shows that the experimental result does *not* depend on one determination since it illustrates the line of best fit for the various determinations. It provides the ratio:

$$\frac{\text{mass of magnesium used, in grams}}{\text{mass of oxygen used, in grams}}$$

Figure A11.21 shows theoretical lines which may account for the possibilities shown in figure A11.18. Thus, if magnesium oxide were of composition Mg_2O, the line corresponding to this composition should yield the ratio:

$$\frac{\text{mass of magnesium used, in grams}}{\text{mass of oxygen used, in grams}} = \frac{24 \times 2}{16 \times 1} = \frac{3}{1}$$

using the relative atomic masses of magnesium and oxygen.

Similarly, for the composition MgO_2, the line would yield the ratio:

$$\frac{\text{mass of magnesium used, in grams}}{\text{mass of oxygen used, in grams}} = \frac{24 \times 1}{16 \times 2} = \frac{3}{4}$$

Figure A11.21 shows that the experimentally determined data fall close to the line corresponding to the composition MgO. This presentation avoids arithmetical difficulties which can be an *initial* barrier to the interpretation of experimental results.

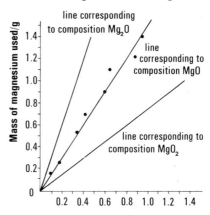

Figure A11.21
Reaction of magnesium with oxygen, with lines corresponding to the compositions of Mg_2O, MgO, and MgO_2. (OHP 8 does not include the experimental points.)

Alternative route

To find the formula of black copper oxide

Apparatus

Each pupil (or pair) will need:

Experiment sheet 45

Hard-glass test-tube, 125 × 16 mm, with small hole near the closed end

Length of rubber tubing and glass delivery tube, mounted in a cork to fit the test-tube

Bunsen burner and heat resistant mat

Access to two gas-taps

Access to balance

Stand, boss, and clamp

Note to teachers
It is important to use analytical grade copper oxide for this experiment. 'Technical' copper oxide contains appreciable quantities of copper and gives poor results. For the best results, heat the copper oxide before the lesson in an open dish at 300–400 °C for several minutes to drive off any absorbed water vapour. Allow the dish to cool in a desiccator containing either anhydrous calcium chloride or silica gel.

Procedure
Town gas or natural gas (see Table A11.2) is used for the reduction of copper oxide. Remind pupils that when the reduction is complete, it is essential to keep a small stream of gas passing through the apparatus until it is quite cold: this prevents air from entering the apparatus and reoxidizing the copper. Note that each group will need to be told how much copper oxide to use. It is suggested that between 0.5 g and 4.5 g is appropriate.

(Continued)

Pure dry black copper(II) oxide (use between 0.5 and 4.5 g – analytical grade preferred)

Desiccator in which to store pure dry copper(II) oxide

Experimental details are given in *Experiment sheet* 45, see page 59.

The argument and calculation follows that used in the previous experiment. Theoretically determined graphs corresponding to the projected formulae for black copper oxides, Cu_2O, CuO, and CuO_2 are shown in figure A11.22 and typical experimental results can be seen to fall around the line corresponding to the formula CuO.

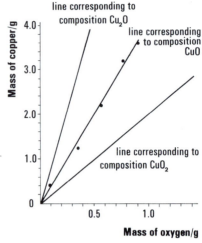

Figure A11.22
Reaction of copper with oxygen, with lines corresponding to the compositions Cu_2O, CuO, and CuO_2. (OHP 9 does not include the experimental points.)

Experiment A11.4c

Alternative route
What is the formula of mercury chloride?
This experiment must be done by the teacher.

Apparatus

The teacher will need:

$100 cm^3$ beaker

Water-bath or large beaker

Filter paper

Access to balance and oven

Bunsen burner, tripod, gauze, and heat resistant mat

Measuring cylinder, $25 cm^3$

Glass rod

Acetone (propanone)

Mercury(II) chloride

Hypophosphorous acid (phosphinic acid) solution (about 50%)

Procedure
Transfer a known mass (about 5 g) of mercury(II) chloride to a $100 cm^3$ beaker of known mass. Add $30 cm^3$ of distilled water and heat in a water bath. Add about $10 cm^3$ of hypophosphorous acid solution. The mercury(II) chloride is soon reduced to the metal which, on further heating and stirring, collects into one or more globules. When this stage has been reached, the metal is washed by decantation with water and acetone (propanone) in succession, the last few drops being removed by filter paper. Allow to stand for two or three minutes and then find the total mass of the beaker and mercury.

Treatment of results

Mass of beaker empty	m_1 g
Mass of beaker and mercury(II) chloride	m_2 g
Mass of beaker and mercury	m_3 g
Mass of mercury(II) chloride taken	(m_2-m_1) g
Mass of mercury formed	(m_3-m_1) g

Topic A11 Atoms in Chemistry

The mass of mercury $(m_3 - m_1)$ g formed is compared with the mass of chlorine present in mercury chloride, $(m_2 - m_3)$ g. These masses are then used as a ratio and compared with various theoretical mass ratios, as in the previous two examples.

After completing the demonstration, it is suggested that the teacher spends some time revising the procedure for determining the formula of a simple compound. Thus,

2.3 g of sodium combines with 0.8 g of oxygen

Therefore:

$$\frac{\text{mass of sodium/g}}{\text{mass of oxygen/g}} = \frac{2.3}{0.8} = 2.9 \text{ (correct to 1 place of decimals)}$$

It may be appropriate to question this approximation before going on to interpret the result. We try various theoretical possibilities:

1. If one sodium atom combines with one oxygen atom to form sodium oxide, then:

$$\frac{\text{relative atomic mass of sodium}}{\text{relative atomic mass of oxygen}} = \frac{23}{16} = 1.4$$

2. If two atoms of sodium combine with one atom of oxygen to form sodium oxide, then:

$$\frac{\text{relative atomic mass of sodium} \times 2}{\text{relative atomic mass of oxygen}} = \frac{2 \times 23}{16} = 2.9$$

It follows that the formula Na_2O yields a similar result to that found by experiment.

Other examples may follow as appropriate for a given class.

Note. Whichever route has been followed in A11, pupils now go on to Topics A12 and A13. Section A14.1 is only used if the alternative route has been followed, and it introduces the mole of atoms.

It is suggested that the teacher concludes the Topic (both routes) by tabulating the formulae of some simple compounds as a teaching aid. This list can be extended from time to time. A suitable table might use the following headings:

Element	Relative atomic mass	Symbol	Oxide	Chloride	Bromide	Iodide
Sodium	23	Na	Na_2O	NaCl	NaBr	NaI
Phosphorus	31	P	P_2O_5	PCl_5	—	—
Silver	108	Ag	Ag_2O	AgCl	—	—
Magnesium	24	Mg	MgO	—	—	—

The Topic can be rounded off with a discussion of the terms analysis and synthesis using various experiments which have either been mentioned in the Topic or have been used as examples when constructing the above table.

Suggestions for homework

1. Make a summary of what you have learnt in this Topic.
2. In each of the following examples, calculate the formula of the compound:
a. The hydride of nitrogen where 1.4 g of nitrogen combines with 0.3 g of hydrogen.
b. The nitride of magnesium where 0.7 g of magnesium combines with 0.28 g of nitrogen.
c. An oxide of phosphorus where 0.62 g of phosphorus combines with 0.48 g of oxygen.
d. An oxide of lead where 0.69 g of lead with 0.064 g of oxygen.
e. Silicon oxide where 3.5 g of silicon combines with 4.0 g of oxygen.
f. Manganese oxide where 1.65 g of manganese combines with 0.64 g of oxygen.
g. An oxide of nitrogen where 4.2 g of nitrogen combines with 12 g of oxygen.
h. Chromium chloride where 2.6 g of chromium combines with 5.3 g of chlorine.
i. An iron oxide where 1.68 g of iron combines with 0.72 g of oxygen.
j. Sodium chloride where 2.3 g of sodium combines with 3.55 g of chlorine.
3. *(For those students who have used moles of atoms (gram-atoms) in this Topic).* Calculate the number of moles of atoms (gram-atoms) of the elements combining with 1 mole of atoms (gram-atom) of oxygen in the following list and hence deduce the formula of the oxide.
a. 0.69 g of sodium formed 0.93 g of sodium oxide.
b. 10.35 g of lead formed 11.15 g of lead oxide.
c. 0.030 g of nitrogen formed 0.094 g of an oxide of nitrogen.
d. 5.6 g of iron combined with 2.4 of oxygen.
e. 1.22 g of phosphorus combined with 0.96 g of oxygen.

Summary

During this Topic, pupils should have looked at some of the evidence available to support the particulate nature of matter and have some idea of the size of such particles. In addition, they will have considered the possibility of counting particles by mass measurement and used the concept of the mole of atoms (gram-atom) in the determination of the formulae of simple compounds.

The alternative route to formulae may have been used. This does not require the mole concept but uses Daltonian-type simplification rules.

They should now know how to find the formula of copper oxide, magnesium oxide, and mercury chloride, and be able to work out the formulae of similar compounds from experimental data. They should understand how to write the formula of a compound using symbols.

Investigation of 'salt gas'

Purposes of the Topic	1. To provide an opportunity to do some qualitative investigations. 2. To introduce the properties of an important compound, hydrogen chloride. 3. To use the ideas developed in Topic A11 and to determine the formula of hydrogen chloride. 4. To provide an opportunity to discuss the terms 'element' and 'compound'.
Contents	A12.1 What are the properties of 'salt gas'? A12.2 What elements are present in 'salt gas'? A12.3 Is 'salt gas' a compound of hydrogen and chlorine only? A12.4 What is the formula of hydrogen chloride?
Timing	Two to three weeks will be required.
Introduction to the Topic	'Salt gas' is made by the addition of concentrated sulphuric acid to common salt – hence its provisional name. Simple investigations lead to the fact that salt gas contains both hydrogen and chlorine. The composition of the gas is confirmed experimentally and its formula is deduced. The opportunity is taken to revise the meaning of the terms 'analysis' and 'synthesis' and the Topic concludes with a simple discussion of the meanings of the terms 'element' and 'compound'. In this way, a number of ideas introduced in Stage I can be revised and extended. Pupils can be told that the work on elements originated from that of Boyle, Lavoisier, and Davy.
Alternative approach	The salt-gas problem appears in Topic B11 as part of a general survey of elements and compounds. The quantitative investigation (Experiment A12.4) leading to the determination of the formula of 'salt gas' is deferred until Topic B18.
Background knowledge	Pupils who followed Stage I will have had plenty of experience with hydrogen and will have met chlorine in A10.2 and A10.4 or in B6.2, B6.3, B8.2, B10.4 and B10.5. Hydrogen chloride has been used in a diffusion experiment in the previous topic (Experiment A11.1b), but it was probably used there under its correct name, so that it would be as well not to refer to this experiment until the composition of the gas has been established. Pupils will have met and used the terms element and compound during Stage I (for example, A4, A5, A6, and A7, or B7, B8, and B10), and should be able to contribute to a discussion on the meaning and usage of these terms.

Further references

for the teacher

Rogers, M. J. W. (1970) *Gas syringe experiments*, Heinemann Educational.

Supplementary materials

Film loop
2–16 'The formula of hydrogen chloride'

Reading material

for the pupil

Handbook for pupils, Chapter 4 'Formulae and equations'.
Chemists in the world, Chapter 4 'Davy and Faraday'.

A12.1
What are the properties of 'salt gas'?

In this section pupils investigate the gas given off when concentrated sulphuric acid is added to common salt, and find out some of its properties.

A suggested approach

The starting point for this investigation is the action of concentrated sulphuric acid on common salt. The pupils perform this reaction in a test-tube and see that a fuming gas, called 'salt gas' is evolved. They then investigate its properties.

Objectives for pupils

1. Awareness of an important compound and some of its properties
2. Ability to frame a simple hypothesis and to devise experiments to test it.

Experiment A12.1a

Investigating some properties of 'salt gas'

Apparatus

Each pupil (or pair) will need:

Experiment sheet 46

4 test-tubes, 100 × 16 mm

Cork, with right-angled delivery tube, to fit test-tube

Small trough or basin to contain water

Spatula

Splint

Full-range Indicator paper

Sodium chloride, spatula measure

Concentrated sulphuric acid, few drops

Procedure
Experiment sheet 46 is reproduced below. The solubility test is unlikely to be successful and a discussion of the problem can lead to the fountain experiment (A12.1b).

Experiment sheet 46
Place a spatula measure of sodium chloride in a test-tube and add a few drops (about 5 or 6) of concentrated sulphuric acid to it. **(Concentrated sulphuric acid is a very corrosive liquid, so handle it carefully. If any of it is accidentally spilled on your skin, wash it off IMMEDIATELY with a stream of water from a tap, and then report the mishap to your teacher.)** Describe all that happens, paying special attention to what you see inside the test-tube and what you see just outside the open end.

Devise and carry out tests on the gas to find out:
1. whether it is acidic, neutral, or alkaline;
2. whether it burns or allows a lighted splint to burn in it;
3. whether it is soluble in water.
(Your teacher will probably give some help in planning these tests.)
Describe below what you do, what happens, and the conclusions that you reach.

You may be shown another way of finding whether 'salt gas' is soluble in water. If so, draw a diagram of the apparatus used.

The fountain experiment

It is best for this experiment to be demonstrated by the teacher, as a spectacular result is only achieved if the flask is very thoroughly filled with salt gas. The filling must be done in a fume cupboard and inevitably involves the escape of a large quantity of gas.

Apparatus

The teacher will need:

Gas generator for 'salt gas': Conical flask or filter flask, 500 cm^3

Tap funnel

Delivery tube and rubber connector

Bung

Stand, boss, and clamp

Dry round-bottom flask (preferably 1000 cm^3), filled as shown in figure A12.1

Glass trough

Cotton wool

Litmus solution

Bunsen burners and heat resistant mats

Rock salt (lumps, not powder – to prevent frothing)

Concentrated sulphuric acid

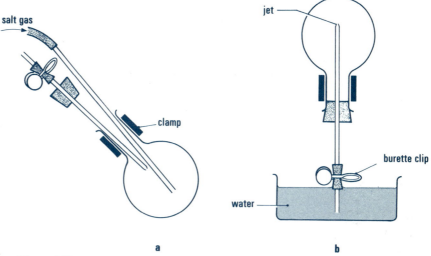

a   b

Figure A12.1
Fountain experiment: (a) arrangement for filling the flask; (b) apparatus ready for the demonstration.

Procedure

Fill the dry flask with salt gas by downward delivery as shown in figure A12.1a. The difficulty lies in displacing all the air from the flask. A separate delivery tube which leads to the bottom of the flask is therefore to be preferred to the use of the central jet with the cork loosened. Allow a great deal of gas to escape before carefully withdrawing the delivery tube and inserting the cork fitted with the jet. Invert the flask over a trough of water which has been coloured blue with litmus solution and one or two drops of alkali (figure A12.1b).

In order to start the 'fountain', warm the flask carefully with an almost luminous Bunsen flame until a few bubbles of gas (mainly air from the tube) have escaped as the result of expansion; then cool the flask with a wet cloth and, as the gas contracts, the water will rise in the central tube. Once water enters the flask through the jet the 'fountain' will operate without further assistance. The acidic solution which forms in the flask will be coloured red by the litmus, which therefore heightens the effect of the demonstration. Discuss the reason for the 'fountain', bringing out the point that all the gas dissolves in the drop of water in the tube, thereby reducing the pressure in the flask.

Discuss the question of how one might prepare a solution of salt gas and point out the problem of 'sucking back' if a normal delivery tube is used. Quickly illustrate the points discussed by showing the pupils a convenient laboratory procedure.

Experiment A12.1c	**Preparing a solution of 'salt gas'** This experiment should be done by the teacher.

Apparatus

The teacher will need:

Gas generator used in A12.1b

either 2 filter tubes
(150 × 25 mm)
or 2 Drechsel bottles, fitted as
shown in figure A.12.2 *a* and *b*

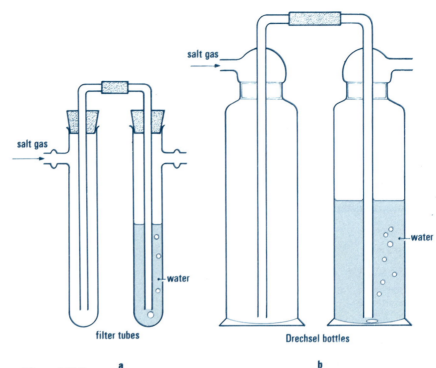

salt gas

salt gas

water

water

filter tubes

Drechsel bottles

a

b

Figure A12.2

Procedure
Pass salt gas through the apparatus shown in figure A12.2 *a* or *b* so as to obtain a solution of the gas in water.

After these demonstrations, pupils can be asked to devise their own experiments to decide what substance causes the steamy fumes made when salt gas comes into contact with the air. In answer to the question in *Experiment sheet* 47 they will probably suggest carbon dioxide and water as possible substances to detect, and the apparatus listed for the experiment is based on this assumption.

Experiment A12.1d	**Finding out what causes the steamy fumes when 'salt gas' is exposed to air**

Apparatus

Each pupil (or pair) will need:

Experiment sheet 47

6 test-tubes, 100 × 16 mm

2 corks to fit test-tubes

Cork with right-angled delivery tube to fit test-tubes

Teat pipette

(Continued)

Procedure
Experiment sheet 47 is reproduced below and gives full details. Pupils may require assistance with these investigations and a demonstration of the effect of dry carbon dioxide and of steam on salt gas may be necessary.

> **Experiment sheet 47**
> You will have noticed that 'salt gas' appears clear and colourless until it comes out into the air, when it forms steamy fumes. These fumes increase greatly if the gas comes into contact with your breath. What substances are there in your breath which also occur, but in smaller proportions, in

Spatula

Sodium chloride, one spatula measure

Concentrated sulphuric acid, few drops

Small marble chips, few chips

Dilute hydrochloric acid

Cotton wool (previously dried in oven)

ordinary air? You can probably think of two such substances and can now try to devise simple experiments which will enable you to decide whether one or both of them are responsible for the fuming.

You will have to decide how to produce separate specimens of the two substances that you wish to test. It may help you to know that a plug of oven-dried cotton wool, placed in the mouth of a test-tube in which a gas is being produced, will dry the gas as it passes through.

Write down the names of the substances, in the air and in your breath, which you think may be responsible for the fuming of 'salt gas'.

Describe the experiments you try and the results you obtain.

What are your conclusions from these experiments?

Summary

Pupils now know that when concentrated sulphuric acid is added to common salt, a gas, provisionally called salt gas, is obtained. They have found that salt gas is acidic, that it does not burn, has a high solubility in water, and fumes in air owing to the presence of water vapour.

A12.2
What elements are present in 'salt gas'?

A suggested approach

Objectives for pupils

Recognition of 'salt gas' as a compound of hydrogen and chlorine

After their introduction to this new compound, pupils should be encouraged to consider ways of finding out what elements it contains. Where do we start? A hint may be taken from the acidity of the gas. What properties of acids have been met with before? Many pupils will remember that some metals (particularly magnesium) react with acids to give hydrogen. Perhaps salt gas contains hydrogen. Discuss ways of finding out about this, and lead the pupils to appreciate that no test for hydrogen is valid if water is present, since water itself contains hydrogen. A simple demonstration experiment in which salt gas is passed over heated iron is described below. Remind pupils of the fountain experiment and point out the danger of sucking-back, which is prevented in this experiment by the use of a simple valve (a Bunsen valve).

Experiment A12.2a

Apparatus

The teacher will need:

Buchner flask fitted with cork and tap funnel to serve as apparatus for generating salt gas

Combustion tube, 125 × 16 mm, with end delivery tube

(Continued)

Finding one of the elements in 'salt gas'
This experiment should be done by the teacher.

Procedure
Place three or four spatula measures of iron powder in the combustion tube and assemble the apparatus as shown in figure A12.3 with the Bunsen valve under water in the basin. Drive the air out of the apparatus with a steady stream of salt gas, when the bubbles will be seen to escape through the valve. As soon as most of the gas appears to be dissolving, heat the iron powder with a moderate flame and collect the bubbles of gas which then begin to escape again through the valve. The way in which this gas burns, when the tube is held to a flame, shows that it is hydrogen. Near-colourless crystals of iron(II) chloride will be seen to form in the combustion tube.

Delivery tube fitted with a Bunsen valve (see figure A12.3: a Bunsen valve can be made from a piece of rubber tubing which has a short slit in it and a piece of rod)

Small trough or basin

Test-tube 100 × 16 mm

Bunsen burner and heat resistant mat

Spatula, stand, boss, and clamp

Rock salt

Concentrated sulphuric acid

Iron powder

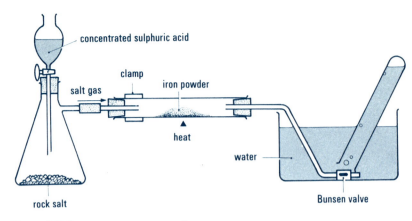

Figure A12.3

From the results of this experiment it should be clear to pupils that the hydrogen produced must have come from the salt gas. 'What else is there in salt gas?' 'If we remove the hydrogen what will we find?' The problem, then, is to remove the hydrogen. They have discovered already that salt gas does not burn and so burning cannot be used to remove the hydrogen. 'How else can hydrogen be removed?' By combining it in some way with oxygen to form water? Perhaps the oxygen can be supplied by a compound? They will have come across a number of oxygen-containing compounds in Stage I: copper(II) oxide, manganese(IV) oxide, etc. The pupils can now try some of these chemicals themselves to see whether another substance is produced.

Apparatus

Each pupil (or pair) will need:

Experiment sheet 48

Test-tube, 100 × 16 mm

Full-range Indicator paper

Splint

Bunsen burner and heat resistant mat

Teat pipette

Manganese(IV) oxide, half a spatula measure

Copper(II) oxide, half a spatula measure

Potassium permanganate (manganate (VII)), half a spatula measure

Supply of concentrated hydrochloric acid labelled 'solution of "salt gas" in water'

To find out what other substance is present in 'salt gas'

Warning. On no account should the following experiment be carried out on the open bench by pupils in a badly ventilated laboratory or in an inefficient fume cupboard.

Furthermore, it is essential that the quantities of manganese(IV) oxide, copper(II) oxide, and potassium permanganate taken by pupils should be very strictly supervised.

Procedure
Throughout this experiment, **pupils should be warned not to use larger quantities than stated, and not to breathe the gas formed.** If anyone accidentally inhales some chlorine, he should gently sniff the bottle of dilute ammonia solution. Details are given for the investigation in *Experiment sheet* 48.

Experiment sheet 48
You have seen an experiment which shows that constituent of salt gas is hydrogen. If you can find a way of removing hydrogen from salt gas you may be able to release the other substance present. You will have discussed this in class and prepared a list of substances that might be needed for this purpose. Test each of these substances as follows:

Add not more than half a measure of the substance to about one cm³ of a solution of salt gas in water in a test-tube. If there is no sign of anything

---

Topic A12 Investigation of 'salt gas'

happening when the mixture is cold warm it gently. Test any gas evolved with damp indicator paper – remember that unchanged salt gas may be escaping from the mixture.

(Do not use larger quantities than those stated and keep the test-tube well away from your nose and eyes.)

Record your results and conclusions below.

Which of the substances releases a gas from salt gas which behaves differently from salt gas towards indicator paper?

What is the name of this gas?

Suggestion for homework

How would you show that 'salt gas' is composed of hydrogen and chlorine *only*?

Summary

As the result of these investigations, pupils should know that 'salt gas' contains hydrogen and chlorine. They will have been able to detect chlorine by its colour (greenish-yellow) and by its bleaching action on moist Full-range Indicator paper. Chlorine is formed when strong salt-gas solution is used with either manganese(IV) oxide or potassium permanganate. However, pupils may be confused by smells and by the greenish-yellow colour of copper(II) chloride solution formed when copper(II) oxide is used – pupils may report that chlorine is formed too – unless the need for a positive bleaching test has been insisted on.

A12.3
Is 'salt gas' a compound of hydrogen and chlorine only?

In this section hydrogen and chlorine are made to react together, and the product is identified as salt gas. This makes it possible to rename salt gas as hydrogen chloride, and to mention the term 'synthesis'.

A suggested approach

Objectives for pupils

1. Knowledge that 'salt gas' is a compound of hydrogen and chlorine *only*, and has the name hydrogen chloride
2. Recognition of synthesis as proof of composition

Is salt gas a compound of hydrogen and chlorine only? Discuss ways of finding this out with the class and discuss the results of the homework if the suggested homework for the last section was set. One simple way is to burn hydrogen in chlorine. Is salt gas formed? Details of an experiment to demonstrate this are given below.

Experiment A12.3

Apparatus

The teacher will need:

Full-range Indicator paper

(Continued)

Burning hydrogen in chlorine
This experiment *must* be done by the teacher.

Procedure
Prepare one or two gas jars of chlorine.

Chlorine may be most conveniently prepared by the dropwise addition of concentrated hydrochloric acid onto potassium perman-

Splints

Delivery tube for burning hydrogen in a gas jar

Rubber connection tubing

Cylinder of hydrogen

Gas jar full of chlorine (see Experiment A13.2a for laboratory preparation)

Access to chlorine generator

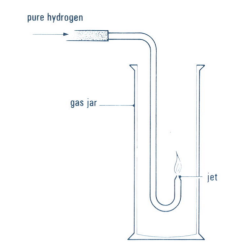

Figure A12.4

ganate and collected by the displacement of air upwards.

Use a steady stream of hydrogen from a cylinder. Light the hydrogen at the jet, adjust the pressure of the gas to provide a flame about 2–3 cm high, and lower the burning hydrogen jet into a gas jar of chlorine.

The hydrogen will be seen to continue to burn, but with a white flame. The colour of the chlorine will go, and its place will be taken by steamy fumes of hydrogen chloride, salt gas. Test this with damp Full-range Indicator paper, and a lighted splint.

This experiment shows that salt gas is a compound of hydrogen and chlorine only, and it may therefore be given the more explicit name of hydrogen chloride. Pupils should by now realize the value of synthesis as a proof of composition. The meaning of the terms 'analysis' and 'synthesis' should be discussed at this point.

A12.4
What is the formula of hydrogen chloride?

Now that salt gas is known to consist of hydrogen and chlorine only, the final problem is to determine its chemical formula, that is, to find the proportions in terms *either* of moles *or* of atoms in which the elements combine together.

Although the most direct method is the combination of hydrogen and chlorine as described in the first edition, it is omitted here because of the chance of explosion. While it is true that many teachers have carried out the experiment repeatedly without incident, others have experienced explosions and it has been decided to replace the experiment by a rather less direct method. This is described below.

Teachers who nevertheless wish to carry out the original experiment will find full details in *Collected Experiments*, Experiment E12.22, and in Rogers, M.J.W. (1970) *Gas syringe experiments*, Heinemann Educational. Alternatively the film loop 2–16 'The formula of hydrogen chloride' may be used.

1. Awareness of a technique for measuring volumes of gases taking part in chemical reactions
2. Knowledge of the formula of hydrogen chloride
3. Awareness of the work of Gay-Lussac on the combining volumes of gases

Experiment A12.4

Apparatus

The teacher will need:

2 syringes, $100 \, cm^3$, in holders

Three-way stopcock with capillary tubing

Piece of hard-glass tubing, 25–30 cm × 7 mm o.d.

Some glass beads to fit inside the hard-glass tubing

Thick-walled plastic connecting tubing

Asbestos paper or equivalent material

Test-tube, rubber tubing, small trough of water, wooden splint

Delivery tube fitted with a Bunsen valve

Apparatus for generating hydrogen chloride (as in A12.1b)

Zinc foil, 0.8 g (cut into narrow strips)

The problem now is to find the formula of hydrogen chloride. If we can find the masses of the two elements which combine together, and we know the relative atomic masses of the elements, we can work out the formula in the same way as we have already done for several other compounds in section A11.4. In chemical reactions it is sometimes easy to find directly the masses of gases involved but it is generally easier to measure the volumes of gases.

What is the formula of hydrogen chloride?
This experiment should be carried out by the teacher.

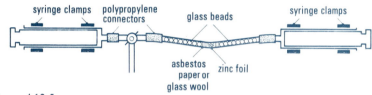

Figure A12.5

Procedure

Heat the piece of hard-glass tubing at its middle point in a Bunsen flame and bend it through an angle of about 140°. Bend the strips of zinc foil (about 0.8 g) small enough to slip into the tube. Then pack the tube with glass beads such that the zinc is in about 2 cm of clear space in the middle, and add small pads of asbestos paper* to keep the glass beads away from the metal. These glass beads reduce the volume of air which would otherwise be in the tube. Additional pads of asbestos paper may be used to secure the beads in the tube before assembling the apparatus as shown in figure A12.5. Everything must be dry.

Admit hydrogen chloride to syringe 1. Make sure that all air has first been displaced from the generating apparatus and flush out the syringe and stopcock tubing once or twice with the gas before filling. Admit a little more than the required volume of the gas, then disconnect the generator and expel the extra gas until the syringe is exactly at the $50 \, cm^3$ mark, with the gas at atmospheric pressure. Then turn the stopcock to connect the syringe with the hard-glass tube. Heat the tube under the zinc with a moderate, wide, but mobile Bunsen flame, and as soon as the metal melts, *GENTLY* pass the gas from syringe 1 through the tube into syringe 2. Pass the gas *GENTLY BACK AND FORTH between the syringes until there is no further reduction in volume.* If the reaction is slow, raise the temperature of the metal, *but avoid overheating*, which will cause excessive volatilization of the zinc chloride. (The wide flame is intended to keep the central pads of asbestos paper hot so that the zinc chloride

*Equivalent materials may be used, such as Rocksil or glass wool.

does not solidify in them and block the tube. The bend in the tube should suffice to keep the molten zinc in place, but there is some risk of its being blown on to one of the pads, and perhaps penetrating through it, if the gas is not passed *gently* enough through the tube.)

The final volume of hydrogen left in the tube should be very close to 25 cm^3 and a glass syringe is likely to be sufficiently free-running for the volume to be read off without using a manometer to check the pressure. The residual gas can be expelled from the apparatus and collected over water in a small test-tube. The gas can be ignited to convince the class of its identity.

Treatment of the result
The experiment shows that 50 cm^3 of hydrogen chloride contains 25 cm^3 of hydrogen. So 1 dm^3 of hydrogen chloride contains 500 cm^3 of hydrogen.

From the Reference section of the *Handbook for pupils*:
At room temperature and pressure,
Mass of 1 dm^3 of hydrogen chloride $= 1.50 \text{ g}$
Mass of 500 cm^3 of hydrogen $= 0.04 \text{ g}$
So, by difference: mass of chlorine present $= 1.46 \text{ g}$
0.04 g of hydrogen combines with 1.46 g of chlorine.
Hydrogen has a relative atomic mass of 1.0. It is convenient to calculate how much chlorine will combine with this mass of hydrogen.

1.0 g of hydrogen combines with $\frac{1.46}{0.04} \text{ g}$ of chlorine

$$= 36.5 \text{ g of chlorine}$$

This figure is very close to the relative atomic mass of chlorine (35.5) given in the *Handbook for pupils*. So, given this sample result, it seems reasonable to suggest that one relative-atomic-mass-worth of chlorine reacts with one relative-atomic-mass-worth of hydrogen. The formula of hydrogen chloride may thus be written as H_1Cl_1, or more simply, as HCl.

After the experiment, draw attention to the simple relationship between the initial and final volumes of gases. Mention that in the early nineteenth century Gay-Lussac noticed gases reacted in volumes which were in a simple ratio and that he proposed a 'law' to that effect. It was only later that an explanation was given for this law. (The explanation will be discussed in the course of section A14.9.)

It can be helpful towards the end of this lesson to recapitulate the evidence for saying that hydrogen and chlorine are elements and hydrogen chloride is a compound. If time allows list some of the other examples of elements and compounds which the pupils have met during Stage I.

The film loop 2–16, 'The formula of hydrogen chloride' could be shown. The apparatus used is that described in *Collected Experiments*, Experiment E12.22.

1. Make a short summary of this Topic.
2. Extend the table of formulae begun as the result of homework set during Topic A11.
3. Make a list of the elements and compounds met during the course so far.

Summary

By the end of the Topic, pupils should have acquired some knowledge of the process used to identify an 'unknown compound'. In the last section, a quantitative experiment is used to show that hydrogen chloride contains hydrogen and chlorine only and that it has the formula HCl. They should be clear in their minds as to how this formula has been established, and what is meant by the terms element, compound, analysis, and synthesis.

Looking at the elements in the light of the Periodic Table

Purposes of the Topic

1. To provide a historical background to the development of the Periodic Table.
2. To demonstrate some of the simpler patterns and trends in the Periodic Table.
3. To provide some knowledge of the properties of the alkali metals, the halogens, the inert gases, the transition metals, and carbon and silicon.
4. To show the value of relating the properties of any element to its position in the Periodic Table.
5. To provide practice in the use of tables of data.

Contents

A13.1 The story of the Periodic Table
A13.2 The alkali metals
A13.3 The halogens (and the inert gases)
A13.4 The transition metals
A13.5 Carbon: an element found in living things
A13.6 Silicon: an element found in many rocks

Timing

Each section of this Topic requires a double period, and if there is to be enough time for discussion most sections will require a single period as well. It is suggested that about six weeks be allowed for the Topic as a whole.

Introduction to the Topic

The Topic begins by reviewing the properties of elements the pupils have met so far. The classification of elements as metals and non-metals is discussed and shown to be unsatisfactory. The developments which led to the Periodic Table are then discussed. The Periodic Table is introduced as a list of elements arranged in the order of their relative atomic masses, such that elements of similar properties are near to one another in the table.

We might then ask 'Where are the metals and non-metals in this arrangement'? and 'How are the elements which are in vertical groups related to one another'? Such questions can be answered partly from the experimental work of Stage I, and partly by the investigations in this Topic, which include the properties of the alkali metals, the halogens, and briefly the group of inert (noble) gases. Other questions arise – such as 'How are elements in the horizontal periods related to one another'? Elements from other parts of the table are examined, and include copper, nickel, and iron (as examples of transition elements). Elements from the middle groups of the Table are also studied, i.e. carbon and silicon. We note that carbon is an element which is found in many living things and whose chemistry receives further study in Topic A21, and that silicon is a constituent of many rocks.

Reference to periodicity is made in many other Topics in Alternative IIA, especially in Topic A24. The theme of periodicity is taken up in Option 7 in Stage III.

The material used in this Topic is presented in a different way in Alternative IIB. In Topic B16 a simulation exercise is used to illustrate the development of the Periodic Table. In Topic B24, the theme is developed further and indeed overlaps some of the content of Option 7 in Stage III.

This Topic follows from much of the work of Stage I, and in particular from the following Topics:

(Alternative IA) Topic A5, 'The elements'. Here the idea of an element is introduced, and the elements are divided into metals and non-metals.

Topic A6, 'Competition among the elements'. The idea of an order or reactivity or displacement series is introduced during this Topic.

Topic A10, 'Chemicals from the sea'. The halogens are considered as a group of similar elements in this Topic.

(Alternative IB) Topic B7, 'The elements'. Here the idea of an element is introduced and metals and non-metals are distinguished.

Topic B8, 'Further reactions between elements', places emphasis on trends in the properties of oxides, sulphides, and chlorides.

Topic B10, 'Competition among the elements', introduces the idea of an order of reactivity or displacement series, and the halogens are considered as a group of similar elements.

The previous Topics in Stage IIA (A11 and A12) set the scene for Topic A13.

Collected experiments, Chapter 6, 'Elements, their classification, and their differences from compounds', contains further experiments for this Topic.
Handbook for teachers, Chapter 3, 'Making a start, the route to elements and the Periodic Table'
Partington, J. R. (1937) *A short history of chemistry*, Macmillan, contains an authoritative account of the history of the Periodic Table.
Spiers, A., and Stebbens, D. (1973) *Chemistry by concept*, Heinemann Educational, has a number of examples of the use of bar charts when comparing physical properties of elements and in other situations.
Nuffield Advanced Chemistry Students' books I and *II*, Penguin, (Topics 2, 4, 5, 6, 16, and 19) and *Nuffield Advanced Physical Science Students' books I* and *II*, Penguin, (Sections 2, 5, 10, and 11) provide extended treatments of this Topic.

Some of the Nuffield Advanced Chemistry overhead projection originals may be used in this Topic.

Overhead projection original
10 Common elements and their abundance in the Earth's crust and in the human body *(Table A13.1)*

Film
'Metals and non-metals' Encyclopaedia Britannica

A specially designed Nuffield Periodic Table is available in a small size for use by pupils.
Handbook for pupils, Chapter 1 'Periodicity'.

A13.1
The story of the Periodic Table

The historical developments which led to the Periodic Table are described and the position in that table of those elements which the pupils have already met during the course are discussed.

A suggested approach

Objectives for pupils

1. Recognition of the meaning of the term element
2. Awareness of the possibility of classifying elements as metals and non-metals, and the limitations of this classification
3. Knowledge of the history of the development of the periodic classification of elements
4. Recognition of the Periodic Table as a list of elements in order of their atomic masses and arranged so as to group together elements with similar properties
5. Ability to design experiments to compare the properties of of elements

The way to the Periodic Table has been prepared by the gradual evolution of ideas in Stage I and in particular by the use of the term element in Topic A5 or B7. Topic A11 introduces the notion of relative atomic mass and this is used in Topic A12 which also reviews ideas about elements and compounds. This Topic can begin very profitably by classifying elements according to whether they are metals or non-metals. The limitations of this classification may then be exposed and the search for other classifications mentioned: thus, the treatment of the Periodic Table at this point in the course can be historical.

The Periodic Table was developed without a knowledge of atomic structure and with an understanding of chemistry which closely approximates that which the pupils have reached.

Early attempts to find relationships between relative atomic masses (formerly atomic weights) and chemical properties, such as those of Döbereiner (1817, 1829), Dumas (1859), Newlands (1863, 1865), and others, may be mentioned. The scene is now set to reveal the almost simultaneous and quite independent discoveries by Lothar Meyer (1869) in Germany, and by Mendeleev (1869) in Russia, of the Periodic Law. The key point for pupils to know is that the Periodic Table is a list of elements arranged in order of their relative atomic masses. (It is helpful to avoid the exceptions to this rule for the time being.) The elements are arranged so that elements with similar properties are in vertical groups, or in horizontal periods – as in the case of transition metals.

It is convenient at this point in the lesson to issue copies of the Periodic Table. The symbols used may require a brief explanation. The pupils could start by marking all of those elements that they have met so far during the course. They will see that the metals are on the lefthand side of the table and the non-metals are on the right. On the righthand side, they will recognize the halogen group: fluorine, chlorine, bromine, and iodine, which is the first group of elements with similar properties that they met.

Later on in this Topic they will look more closely at these elements.

Mendeleev claimed that in his periodic classification, elements arranged in vertical groups had similar properties.

In addition, he correctly predicted the properties of elements which had not at that time been discovered. How can *we* test Mendeleev's claim? Can we see if the elements within a group are really similar? Can we find out how the properties of the elements vary within a particular group? These questions can form part of a discussion and lead to the next set of experimental investigations for pupils.

Either section A13.2 or A13.3 could follow immediately from this introduction.

Suggestions for homework

Read Chapter 1 'Periodicity' of the *Handbook for pupils*.

A13.2
The alkali metals

In this section the first group of the Periodic Table, the alkali metals, is studied. The properties of sodium and potassium are investigated. The pupils are asked to predict whether lithium will be more or less reactive than sodium and potassium, and then to test their prediction. Chemical reactions are summarized using the word equations, and the properties of the alkali metals are compared using bar charts.

A suggested approach

Pupils will have met sodium and potassium salts if they followed Stage I, and may also have seen sodium metal. It is suggested that the teacher starts by finding out what the pupils remember about these chemicals. The reactions of sodium and potassium with air, chlorine, and water can then be demonstrated.

Objectives for pupils

1. Ability to identify a pattern of reactivity and to make predictions from the pattern
2. Knowledge of some reactions and the comparative reactivity of the alkali metals
3. Ability to summarize information about chemical reactions using 'word equations'
4. Ability to compare data for different elements

Experiment A13.2a

The action of (1) air; (2) chlorine on sodium and potassium
This experiment **must** be done in a well-ventilated fume cupboard by the teacher.

Apparatus

The teacher will need:

Safety spectacles

Chlorine generator (set up in the fume cupboard) made from a filter flask, and fitted with a bung carrying a tap funnel

Delivery tube and rubber tubing connector

(Continued)

Procedure
It is recommended that the teacher wear safety spectacles for these demonstrations.

Warning. Only very small pieces of sodium and potassium should be used for these experiments.

Part 1: remove a small piece of sodium from the bottle, and free it from protective paraffin oil by dipping it first into hexane or petroleum spirit, and then gently pressing dry between filter papers. Cut

Stand, boss, and clamp

2 gas jars and covers

2 combustion spoons for use in gas jars

Bunsen burner and heat resistant mat

Filter paper

Asbestos paper or equivalent material

Long-handled laboratory knife

Concentrated hydrochloric acid

Potassium permanganate (manganate(VII))

Vaseline

Sodium

Potassium

Hexane or petroleum spirit (Caution! Keep away from flames)

away any surface coating of oxide and place the prepared piece of sodium (a cube of side 1–2 mm is suitable) on a small piece of asbestos paper in a combustion spoon. Heat the spoon in a Bunsen burner flame and observe how vigorously the sodium burns. Repeat the experiment using potassium, which is more reactive than sodium.

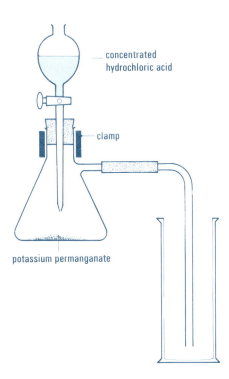

concentrated hydrochloric acid

clamp

potassium permanganate

Figure A13.1

Warning. Chlorine gas is extremely dangerous if inhaled.

Part 2: Chlorine (sufficiently pure for this experiment) can be obtained by the dropwise addition of concentrated hydrochloric acid on to potassium permanganate crystals, using a filter flask as a gas generator (figure A13.1). Collect several gas jars full of chlorine gas and cover them with appropriately vaselined gas-jar lids. Obtain another small sample of sodium, clean it as in Part 1, and place it on asbestos paper on a combustion spoon as in the earlier experiment. Heat the spoon carefully in a Bunsen burner flame and as soon as the sodium shows signs of burning, plunge the spoon and its contents into a jar of chlorine. The sodium will continue to burn. If the sample of sodium has been correctly cleaned, white fumes and powder of sodium chloride will be seen to be formed. Repeat the experiment using a small sample of potassium, remembering that this is much more reactive than sodium.

The section should be concluded by reviewing the reactions studied and constructing suitable word equations.

Note. Small pieces of alkali metal which may be left over from the experiment should not be returned to the stock bottle, and must on no account be put into sinks or rubbish boxes. They can be dealt with by putting them into an excess of industrial methylated spirits in a beaker until the effervescence stops. The resulting solution can be diluted and flushed down the laboratory sink with water.

Experiment A13.2b

Apparatus

The teacher will need:

Safety spectacles

Plastic safety screen

2 glass troughs

Long-handled laboratory knife

Filter papers

Beaker

Sodium

Potassium

Full-range Indicator solution

Hexane or petroleum spirit (Caution! Keep away from flames)

The action of sodium and potassium on water
This experiment should be done by the teacher.

Procedure
It is recommended that the teacher wears safety spectacles for this demonstration.

Stand two clean grease-free glass troughs side by side on the demonstration bench, each half filled with water. Place a safety screen between the troughs and the pupils. Remove a piece of sodium from the stock bottle and free it from adhering paraffin oil by dipping it (using the penknife) into the beaker containing hexane or petroleum spirit and then drying it by means of filter paper. Cut a small piece of sodium free from surface corrosion. A cube of side 1–2 mm is suitable. Place the sodium on the water in one of the troughs.

Repeat the experiment with potassium, using the other trough.

The sodium melts and skates about on the surface of the water; potassium melts, skates about on the surface of the water, and a flame may appear.

Pupils should not come too near the bench in case pieces of molten metal or molten metal hydroxide 'jump out' of the troughs (hence the recommended precautions).

Finally, put a little Full-range Indicator into each trough of water and note the result.

After these experiments have been performed, the pupils can summarize, perhaps in a table, their knowledge of the properties of sodium and potassium.

Property	Sodium	Potassium
Action of air		
Action of chlorine		
Action of water		

This summary is then completed by adding word equations for the reactions studied.

After these experiments, ask the pupils to find the other elements in the alkali metal group, using their own copies of the Periodic Table. They will see that lithium, rubidium and caesium are in the same vertical group as sodium and potassium. Ask the pupils what properties they would expect them to have. How, for instance, would they expect these metals to react with air? What about the order of reactivity? Would they expect these metals to be soft or brittle?

Would they expect these metals to have a high or a low melting point? After a discussion, Experiment A13.2c can be carried out; teachers should note that the first part of the experiment is a teacher demonstration. Pupils will need the *Handbook for pupils* for the last part of the experiment.

Note. Details for the treatment of *small* pieces of unused alkali metals appear at the end of Experiment A13.2a.

What are the reactions of lithium?

Note to teachers
Lithium should be purchased in the form of 'shot' of high purity. Normally, lithium shot is kept in liquid paraffin. The traces of liquid paraffin can be removed by using hexane or petroleum spirit and pieces of filter paper – see Experiment A13.2a – but it is suggested that pupils should not be allowed to clean the pieces of lithium for themselves.

Procedure
Tell pupils to wear safety spectacles for this investigation. The procedure is described in *Experiment sheet* 49. It is important to note that the first part of the experiment is a demonstration.

Experiment A13.2c

Apparatus

The teacher and each pair of pupils will need:

Experiment sheet 49

Safety spectacles

Beaker, 100 cm^3

Tripod

Asbestos paper or equivalent material

Pipeclay triangle

Tongs

Bunsen burner and heat resistant mat

Handbook for pupils

Full-range Indicator solution

Small piece of lithium (1 mm^3)

Distilled water (wash bottle)

Experiment sheet 49
You will be required to wear safety spectacles when doing this experiment.

1. *Your teacher will first demonstrate the effect of heating a small piece of lithium on dry asbestos paper.*

Does the lithium burn in air?

What is the colour of the flame?

Record the appearance of the residue.

Your teacher will then transfer the residue into a test-tube using distilled water and add drops of Full-range Indicator to the solution.

Record what happens.

2. Half fill a 100 cm^3 beaker with water and add a rice-grain size piece of lithium. What happens?

Add 10 drops of Full-range Indicator solution to the contents of the beaker.

What is the pH of the solution?

Write word equations for the two reactions you have studied.

lithium + oxygen $\longrightarrow$
lithium + water $\longrightarrow$

Find out as many similarities and differences as you can between the properties of lithium, sodium, and potassium, using the Reference section of the *Handbook for pupils*. Tabulate your results using the headings below.

Property	Lithium	Sodium	Potassium

Topic A13 Looking at the elements in the light of the Periodic Table

After the experiments and demonstrations, discuss the results. The question may then be put, 'To what extent were your predictions borne out in practice?' Pupils will have seen that in their experiments lithium is like sodium and potassium but that in general it is less reactive. In this group, the elements become more reactive as their relative atomic masses get bigger. During this final discussion pupils should be told that the formulae of the chlorides of these elements are LiCl, NaCl, and KCl. They should understand that just as they were able to work out the composition of magnesium oxide in Topic A11, somebody had to work out, for example, that lithium chloride has a composition which can be represented by the formula LiCl.

Teachers should use their discretion about giving the formulae of the other compounds mentioned in this section. The principle oxides, for example, do not all have the same kind of formula. Thus, for lithium the principal oxide is Li_2O; for sodium, Na_2O_2; and for potassium, KO_2. The increasing oxygen content might be used to illustrate the increasing affinity for oxygen. Although the hydroxides (formed when the metals are put in water) all have formulae of the type MOH, they are not, of course, binary compounds, and it is therefore less easy for pupils at this stage to see how the formulae might be found.

In this discussion of the properties of these elements the Reference section of the *Handbook for pupils* can play a large part. The pupils can use it to look up the melting points and other physical properties of the alkali metals, and some of this could well be done for homework (see 'Suggestions for homework' at the end of this section). The use of bar charts can give a real feeling for data, as explained below. There are a number of examples of bar charts used in this way in the book *Chemistry by concept*. (See 'Further references for the teacher', at the start of this Topic.)

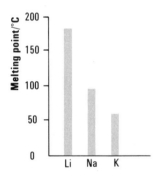

Figure A13.2
Displaying melting points by means of a bar chart.

The use of bar charts
There are several reasons for using bar charts rather than 'graphs'. Some examples are shown in figure A13.2. Not only does a bar chart give the pupils more feeling for the relative magnitudes of the values displayed, it also emphasizes the fact that data values for an arbitrarily chosen set of elements (such as lithium, sodium, and potassium) are being compared.

It is, of course, true that there are chemical relationships between lithium, sodium, and potassium, but there is no simple *mathematical* connection between the melting points of these elements as the horizontal 'axis' of the 'graph' of figure A13.3 misleadingly suggests. The lines joining the points in the graph merely serve as a guide to the eye from one point to the next and indicate trends. However, the slopes or gradients of such graphs are quite meaningless.

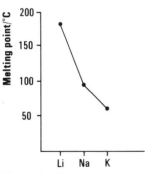

Figure A13.3
A graph of melting points of lithium, sodium, and potassium.

For large-scale display for use in class discussions, bar charts can conveniently be made from large strips of coloured paper or card. Alternatively, solid models can be useful (see figure A13.4).

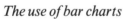

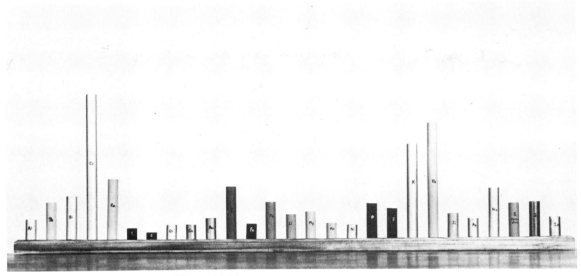

Figure A13.4
Solid models for displaying data relating to the elements. (*Note:* such models can be arranged to show the periodicity of the property of atomic volume.)

1. Using the Reference section of the *Handbook for pupils*, construct bar charts showing (*a*) melting points, (*b*) boiling points, (*c*) densities, for the alkali metals. Report on the trends in these properties.

(*Note for teachers*. Data for lithium, sodium, potassium, rubidium, and caesium, are provided in the *Handbook for pupils*. The extent to which this is used is, of course, at the teacher's discretion.)

2. Use the densities of the alkali metals to work out the volume of those quantities of the elements which contain the same number of atoms. Plot the results on a bar chart. How do the results compare with one another, and with values for other metals – such as iron and copper?

Hint. The relative atomic masses of elements expressed in grams each contain the same number of atoms.

Summary

3. In what ways would you expect rubidium and caesium to be different from or similar to the alkali metals?

By the end of this section pupils should know that the alkali metals have similar chemical properties, but differ in the degree of reactivity. An increase in reactivity occurs with an increase of relative atomic mass. Pupils should know that the metals burn, that they combine with chlorine to give chlorides of the general formula MCl, and that they react with water to give alkaline solutions. Pupils should have some experience in writing word equations as summaries for the reactions studied, and in comparing physical properties of these elements by the use of bar charts or by other means.

A13.3
The halogens (and the inert gases)

In this section, the halogens are examined more thoroughly than in Stage I. The pupils look for a 'reactivity trend' and find that in this case the elements of lower relative atomic mass are more reactive. Mention is also made of the inert (noble) gases.

Objectives for pupils

1. Ability to identify a pattern of reactivity and make deductions from the pattern
2. Knowledge of some reactions and comparative reactivity of the halogens
3. Practice in writing reaction summaries and word equations
4. Ability to compare data for different elements

Begin by asking the pupils to look at the righthand side of their Periodic Tables. On the extreme right are the inert (or noble) gases. A brief mention may be made of the occurrence, properties, and uses of these gases (e.g. occurrence in the atmosphere; isolation from liquid air; use in gas-filled electric lamps, etc.). The main purpose of the section is to examine the halogens and to see how their properties vary within the group. The pupils will know something of their properties already through the work done in Stage I, but this work did not go far beyond simple description, except in Alternative IB in which pupils saw a trend in reactivity for the halogens. Discuss ways of finding out how to determine the relative reactivity of the elements and then let the pupils try Experiment A13.3b. Demonstrate A13.3a yourself.

Experiment A13.3a

Apparatus

The teacher will need:

5 hard-glass test-tubes, 150 × 25 mm, arranged in a test-tube rack

Corks to fit the 150 × 25 mm test-tubes

Hard-glass test-tube, 125 × 16 mm, with small hole near the closed end and fitted with a cork carrying a straight delivery tube

Stand, boss, and clamp

Teat pipette

Beaker, 250 cm³

Bunsen burner and heat resistant mat

Apparatus for the production of chlorine (see Experiment A13.2a)

Delivery tube

Access to fume cupboard

(Continued)

Some properties of the halogens

This experiment should be performed by the teacher. All of the reactions carried out should be in or near a well-ventilated fume cupboard.

Procedure
First demonstrate the action of chlorine and bromine on water. Show that chlorine dissolves in water by bubbling chlorine through a delivery tube into a 150 × 25 mm test-tube half full of water. (This experiment should be done in a fume cupboard.) Demonstrate the action of bromine on water by shaking up a little bromine in a 150 × 25 mm test-tube half full of water. Test the chlorine and bromine solutions with Full-range Indicator paper and then add sodium hydroxide solution.

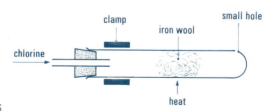

Figure A13.5

Demonstrate the action of chlorine on iron wool by passing chlorine from the generator through a 125 × 16 mm test-tube with a small hole near the closed end (see figure A13.5). (This experiment should be performed in a fume cupboard.) Heat the iron wool to start the

Bromine

Indicator paper – Full-range Indicator or litmus

Iron wool

Sodium hydroxide solution

reaction and then remove the Bunsen burner flame. The iron wool glows red hot and iron(III) chloride is formed.

Demonstrate the action of bromine on iron wool by putting two or three drops of liquid bromine into a test-tube and pushing a tuft of iron wool halfway down the tube. Heat the iron wool until a reaction can be seen to take place.

Discuss the results of each experiment with the class. Make the point that bromine and chlorine react in a similar way but with a different degree of vigour. Ask them which they think more reactive. Pupils are now in a position to predict from the Periodic Table how they would expect iodine to react with water, sodium hydroxide solution, and iron wool. Let the pupils make their predictions and guess whether iodine is more or less reactive than the other two halogens.

Their previous experience with lithium (which is above sodium and potassium in the Periodic Table) is that it is less reactive than the other two metals. 'But what about iodine which is below chlorine and bromine?' After such a discussion let the pupils try the experiments with iodine which are described in Experiment A13.3b.

Experiment A13.3b

Investigating some of the properties of iodine

Apparatus

Each pupil (or pair) will need:

Experiment sheet 50

2 hard-glass test-tubes, 100×16 mm

Small crystals of iodine, about 5 small crystals

Iron wool, small tuft

Sodium hydroxide solution, $5 \, \text{cm}^3$

Procedure
Details of the procedure are given in *Experiment sheet* 50 which is reproduced below.

Experiment sheet 50
You will have seen a number of experiments with the elements chlorine and bromine. The following experiments with iodine will enable you to find out whether it resembles chlorine and bromine in its properties and reactions.

1. Add a small crystal of iodine to a test-tube half filled with water; shake the mixture well.

Does the iodine appear to dissolve?

Test the liquid with Full-range Indicator paper. What happens?

2. Shake a small crystal of iodine with a quarter of a test-tube full of sodium hydroxide solution. **(Be careful, sodium hydroxide solution attacks the skin – do not put your thumb over the open end of the tube to shake it.)**

What happens?

3. Place a crystal of iodine at the bottom of a dry test-tube and wedge a small tuft of iron wool about half-way down the tube. Support the test-tube horizontally in a clamp or holder and heat the iron wool strongly. Enough heat will probably reach the iodine to vaporize it so that purple iodine vapour will surround the iron wool. If this does not happen, warm the iodine for a second or two. Is there any evidence of a reaction between the iron and the iodine? If so, describe it.

After these experiments, ask the pupils again whether they think iodine is more or less reactive than bromine and chlorine. Their results should enable them to see that it is less reactive. Continue by telling them that the reactions with iron produce compounds with formulae $FeCl_3$, $FeBr_3$, and FeI_2. Finally draw the pupils' attention to some physical properties of the halogens using the Reference section of the *Handbook for pupils*. As in the previous section, the use of bar charts may prove helpful. The reactions of the halogens should now be summarized using word equations for as many reactions as would seem appropriate for the ability of the group. (The essential ones are those concerned with the formation of the halides of iron, which may also be summarized by chemical equations.)

Suggestions for homework

1. Draw up a table of the chemical properties of chlorine, bromine, and iodine that have been observed. Indicate any trends that have been noted.
2. Use the Reference section of the *Handbook for pupils* to construct bar charts showing (*a*) melting points, (*b*) boiling points, of (*i*) the halogens; (*ii*) the inert gases.
Report on any trend that can be noted.
3. Extend the table of formulae begun in Topics A11 and A12.

Summary

By the end of this section, pupils should know that the halogens have similar chemical properties, but differ in degree of reactivity. A decrease in reactivity occurs with an increase of relative atomic mass. Pupils should know that the halogens dissolve in water to a limited extent, that they bleach indicators, that they react with alkalis to give colourless solutions and that they each react with iron to form a halide. Pupils should have also gained some experience in writing word equations, as summaries for the reactions studied. They will have also compared the physical properties of these elements by the use of bar charts or by other means. Pupils should also be aware of the occurrence, isolation from liquid air, and uses, of the inert gases.

A13.4
The transition metals

The transition elements occur between the two extremes which have been studied so far in the Periodic Table. These elements are examined and their properties compared with those of the alkali metals.

Teachers should note that pupils who have followed Alternative IB have not met the sequence copper carbonate to copper oxide to copper (which was used in the study of malachite) nor the action of heat on limestone (calcium carbonate). Such pupils will therefore have less background knowledge in this section than those who have followed Alternative IA.

A suggested approach

Reference should be made to the Periodic Table, and the location of various groups of elements which have been studied so far. Use the

1. Ability to identify a pattern of reactivity and to make deductions from the pattern
2. Knowledge of some reactions and comparative reactivity of some transition metals (for example, copper, nickel, and iron)
3. Practice in writing reaction summaries in the form of word equations
4. Ability to compare data for different elements

long form of the Periodic Table and point out the large group of elements in the central section known as the transition metals. Particular attention should be drawn to the top row of transition elements, i.e. the elements scandium to copper. Tell the pupils that these metals have very similar properties, but are unlike the halogens and the alkali metals. These elements are 'grouped' horizontally – in a period – rather than vertically in a group like other elements in the table.

It helps to show pupils samples of both the transition metals and their salts. What do they notice about the metals?

Extend the range of possibilities by showing as many samples of the elements in the first transition series as practicable. Describe their properties and uses. What do pupils notice about transition metal salts compared with alkali metal salts? They are coloured whereas the alkali metal salts are colourless (or white). Now let pupils do some simple experiments on the general properties of transition metals and their salts (see *Experiment sheets* 51 and 52 reproduced below).

Pupils should see that transition metals have coloured salts, are less reactive than the alkali metals (for example they do not react violently – if at all – with cold water), and are ductile, 'stronger', and harder.

Experiment A13.4a

Apparatus

Each pair of pupils will need:

Experiment sheet 51

3 hard-glass test-tubes, 100×16 mm

Samples of copper foil, nickel foil, iron wire

Looking at the properties of some transition metals

Procedure
Details of the procedure are given in *Experiment sheet* 51 which is reproduced below.

Experiment sheet 51
Using the samples of copper foil, nickel foil, and iron wire provided, investigate their hardness and ability to bend without breaking. Their hardness may be described as very hard (like stone), quite hard (like wood), or soft (like cheese). Describe your results below.

Now investigate the effect of cold and hot water on the three metals.

How do the properties of the transition metals which you have studied compare with those of the alkali metals (lithium, sodium, and potassium) which you have studied earlier in this Topic?

After the experimental work is complete, discuss the results. Relate the experimental results to the uses of these metals. The pupils' knowledge of transition metal compounds is extended in the following experiment.

Experiment A13.4b

Looking at the properties of some compounds of certain transition metals

These investigations will require an introduction by the teacher. For example pupils will need to be reminded of the test used to detect carbon dioxide, and of ways of carrying out this test. The

---

Apparatus

Each pupil (or pair) will need:

Experiment sheet 52

8 hard-glass test-tubes, 100×16 mm

Bunsen burner and heat resistant mat

About 1 g of each of:

Copper(II) carbonate

Nickel(II) carbonate

Iron(II) carbonate

Sodium carbonate

4M hydrochloric acid, 10 cm³

2M ammonia solution, 20 cm³

The teacher will need as many solutions of transition metal compounds as are conveniently available.

technique of heating a powder or small crystals in a test-tube may also need demonstrating.

Experiment sheet 52

You will be provided with samples of carbonates of some transition metals; also with sodium carbonate, which is *not* a transition metal carbonate but is included so that you can compare its properties with those of the transition metal carbonates. Investigate the properties of these carbonates as follows:

1. Put one measure of carbonate in a test-tube and add about a 2 cm depth of dilute hydrochloric acid. If a gas is evolved, try to identify it. If the carbonate dissolves, note the colour of the solution and compare it with that of the carbonate. When no further change appears to be taking place, add ammonia solution to the contents of the test-tube, a little at a time with shaking, until the mixture is strongly alkaline (test with Full-range Indicator paper). Describe all that happens. To record your results make a table on a separate sheet of paper, using the headings:

Name of carbonate	Gas evolved	Colour of solution	Effect of ammonia solution

2. Put one measure of carbonate in a *dry* test-tube. Heat gently using a small Bunsen flame. Identify any gas evolved and note the colour of the residue. Use the following headings to record your results on a separate sheet of paper.

Name of carbonate	Gas evolved	Colour of residue

Discuss the results of their experiments, and encourage pupils to interpret their findings in a simple way. With able children, it is possible to get them to write word equations and/or chemical equations as summaries for these reactions. Properties of several transition elements can then be compared using the Reference section of the *Handbook for pupils* (this suggestion is included as a possible homework).

Suggestion for homework

Use the Reference section of the *Handbook for pupils* to make a comparison of the physical properties of three transition metals with those of the alkali metals.

Summary

By the end of this section, pupils should know the position of the transition metals in the Periodic Table, and some of the characteristic properties of these elements and their common compounds as exemplified by the first series.

This knowledge of the transition elements should include their physical characteristics; that the solutions of crystals of their salts are coloured; that different colours may be formed on the addition of aqueous ammonia to solutions of these salts; and that on heating or on the addition of dilute acid, the carbonates of these elements decompose and that carbon dioxide is one of the products of these reactions.

Pupils will also have had practice in organizing the results of a large number of experiments. The homework section will ensure that experience has been gained in comparing properties of elements using bar charts or by some other comparable method.

A13.5
Carbon: an element found in living things

The pupils now take one element from the centre of the Periodic Table, carbon, and find out more about it.

A suggested approach

Objectives for pupils

1. Awareness of carbon as the key element in organic compounds
2. Awareness of hydrogen as a common constituent of organic compounds
3. Knowledge of a test for carbon and hydrogen in substances

The Group I metals, the transition metals, and the non-metals in Group VII have now been examined. Draw the attention of the class to the middle of the Periodic Table: there are clearly a large number of elements between these two extremes and we take one element, carbon, for further examination. We cannot say that this element is typical of those in the middle of the Periodic Table because they vary so greatly in their properties. It is only *one* example of an element from the middle of the Table.

Ask the pupils what they know about carbon already. They should know from experiments with electrolysis in Stage I that carbon rods (graphite) can conduct electricity. (Note that the allotropy of carbon is not discussed until section A14.8.) At this point a carbon rod may be tested with a simple conductivity apparatus (e.g. 6 volt battery, bulb, connecting wire fitted with crocodile clips). The pupils also know that carbon burns in oxygen to produce a gas, carbon dioxide, which dissolves in water to give an acidic solution. In this last property carbon behaves like other *non*-metals. To some extent then it has 'in-between' properties.

Ask the pupils where they expect to find carbon. They may know that charcoal is an impure form of carbon; or they may realize, to take a more homely example, that the black stuff on burnt toast is carbon. Both these examples show that carbon is connected with living things; the charcoal was made from wood, the bread from wheat. Tell pupils that carbon is found in all living things and is thus apparently necessary for life.

Can they suggest how to identify carbon in carbon compounds (organic compounds)? Discuss ways of doing this with the class. They should be able to suggest that heating the substance may produce carbon and that if so, the substance will go black. Now give the pupils a selection of substances to test. If there is not time for each pair of pupils to test every substance, divide the substances among the pupils and get them to pool their results at the end of the experiment.

To find out if carbon is formed when substances from plants and animals are heated

Each pupil (or pair) will need:

Experiment sheet 53

Piece of broken crucible or a crucible lid

Tongs

Bunsen burner and heat resistant mat

Teat pipette

2 hard-glass test-tubes, 100×16 mm

Test-tube holder

Samples of substances from plants or animals as listed below:

Sugar, starch, fat (e.g. butter or lard), rice, custard powder, dried leaves or grass, coal, polystyrene

Procedure
Details of the procedure are given on *Experiment sheet* 53 which is reproduced below.

Experiment sheet 53
A selection of substances which are obtained from plant or animal sources is provided for this experiment. Place a small piece of one of them, about the size of a rice grain, on a piece of broken porcelain held in tongs. Heat the substance, gently at first and then more strongly. Record all that happens in the table below, looking for any evidence that the substance contains carbon. When no more changes occur, allow the porcelain to cool on the asbestos square, clean it if necessary and use it to heat another substance. Repeat this procedure with each substance in turn.

Name of substance heated	Changes which take place during heating

---

After the experiment discuss the results with pupils. They should have seen a black substance (carbon) produced in each case. Ask them if there were any cases in which carbon was not produced on heating. You may then broaden the subject of carbon chemistry by telling the pupils that carbon is also present in a large number of compounds which are not found in nature. In order to demonstrate this, give each pupil a small piece of polystyrene and ask him to heat it and observe the change. Can they suggest a reason for the change in colour?

In the experiment the presence of carbon was inferred in a large number of substances, since they turned black on heating. However this is not completely satisfactory, as 'blackness' need not necessarily indicate the presence of carbon. A more convincing method of verifying the presence of carbon must be found. Discuss the matter with the class. They may suggest burning the black stuff and testing the products of combustion for the presence of carbon dioxide using lime water.

Not all the original materials burn readily in air. Is there any other way of supplying oxygen to a compound which is suspected of containing carbon? What about a solid oxide? From their knowledge of the reactivity series the pupils should be able to suggest that carbon can remove oxygen from certain oxides of metals. They may remember that carbon is capable of removing oxygen from copper(II) oxide. Suggest that pupils try heating some of the organic compounds with copper oxide.

To confirm the presence of carbon in various substances

Procedure
Details of the procedure are given in *Experiment sheet* 54 which is reproduced overleaf.

Apparatus

Each pupil (or pair) will need:

Experiment sheet 54

Hard-glass test-tube, 100×16 mm

Hard-glass test-tube, 75×10 mm

Teat pipette

Spatula

Test-tube holder

Bunsen burner and heat resistant mat

Dry copper(II) oxide, 5 g

Anhydrous copper sulphate, 1 g

Limewater, $20\ cm^3$

Selection of carbon compounds (the same materials may be used as in Experiment A13.5a)

Experiment sheet 54

Mix a measure of the substance (in powder form if possible) with a measure of dry black copper oxide in a dry test-tube. Put another measure of copper oxide on top of the mixture. What is the function of the copper oxide?

What is the effect of heat on copper oxide alone?

Have ready a small test-tube containing a 1 cm depth of limewater and a teat pipette.

Heat the test-tube containing the substance mixed with copper oxide, gently at first, then more strongly. Remove a sample of the gas from just above the mixture using the teat pipette and bubble it through the limewater in the small test-tube. (Your teacher will show you how to do this.) What happens?

What do you see on the cool upper part of the test-tube in which the mixture was heated?

What do you think this is?

Allow the test-tube to cool and then carry out a test to confirm your suggestion. Describe what you do and what happens.

What other element does this test show you to be present in the substance that you started with?

Repeat the above procedure with other substances. Make a table to record the substances used and the elements which you have shown them to contain.

After the experiment, discuss the results obtained.

Take a piece of transparent tracing paper and cut it to the size of your Periodic Table. Choose a physical property, such as melting point, and write the melting points of all the elements you have met on the transparent paper so that when you place it over the Periodic Table the melting points are just below the elements to which they refer.

Summary

Pupils should now know that carbon is present in compounds from plant and animal sources and also in compounds not found in nature; that such compounds usually give a residue of carbon (on heating); and that carbon dioxide and water are usually formed when such materials are heated with copper(II) oxide.

A13.6
Silicon: an element found in many rocks

In this section the element of the rocks, silicon, is studied together with one of its compounds.

A suggested approach

Point out that below carbon in the Periodic Table is another element, silicon, which we would expect from previous experience to have similar properties to those of carbon. In fact it is not found in living creatures as carbon is, but in rocks. Silicon is known as the element of the rocks. If this is so, how can we try to get a sample of pure silicon? One source of silicon is sand, which is one of the products of the breakdown and weathering of rocks. Show pupils some sand and ask them to suggest ways of getting silicon from it. Tell them that sand is mainly composed of silica, that is, silicon oxide. How can we remove the oxygen from the silica? Reference to the reactivity series will suggest that a reactive metal or non-metal is needed. It might be a good idea to allow the pupils to make their own choice in the first instance. If someone suggests carbon, let them try heating the carbon and silica together in a hard-glass test-tube. It becomes obvious that a more reactive element is required. In fact magnesium is sufficiently reactive to remove the oxygen from the silicon. Then carry out the experiment described below.

1. Ability to design an experiment
2. Knowledge of the simple chemistry of silicon
3. Awareness of the similarities and differences in properties of carbon dioxide and silicon dioxide

Experiment A13.6

The reaction between magnesium and silica
This experiment MUST be done by the teacher.

Caution. As explosions have been known to occur during this experiment, make sure that the sand and tube are absolutely dry.

Apparatus

The teacher will need:

Safety spectacles

Plastic safety screen

Hard-glass test-tube, 100×16 mm

Test-tube holder

Bunsen burner and asbestos square

Beaker, 100 or 250 cm^3

Watch-glass big enough to rest on top of the beaker

Tripod and gauze

Funnel and quick filter paper

Funnel stand

Magnesium powder

Sand

2M hydrochloric acid

Procedure
It is recommended that the teacher wear safety spectacles for this demonstration. Some teachers use a plastic safety screen as well.

First make an intimate mixture of magnesium powder (1 part by volume) and *dry* purified sand (2 parts by volume). About 2–3 g of the mixture are required.

Place the mixture in a hard-glass test-tube, and clamp it horizontally. Then heat the mixture at the end nearest the mouth of the tube with a Bunsen burner. As the reaction proceeds and the mixture glows, follow the glow with the flame to the bottom of the tube. A sample of impure silicon may be obtained as follows: when the tube is cool its contents should be shaken onto an asbestos square for examination and then transferred to a beaker containing about 20 cm^3 of 2M hydrochloric acid. This will react with the magnesium oxide and any magnesium silicide that may be present: the latter will form gaseous hydrides of silicon and a few harmless explosions will occur as they ignite on contact with air. The teacher should be prepared for this possibility.

Place the beaker on a tripod and gauze and heat the contents just to boiling point with a Bunsen burner. Filter the hot mixture rapidly and wash the silicon on the filter paper with a little hot hydrochloric acid, followed by a quantity of water. The paper plus silicon can then be transferred to a watch-glass and dried.

Finally, examine some of the properties of the silicon prepared, pointing out its appearance, insolubility in water, and lack of action with dilute hydrochloric acid, sodium hydroxide solution, and lime water.

After the experiment, compare some of the properties of silica and carbon dioxide. Although carbon dioxide is a gas and silica is a solid with a very high melting point similarities do exist. Analysis of silica shows that the ratio of numbers of atoms of silicon to those of oxygen is 1 : 2. Its formula is thus SiO_2. This is similar to the formula of carbon dioxide, CO_2. The pupils may remember that magnesium is also able to remove oxygen from carbon dioxide and to that extent this is another similarity.

The pupils should have practice in using word equations as summaries for chemical changes in this Topic.

Pupils may be interested to know why two oxides of elements next to each other in the Periodic Table which have similar formulae have such different physical properties. This is a good moment to explain that it is not the formula which determines the physical properties but the way in which the atoms are joined together. The question of how atoms are joined together will be taken up in the next Topic, A14.

Common elements and their abundance in the Earth's crust and in the human body.
At the end of this Topic it may be helpful to provide some factual information about the abundance of elements. The table below does this for the Earth's crust and the human body.

Element	Symbol	Abundance in the Earth's crust (percentage by mass)	Abundance in the average human body (percentage by mass)
Aluminium	Al	7	—
Calcium	Ca	3	2
Carbon	C	very small*	18
Chlorine	Cl	very small*	0.15
Hydrogen	H	1	10
Iron	Fe	4	very small
Magnesium	Mg	2	—
Nitrogen	N	very small*	3
Oxygen	O	50	65
Phosphorus	P	very small*	1
Potassium	K	$2\frac{1}{2}$	0.4
Silicon	Si	26	—
Sodium	Na	$2\frac{1}{2}$	0.15
Sulphur	S	very small*	0.3

*The remaining two per cent of the Earth's crust is made up of these and 78 elements.

Table A13.1 Common elements and their abundance in the Earth's crust and in the human body [OHP 10]

Suggestions for homework

1. Find out what you can about the way in which rocks were first formed.
2. Make another transparent 'overlay' for your Periodic Table. (See suggested homework for section A13.5.) For example, plot the densities.
3. Make a short summary of your work in this Topic.
4. Write a simple comparative survey of physical and chemical properties of some elements.

5. Refer to *Chemists in the world*, Chapter 8, 'Ceramics and glass' and write a short note on the importance of silicon compounds in everyday life.

Summary

By the end of the Topic, pupils should be aware of the historical development of the Periodic Table and of some of the simpler patterns and trends. For example the pupils should know that metals appear to the lefthand side and in the centre of the table whereas non-metals appear to the righthand side. They will have gained some personal knowledge of the properties of elements and of relating those properties to the position of an element in the Table. The alkali metals and the halogens are used to illustrate 'vertical' relationships in the table. Thus, the alkali metals show an increase in reactivity with increasing relative atomic mass whereas the reverse occurs with the halogens. The inert gases are mentioned.

Pupils should also have become aware of the importance of carbon in organic materials. They should know that silicon is an abundant element, present in many rocks, and that it can be obtained from sand by heating with magnesium. Pupils should have compared the properties of carbon dioxide and silicon dioxide. An indication of the abundance of the elements is also included in the Topic.

Finding out how atoms are arranged in elements and other substances

1. To introduce and use the term mole of atoms (gram-atom).
2. To consider the volumes containing 1 mole of atoms (gram-atom) of the elements, both solid and gaseous.
3. To give an appreciation of the regularity of crystal shape.
4. To introduce the technique of X-ray diffraction for the determination of crystal structure and of the Braggs' original discoveries in this field.
5. To relate the physical properties of metals to their structures.
6. To introduce carbon and sulphur as elements showing allotropy and to examine their structures.
7. To introduce the terms giant structure, molecule, and molecular structure.
8. Where the term gram-atom is used, to introduce and use the terms gram-formula, gram-molecule, and (for gases) gram-molecular volume.
9. To give practice in the use of tables of data.

Contents

A14.1 Alternative route: comparing numbers of atoms by mass measurement. (*Only* to be followed if the Alternative route for Topic A11 was chosen.)

A14.2 Looking at the volumes occupied by 1 mole of atoms (gram-atom) of each of several solid elements.
A14.3 Watching crystals grow.
A14.4 The problem of trying to investigate things too small to be seen.
A14.5 How are atoms arranged in metals?
A14.6 The importance of the structure of metals to engineers.
A14.7 The effect of packing sulphur atoms in different ways.
A14.8 Another element which has different forms: carbon.
A14.9 What can we find out about the structure of gases?
A14.10 Differences between giant and molecular structures.
A14.11 The arrangement of atoms in compounds, such as magnesium oxide and sodium chloride.
A14.12 Starting to write chemical equations.
Appendices
1. Laboratory models of moles of atoms (gram-atoms) of a series of solid elements (see section A14.2).
2. Construction details for a perspex funnel (Experiment A14.3b).
3. Model crystal for use in a ripple tank (Experiment A14.4a).

4. Construction of the triangle for use in the close packing of spheres (Experiment A14.5d).

5. Protractor for marking angles on polystyrene spheres (Experiment A14.8).

6. Template for magnesium oxide model (Experiment A14.11b).

7. Template for caesium chloride model (Experiment A14.11b).

8. Magnetic blackboard as an alternative form of presentation of chemical reactions (see section A14.12).

Timing

If the Alternative route was used in Topic A11 (see page 66), then A14.1 has to be covered first. This should take about a double period. A single period should be enough for A14.2 but the remaining sections need at least a double period and A14.5 and A14.6 will take longer. No more than 8 or 9 weeks should be allotted to the Topic.

Introduction to the Topic

Earlier Topics introduced the idea that matter is particulate and that the particles, or atoms, of which elements are made have definite masses. Pupils should have some idea of the size of atoms and will also have had some experience of using the mole of atoms (gram-atom) concept if the main route through Topic A11 was followed. If the alternative route through Topic A11 was followed, A14.1 introduces the mole concept. Then all pupils start A14.2 with the same background knowledge of the mole concept.

By looking at the volumes occupied by moles of atoms (gram-atoms) of solid elements pupils will see that these volumes are not in proportion to the relative masses of the atoms. This observation is used to introduce the subject of the packing of atoms in solid elements.

To emphasize the reasonableness of the idea that there is regularity in at least some solid structures, the Topic continues with the study of some crystals. 'Is the regularity in the shapes of the crystals a sign that the particles of which they are composed are also regularly arranged?' The ways in which atoms are arranged in elements is considered by referring to X-ray diffraction patterns. The approach is novel and should be considered carefully. (The object is not to involve pupils in the intricacies of X-ray analysis but to make the idea seem plausible.) Analogues of X-ray diffraction (e.g. Nuffield diffraction grids) are used to demonstrate the origin of diffraction patterns. These diffraction patterns provide information about the arrangement of dots photographed on the transparent slides used. After some exercises with these 'grids' the pupils should see that it is reasonable to believe that the arrangement of atoms in solids can be found by means of a technique using X-rays.

To illustrate the structure of elements, pupils grow some metal crystals and make models of their structures. The structure of elements may be important in laboratory work, but has it any practical implications? This question is briefly considered in section A14.6, 'The importance of the structure of metals to engineers'. Bubble rafts and other analogues are introduced to give the pupils a picture of the effect of dislocations and foreign atoms on the regular structure of the metals.

Further consequences of the arrangement of atoms in elements are then considered with special reference to the properties of sulphur and carbon. Allotropes of sulphur are prepared and their properties explained by references to 'what the X-ray crystallographer tells us about the arrangement of atoms in each allotrope'. The allotropes of carbon are also examined.

The volumes occupied by elements in the gaseous state are studied next, and it is found that 1 mole of atoms (gram-atom) of most gases occupy (under normal conditions) about $12 \, dm^3$. An exception occurs in the case of inert gases, 1 mole of atoms (gram-atom) of which are found to occupy about $24 \, dm^3$. In order to explain this difference the term molecule is introduced and used with Avogadro's hypothesis to explain the different values found for volumes of gaseous elements.

In section A14.10, the properties of molecular substances and giant structures are contrasted using iodine and graphite as examples. Iodine (a molecular substance) can be vaporized fairly easily whereas this is not the case with graphite (a giant structure). This difference is explained by supposing that the forces holding the atoms together in graphite are very strong compared with those holding the molecules together in iodine crystals.

Attention is turned in section A14.11 to compounds such as magnesium oxide and sodium chloride. Such substances are found to have high melting points and this suggests that they too have giant structures. Pupils are told that X-ray diffraction methods have enabled the arrangement of the atomic particles to be worked out. (*Note:* The term ion is *not* used during this Topic – it is introduced for the first time during Topic A16.)

Pupils should understand the differences between a giant structure and a molecular structure, and, on the basis of boiling point, be able to make a reasonable guess as to which structure any particular substance has. The Topic is brought to a close by representing the particles which constitute the reactants and products of a chemical change by models. For example, the reaction of hydrogen and oxygen to form water may be represented as shown in figure A14.1.

Figure A14.1
Scale models used to represent the formation of water from hydrogen and oxygen.

The emphasis given in this Topic to the writing of equations is on *visualizing* a regrouping of particles, *not* on balancing complicated chemical equations.

Subsequent development

The ideas introduced in Topic A14 are used throughout the rest of the course: they include knowledge of what is meant by molecules, molecular structures, giant structures, the relationship between structure and properties, the mole of atoms, gram-formulae, and moles of molecules (gram-molecules). Option 4 'Metals and alloys' of Stage III extends some aspects of this Topic.

Alternative approach

Topic B17 is an introduction to structure and two approaches are considered. In essentials, the content is the same although greater emphasis is given to applied aspects. Also, Topic B17 does not assume knowledge of the mole (gram-atom).

Background knowledge

Crystals are considered in a number of Topics during Stage I. If the pupils are following Nuffield Physics, it should be noted that the structure of crystals is discussed in a very elementary way during Year 1.

The mole (gram-atom) concept is required for sections A14.2 onwards. Pupils should either have met the mole (gram-atom) concept in Topic A11 or have followed the alternative route (page 66) in which case they meet the mole (gram-atom) in section A14.1.

Note on mathematics
Some school mathematics courses include the properties of polyhedra which 'fill space', for example prisms, cubes, rhombuses, dodecahedra, truncated octahedra.*

Such work is obviously related to the problem of the packing of spheres so as to minimize 'unfilled' space and some teachers may be interested in pursuing the links between the two subjects, chemistry and mathematics, during this Topic and in particular during section A14.5 onwards.

Further references

for the teacher

See *Collected experiments*, Chapter 13, for additional experiments on this Topic.
See *Handbook for teachers*, Chapter 9 to 12, for more detailed discussion of the structure of elements and compounds.
Nuffield Advanced Physics (1971) Teachers' guide Unit 1 *Materials and structure*.
Nuffield Advanced Chemistry (1971) *Metallurgy* – a Special study.
Nuffield Advanced Physical Science (1974) *Teachers' guide III*, the materials options M1, M2, and M3.
Farrar, R. A. (1971) *The mechanical properties of materials*. Methuen Educational.
Jenkins, E. W. (1973) *The polymorphism of elements and compounds*. Methuen Educational.
Mander, M. and Pargeter, F. W. J. (1974) *Metals and alloys*. Heinemann Educational.
Martin, J. W. (1969) *The elementary science of metals*. Wykeham Publications.
Schools Council (1972) Project Technology Handbook 3 *Sample materials testing equipment*. Heinemann Educational.

*For example, see *School mathematics project: Book B with Teachers' guide*, Chapter 2 (1969), Cambridge University Press and *Book E, with Teachers' guide*, prelude (1970). Cambridge University Press.

Films
'Considering crystals' Unilever Film Library
'Metallurgy – a special study' ESSO. Films for Science Teachers No. 28
'Exploring chemistry' (a film for teachers) Unilever Film Library
'X-ray diffraction' EFVA
'Crystal structure' ICI Film Library
Film loops
2–2 'Gram-atoms (The mole of atoms)'
2–3 'Sulphur crystals'
Special teaching aids
The following figures are available in a series of slides:
Figure A14.3   Photographs of various crystals
Figure A14.8   Simplified representation of an X-ray diffraction apparatus
Figure A14.9   X-ray diffraction by the von Laue method of a crystal of beryl
Figure A14.12 Photographs to illustrate the crystalline nature of metals
Figure A14.23 A bubble raft
Figure A14.29 (a) Graphite. (b) Diamond in rock. (c) Diamond in jewellery and a natural diamond.
Sets of Nuffield Diffraction Grids are available which give an optical analogue for X-ray diffraction patterns. Cards showing X-ray diffraction patterns are included with these sets (see figure A14.10).
Construction details for a perspex funnel, crystal analogue, wooden triangle, protractor, templates, etc., are provided in the Appendices in this Topic.
Overhead projection originals
11 Spheres arranged to show (a) ABCABC and (b) ABAB packing *(figure A14.14)*
12 Stacks of spheres showing packing *(figure A14.15a)*
13 Stacks of spheres showing packing *(figure A14.15b)*
14 Stacks of spheres showing packing *(figure A14.15c)*
15 Models of the three common ways in which atoms are arranged in metals *(figure A14.16)*
16 Changes in the arrangement of sulphur atoms on heating *(figures A14.26 & .27)*
17 Packing of S_8 rings *(figure A14.28)*
18 Avogadro's law and gases *(figure A14.32)*
19 An equation using molecular models *(figure A14.41)*
20 The magnesium–oxygen reaction *(figure A14.44)*

Handbook for pupils, especially Chapter 3 'Atomic structure and bonding' and Chapter 9 'The structure of elements and compounds'.
Chemists in the world, Chapters 6 'Polymers from Petroleum' and Chapter 8 'Ceramics and glass' illustrate applications of studies of structure.

A14.1
Alternative route: Comparing numbers of atoms by mass measurement

This section is only for those who have followed the Alternative Route through Topic A11 (see page 66).

The purpose of this section is to introduce the mole of atoms (gram-atom) and the Avogadro constant.

Begin the lesson by reminding pupils of the ideas that matter is particulate, that elements are made up of atoms, and that such particles have distinctive properties. The Periodic Table was introduced in Topic A13 as a particular arrangement of the elements in the order of their relative atomic masses. Show pupils bottles which contain the relative atomic mass of as

1. Revision of the particulate theory of matter
2. Knowledge of the symbols of some common elements
3. Revision of the term relative atomic mass of an element
4. Understanding the meaning and use of the terms mole of atoms and the Avogadro constant
5. Ability to use tables of data

many solid elements as possible. For example, show them 12 grams of carbon, 24 grams of magnesium, 32 grams of sulphur, etc. Remind the pupils of the meaning of these numbers: carbon has an atom which is 12 times heavier than a hydrogen atom; magnesium 24 times, sulphur 32 times, and so on.

It is not necessarily obvious to pupils that each of these quantities of material contains the same number of atoms. It is therefore suggested that the teacher uses a series of related articles such as ballasted polystyrene balls, wooden balls, and table tennis balls, each of known relative mass. Use three paper bags and place ten balls of one kind in each bag. Ask the pupils to predict the relative masses of the three bags. Finally, without showing the pupils, put simple proportionate numbers of balls into the bags and weigh them. Ask the pupils to predict the number of balls present in each of the bags knowing only the relative masses of the balls. Explain that this is a process of counting out by weighing the bags. Banks do this when they weigh out £5 of 10p pieces. The film loop 2–2 'Gram-atoms (The mole of atoms)' recapitulates many points raised in this discussion. To bring home the physical reality of moles of atoms (gram-atoms), pupils could weigh out different proportions of moles of atoms (gram-atoms) (Experiment A14.1).

Pupils may be curious to know the actual number of atoms in a mole of atoms (gram-atom) of an element. Although it is not important to the argument at this point many teachers have found that this knowledge helps the pupils' understanding. Pupils may therefore be told that it is a very large number, about 6×10^{23}; that it can be found experimentally by a number of independent methods too difficult to explain at this point; and that it is known as the Avogadro constant after a famous nineteenth-century Italian scientist.

Experiment A14.1

Alternative route: exercises in using moles of atoms (gram-atoms) of various elements

Apparatus

The following apparatus will enable all the suggested exercises to be carried out by the pupil:

Copies of the *Handbook for pupils*

Access to a rough balance (measuring mass to the nearest gram)

Supply of paper squares to place on balance pans

(Continued)

Procedure
Pupils should be asked to use the Reference section of the *Handbook for pupils* as a source of the relative atomic masses. Ask them to obtain the quantities suggested below on paper squares using a rough balance. They place these weighed amounts in beakers – suitably labelled – to be kept until the end of the lesson and compared.

Exercises
1. Obtain 1 mole of atoms (gram-atom) of sulphur.
2. Obtain $\frac{1}{2}$ mole of atoms (gram-atom) of copper.
3. Obtain $\frac{1}{10}$ mole of atoms (gram-atom) of zinc.
4. Obtain as many atoms of carbon as there are in 16 g of sulphur.

Supply of 100 cm^3 beakers in
which to place known masses
of the elements

Labels

Carbon (graphite, *not* charcoal)

Copper (metal turnings)

Sulphur

Zinc (metal granulated)

5. Weigh twice as many atoms of carbon as there are atoms in
64 g of copper.

In doing these exercises, pupils will begin to appreciate the
relationship between mass in grams and moles of atoms
(gram-atoms). Thus:

$$\text{number of moles of atoms (gram-atoms) of an element} \equiv \frac{\text{number of grams of the element}}{\text{relative atomic mass of the element}}$$

i.e.

$$\text{the amount of an element expressed in moles of atoms (gram-atoms)} \equiv \frac{\text{mass of the element in grams}}{\text{relative atomic mass of the element}}$$

Note. The numbers involved in the exercises have deliberately
been kept simple. In the experimental situations the numbers
used will *not* necessarily be simple (see the experiments in
Topic A11, Alternative route).

By this time, pupils should know that:
1. Elements are considered to be made up of atoms; all atoms
of one element are alike but differ from the atoms of every
other element.
2. It is possible to compare the masses of atoms of different
elements and so build up a list of relative atomic masses.
3. The amount of an element having the same number of grams
as the numerical value of the relative atomic mass is known as
the mole of atoms (gram-atom) of that element.
4. 1 mole of atoms (gram-atom) of all elements contains the
same number of atoms. This number, 6×10^{23}, is known as
the Avogadro constant.

In addition, pupils should be able to use data tables to look up
relative atomic masses, and should understand the relationship between quantities expressed in grams and those expressed
in moles (gram-atoms). *They may revise the symbols for the
elements and learn that these may be used to represent the
quantity of the element expressed in moles of atoms
(gram-atoms).*

This change in use of symbols requires emphasis. Ask the
pupils if they can see any advantage in making this change.
Show them a series of bottles each containing 1 mole of atoms
of a solid element. What do they notice about these quantities?
All contain the same number of atoms but occupy different
volumes. Why? The answers to this question form the theme
for section A14.2.

Suggestion for homework

See section A11.3, page 50 for suggestions for homework.

A14.2
Looking at the volumes occupied by 1 mole of atoms (gram-atom) of each of several solid elements

Pupils calculate the volumes occupied by a mole of atoms (gram-atom) of solid elements and investigate reasons for differences between them.

A suggested approach

Objectives for pupils

1. Knowledge of the term gram-atomic volume and its relationship to the density of an element
2. Appreciation of the fact that the gram-atomic volume depends on the sizes and on the arrangement of the atoms

Begin by showing pupils scale models of 1 mole of atoms (gram-atom) of each of several *solid* elements. A typical collection of full-scale models arranged in order of increasing relative atomic mass is shown in figure A14.2, and instructions for making these models are given in Appendix 1, page 159. Teachers may recall that the idea shown in this model was presented in film loop 2.2 'Gram-atoms (The mole of atoms)'. Teachers may also find it helpful to use bar charts (see Topic A13, page 93).

Pupils will see from the models that there is no simple relationship between relative atomic masses and gram-atomic volumes. They might have expected that the volume would be proportional to the mass of an element, but this is not so. Pupils should be shown that the volume of the mole of atoms (gram-atom) can be calculated by dividing the relative atomic mass by the density of the solid element. Since moles of atoms (gram-atoms) of different elements all contain the same number of atoms, there must be considerable differences in the *sizes* of the atoms of the different elements. It is possible that the pupils' introduction to the Periodic Table in Topic A13 may lead them to notice a periodic repetition of large and small volumes. They might notice, for instance, the large volumes occupied by a gram-atom of each of the alkali metals. Then draw attention to the fact that the gram-atomic volumes of graphite and

Figure A14.2
Wooden models of moles of atoms.

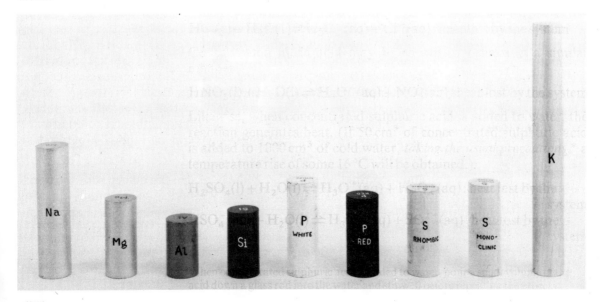

diamond are not identical, although these solids are forms of the same element – carbon – and so must be made up of identical atoms. One possible explanation is that the atoms are arranged differently in the two solids. This focuses attention on the arrangement or packing of atoms which is the main theme of this Topic.

Suggestions for homework

1. Using the data given in the Reference section of the *Handbook for pupils*, calculate the volumes of 1 mole of atoms (gram-atom) of the following elements: Li, Be, B, C, Na, Mg, Al, Si, K, Ca. Plot a bar chart showing your results in this order. You know that lithium (Li), sodium (Na), and potassium (K) have similar chemical properties. Can you see any similarities in the values of the volumes of moles of their atoms when compared with those of neighbouring elements?

2. Calculate the volumes of one mole of atoms (gram-atom) of the two forms of phosphorus – red and white. Suggest a reason for the two volumes being different. Also suggest possible reasons for both being larger than the value for aluminium ($10\,cm^3$).

Summary

Pupils should now be able to calculate the gram-atomic volume of a solid element and should recognize that this volume depends on some other factor apart from the size of individual atoms. They should understand that the arrangement, or packing, of the atoms in the solid element is likely to be important.

A14.3
Watching crystals grow

In this section pupils observe the growth of crystals. The regularity of their shape is used as evidence for the orderly arrangement of their constituent particles.

A suggested approach

Objectives for pupils

1. Awareness of the regularity of shape of the crystals of a given substance
2. Understanding that regularity of shape strongly suggests that the particles in a crystal are arranged in a regular manner

Are atoms in solids all jumbled up, or are they arranged in an orderly manner? This question might be put to the pupils to see if they can think of any evidence to support one view or the other. Obviously indirect evidence will be required because atoms are far too small to be seen! The most convincing evidence for orderliness in certain substances rest on the remarkable regularity of their crystals. (Since the easiest crystals to show are often those of compounds, pupils need to be reminded that the 'building blocks' used are not necessarily atoms. Accordingly the term particle will be used in such cases.)

Pupils should be provided with some striking evidence of the regularity of crystal shapes. A good method is to let pupils observe crystallization taking place in their own experiments (Experiment A14.3a).

Alternatively, the teacher can use a micro-projector to show the formation of crystals. (This method was described in the first edition and is given in *Collected experiments* (Experiment E13.3).)

After pupils have seen crystallization taking place, they should have the opportunity of looking at some really large crystals. Include some naturally occurring crystals in the collection to dispel any idea that crystals are some kind of laboratory curiosity.

Experiment A14.3a

Watching crystals grow

Note: crystals for display
A variety of large crystals are needed for display purposes, for example, various alums, copper(II) sulphate, calcite, quartz, mica, and sulphur.

It is suggested that good samples obtained as the result of other experimental work (for example from the preparation of salts) are stored. Fragile crystals may be protected from rough handling by mounting them in transparent embedding plastic. (*Note:* some laboratory suppliers stock suitable kits.)

Large naturally occurring crystals may also be purchased. The following specimens are suggested as a minimal requirement: fluorite (calcium fluoride); calcite, a variety of Iceland spar (calcium carbonate); quartz (silicon dioxide); gypsum (calcium sulphate-2-water); mica; and sodium chloride.

Procedure
Before the lesson prepare warm saturated solutions of the salts and check that the solutions deposit crystals in a reasonable time when poured onto a watch glass. These solutions (except for the acetone solution) should be kept just below the boiling point during the introduction to the lesson. When practical work commences the Bunsen burners should be extinguished and the solutions placed in accessible positions around the laboratory. Each beaker should be clearly labelled with the name of the solution it contains, and a (previously warmed) teat pipette should be placed in the solution to enable pupils to withdraw samples. Other details are given in *Experiment sheet* 55 which is reproduced below.

Apparatus

Each pupil (or pair) will need:

Experiment sheet 55

3 watch glasses

Access to a microscope and/or hand lenses

Microscope slides

The teacher will need:

7 beakers 250 cm³

7 teat pipettes

7 test-tubes, 16 × 150 mm

Test-tube rack

6 Bunsen burners, tripods, and heat resistant mats

Hot saturated solutions of:

Ammonium chloride

Ammonium nitrate

Potassium chlorate

Potassium nitrate

Potassium chromate

Sodium hydrogen sulphate

Warm solution of acetamide in acetone (propanone)

Supply of distilled water

Experiment sheet 55

You will have grown crystals earlier in the course. In this experiment you will concentrate on watching crystals grow rather than on the crystals that are formed.

Your teacher will provide hot, saturated solutions of a variety of substances. Using the teat pipette in one of the solutions transfer about 2 cm³ of it to a small watch glass. Stand the watch glass on the bench and observe carefully the formation of crystals as the solution cools, noting any points of interest below. Make a sketch to show the appearance of one or two of the crystals.

Repeat the observations and sketches for other solutions, using a clean watch glass in each case.

The experiments show the following results:

Ammonium chloride	– dendrites
Potassium chromate	– 'fluffy' clusters
Ammonium nitrate	– needles

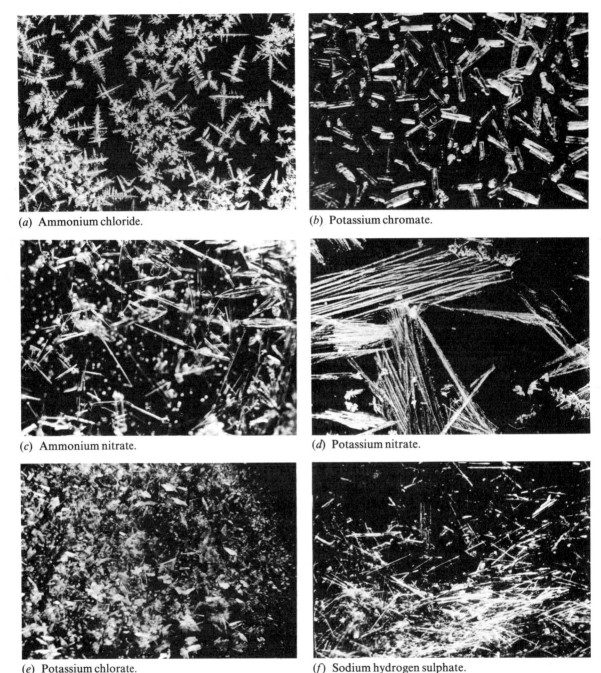

(a) Ammonium chloride.

(b) Potassium chromate.

(c) Ammonium nitrate.

(d) Potassium nitrate.

(e) Potassium chlorate.

(f) Sodium hydrogen sulphate.

Figure A14.3
(Available in a series of slides.)

Potassium nitrate	– needles
Potassium chlorate	– thin plates
Sodium hydrogen sulphate	– needles
Acetamide	– needles

Photographs which illustrate these descriptions appear in figure A14.3.

After the practical work has been completed, pass round some large crystals for inspection. Pupils are fascinated by the shape and colour of crystals and need little encouragement to make sketches of them in their notebooks. Finally, let the pupils look at some crystals under a microscope – if their practical work has not required this.

It does not seem possible that such regular shapes could be made from a jumble of particles, but only from particles which are arranged in an orderly manner. Pupils may think that the different shapes of the various crystals arise because of the different shapes of the constituent particles. Remind them of graphite and diamond, in which the 'building blocks' are identical but the structures are different.

In the next experiment, a model of a crystal is made in which spherical particles are shown to form a regular pattern.

Figure A14.4
Pouring spheres into a funnel. (*Note:* the spheres immediately arrange themselves into a regular pattern.)

(*Apparatus overleaf*)

| Experiment A14.3b | Watching growth in a model of a crystal |

Procedure
Support the funnel in a vertical position with the wide end uppermost. Place the large tray underneath the funnel. Now pour the spheres from a container into the funnel, tapping the funnel from

The teacher will need:

Large perspex funnel (construction details are given in Appendix 2, page 159)

Supply of small spheres (constant diameter, say 25 mm) – large number required

Large tray (0.5 m × 0.5 m)

Container from which the spheres can be poured

Summary

time to time to assist the packing process. When the funnel is full, continue to pour the spheres to enable them to build into layers above the edge of the funnel. (Any loose spheres will collect in the tray.) Pupils should now observe how the building up of a 'crystal' occurs naturally – especially from the 'growth' seen above the edge of the funnel.

This experiment can lead to a brief discussion on the mechanism of growth of real crystals. Growth occurs through the attachment of fresh particles onto an existing crystal face, thereby building up additional layers. The main emphasis of this discussion should be on the regularity of the structure attained.

Pupils should now be aware that the regularity of shape of crystals is very likely due to the regularity of the arrangement of particles which make up the crystal.

A14.4
The problem of trying to investigate things too small to be seen

The solution to the problem of investigating things too small to be seen was found by Max von Laue, and Sir William Henry Bragg and Sir William Lawrence Bragg.

Their work is discussed in this section and optical analogues are used to explain the use of X-ray diffraction in the determination of crystal structures.

A suggested approach

Objectives for pupils

1. Awareness of X-ray diffraction as a way of investigating crystal structures
2. Knowledge of the part played by M. von Laue, and Sir W. H. and Sir W. L. Bragg in developing the X-ray diffraction method of structure determination
3. Knowledge of the regularity of atomic arrangement in some structures which have been determined by X-ray methods

In the previous section pupils have been led to believe, from their observation of crystal shapes, that the particles in crystals are arranged in a regular way. The question now arises, 'How do we know the precise nature of this regular arrangement, since the particles are too small to be seen?'.

This problem was solved by the work of von Laue and the Braggs and aspects of their work are described in the *Handbook for pupils* and in *Chemists in the world*. The main point is that the Braggs were able to deduce the arrangement of the atoms in a crystal by passing a beam of X-rays through it. It is not essential for pupils to be aware of the details of the experimental procedure at this stage. (However, teachers may wish to refer to the electromagnetic radiation spectrum and to indicate the location of visible light, X-rays, radiowaves, and so on. The choice of X-rays for the investigation of crystal structure was made because the wavelength of the radiation used had to be about the same as the distance which separated planes of atoms in a crystal.)

In order to give pupils some insight into the principle of X-ray diffraction, the following experimental analogues may be used. The

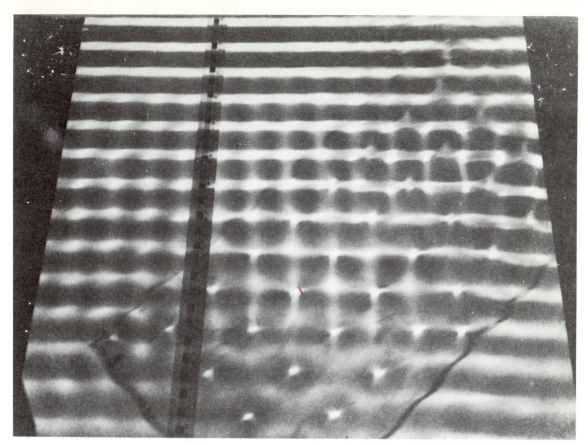

Figure A14.5
Ripple tank crystal analogue.
Photograph, Miss J. Henshelwood, Department of Physics, King's College, London.
Crystal analogue by G. R. Pierce, Department of Chemistry, King's College, London.

first is a simple demonstration using a ripple tank and a model crystal made from some nuts and bolts arranged in a regular pattern in a piece of perspex. This can be used to show the effect of varying the wavelength of the 'radiation'. The second analogue requires special Nuffield diffraction grids.

Experiment A14.4a

Crystal structure analogy using a ripple tank and a 'model' crystal

Apparatus

The teacher will need:

Ripple tank fitted to demonstrate linear waves

Ripple-tank illumination source

Ripple-tank power pack

'Model' crystal (for details of the construction of this model see Appendix 3, page 160)

Procedure
Fill the ripple tank with water and illuminate the tank with the special light source. Place the model crystal in the tank (see figure A14.5) and adjust the level of the water until bright spots of light appear in the image seen on the screen below the tank. This effect is usually obtained when the bolts of the 'atoms' in the crystal just 'sit' on the water surface in the ripple tank. Arrange for the production of linear waves and show the effect of varying the speed of the ripple-tank motor. Alter the position of the crystal to obtain the conditions for Bragg reflection (see figure A14.5). The condition is given by:

$$n\lambda = 2d \sin \theta$$

A14.4 The problem of trying to investigate things too small to be seen   **119**

where λ is the wavelength of the wave motion
d is the distance between the regular layers of atoms
$(90-\theta)$ is the angle of the wave front to the regular array of atoms in the crystal.

If $\theta = 60°$, then the minimum distance d is $\dfrac{\lambda}{2}$.

This relationship is not intended for the pupils, but to help the teacher to achieve the condition for Bragg reflections.

After showing the effect of using different wavelengths, show the effect of using a wavelength related to the distance between 'atoms' (see figure A14.5). Tell the pupils that this can lead to an estimate of the distance between atoms in such a crystal. In what ways is this analogue similar to X-ray diffraction? This model system is very simple and could be improved. Light-waves would be a better analogue for X-rays than water-waves, as the wavelengths of X-rays and light are so much more alike. Instead of the pattern of nuts and bolts in perspex, a pattern of very small dots on a piece of film might form a better crystal model. This is the basis of the Nuffield diffraction grids used in Experiment A14.4b.

Experiment A14.4b

Apparatus

The teacher will need:

Piece of material with a regular structure, for example, cotton handkerchief or piece of nylon or terylene net.

Set of Nuffield diffraction grids (see figure A14.6)

Small light source, for example, 6 V battery with holder and bulb

Low power microscopes or hand lenses for pupils' use

Photographs of X-ray diffraction patterns (see figure A14.10)

1 or 2 models of crystal structures, for example, sodium chloride, diamond

An optical analogue for X-ray diffraction patterns

Procedure
Part 1: to explain the nature of a diffraction pattern the teacher will need a point source of light. A bright torch bulb will do. It is important that the bulb should be bright for this experiment so that blackout is unnecessary. Pupils look at the bulb through a stretched handkerchief. Ask them what they see.

Ask them to try the same kind of experiment using a piece of nylon or a piece of terylene net. What happens to the pattern when the handkerchief is turned through 90°? It also turns. Clearly, the pattern is related to the structure of the handkerchief. Hand lenses should now be made available so that pupils can look at the structure of the handkerchief used. What happens when the handkerchief is stretched? The pattern changes. Finally, what happens when the handkerchief is tilted? They will see that the spots in the pattern get further apart. Some pupils may realize that this demonstrates that the nearer the strands of the handkerchief are together the further the spots are apart. Thus the distance between the spots in the pattern varies inversely with the distance between the holes of the structure used.

Part 2: Nuffield diffraction grids may now be introduced for class use. Tell the pupils that these have three types of pattern – shown in figure A14.6.

Give the pupils some of the marked cards to examine first. They should hold the cards in one hand, close to their faces, with the card covering the one eye as shown in figure A14.7. They can then look through the grid towards the light source with their left eye

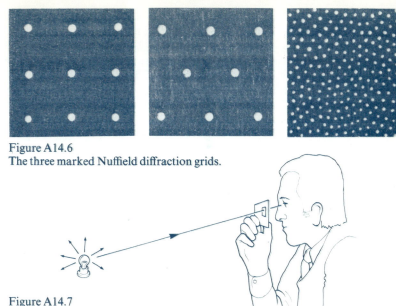

Figure A14.6
The three marked Nuffield diffraction grids.

Figure A14.7
Method of using Nuffield diffraction grids.

without having to close the right eye, or vice versa. With an ordinary 'white' source, such as a torch bulb, there will be considerable dispersion and coloured streaks instead of white spots. The pattern is still quite clear, but if a sufficiently bright source is used a piece of coloured glass or film may be placed in front of it to act as a filter. The pupils should look at each of the three kinds of grid. They will realize that the pattern that they can see tells them about the arrangement of spot on the film which they cannot see. Once again they will see the pattern move when they tilt or turn the card. Now provide pupils with some unmarked cards. Can they tell the arrangement of dots on each card by looking through the grid at the small light bulb? Let them check their ideas by examining the grid with a hand lens or low powered microscope, or by comparison with a marked card.

The various Nuffield grids may be shown to a class by projection. To do this effectively, the room must be blacked out and a standard slide projector employed. Place a fine piece of foil with a pin hole in it in the slide position and hold the Nuffield grid in front of the projector some 15 to 30 cm away from the projector bulb. A diffraction pattern will be seen on the screen. (This demonstration requires careful preparation – before the lesson begins!)

After the experiment with the grids, show the class an example of a von Laue-type X-ray photograph and explain that a beam of X-rays passing through a crystal produces a pattern similar to those seen through the grids (figures A14.8 and A14.9). The contribution of the Braggs was to provide a relatively simple mathematical analysis which allows the position of the atoms in a crystal to be calculated. The details need not concern the pupils at this point – only the results. Figure A14.10 is included in the set of Nuffield diffraction grids.

Figure A14.8
Simplified representation of an X-ray diffraction apparatus. Note the similarity between this arrangement and the use of Nuffield diffraction grids. (Available in a series of slides.)

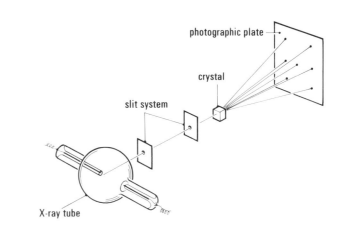

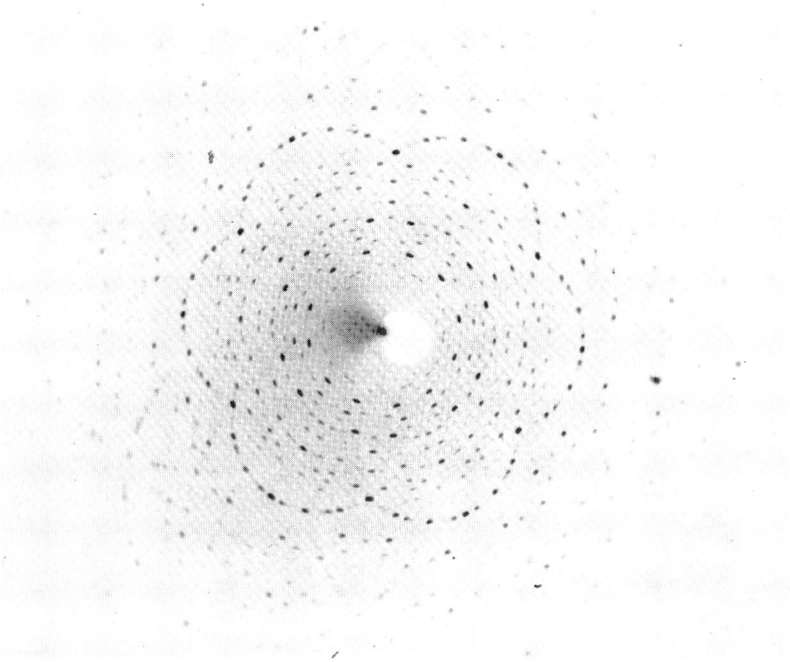

Figure A14.9
X-ray diffraction by the von Laue method of a crystal of beryl. (Available in a series of slides.)
Photograph, The Royal Institution.

Figure A14.10
Photographs of X-ray diffraction patterns. (Included with the Nuffield diffraction grids.)
Photographs: diamond, Dr H. Judith Milledge, University College, London. Aluminium, Dr G. I. Williams, Fulmer Research Institute.

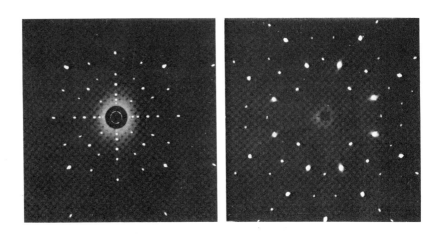

Then show some models of crystal structures which have been determined by X-ray analysis and point out the regularity of arrangement of the atoms. Such an arrangement would be expected from the regularity of the crystal shape. Suitable examples are diamond and sodium chloride structures. The models used should be of the same kind ('tangential contact' or 'ball and stick') so as to avoid the need for a discussion of model types at this point of the course.

Suggestion for homework

Find out all you can about the Nobel prize winners who did their research using the technique of X-ray analysis (for example, *Chemists in the world*, Chapter 3 'The way of discovery').

Summary

Pupils should now be aware that the use of X-ray diffraction methods allows us to determine the arrangement of the atoms in crystals. They should know that the methods were developed by Sir W. H. and Sir W. L. Bragg, following an original discovery by Max von Laue. They should have seen some examples of regular atomic arrangements in models of structures which have been determined by the X-ray diffraction method.

A14.5
How are atoms arranged in metals?

In this section crystals of various metals are prepared by the pupils and models of metal structures are made from polystyrene spheres.

A suggested approach

Objectives for pupils

1. Awareness that metals have a crystalline structure
2. Knowledge of the existence of different patterns for the close-packing of spheres, namely, 'ABAB' and 'ABCABC', and the less closely packed body-centred arrangement
3. Knowledge that most metals crystallize with their atoms in one of these three arrangements

The teacher will need:

Sheet of galvanized iron

Piece of tin-plated iron sheet (unpolished)

In the last section, it was shown that the arrangement of particles in a crystal can be determined by X-ray diffraction. In this section we consider the structures of metals. The first point to make is that metals are crystalline. 'But a piece of copper wire does not look like a crystal.' Tell the pupils that in a piece of metal the crystals are jammed together and are often very small and cannot be seen directly. Sometimes the original shapes can be seen in the surface of a piece of metal. This can be demonstrated by showing the class some galvanized iron or a sheet of tinned iron. Ordinary galvanized corrugated iron sheeting shows the crystalline structure of zinc quite clearly. The tin-plated iron sheet will show crystalline structure only if the tin has not been polished. However, the inside of an old tin can *may* show the structure as the result of etching by its former contents. Alternatively a piece of rapidly cooled tin plate may be etched using hydrochloric acid (Experiment A14.5c). Teachers will recall that pupils who followed Alternative IB saw crystals of metals in Topic B10. Pupils now carry out some experiments to obtain some metal crystals.

Growing crystals of metals

Apparatus

1. Lead crystals

Each pupil (or pair) will need:

Experiment sheet 56

Test-tube, 100×16 mm

Access to a hand lens

0.1M lead acetate solution, 10 cm^3

Strip of zinc foil, 100×10 mm

2. Silver crystals

Each pupil (or pair) will need:

Experiment sheet 56

Test-tube, 100×16 mm, and a loosely fitting cork

Access to a hand lens

15 cm length of copper wire (about 22 s.w.g.) or a strip of copper foil, 100×10 mm

0.1M silver nitrate solution, 10 cm^3

Procedure
Details are given in *Experiment sheet* 56, reproduced below. Warn the pupils about the need for caution when using silver nitrate solution.

Experiment sheet 56

You may not normally think of metals as being crystalline because, as we usually see them, metals are highly polished so that the crystal boundaries are obscured. It is fairly easy, however, to grow crystals of some metals. In this experiment you will try to do this for lead and silver.

1. *Lead crystals*
Half fill a test-tube with lead acetate solution. Bend over the top 1 cm of a strip of zinc foil, so that it can be supported over the rim of the test-tube, and place it in the solution. Allow the test-tube to stand in a rack. Describe what happens, drawing a sketch if this will help.

2. *Silver crystals*
These can be made using silver nitrate solution and copper foil or copper wire. (Silver nitrate solution should not be allowed to remain in contact with skin or clothing, as it produces black stains which are difficult to remove; if any is accidentally spilled, wash it off at once with plenty of water. **So, be careful when using silver nitrate solution!**)

If copper foil is used, the same procedure is followed as for lead crystals, see (1) above. If copper wire is used, shape the wire into a helix by winding it round a pencil. Open up the coil slightly so that each turn is separate from its neighbour. Place the wire in the silver nitrate solution and bend the top end over the edge of the test-tube. Allow the tube to stand in a rack.

Describe what happens.

---

3. Silver crystals

The teacher will need:

Test-tube, 100×16 mm

0.1M silver nitrate, 10 cm^3

Mercury, 1 drop

3. This experiment *must* be carried out by the teacher.

Procedure
Transfer a drop of mercury to a test-tube and then half fill the tube with silver nitrate solution. Needlelike crystals of silver will begin to grow from the mercury after a few minutes.

Apparatus

The teacher will need:

Safety spectacles

Crucible (fused silica)

Flat smooth surface – such as a piece of copper sheet or stainless steel

Bunsen burner and heat resistant mat

Crystallizing dish (10 cm diameter)

(Continued)

How do crystals form from molten lead?

This experiment may be shown to pupils by the teacher (or by one or two groups of pupils). Lead is easily melted and the shapes of the crystals formed on solidification are revealed by treatment with dilute nitric acid.

Procedure
Caution is necessary – molten metals can be hazardous! Teachers are recommended to wear safety spectacles.

Melt the lead in the crucible but do not overheat. If the metal is dirty, a scum will collect on the surface but this can be removed using a nickel spatula, preferably after adding a little powdered charcoal. Pour the molten lead onto the flat surface so that it forms a pool some 30 or 50 mm in diameter.

Figure A14.11
Crystals in a lead pancake. (See also figure B15.17.)

Immerse the cold lead 'pancake' in dilute nitric acid. Remove the specimen from the acid when the metal crystals reflect light like jewels. Wash the specimen with water and examine it using a magnifying glass.

The use of acids to highlight the structure of metals is known as *etching*. The acid attacks the grain boundaries first. Deep etching results in some grain boundaries appearing darker than others.

Two types of crystal will be seen in the sample.
a. Elongated crystals which have grown from the outer edges of the specimen towards the centre (*columnar* crystals). Such crystals are produced by a steep temperature gradient.
b. Small crystals which are formed in the centre of the specimen where the rate of cooling is approximately the same over a comparatively large area (*equi-axed* crystals).

Pipeclay triangle and tripod

Magnifying glass

Tongs

Nickel spatula

30–40 g lead

2M nitric acid

Powdered charcoal

How do crystals form from molten tin?

This experiment may be done by the teacher or by the pupils.

Apparatus

The teacher (or each pair of pupils) will need:

Micro-burner

Heat resistant mat

Tongs

Teat pipette

Access to sink

Piece of good quality tin plate – about 150 mm square

Concentrated hydrochloric acid

Procedure

Hold the piece of tin plate with tongs over the flame of a micro-burner for about 5 seconds to heat a small area about 10–15 mm across. Remove the plate from the flame and allow the tin to cool. Pour a little concentrated hydrochloric acid over the heated surface of the plate and wash the acid off with water.

Repeat the experiment but this time heat a similar area in one of the corners of the piece of metal.

Record and discuss the results obtained. The rate of heat loss from the molten tin varies with direction and this affects the growth of the crystals.

After these experiments, the teacher should show pupils some metal crystals that have been prepared previously, or have been purchased as specimens. Any available metal crystals may be used, the important point being to convey orderliness and regularity of the crystal shapes. Photographs showing the crystalline nature of metals (such as figure A14.12) may be shown to the pupils. The effects shown in figure A14.12b and c could be demonstrated.

Figure A14.12
Photographs to illustrate the crystalline nature of metals. (Available in a series of slides.)

(*a*) See opposite below.

(*b*) Dendrites in antimony (× 1)

(*c*) See opposite.

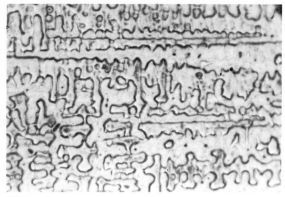

(*d*) Cast bronze (5% Sn; 95% Cu), polished and etched (× 400). Note dendrite-type structure (fern-like). If bronze is annealed, the structure looks like that shown for brass.

(*e*) Brass (70% Cu; 30% Zn), polished and etched (× 100).

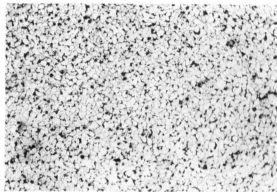

(*f*) Steel (0.12% C, × 100). Note ferrite grains and small dark regions of pearlite in the grain boundaries.

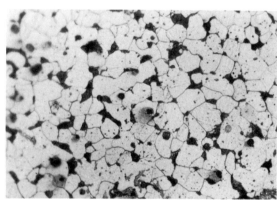

(*g*) Steel (0.12% C, × 400).

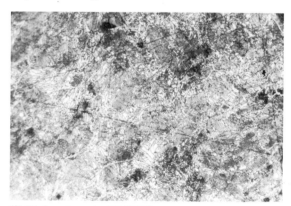

(*h*) High carbon steel (1.2% C), polished and etched (× 400).

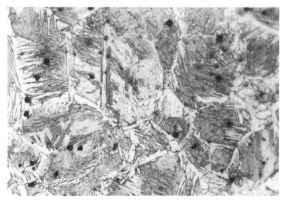

(*i*) Steel (0.12% C), quenched (× 400). Note change in structure. The dark regions are oxide.

(*a*) Silver crystals: the tree-like effect was obtained by placing a shaped copper wire in silver nitrate solution.

(*c*) Lead crystals: the effect was obtained by placing a strip of zinc in lead acetate solution.

Photographs: b, *Science Museum, London;* d–i, *Dr. A. A. Smith, King's College, London.*

A14.5 How are atoms arranged in metals?

Discuss the results of the experiments and ask 'How are atoms arranged in these crystals?' The pupils know that X-ray diffraction patterns will provide the answer to the question. Before telling pupils about the three main types of metal structure, ask them to suggest possible arrangements of atoms in these crystals. Polystyrene spheres may be used as models of atoms. It is important to stress the fact that these spheres are only models of atoms and that they are useful in discussions about packing arrangements but not for other problems to do with atoms.

Experiment A14.5d

Apparatus

Each group of pupils will need:

Bag containing 35 polystyrene spheres (preferred size 25 mm diameter) *or* rafts of close-packed spheres

Wooden triangle (the size will depend on the size of the spheres selected; for construction details see Appendix 4, page 160)

Thin glass rod

Tracing paper

Supply of 2p pieces

Figure A14.13
Model to show the close packing of spheres.

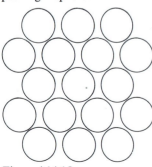

Figure A14.15
(OHPs 12, 13, and 14)

How might atoms be arranged in metals?

Procedure
Suggest that pupils first arrange the spheres as close together as possible in the triangle – there is one obvious arrangement, and this is called a 'close-packed layer'. Ask them to arrange another layer on top. Is this layer like the other one? Yes, but the spheres now lie over the gaps between spheres in the layer below. What about another layer on top of the last one? Pupils should see that they can place their spheres either directly over the spheres in the first layer ('ABAB' arrangement) or in a new position ('ABCABC').

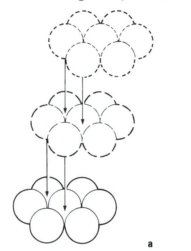
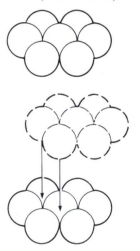

a   b

Figure A14.14
Stacks of spheres showing two kinds of close packing.
(*a*) Close-packed spheres in ABCA sequence;
(*b*) Close-packed spheres in ABAB sequence. (OHP 11)
From Roberts, I. F. (1974) Crystals and their structures. Methuen.

Tell pupils that the structures of all metals have been identified using X-ray diffraction and that many metals are found to have one of these two structures, e.g.

ABAB   zinc, magnesium
ABCABC   copper, aluminium

If time allows tell the pupils to use 2p pieces and to draw several rows of circles on a piece of tracing paper (figure A14.15). Each group will require at least three prepared tracings. Then ask the pupils to 'stack' their sheets of tracings. Can they stack them in two different ways – as they found with the spheres?

The terms 'cubic close-packed' for ABCABC and 'hexagonal close-packed' for ABAB could be introduced *but are not essential*. It is *not* intended that any special emphasis be placed on the cubic symmetry which develops in ABCABC structures. However, some pupils may find it interesting that ABCABC structures made of spheres have close-packed layers in more than one plane, related by symmetry, which will all be possible planes of slip when the sample of metal is under stress.

However, there is another sort of arrangement. Some pupils may have discovered body-centred cubic for themselves – most will need to have it demonstrated. Have ready some models of this structure (it is difficult for pupils to make). The following metals have body-centred cubic structures:

sodium, potassium, chromium, iron.

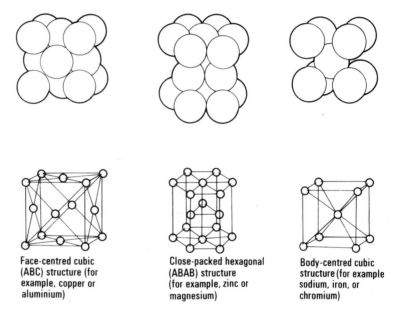

Face-centred cubic
(ABC) structure (for
example, copper or
aluminium)

Close-packed hexagonal
(ABAB) structure
(for example, zinc or
magnesium)

Body-centred cubic
structure (for example
sodium, iron, or
chromium)

Figure A14.16
Models of the three common ways in which atoms are arranged in metals. (These should be on display in the laboratory.) (OHP 15)

Ask pupils to count how many nearest neighbours each atom has in the three structures (12 in the ABAB and ABCABC arrangements, 8 in the body-centred cubic arrangement). However, do *not* allow pupils to think that metals with this last structure necessarily have a low density, or are softer.

Summary

By the end of this section, pupils should be familiar with the idea that metals possess a crystalline structure. They should know that three types of packing of atoms are commonly found in metals. They should understand that the difference between the two close-packed structures rests in the positioning of the successive layers, and should also have met the body-centred cubic structure. Pupils should also understand that atoms are less closely packed in the body-centred cubic structure than in the other two arrangements.

A14.6
The importance of the structure of metals to engineers

Some physical properties of metals are examined and accounted for in terms of their structures. There is an opportunity to show the relevance of metallic structure to everyday life by discussing the uses to which metals of given properties (and hence of given structure) can be put. Experimental work must necessarily depend on the resources available to the teacher.

Objectives for pupils

1. Awareness of the fact that the strength and ductility of a metal are related to its structure
2. Understanding of the terms dislocation, crystal grain, and giant structure
3. Awareness of the relevance of the structure and properties of metals to everyday life

Four experiments (A14.6a, b, c, d) on the properties of metals are described here, but it is not essential for pupils to see or do them all. The teacher should select experiments on the basis of the interest of the class. The examiners will not expect pupils to have done all the experiments and will not ask specific questions on them, although general questions on properties of metals may be set. The next two experiments (A14.6e, f) relate properties and structure, and form part of the basic course.

The first three experiments look at important properties of metals – strength, ductility, and hardness. As wide a range of metals as possible are tested, and it is worthwhile comparing a hardened form of a metal with the soft form, e.g. hard-drawn copper and copper wire. The next experiment is on the effect of heat on steel, and there could be a short discussion of the effect of carbon content. These four experiments should give an idea of the range of values of different properties for different metals, and also the effect the previous treatment of the metal has on its properties.

Experiment A14.6a

Apparatus

The teacher will need:

Standard lengths of 26 s.w.g. wire of some or all of the following:
Copper, brass, aluminium, steel, iron, nickel, nichrome, tin (fuse wire), etc.

Slotted weights

Wire-stretching apparatus (see figure A14.17)

Pad of foam rubber

2 G-clamps

How strong are metals?

Procedure
Figure A14.17 shows the arrangement of the apparatus. A standard length of wire (30 cm is a convenient length) of 26 s.w.g. is required for this experiment. Slotted weights are added to the wire and a pad of foam rubber placed on the floor or bench underneath the slotted weight to prevent damage. Tests should be carried out on at least two samples of metal.

After the experiment discuss the results. 'If you were making a bridge/an aeroplane/a saucepan/the buzzer of an electric bell/the axle of a car, which metal would you choose?' Many other properties as well as strength are involved. Density, for example, is important when metals are used in aircraft construction but not particularly significant for railway lines.

It also effects price, as metals are bought by weight (copper saucepans are heavy and more expensive than light aluminium ones of the same size). Malleability is important for some uses (toothpaste tubes) but not for others (tin cans). Corrosion resistance, thermal conductivity, electrical and magnetic properties, durability, fatigue properties, and many others might be mentioned in discussion, as can the concept of 'cost effectiveness'. The next two experiments investigate other properties.

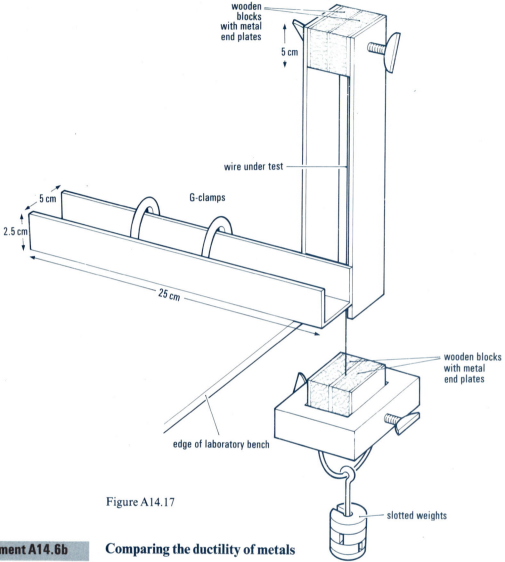

wooden blocks with metal end plates

5 cm

wire under test

5 cm

2.5 cm

G-clamps

25 cm

wooden blocks with metal end plates

edge of laboratory bench

Figure A14.17

slotted weights

Comparing the ductility of metals

Apparatus

The teacher will need:

Strips of different metals of the same thickness

Metal-bending apparatus (see figure A14.18) or a metalwork vice

Procedure

Clamp a metal strip firmly in the metal-bending apparatus. Move the handle backwards and forwards and count the number of bends needed for the metal to fracture. Repeat the experiment using a different metal strip of the same thickness and size as the first one. Tabulate the results.

An alternative procedure is to use the metalwork vice to hold each metal strip in turn and to bend the strip backwards and forwards about the point at which it is held. To ensure that bending does not occur over the length of the metal strip a piece of rigid steel needs to be strapped to the metal under test at a point just above the clamp and secured along the entire length of the metal strip.

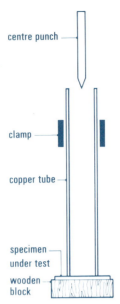

centre punch

clamp

copper tube

specimen under test

wooden block

Figure A14.19

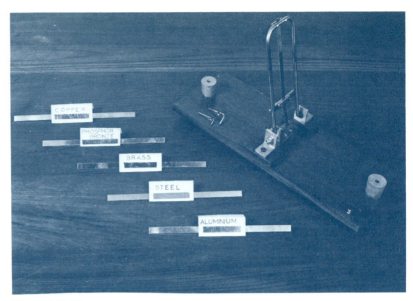

Figure A14.18
(See also figure B15.13.)

(See also figure B15.13.)

Experiment A14.6c

Apparatus

The teacher will need:

Small pieces of various metals of approximately the same thickness

Apparatus for testing the hardness of metals (see figure A14.19)

Small wooden block

Retort stand, boss, and clamp

Hand lens or microscope

Investigations on the hardness of metals

Procedure
1. Scratch hardness
The selection of metals may be arranged in order of decreasing hardness by comparing the ease with which one metal will scratch another. This test, although qualitative, can yield useful results.

2. Using an apparatus for making indentations
Arrange the apparatus as shown in figure A14.19. Allow the hardened centre punch to fall from a standard height onto the metal sheet resting on a piece of wood. It is important to control the dropping of a punch by using the metal tube – as shown in the diagram. Examine the indentations produced in the samples of metals using the small hand lens and compare the hardness of the samples qualitatively.

Contrast the results obtained by the two methods.

The discussion of these experiments could be led towards ways in which an engineer, rather than just choosing a metal, can change the properties of a material to suit a particular need. The effect of heating steel and cooling it in different ways is investigated in the next experiment.

The effect of heat treatment on steel

Apparatus

The teacher will need:

Bunsen burner

Some steel pins

Beaker (250 cm³)

Supply of water

Tongs

Procedure

Heat one of the pins in a Bunsen flame and allow it to cool in the air. The pin becomes very soft and can be easily bent. If a second pin is heated in a Bunsen flame and then quenched rapidly in a beaker of water, it will be found that the pin cannot be bent in the same way. Indeed, it will break in a brittle manner. The ductile state can again be induced by reheating the quenched pin at a temperature which just causes a blue oxide colour to appear again on the polished surface. (That is, the metal is tempered.)

After the experiment, the importance of iron and steel could be discussed. The effect of carbon content on the properties of steel might be mentioned briefly.

The next two experiments are on the relationship between the structure of a metal and its properties.

Pupils have learnt in A14.5 that a piece of solid steel is made up of lots of crystals. The crystals may be too small to see or quite large, all the same size and shape or different sizes and shapes within a particular piece of metal.

Why are some metals stronger than others? Can we use the polystyrene sphere model to explain the strength of metals? We now look at what happens to the crystals when a piece of metal is compressed. Pupils saw earlier how to prepare and etch the surface of a metal to show the crystals (see Experiments A14.5b and A14.5c). Explain that metallurgists call each crystal a 'grain', and the interface between crystals a 'grain boundary'.

The preparation of a metallic surface to show slip lines

Apparatus

The teacher will need:

Several similar pieces of copper

Emery cloth (various grades)

Sheet of plate glass

Metal polish

Crystallizing dish

Tongs

A solution containing 5% iron(III) chloride, 2% hydrochloric acid in ethanol

Metalwork vice

Magnifying lens (× 10) or metallurgical microscope

Procedure

Polish a freshly cut piece of copper with an emery cloth resting on a plate glass surface. Use various grades of emery cloth and complete the task using metal polish. Etch the polished face for between 15 and 30 seconds in the iron(III) chloride–hydrochloric acid–ethanol solution. Repeat the process using other similar pieces of copper. Compress one piece of copper using the metalwork vice until a slight crushed effect can be seen on the etched surface. Compare the polished and etched faces of two pieces of copper by viewing them either with a powerful hand lens or with a metallurgical microscope. Figure A14.20 shows typical results for this experiment.

After the experiment, discuss the results. Point out that the slip is *within* the crystal, not the slipping of one crystal over another. The more slip, the more ductile the metal is; less slip means the metal is harder. Pupils have studied the structures of crystals in A14.5. Can they suggest how slip can occur in the polystyrene sphere model? Take two close-packed layers of spheres and demonstrate how the

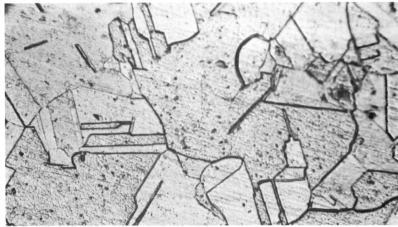

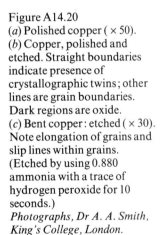

Figure A14.20
(*a*) Polished copper (× 50).
(*b*) Copper, polished and
etched. Straight boundaries
indicate presence of
crystallographic twins; other
lines are grain boundaries.
Dark regions are oxide.
(*c*) Bent copper: etched (× 30).
Note elongation of grains and
slip lines within grains.
(Etched by using 0.880
ammonia with a trace of
hydrogen peroxide for 10
seconds.)
*Photographs, Dr A. A. Smith,
King's College, London.*

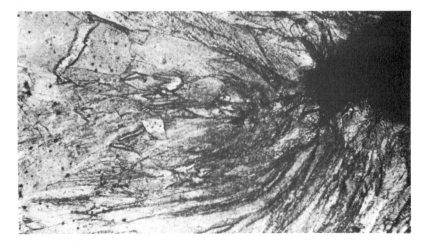

top layer slips over the other one in a zig-zag way. (It is also easy to
see why a close-packed layer does not slip back again.) In a real
metal, there are always imperfections in the crystal, and these make
it easier for atoms to move. Diagrams showing vacancies in crystal
structures and the mismatching of crystal planes make this clearer,
see figure A14.21. A bubble raft can also be used to show some of
these features.

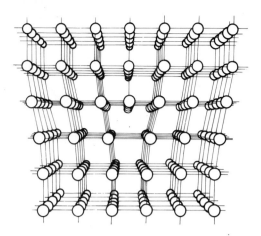

Figure A14.21
Model of an edge dislocation.

Apparatus

Each pupil (or pair) will need:

Beaker, 100 cm³

Petri dish

Glass tube drawn out to a fine jet in the way shown in figure A14.22

50 cm³ syringe and rubber tubing to connect the syringe to the glass tube

Small quantity of a solution of Teepol in water (10 cm³ to 1 dm³ of solution)

Making a bubble raft

There are several ways of performing this experiment. The original method due to Bragg and Nye (1947)* is somewhat complex and is not recommended for routine use. What follows is a very simple way of conducting the experiment successfully.

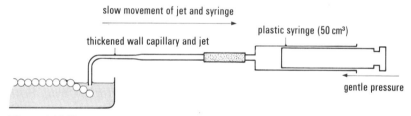

Figure A14.22

Procedure
Fill the Petri dish with Teepol solution. Blow a large number of very small similarly sized bubbles by moving the jet slowly below the surface as shown in figure A14.22.

Note. It is inadvisable to replenish the volume of air in the syringe without disconnecting the syringe from the glass tube. Alternatively, when no plastic syringes are available, connect the glass tube to the laboratory gas supply and adjust the gas flow to yield the convenient size of bubble. (**Warning.** If gas is used, make sure that the laboratory is well ventilated and that the gas is turned off as soon as the rafts have been made.)

It is convenient to demonstrate this technique before asking the pupils to carry out the experiment. Figure A14.23 illustrates a typical bubble raft obtained by this procedure.

In the bubble raft it should be possible to see large areas of regular patterns of bubbles. It is often helpful to stand back and view the dish from several feet away to see this effect. Closer inspection

*Bragg, W. L., and Nye, J. F. (1947) *Proc. Roy. Soc.* A190, 474–481.

Figure A14.23
A bubble raft. (Available in a
series of slides.)

shows that these large areas are separated by zones where bubbles
have got out of their regular positions – perhaps because of an oversized bubble or missing bubbles. The area boundaries are made up
of a large number of such 'dislocations'. In a metal crystal there
may be as many as a thousand million dislocations in each cubic
centimetre of crystal, but this works out at something less than one
dislocation per ten thousand atoms in any line through the crystal.

What is the effect of these dislocations? A line of atoms slips across
another line until it reaches a dislocation. There it stops. In the
copper specimen, different grains had slip lines at different angles.
This means that it is *more* difficult to distort than if all the grains
slipped at the same angle. Tell pupils that single crystals of metals
are found to compress more easily than polycrystalline metals, and
this has important implications for engineering. Metals that cool
rapidly form small crystals, and are therefore harder. (Can pupils
see how this explains the results of Experiment A14.6d?)

Grain boundaries in the bubble raft may be due to oversized bubbles.
We introduce 'larger bubbles' into a metal by adding small proportions of another metal. Soft metals like copper can be made
much harder by adding metals such as zinc (brass), aluminium, and
tin.

Finally, point out to pupils that even small crystals of metals are
made up of thousands of atoms. Tell the pupils that metals possess
the type of structure known as a giant structure, whatever the form
of detailed packing of metal atoms.

Imagine that a way of producing metal structures is discovered which results in making all metals and their alloys ten times as strong (and no more expensive!). What difference would it make to the way in which things are made; for example, bridges, aircraft, buildings?

Summary

Pupils should be aware of the grain structure of the metals which they have seen and handled. They should know that when strength is important in a metal object, special steps have to be taken to keep the crystal grains small. They should also understand the meaning of the term giant structure. The pupils will have seen the results of various experiments involving 'models' of metal structures. They will have seen that these models can be used to explain the physical properties of metals and should appreciate the relevance of the structure of metals to engineers.

A14.7
The effect of packing sulphur atoms in different ways

The relationship between micro-structure and macro-properties is studied again, this time with respect to the allotropes of sulphur.

A suggested approach

In the previous section the properties of metals were shown to be related to the arrangement of the metal atoms. This same conclusion will now be reached by quite a different route using the non-metallic element sulphur.

Objectives for pupils

1. Familiarity with the relationship between physical properties and structure
2. Knowledge of the allotropes of sulphur
3. Recognition of the idea that atoms of elements can join together in small numbers to form units called molecules
4. Recognition of the differences between molecular and giant structures

Experiment A14.7a

Making various allotropes of sulphur

Apparatus

The teacher will need:

Test-tube, 100×16 mm

Watch glass

Spatula

Filter paper

Powdered roll sulphur or a pestle and mortar and a small lump of roll sulphur

(Continued)

Note. The first two preparations that follow should be done by the teacher and not by the class since the solvents used have dangerous properties. Powdered roll sulphur must be used. 'Flowers of sulphur' is *not* suitable since it contains a large proportion of an insoluble amorphous form of sulphur.

1. Preparation of rhombic sulphur

Procedure
Powder a piece of roll sulphur about the size of a pea, or take a spatula measure of powdered roll sulphur, and add to it about 2 cm depth of carbon disulphide in a test-tube. Agitate the mixture (but do not warm it). When most of the sulphur has dissolved, decant

the solution onto a watch glass in a fume cupboard. Allow the solution to evaporate slowly. This may be done by covering the watch glass with a sheet of filter paper. Crystals of rhombic sulphur will form after about 10 to 20 minutes.

Caution. Carbon disulphide is dangerously flammable, extremely toxic, highly volatile, and has an unpleasant smell. It should be used with care.

Apparatus

The teacher will need:

Test-tube, 150 × 25 mm

Spatula

Bunsen burner and heat resistant mat

Test-tube holder

Post card

Powdered roll sulphur or pestle and mortar and a small lump of roll sulphur

Xylene (*Note.* Toluene is a suitable alternative solvent but it is a little more volatile)

(Alternative apparatus is shown in figure A14.24)

2. Preparation of monoclinic sulphur from a solution of sulphur

Procedure

Powder a piece of roll sulphur about the size of a pea, or take a spatula measure of powdered roll sulphur, and add to it about 3 cm depth of xylene in a test-tube. Place the test-tube in a test-tube holder and then hold the tube cautiously over a low Bunsen burner flame and heat the tube until the sulphur just dissolves. The liquid should NOT be allowed to boil. Carefully place the test-tube in a test-tube rack away from the flame of the Bunsen burner and allow the solution to cool. Crystallization must start above 96 °C if monoclinic crystals are to be obtained: if not, add more sulphur and repeat the experiment.

Caution. Xylene is flammable but it is not very volatile and the vapour is dense and tends to stay in the test-tube – *provided* a low Bunsen burner flame is used when the solution of sulphur is formed. In the event of fire, turn the burner out at once, and cover the mouth of the test-tube with a piece of card to extinguish the flames. If by any chance the test-tube should break, use sand, not water, to extinguish the flames.

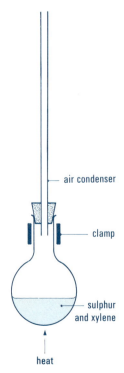

air condenser

clamp

sulphur and xylene

heat

Figure A14.24

Alternatively use the apparatus shown in figure A14.24. The mixture of xylene and sulphur is heated in the flask and the solution is allowed to crystallize. Monoclinic sulphur separates out from solution if a concentrated solution of sulphur begins to crystallize above 96 °C.

Apparatus

Each pupil (or pair) will need:

Experiment sheet 57

Hard-glass test-tube,
100×16 mm

Beaker, 250 cm^3

Test-tube holder

Paper clip

2 filter papers

Tongs

Bunsen burner and heat resistant mat

Powdered roll sulphur

3. Preparation of monoclinic sulphur from liquid sulphur

This experiment and Experiment A14.7b, may be carried out by pupils at the discretion of the teacher. A well-ventilated laboratory is necessary since it is likely that some pupils will allow their samples of sulphur to catch fire, particularly during the second experiment.

Procedure

This is described in *Experiment sheet* 57 which is reproduced below. It is *very important* for the teacher to emphasize the need for gentle and uniform heating of the sample during the experiment. Sulphur is a bad conductor of heat and five or more minutes of heating will be necessary. (The practice in technique during this experiment will be a good preparation for the experiment which follows.)

Experiment sheet 57
In this experiment sulphur crystals are obtained by allowing liquid sulphur to cool.

Three-quarters fill a test-tube with powdered roll sulphur. Fold a filter paper in the usual way and fasten it with a paper clip.

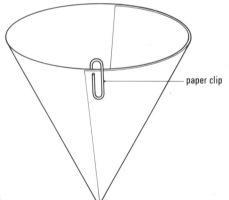

Figure A14.25

Have ready a beaker of water.

Hold the test-tube in a holder and heat the sulphur very gently, keeping the test-tube moving all the time. It is important that the sulphur is only *just* melted, to a clear, amber-coloured liquid.

When all the sulphur has melted, hold the filter paper by the rim with a pair of tongs and pour the molten sulphur into it.

Allow the sulphur to cool until a crust has about half covered the surface and then pour out the remaining liquid sulphur into the water in the beaker. Immediately open the filter paper (be careful not to burn your fingers) and inspect the monoclinic sulphur crystals contained in it. Draw one or two of them.

Observing what happens when sulphur is heated: making plastic sulphur

Apparatus

Each pupil (or pair) will need:

Experiment sheet 58

Test-tube, 100×16 mm

Beaker, $250 \, cm^3$

Test-tube holder

Bunsen burner and heat resistant mat

Post card

Powdered roll sulphur

Procedure

This is described in *Experiment sheet* 58. The teacher should be prepared with a piece of card to hold over the mouth of any test-tube in which the sulphur catches fire. Again, emphasize the need for gentle uniform heating – until, on this occasion, the sulphur boils.

Experiment sheet 58

Gently heat a test-tube which is three-quarters full of powdered roll sulphur until *all* the sulphur melts, holding the test-tube in a holder. Observe carefully all that happens during the slow heating process and continue this until the liquid sulphur just boils. The vapour may catch fire, so be careful. (Your teacher will show you how to extinguish the flame if this does happen.)

Pour the liquid obtained into a beaker half-filled with cold water. Leave the sulphur in the water for a few minutes to cool and occupy the waiting time by describing below what you saw during the heating.

Now remove the sulphur from the water and examine its properties. It is called *plastic sulphur*. Knead a small piece between the finger and thumb for a few minutes and note any changes that take place. Record your observations below.

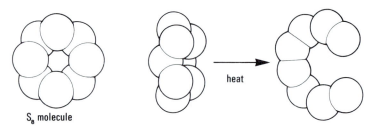

S_8 molecule   heat

Figure A14.26 (OHP 16A)

After these experiments show pupils a model of an S_8 sulphur ring. Tell them that X-ray diffraction studies have shown that rhombic and monoclinic sulphur are made up of S_8 rings. The atoms in these rings are tightly bound together – it is possible to break them open but it is not at all easy. Pupils saw in the last experiment how roll sulphur melted to form first a fairly free-flowing liquid. In this state, the sulphur atoms are still bound together in S_8 rings which can flow easily over each other. If liquid sulphur in this condition is cooled (see *Experiment sheet* 57) it forms monoclinic sulphur, made up of S_8 rings. At a higher temperature the liquid sulphur became very sticky and viscous. This can be explained by suggesting that more and more rings break open – here you could break the model S_8 ring. The atoms at each end of the broken ring are reactive and immediately join up with atoms from other broken rings to form long chains. These chains become intertwined and are unable to slide over each other easily. This process of polymerization is greatest just below $200 \,°C$, when the longest chains are estimated to contain a million atoms. (Even at this temperature, however, there are still some unbroken S_8 rings left.)

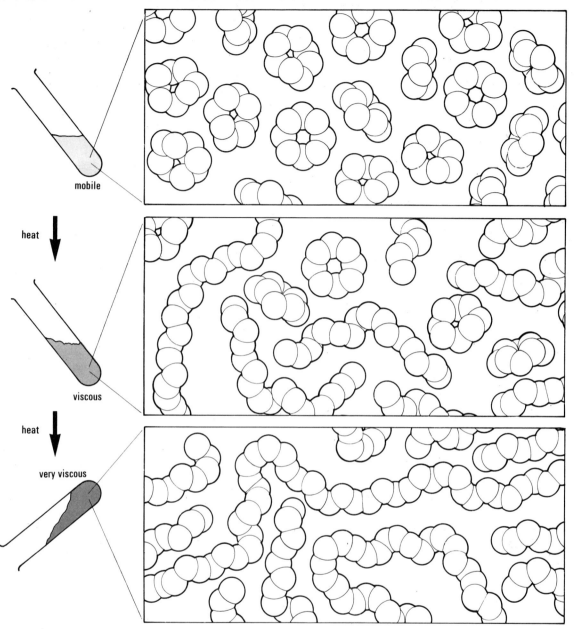

Figure A14.27
Changes in the arrangement of sulphur atoms on heating (OHP 16B)

Between 200 °C and the boiling point of sulphur (444 °C), the liquid becomes less viscous. The chains break into shorter lengths and the viscosity gradually falls.

When the near-boiling sulphur is poured into cold water (*Experiment sheet* 58) it enters the water as a mixture of rings and chains of sulphur atoms. This arrangement is 'frozen' as there is no time for a general rearrangement of the atoms into S_8 rings as happens during slow cooling. Plastic sulphur is therefore a mixture of rings and long chains. Any change from chains back into S_8 rings must involve changes in the positions of the atoms. Such a process is

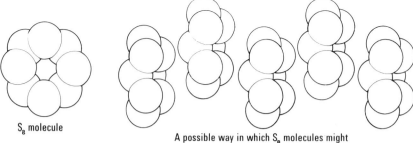

S$_8$ molecule

A possible way in which S$_8$ molecules might be packed together in a crystal

Figure A14.28
Packing of S$_8$ rings. (OHP 17)

difficult in the solid state and so plastic sulphur changes back into the rhombic form extremely slowly, taking months or even years to achieve the transformation.

The changes which occur when sulphur is heated are explained in the *Handbook for pupils*, Chapter 9. The illustrations shown in figures A14.27 and A14.28 are available as overhead projection originals which may be used in class discussions.

If pupils ask about the structures of rhombic and monoclinic sulphur, explain that both are made up of S$_8$ rings. Figure A14.28 shows how S$_8$ rings can be packed together. Pupils should appreciate that S$_8$ rings (show them the model again) can be fitted together in a number of ways. (In fact, the structure of monoclinic sulphur was not known until 1965.*)

Summarize this section by pointing out the differences between the structures of rhombic and monoclinic sulphur and the giant structures of metals. Giant structures are made up of any number, up to millions, of atoms arranged regularly. Rhombic and monoclinic sulphur are made of S$_8$ rings packed together regularly. We call a small grouping like the S$_8$ rings a *molecule*. A molecule is a group of atoms which cling together and are quite hard to break apart. When a group of molecules are packed together in a regular arrangement in a crystal it is called a molecular structure. What are rhombic and monoclinic sulphur? Is plastic sulphur a molecular substance?

Molecules are found in liquids and gases as well as in solids. There is evidence for the presence of S$_8$ molecules in boiling sulphur and in sulphur vapour.

Suggestion for homework

Industrial countries use millions of tonnes of sulphur each year. Try to find out where this sulphur comes from and what it is used for.

Summary

By the end of this section pupils should have a knowledge of the properties of the allotropes of sulphur and explanations for these properties in terms of structure.

They should also know that in some substances atoms are joined together in small groups and that these groups are called molecules.

*Sands, D. E. (1965) 'The crystal structure of monoclinic (β) sulphur' *J. Amer. Chem. Soc.*, **87**, 1395.

A14.8
Another element which has different forms: carbon

The allotropic forms of carbon are studied and their properties are related to their structures.

This section may be opened by reminding pupils that in Topic A13 they found that carbon is an element which occurs in all living things. They met it then as charcoal, formed by strongly heating sugar or various other materials. Tell them that the free element occurs naturally in two very different forms – graphite and diamond. If specimens are available, demonstrate some of the differences. Figure A14.29 (available as slides) can also be used.

Figure A14.29
(*a*) Graphite.

(*b*) Diamond in rock

(*c*) Diamonds in jewellery and a natural diamond. (Available in a series of slides.)

The main differences between graphite and diamond are listed in Table A14.1.

Graphite	Diamond
Conducts electricity	Does not conduct electricity
Opaque to light	Transparent to light
Soft enough to mark paper and can be split easily into flakes parallel to one plane only	One of the hardest substances known; but it can be cleaved with difficulty parallel to certain planes
1 mole of atoms (gram-atom) occupies 5.3 cm^3	1 mole of atoms (gram-atom) occupies 3.4 cm^3

Table A14.1 Properties of the allotropes of carbon: graphite and diamond

Ask pupils how they would prove that graphite, diamond, and charcoal contain only carbon. It may be necessary to demonstrate burning charcoal in oxygen.

The important point is that charcoal, graphite, and diamond are all forms of the same element, as shown by the fact that equal masses of all three forms give exactly the same quantity of the same product – carbon dioxide – when burned in an excess of oxygen.*

Differences between these allotropes must be due to differences in the arrangement of the atoms. Point out that charcoal does not give *easily recognizable* crystals. We say that it is 'amorphous'. In graphite, X-ray diffraction methods have shown that the atoms are arranged in layers made up of huge numbers of atoms. In diamond, the atoms are joined together tetrahedrally, each atom being joined to four others to give an unending three-dimensional network. Show models of the structures of graphite and diamond. Pupils might then make their own models as described below.

Experiment A14.8

Apparatus

Each group of pupils will require:

Pipe-cleaners

Access to wire cutters

Dividers or other instrument with a point

For graphite: 13 expanded polystyrene spheres, 25 mm diameter

(Continued)

Making models of graphite and diamond structures
This may be a more useful exercise for a science club than for a normal lesson.

Procedure
Graphite model. Mark the spheres at 120° round the equator. Pierce them and join them together in rings of six carbon atoms using short lengths of pipe-cleaner. With 13 spheres, one layer of these rings can be made.

Diamond model. Mark the spheres tetrahedrally. Pierce them and join them together, using short lengths of pipe-cleaner.

The *Handbook for teachers* (page 155) provides details of diamond and graphite models.

If groups of pupils prepare graphite layers, these may be assembled to make a large model by using stiff struts of wood or wire (but not

*Teachers wishing to demonstrate this point experimentally may like to refer to Rogers, M. J. W. (1970) *Gas syringe experiments* Heinemann Educational.

For diamond: 14 expanded
polystyrene spheres, 25 mm
diameter

Protractor: details for making
this instrument appear in
Appendix 5, page 161

pipe-cleaner) to hold the parallel layers the correct distance apart. If polystyrene spheres of 25 mm diameter are used, the distance between the centres of the spheres is 59 mm.

The diamond structure is more difficult to make but with a competent group of pupils it is possible to assemble a larger portion of the structure than that shown in figure A14.30. The diagram is drawn with a large space between the spheres so that the connections can be seen clearly. It is intended that the model should be of the tangential-contact type.

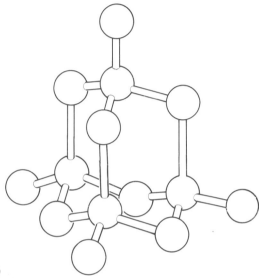

Figure A14.30

Remind pupils of the giant structure of metals. They should see at once that the same description applies to the structure of diamond, but may need more convincing that it applies to graphite. Pupils should now look more closely at models of graphite. They can see that each layer is made up of interlocking hexagons of carbon atoms, forming a two-dimensional giant structure. The distance between atoms within the layers is much less than the distance between the layers. The forces holding the atoms in layers are very strong whereas the forces holding the layers together are weaker, explaining the ease with which layers slip over one another and why graphite flakes so easily. In metals there are close-packed layers which slip across each other – but it is obviously much easier for the graphite layers to slip than the layers of metal atoms.

In the diamond structure the atoms are joined together tetrahedrally, each carbon atom being joined to four others. Pupils may ask why diamond is so very hard, especially compared with single crystals of metals. At this stage we can only say that the attractive forces between atoms in diamond are very strong, and remind them of the strength of the attractive forces between sulphur atoms in S_8 rings.

The reason for the difference between the volumes occupied by 1 mole of atoms (gram-atom) of graphite and of diamond should be obvious from the models.

The section could be completed by going back to the Periodic Table, and discussing the differences between metals and non-metals.

Read *Chemists in the world*, Chapter 5, 'Energy and chemicals', the section on diamonds.

Summary

By the end of this section pupils will have studied some properties of two allotropes of carbon, diamond and graphite, and related their properties to their respective structures. They will have been introduced to the idea that the attractive forces between atoms in different elements can be of different strengths (compare diamond and metals) and that within the same substance there may be different attractive forces (for example, much stronger forces holding atoms together in S_8 rings than hold the S_8 rings together in a crystal; similarly, the arrangement of atoms in layers in graphite).

A14.9
What can we find out about the structure of gases?

The term *molecule* was introduced at the end of A14.7 to describe the S_8 rings and was mentioned again in the previous section when discussing the structure of graphite and diamond.

In this section it is suggested that teachers use book values for the densities of gases to lead pupils to the idea that the common gaseous elements exist at room temperature as molecules of two atoms each.

A suggested approach

Objectives for pupils

1. Knowledge of the general relationship between the volumes occupied by moles of molecules of gases and its expression in Avogadro's hypothesis
2. Recognition of the molecule as the important unit in the structure of gases and understanding of the meaning of such terms as atomicity, monatomic, diatomic, etc.

Review briefly what pupils have learnt about the three states of matter. Remind them of their studies on the structure of substances starting with a comparison of the volumes occupied by the mole of atoms (gram-atom) of a large number of solid elements (A14.2). Display again the models which show these volumes. Then ask pupils to look up the densities of gaseous elements in the Reference section of the *Handbook for pupils* and show them how to work out the volume occupied by 1 mole of atoms (gram-atom) of each element. Construct a table on the blackboard and fill in the details. Their answers will be approximately $12 \, dm^3$ for hydrogen, oxygen, nitrogen, chlorine, and fluorine, but approximately $24 \, dm^3$ for helium, neon, argon, krypton, and xenon. The constancy of these results is a little surprising, particularly in the case of the inert gases with their wide range of atomic masses. The difference between the two sets of values also demands some explanation.

Corresponding calculations need to be made for other gases but since these are compounds we add the relative atomic masses and obtain a total formula mass for the compound concerned. It is soon found that the gram-formula of each gas occupies approximately $24 \, dm^3$.

Look at those gases which give a value of $24 \, dm^3$. First there is the group of inert gases. The name 'inert' arose from the fact that it is hard to get these gases to react. Pupils may guess that the inert gases consist of atoms. Then there is the group of compounds. These substances are made of molecules, that is of particles which behave like atoms. Remind pupils of the S_8 molecules considered in section A14.7.

What about the gases which give a value of $12 \, dm^3$? Pupils may be able to guess that they form molecules of pairs of atoms.

An Italian scientist, Amedeo Avogadro, formulated a relationship to explain these findings as long ago as 1811. We state Avogadro's hypothesis now as 'Equal volumes of all gases contain the same number of molecules – given the same conditions of temperature and pressure'.

Relate Avogadro's hypothesis to this present problem and explain that those substances which have two atoms per molecule are said to be diatomic, those which have three atoms are triatomic, etc. Electron diffraction studies of gases can be used to discover the atomicity of gases in much the same way as X-ray diffraction is used to investigate the structure of crystals.

Show pupils models of atoms and molecules (see figure A14.31). Overhead projection originals reproduced in figures A14.32 and A14.33 are also helpful in this presentation.

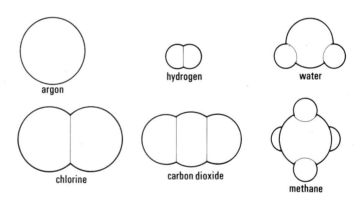

Figure A14.31
Models of atoms and molecules of common gases.

Before concluding this section, it is worthwhile reminding pupils that when they worked out the volumes occupied by moles of atoms (gram-atoms) of gaseous elements the answers were not all *exactly* the same. Show them on an overhead projector the exact values for the mole of molecules (gram-molecules) – after multiplying by two where necessary. They will see why we say that at room temperature a mole of molecules (gram-molecule) of any gas occupies *approximately* $24 \, dm^3$. The teacher will need to remind pupils that the reverse of Avogadro's law is also true; equal numbers of molecules of gases occupy the same volume, given the same conditions of pressure and temperature.

An alternative way of using the densities of gases to verify Avogadro's hypothesis is shown on Experiment sheet 59 which is reproduced below.

Experiment A14.9

What volumes are occupied by one mole of molecules (gram-molecule) of various gases?

Requirements

Each pupil will need:

Experiment sheet 59

Handbook for pupils

Experiment sheet 59
Use Tables 1, 3, and 5 in the Reference section of the *Handbook for pupils* to compete the following table. To assist you in this, one line of the table, that for oxygen, has been worked out below.

Gas	Assumed molecular formula	Relative molecular mass	Volume of 1 mole of molecules (g-molecule) at 298 K and 1 atm/dm³
Argon	Ar		
Helium	He		
Neon	Ne		
Hydrogen	H_2		
*Oxygen	O_2	32	24.1
Nitrogen	N_2		
Chlorine	Cl_2		
Fluorine	F_2		
Carbon dioxide	CO_2		
Hydrogen sulphide	H_2S		

*For oxygen the relative atomic mass (Table 1) is 16. The molecular formula is assumed to be O_2, and so the relative molecular mass is $16 \times 2 = 32$. The density of oxygen (Table 3) is 1.33 g dm^{-3}; thus 1.33 g oxygen, at 298 K and 1 atm, occupies a volume of 1 dm³. Hence 32 g oxygen (1 mole of molecules or 1 g-molecule) at 298 K and 1 atm, occupies a volume of $\frac{32}{1.33} = 24.1 \text{ dm}^3$ (the gram-molecular volume of oxygen).

What can you say about the volume of one mole of molecules (gram-molecular volume) of these gases?

What does this tell you about the hypothesis put forward by Avogadro?

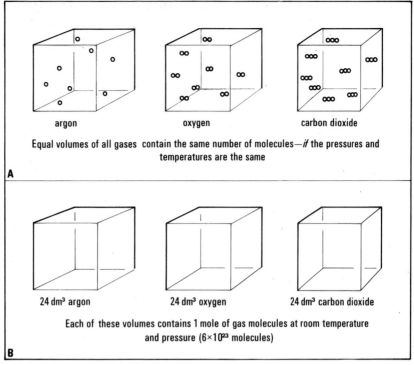

Figure A14.32 Avogadro's law and gases. (OHP 18)

It will be seen that Avogadro's hypothesis fits with the facts. The same information can be conveyed visually using an overhead projection original (figure A14.32a).

Emphasize that moles of molecules (gram-molecules) of all gases occupy approximately 24 dm³ by showing pupils cubes which actually occupy 24 dm³ or by using another overhead projection original (figure A14.32b).

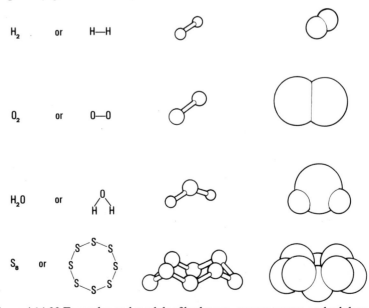

Figure A14.33 Formulae and models of hydrogen, oxygen, water, and sulphur.

Note on the representation of covalently-bonded substances
There are many methods of representing the molecular structures of simple compounds (see figure A14.33) but the two most common methods are:
a. Ball and stick models
b. Space-filling models
Explain to pupils that both of these representations are useful. They are both used in the *Handbook for pupils* (see Chapter 9).

If the pupils have made some models for themselves they will have been of the tangential-contact type and these are particularly convenient for showing the structures of graphite and diamond.

Summary

By the end of this section pupils should know the common property of all gases which is expressed in Avogadro's hypothesis, and the important volume $24\,dm^3$ should be familiar to them. They should recognize the molecule as the important particle in gases and know that the common gaseous elements, apart from the inert gases, have two atoms in their molecules and that their formulae are written H_2, N_2, O_2, etc. Pupils should also understand the meaning of such terms as atomicity, monatomic, diatomic, etc. and be aware of the meaning of gram-molecular volume.

A14.10
Differences between giant and molecular structures

In this section, the properties of substances with giant structures are compared with those with molecular structures.

A suggested approach

Objectives for pupils

1. Knowledge of the structures of molecular substances in the liquid and solid states
2. Knowledge of the strong attracting forces between atoms in giant structures and the weak attracting forces between molecules in molecular substances
3. Understanding of the use of melting point or boiling point as an indication of structure

This section may be introduced by asking the pupils what they think happens when gases such as chlorine and iodine vapour are cooled. The chemical similarity between these two elements suggests that the molecules of iodine should be diatomic and this fact has been confirmed by electron diffraction studies.

We know that gases condense to liquids when cooled and that the volume of the liquid is far smaller than that of the gas. 5 grams of iodine occupy approximately $1\,cm^3$ as a liquid, or solid, but more than $500\,cm^3$ as a gas (that is, above its boiling point). The molecules must be much closer together in the liquid state than in the gaseous state, and the fact that liquids are almost incompressible suggests that the molecules are very close together indeed. The very small volume change which accompanies freezing shows that the spacing of the molecules is much the same in the liquid as in the solid state.

Further cooling of liquid iodine leads to crystallization, and in the crystals we expect to find an orderly arrangement of molecules. Analysis of the structure by X-ray diffraction confirms the existence of I_2 molecules and shows the precise way in which the iodine molecules are packed together.

A specimen of iodine crystals should be viewed and the ideas which have been discussed may be clarified with the help of models.

Models of molecular substances in the liquid and solid states

Apparatus

The teacher will need:

30 small models of the iodine molecule, preferably the space-filling type

Perspex cube large enough to hold these models

Glass rod

Models of the structure of diamond or silicon (these have exactly the same structure) or graphite

Procedure

Show the pupils the large cube containing the models of iodine molecules. Explain that this provides a simple picture of the structure of liquid iodine. (Of course, this analogy has its limitations and it may be necessary to modify it at a more advanced stage.) Stir the models with a glass rod to show that they are able to slide over one another very easily – just as we suppose the molecules can move in liquid iodine (see figure A14.34). Explain that, in fact, the molecules are never stationary but continuously sliding over one another.

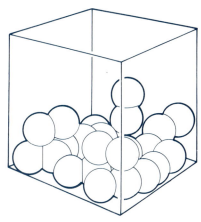

Figure A14.34
Model of iodine molecules in liquid iodine.

Figure A14.35
(*a*) Arrangement of iodine molecules in solid iodine.
(*b*) Successive layers of an iodine crystal, where dots represent atomic nuclei.

Now arrange the models in a regular pattern to illustrate the structure of solid iodine (see figure A14.35*a* and *b*).

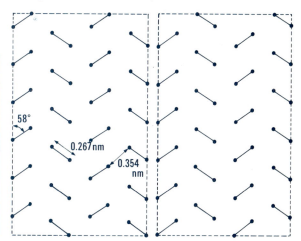

58°

0.267 nm

0.354 nm

Finally, show the pupils models of diamond or silicon or graphite and emphasize the rigidity of the models.

The purpose of the experiment which follows is to show that the forces which hold the molecules together in solid iodine are very much weaker than the forces holding the atoms together in a giant structure such as graphite or silicon. (Silicon is more obviously a giant structure than graphite but a sample of graphite looks very like iodine crystals.)

Experiment A14.10b

Apparatus

Each person will need:

2 test-tubes

Bunsen burner

Graphite or silicon

Iodine

How easy is it to pull atoms apart in silicon or graphite (giant structures) and molecules in iodine (a molecular structure)?
The experiment may be performed by the teacher or by pupils.

Procedure
Take two test-tubes, one containing a little graphite or silicon and the other containing a little iodine. Heat them equally over a Bunsen flame.

The iodine readily melts and vaporizes, whereas there is no change in graphite, or silicon.

Ask the pupils what conclusion they can draw about the ease with which iodine molecules can be separated, remembering how widely spaced they must be in the vapour. It is clear that the forces holding the *molecules* together in the solid iodine are very much weaker than the forces holding the *atoms* together in carbon or silicon with the giant structure. Point out that we knew this in advance.

Finally tell the pupils that they have seen one example of a general rule – that molecular substances have low melting and boiling points whereas substances with giant structures have high melting points and boiling points. This shows that the forces holding molecules together are always relatively weak.

Teachers should note that the distinction between 'forces within molecules' and 'forces between molecules' must be made to avoid the misconception that 'molecular substances are held together by weak forces and giant-structured substances are held together by strong ones'.

Suggestion for homework

Look up melting points in the Reference section of the *Handbook for pupils* and decide which of the substances have molecular structures and which have giant structures.

Summary

Pupils should now know the type of structure to be expected for a molecular substance in the liquid or solid state. They should also know that the forces between molecules in these substances are very weak compared with forces which hold the atoms in a molecule together and that the melting point or boiling point of a substance is usually a good indication as to whether it has a giant or molecular structure.

A14.11
The arrangement of atoms* in compounds such as magnesium oxide and sodium chloride

Objectives for pupils

1. Understanding that substances like sodium chloride and magnesium oxide consist of large numbers of atoms which are strongly held together and would therefore be expected to have giant structures rather than molecular ones
2. Knowledge of the structure of magnesium oxide and sodium chloride

The pupils will know from their earlier work that elements may have giant structures (for example metals) or molecular ones (for example gases, iodine, sulphur). In this section, we consider 'What sort of structure do substances such as magnesium oxide and sodium chloride have?' One way to investigate the problem is to see if these compounds are readily decomposed or affected in any way by the action of heat.

Experiment A14.11a

Apparatus

The teacher will need:

2 test-tubes, 150 × 16 mm

Test-tube holder

Bunsen burner and heat resistant mat

Sodium chloride
Magnesium oxide
Sodium bromide
Tetrabromomethane   } *

*(1 spatula measure each)

Can the atoms in sodium chloride and magnesium oxide be pulled apart readily?

Procedure
Heat a sample of sodium chloride in a test-tube. Repeat the experiment using magnesium oxide. It will be found that there is no evidence of vaporization.

One may conclude from this experiment that the particles of the two substances, magnesium oxide and sodium chloride, are not readily pulled apart and therefore from previous experience, one would expect these substances to have giant structures. A supplementary demonstration using sodium bromide and tetrabromomethane may provide a helpful comparison.

The next question which arises is, 'How are the atoms arranged in such substances?' The pupils have already met X-ray analysis and its use in working out the structure of solid elements (see section A14.4). X-ray diffraction methods have also been used to determine the structures of these two compounds. At this stage one could show pupils models of the structures of several compounds and include sodium chloride and magnesium oxide and perhaps caesium chloride as well.

Directions for making the models are given in Appendices 6 and 7 (pages 161–162).

Experiment A14.11b

Making models of the structure of magnesium oxide, sodium chloride, and caesium chloride
It may be more appropriate to do this as a science club exercise rather than in a normal lesson.

*The term 'atom' is used in this section instead of 'ion', since pupils are not introduced to ions until A16.

Apparatus

Each group will need:

1 or more templates (for dimensions see page 161)

32 polystyrene spheres, 38 mm in diameter

32 polystyrene spheres, 19 mm in diameter (preferably painted silver)

Figure A14.36
One layer of the magnesium oxide model being made.

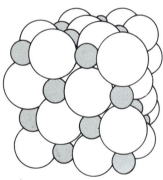

Figure A14.37
Sodium chloride structure (polystyrene spheres in contact).

Procedure

For magnesium oxide, the largest spheres are placed in the holes in the template and the smaller ones positioned between them or in the corners (see figure A14.36). A little quick-drying adhesive on the balls where they touch will hold them together to give one layer of a structure. Different groups could each make one layer and at least four layers should be laid one on top of the other in an appropriate orientation to make the model.

See figures A14.37 to A14.39 for models of sodium chloride and caesium chloride.

These models give very satisfactory representations of the structures concerned. Certain features need to be emphasized in discussion. Thus:

1. The two kinds of atoms are uniformly distributed in a regular pattern, with no special groupings of any kind, and the structures are therefore giant structures. This is also suggested by the negative results of heating magnesium oxide and sodium chloride in Experiment A14.11a.

2. The cubic pattern which can be seen in the models provides an interpretation of the formation of cubic crystals.

3. Each atom is surrounded by six atoms of the other element in the sodium chloride and magnesium oxide structures.

4. The type of structure found is that which gives the closest packing, and depends upon the relative sizes of the atoms. Thus, magnesium oxide and sodium chloride have the same type of structure because the metal atom has only about half the diameter of the non-metal atom. (*Note:* when the two kinds of atom are of a similar size, as in caesium chloride, a different structure is found. The temptation to discuss the caesium chloride structure as body-centred cubic should be resisted, since the atom at the centre of the cube is *not* of the same kind as those at the corners.)

(*Note to teachers:* most of the structures mentioned are also used in a different context in the *Nuffield Advanced Chemistry* course.)

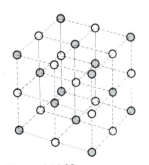

Figure A14.38
Sodium chloride: lattice model.

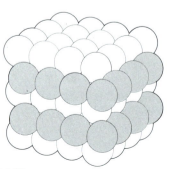

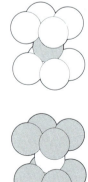

Figure A14.39
Caesium chloride model.

| Summary | By the end of this section pupils should be aware that the atoms in magnesium oxide and sodium chloride are very strongly held together in giant structures. They should be able to sketch the structures of these compounds. |

A14.12
Starting to write chemical equations

Symbols have been used so far to represent both atoms and moles of atoms (gram-atoms). Pupils have *not* been introduced to chemical equations because they had not met the concepts of *molecule* and *giant structure*. Now that these two concepts have been introduced, we may begin to write equations, encouraging the pupils to think in terms of individual atoms or molecules as well as of moles of atoms (gram-atoms) and moles of molecules (gram-molecules).

| A suggested approach |

Objectives for pupils

1. Understanding that chemical equations may be interpreted in two ways:
a. in terms of moles of atoms (gram-atoms), and moles of molecules (gram-molecules), etc.
b. in terms of atoms and molecules (i.e. in terms of particles)
2. Ability to visualize the reactants and products of a chemical reaction in terms of actual samples ('bottles-worth') of moles of atoms (gram-atoms) of elements; moles of molecules (gram-molecules) of simple compounds and so on.

The purpose of this section is to help pupils to appreciate the value of chemical equations in the interpretation of chemical reactions. It is important, therefore, **not** to convey an impression that the writing of an equation is merely an algebraic exercise requiring the juggling of symbols and formulae. Equations *must* be seen to be a convenient method of summarizing ways of thinking about chemical reactions and about the properties of atoms, molecules, etc.

A. Reaction of hydrogen and oxygen to give water
From Topic A11, pupils should know that there is evidence that:

| 2 moles of atoms (gram-atoms) hydrogen | + | 1 mole of atoms (gram-atom) oxygen | → | 1 mole (gram-formula) water |

Tell the pupils that water is a molecular substance. From A14.9, they know that hydrogen and oxygen are in the form of diatomic molecules. Hence we can write:

| 2 moles of molecules (gram-molecules) hydrogen | + | 1 mole of molecules (gram-molecule) oxygen | → | 2 moles of molecules (gram-molecules) water | *1.* |

The pupils also know – although it will probably have to be emphasized more than once – that a mole of atoms (gram-atom) or a mole of molecules (gram-molecule) of any substance contains the same number (6×10^{23}) of particles (atoms *or* molecules). Therefore:

| 2 molecules hydrogen | + | 1 molecule oxygen | → | 2 molecules water | *2.* |

This is an appropriate place to introduce symbols and formulae into the reaction summary and to state that an equation is a convenient way of summarizing equations *1.* and *2.* above.

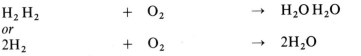

$$H_2 H_2 \quad + \quad O_2 \quad \rightarrow \quad H_2O\, H_2O \qquad 3.$$

or

$$2H_2 \quad + \quad O_2 \quad \rightarrow \quad 2H_2O$$

It is also a good time to introduce state symbols into equations. Hence, we write:

$$2H_2(g) \quad + \quad O_2(g) \quad \rightarrow \quad 2H_2O(l)$$

The danger in writing equations for chemical changes too soon during a course is that the equations become a sort of mathematical juggling and pupils lose sight of what an equation represents. It is suggested that the following may help pupils to interpret equations at two distinct levels:

1. *Using moles of molecules (gram-molecules)*

Show pupils models of moles of molecules (gram-molecules) of hydrogen and oxygen, and two bottles containing 18 g water each (figure A14.40).

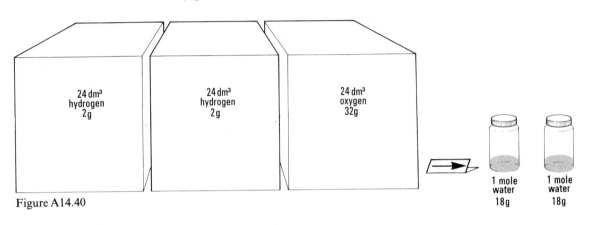

Figure A14.40

2. *Using 'particles'*

Polystyrene spheres or similar model 'atoms', may be used to represent the reaction at the molecular level (figure A14.41).

Figure A14.41
An equation using molecular models. (OHP 19)

Figure A14.42
Using a magnetic blackboard
(*cf* Appendix 8).
*Photograph, R. B. Cowin
(March 1970)* School science
review, *Vol. 51, No. 176,
page 633, figure 4.*

Another way of presenting such information – and possibly the most convenient method of representing reactions at the molecular level – is by mounting half-spheres on a 'magnetic blackboard' (figure A14.42), *or* by placing similar models on the face of an overhead projector.

A number of other similar reactions can be treated in this way including the reaction between hydrogen and chlorine to form gaseous hydrogen chloride.

B. Reaction of magnesium and oxygen to give magnesium oxide
Pupils will know from Topic A11 that there is experimental evidence to show that:

1 mole of atoms (gram-atom) magnesium	+	1 mole of atoms (gram-atom) oxygen	→	1 mole (gram-formula) magnesium oxide

or

2 moles of atoms (gram-atoms) magnesium	+	1 mole of molecules (gram-molecule) oxygen	→	2 moles (gram-formulae) magnesium oxide

The reaction may be considered in two ways by the pupils:

1. *Using moles (gram-atoms), gram-molecules, and gram-formulae*
Show the pupils two bottles each containing one mole (gram-atom) of magnesium, a model of the volume occupied by a mole (gram-molecule) of oxygen, and two bottles each containing one mole (gram-formula) of magnesium oxide (MgO) (figure A14.43).

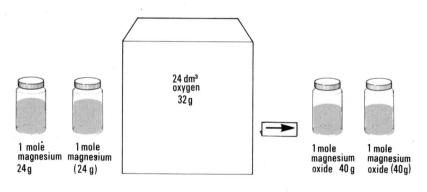

Figure A14.43
The formation of magnesium oxide using mole quantities (gram-formulae).

2. *Using particles*
In this presentation, the problem is how to introduce pupils to writing the chemical equation for a reaction which involves the formation of a giant structure. One might, of course, simply write:

$$2Mg(s) \quad + \quad O_2(g) \quad → \quad 2MgO(s)$$

but it is all too easy for pupils to acquire from such a presentation an image of molecules of magnesium oxide. (There are, of course, no such things!)

It is therefore suggested that the writing of this type of equation be approached very carefully in the following way.

Start by reminding pupils that magnesium has a giant structure and that oxygen forms diatomic molecules. Then show, perhaps on a magnetic blackboard or using an overhead projector, models of the giant structure of magnesium and of some oxygen molecules, *making sure that there are the same number of model atoms of magnesium and oxygen.* Ask the pupils to suggest how the giant structure of magnesium oxide might be represented and lead to the following representation (figure A14.44).

Figure A14.44
The magnesium-oxygen
reaction. (OHP 20)

We might then ask how they would represent this reaction using symbols and formulae. They might suggest:

$$12Mg \quad + \quad 6O_2 \quad \rightarrow \quad 12MgO$$

In this case, one would need to explain that it is usual to represent the reaction even more simply:

$$2Mg \quad + \quad O_2 \quad \rightarrow \quad 2MgO$$

or, using state symbols,

$$2Mg(s) \quad + \quad O_2(g) \quad \rightarrow \quad 2MgO(s)*$$

Another reaction which can be treated in a similar manner is that between sodium and chlorine to form sodium chloride.

Suggestion for homework

Build up a summary of the reaction between copper oxide and hydrogen, *or* between hydrogen and chlorine, using the procedure outlined in this section.

Summary

The structure of a variety of elements and compounds is reviewed in simple terms in this Topic. The work includes the formulation of an explanation using analogues and models, particularly when referring to the structures of metals, sulphur, carbon, and such simple compounds as sodium chloride or magnesium oxide. The terms giant structure and molecular structure are introduced as a means of interpreting the properties of materials. The forces between atoms are contrasted simply with those acting between molecules. The importance of the structure of metals to engineers is reviewed and illustrated by a series of simple tests. The structure of gases is also considered and Avogadro's hypothesis introduced as a common property of gases. Finally, the principles of writing chemical equations are presented in such a way that structural considerations of reactants and products receive attention.

*The state symbol (s) is used to denote all solid substances in this edition. In the first edition (c) was used for those solid substances which were known to be crystalline rather than non-crystalline (amorphous) solids.

Appendices for Topic A14

Appendix 1

Laboratory models of mole of atoms (gram-atom) of a series of solid elements (see section A14.2)

Cylinders of lengths indicated below may be cut from $\frac{3}{4}$ inch dowel, painted realistically, and mounted on a base board by means of metal pegs. Symbols of the elements painted on the cylinder surface (to face the class), cylinder top (to face the teacher), and on the base board help in the correct identification and location in ascending atomic mass by the class and teacher.

(All lengths are in cm)

Aluminium	3.5	Manganese	2.6
Antimony	6.4	Nickel	2.3
Bismuth	7.5	Phosphorus (white)	5.9
Caesium	24.9	Phosphorus (red)	5.0
Calcium	10.5	Potassium	15.8
Carbon (graphite)	1.9	Rubidium	19.6
Carbon (diamond)	1.2	Silicon	4.1
Chromium	2.6	Silver	3.6
Copper	2.5	Sodium	8.3
Gold	3.6	Sulphur (rhombic)	5.5
Iodine	9.0	Sulphur (monoclinic)	5.8
Iron	2.5	Tin	5.8
Lead	6.4	Zinc	3.2
Lithium	4.6		
Magnesium	4.9		

Appendix 2

Construction details for a perspex funnel (Experiment A14.3b)

The funnel may be made from four similar pieces of perspex and a fifth square of perspex which acts as a base (optional – see Figure A14.4). For demonstrations, the funnel may either be stood on its base or supported in a ring attached to a retort stand. A bag containing a large number of small spheres (diameter constant about 25 mm) completes the requirement for this demonstration.

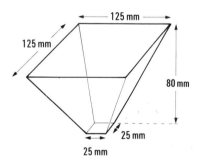

Figure A14.45

Appendix 3

Model crystal for use in a ripple tank (Experiment A14.4a)

Much depends on the design of the ripple tank used. If the ripple tank is of the type recommended for the Nuffield Physics course (see *Teachers' Guide III*) then the 'crystal' can be made from a rectangular piece of sheet perspex of length approximately half to two-thirds of the breadth of the tank. The breadth of the 'crystal' should allow for 4 rows of 'atoms' and for the supporting screws. The 'atoms' and supporting screws need to be arranged in a regular manner such that the distance between the atoms is $\lambda/2$ where λ is one of the longer wavelengths which the tank can produce most easily (when used to give linear waves).

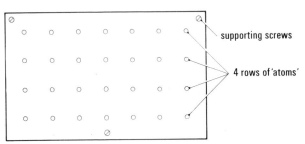

Figure A14.46

The 'atoms' may be made from 4B.A. nuts and bolts such that the length of each bolt just reaches into the surface of the water in the tank.

Appendix 4

Construction of the triangle for use in the close packing of spheres (Experiment A14.5d)

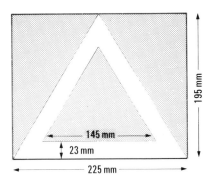

Figure A14.47

A convenient way to make the triangle is to use 18 mm blockboard. The dimensions will depend on the size of the polystyrene spheres used. Those shown relate to spheres of 25 mm diameter. The inner equilateral triangle must accommodate five spheres along each edge (i.e. a total of 15 within the triangle).

Cut off the shaded areas. It is clearly more economical to make several triangles at the same time from one piece of blockboard.

If 38 mm diameter spheres are used, the dimensions are 300 mm × 260 mm with an inner triangle of side 220 mm.

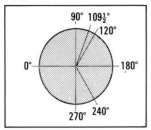

Protractor for marking angles on polystyrene spheres (Experiment A14.8)

For repeated use, this piece of apparatus should be made from stiff cardboard. If pupils make their own protractors for one occasion, then thin card will suffice.

The angles required are shown in figure A14.48. The central shaded area should be cut out. The diameter of the hole should be equal to that of the spheres to be used; a close fit makes for easy handling.

Template for magnesium oxide model (Experiment A14.11b)

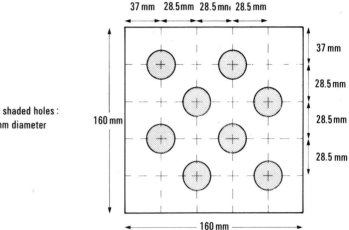

Note. Cut-out shaded holes: all 33 mm diameter

Figure A14.49
Template for magnesium oxide model.

This can be made from wood of thickness 10 mm or more, or from a metal sheet supported at this height. The dimensions shown in figure A14.49 are for 38 mm and 19 mm diameter spheres. No holes are provided for locating the smaller spheres; they rest on the template and are located by their contacts with the larger spheres.

Template for caesium chloride model (Experiment A14.11b)

The making of a useful demountable model of this structure has been described by R. M. Robertson (1971) 'Re-usable crystal models' *School Science Review*, **53** No. 182, 147. It is assembled on a simple template, but vertical walls must be provided to support the pile of spheres. These walls must be of perspex if the structure is to be seen. Dimensions for a model using 38 mm diameter spheres are shown in figure A14.50. Spheres are placed in the holes and then four others, representing the other element, are laid on top of them. With the side walls in position, alternative layers can be added and will remain in position without any adhesive being required. A suitable model uses 35 spheres and has walls 100 mm high made from 3 mm thick perspex sheet.

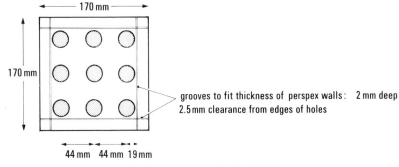

grooves to fit thickness of perspex walls: 2 mm deep
2.5 mm clearance from edges of holes

170 mm

170 mm

44 mm   44 mm   19 mm

Note. Base 1 cm thick (or 1 mm and supported at 1 cm height);
holes are 33 mm in diameter

The pieces of perspex sheet required are two of 170×110 mm, two of 126×110 mm, and four of 19×108 mm. The narrow strips are stuck to the ends of the longer sheets to provide supports at the corners.

If 38 mm diameter spheres are placed in the tray, they may be quickly stuck together and can still accommodate a sphere of 28 mm diameter at the body centre. However in *real* caesium chloride, it must be remembered that the chloride ions are not in contact!

Appendix 8

Magnetic blackboard as an alternative form of presentation of chemical reaction (see section A14.12)

A magnetic blackboard can be put to considerable use in the teaching of chemistry at all levels, but especially when equations are first introduced or in dealing with such topics as ionization and reaction mechanism at a later stage in a chemistry course.

The following references provide some independent reports on the use of this aid.

Maloney, M. J. (1969) 'The magnetic blackboard in chemistry teaching'. *School Science Review*, **50,** No. 170, 126.

Cowan, R. B. (1970) 'Using space-filling models on a magnetic board'. *School Science Review*, **51,** No. 176, 632.

Cowan, R. B. (1970) 'The use of magnetic board and models to illustrate the concept of proton stabilization'. *School Science Review*, **52,** No. 178, 120.

A simple alternative is to use an overhead projector with prepared cut-out atomic and molecular shapes made from acetate sheet *or* from 'half' spheres, etc. to cast shadows on the screen.

Solids, liquids, and gases

Purposes of the Topic	1. To show how kinetic theory applies to the three states of matter. 2. To measure the quantity of energy required to melt solids and to vaporize liquids. 3. To provide information about forces between atoms and molecules (i.e. about bonding). 4. To provide practice in the use of tables of data.
Contents	A15.1 The kinetic theory of matter A15.2 Finding out how much energy is needed to melt solids and to vaporize liquids A15.3 Using data on heats of vaporization to provide information about structure
Timing	About 4 weeks will be required for this Topic.
Introduction to the Topic	Pupils need to be familiar with the particulate theory of matter and with the essential differences between solids, liquids, and gases. Evidence is reviewed which supports the idea that molecules or atoms which constitute gases and liquids are in continual motion. The energy needed to melt a solid or to vaporize a liquid is studied and then measured. The final section of the Topic uses heats of vaporization to distinguish between a giant structure and a molecular structure.
Alternative approach	Differences between solids, liquids, and gases are described in Topic B17, and linked with energy changes in Topic B23.
Background knowledge	Topics A11 and A14 introduce the kinetic theory of matter and some detailed considerations about the the three-dimensional structure of substances. Nuffield Physics Year 1 deals with kinetic theory in elementary terms.
Further references *for the teacher*	Revised Nuffield Physics *Teachers' guide 1* (and Nuffield Physics *Teachers' guide I*). Nuffield Physics *Guide to Experiments I.* Ogborn, J. (1973) *Molecules and motion* (a Nuffield Physics special publication).
Supplementary material	*Film loops* 2–5 'Liquid–gas equilibrium' 2–6 'Solid/liquid equilibrium' 2–7 'Movement of molecules' *Films* 'Change of state' ESSO Film Library 'Molecular theory of matter' Encyclopaedia Britannica 'Properties of matter: Part 1. Solids, liquids, and gases' EFVA 'Properties of matter: Part 2. Atoms and molecules' EFVA

'Energy in chemistry (1). Energy levels, latent heats, and heats of reaction' ESSO. Films for Science Teachers No. 21

Handbook for pupils, Chapter 6 'Energy changes and material changes'.
Chemists in the world, Chapter 5 'Energy and chemicals'.

A15.1
The kinetic theory of matter

In this section, evidence to support a kinetic theory of matter (i.e. the theory that the particles of which matter is composed are continually in motion) is reviewed and extended. Teachers should consult their physics colleagues before they begin this section, as the experiments may already have been done in Physics and a simple reminder may be all that is necessary here.

A suggested approach

Objectives for pupils

1. Knowledge of the kinetic theory as it applies to the three states of matter
2. Awareness that energy changes are associated with changes of state

Pupils will already have some knowledge of the kinetic theory. The teacher might begin by asking what evidence there is to support the view that 'molecules' (as they may now be called) of gases and liquids are in continual motion. They may suggest:
1. evidence based on diffusion
2. evidence based on Brownian motion.

If pupils have *not* already seen Brownian motion (see A11.1c) this is a good time to demonstrate it.

Experiment A15.1a

Apparatus

Each pupil (or pair) will need:

Test-tube, 100×16 mm

Glass rod

Microscope slide and cover slip

Access to microscope with 4 mm objective

Aquadag (colloidal graphite) or toothpaste

Distilled water

Evidence for the movement of molecules in the liquid state

Procedure
Place a speck of Aquadag, about the size of a pin head, in a clean test-tube containing about $5 \, \text{cm}^3$ of water and stir the mixture to disperse the Aquadag. Place one drop of this suspension on a microscope slide and cover it with the cover slip. Illuminate the slide from one side and focus the microscope on the edge of the cover slip before altering the field of view to observe the Brownian movement displayed by the graphite particles. (Toothpaste may be used in place of Aquadag.)

When discussing this experiment and Experiment A11.1c, the teacher should emphasize that pupils are not directly observing individual molecules in motion. They are observing the motion of much larger particles of graphite or, in A11.1c, of smoke, which are being moved by the random bombardment of very much smaller molecules of liquid or gas, respectively. (On occasion it is helpful to offer a simple analogue to make these points realistic – as used in Year 3 of the Nuffield Physics course for gases.)

The film loop 2–7 'Movement of molecules' might be shown. It provides a sophisticated picture of the various types of motion in a molecule of water, namely, translation, rotation, and vibration. The *Handbook for pupils*, Chapter 2, has diagrams which represent the movement of molecules.

The teacher should then extend the use of these ideas to account for the various states of matter. 'If matter is particulate, how could we use these ideas to account for liquids or solids?' 'How can we account for differences between ice – water – steam?'

Suggestions for homework

1. Summarize the evidence for the kinetic theory and use the theory to account for differences between the three states of matter: solid, liquid, and gas.
2. Describe the various forms of motion shown by a water molecule. Include in your answer some sketches of a water molecule to illustrate the points that you make.

Summary

By the end of this section, pupils should have some qualitative understanding of the kinetic theory of gases and of some of the evidence which support it. They should be aware of the existence of quantitative differences of energy requirements for a simple substance (for example, water) depending on whether it exists in the solid, liquid, or gaseous state.

A15.2
Finding out how much energy is needed to melt solids and to vaporize liquids

The previous section leads naturally to measurements of the energy required to melt a solid or to vaporize a liquid. These energy requirements are then related to the nature of the bonding in materials.

A suggested approach

Objectives for pupils

1. Awareness of the fact that (heat) energy must be supplied to melt a solid
2. Knowledge of how to determine the amount of energy (in kJ) needed to vaporize a gram-formula quantity of a liquid
3. Knowledge of energy changes that accompany changes of state and which provide information about the bonding present in a particular material

Pupils will need to be reminded of the differences in the structures of materials (Topic A14) and of the energy needed to effect a change of state. In this section, we estimate the amount of energy needed to bring about melting and vaporization and use such estimates to assess the nature of the bonding in a substance.

In the first experiment, we show that energy is needed to effect a change of state – in this case to form a solid into a liquid. Pupils plot a 'heating curve' and establish that for a period when energy input is constant, there is no change in the temperature of the substance which is being melted.

This experiment is usually carried out by allowing a molten material to cool and plotting a 'cooling curve'. Pupils, however, seem to understand the experiment more readily if the reverse process is adopted, namely supplying heat to melt a solid.

Experiment A15.2a

Is energy needed to melt a solid?

Procedure
Details appear in *Experiment sheet* 60 reproduced below. It may be useful to discuss the procedure before pupils attempt the experiment. Some typical results are shown in figure A15.2.

Apparatus

Each pair of pupils will need:

Experiment sheet 60

Thermometer, -10 to $110 \times 1\,^{\circ}$C, long form

Test-tube, 150×25 mm

Bunsen burner and heat resistant mat

Tripod and gauze

Beaker, $250\ \text{cm}^3$

Clamp, boss, and stand

Stop-clock (or view of clock with a seconds hand)

Spatula

Naphthalene crystals, 5 g

Experiment sheet 60

The solid that you will use in this experiment is naphthalene, which has a melting point of $80\,^{\circ}$C and can therefore be melted in boiling water.

Half fill a beaker with water and place it on a gauze supported on a tripod (with a stand and clamp in position as shown in the diagram below). Heat the water until it boils and then adjust the flame so that the water just keeps boiling. Put naphthalene crystals in a test-tube to a depth of about 3 cm and place a thermometer in the crystals so that the bulb is near the bottom of the tube. Make sure that the thermometer is in such a position that it can be read from this time onwards, and *do not use it to stir the naphthalene*.

Figure A15.1

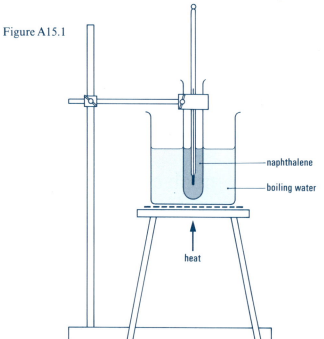

Clamp the test-tube in the water as shown in the diagram and start a stop-clock (or note the time on a clock or watch with a seconds hand). Read the temperature at 15-second intervals and record the results in the table below. Continue the readings until the naphthalene has been liquid for at least 2 minutes.

Time from start/sec	Temp. /°C						
I	II	I	II	I	II	I	II
0							
15							
30							
45							

Plot a graph of temperature (vertical axis) against time (horizontal axis).

Account for the shape of the graph that you obtain.

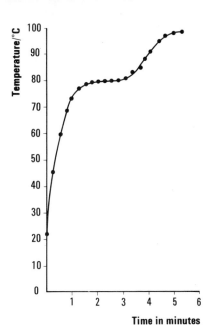

Figure A15.2

When discussing the experiment, make sure that pupils realize that at the melting point, the heat supplied does not raise the temperature of the naphthalene but is needed to turn the solid into liquid. Energy is required for the molecules to break loose from their fixed positions in the crystal lattice and to move about relative to one another. Pupils should be told that the experiment can be adapted to measure the quantity of heat required to melt a solid. How do pupils consider that the results of such a determination should be expressed? The quantity of heat is measured in kilojoules, so we could use the unit kilojoules per kilogram or kilojoules per mole of molecules. Using kilojoules per mole of molecules (gram-molecule) enables comparisons to be made between substances because we are using the same number of particles on each occasion!

The film loop 'Solid/liquid equilibrium' may be used to illustrate ideas raised in discussion.

Suggestions for homework

Complete the table overleaf using Tables 2, 5, and 6 in the Reference section of the *Handbook for pupils* and answer the following questions.

Is there any relationship between melting point and energy of fusion for the materials listed? (*Note:* this need not necessarily be a precise relationship.) Can we use the sort of information given in the table to obtain an indication of the structures of the materials?

Substance	Melting point /°C	Energy of fusion /kJ mol^{-1} (or kJ (gram-molecule)$^{-1}$)	Can the particles of the solid be loosened easily or only with difficulty?
Metallic elements copper zinc magnesium iron sodium			
Non-metallic elements sulphur nitrogen oxygen argon helium			
Compounds methane ammonia			

Attention is now turned to measurements of the energy needed to vaporize the liquid. This can be done by a direct electrical method (see Alternative IIB, Topic B23) or by the calorimetric method which follows.

The pulling apart of the molecules of a liquid to form a gas requires energy in just the same way as transforming solid into a liquid (see Experiment A15.2a). To set the scene for a simple semi-quantitative determination of the heat of vaporization of a liquid, teachers should discuss the units to be used. As mentioned earlier, the kilojoule is used as a unit of thermal energy. It *may* be appropriate to mention calories and then to relate calories to joules and to justify the adoption of the joule before making a statement such as 'The kilojoule is the unit of energy needed to raise the temperature of (approximately) 240 cm^3 of water through 1 °C.' We will be measuring the quantity of energy needed to transform a definite number of particles from one state to another (compare A15.2a) and so the units used for our experiment will be kilojoules per mole of molecules (or kilojoules per mole).

The procedure outlined below may be tried by pupils using water and subsequently demonstrated by the teacher using heptane. Pupils may calibrate their Bunsen burners directly in kilojoules per minute, using 240 cm^3 water for this purpose, so that a rise of 1 °C means that 1 kJ of energy has been applied.

Experiment A15.2b

How much energy is required to vaporize 1 mole of water?

Procedure
Depending on the ability of the class, it may be appropriate for the teacher to demonstrate key points of the procedure beforehand.

Apparatus

Each pair of pupils will need:

Experiment sheet 61

Graph paper

Conical flask, 350 cm³, or 500 cm³

Measuring cylinder, 250 cm³

Thermometer, −10 to 110 × 1 °C

Stirring rod

Bunsen burner and heat resistant mat

2 stands, bosses, and clamps

Stop-clock (or view of clock with a seconds hand)

Experiment sheet 61

Energy is measured in kilojoules, one kilojoule being the amount of energy required to raise the temperature of 240 g water (240 cm³) by 1 °C.

If a known mass of water at room temperature is heated to boiling point by a steady Bunsen burner flame, and the time taken to do this is measured, the number of kilojoules of energy which the flame supplies per minute can be found. If the *same* flame is now used to boil away a known mass of water, the amount of energy needed to change one mole of molecules (gram-molecule) of water at boiling point to steam at boiling point can be calculated. To make the calculation easier, it is convenient to start with 240 g water (i.e. 240 cm³) since one kilojoule is needed to raise the temperature of this by 1 °C.

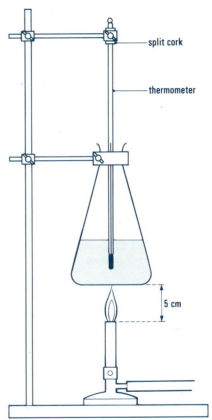

split cork

thermometer

5 cm

Figure A15.3

In this experiment the most important precautions are to keep the size of the Bunsen burner flame constant and to protect it from draughts.

Assemble the apparatus shown in the diagram, with the conical flask empty.

Remove the thermometer. Move the Bunsen burner to one side, light it, adjust the air-hole to give a non-luminous flame, and adjust the gas supply until the flame is about 5 cm high. Replace the burner under the flask and make a final adjustment so that the tip of the flame just touches the bottom of the flask. Remove the burner again and *do not alter the flame for the rest of the experiment*.

As accurately as you can, measure 240 cm³ water into the conical flask, using a measuring cylinder.

Replace the flask in the lower clamp of the apparatus and fix the thermometer (in a split cork) in the upper clamp (care) so that the bulb is about 2 cm clear of the bottom of the flask. Put the Bunsen burner under the flask so that the tip of the flame is under the centre of the flask. Read the temperature on the thermometer and *at the same moment* start a stop-clock (or note the time on a clock or watch with a seconds hand). Take temperatures (after stirring the contents of the flask) every half-minute until the water is boiling steadily, and every minute afterwards. Allow the water to boil for 10 minutes. Remove the Bunsen burner. Allow the flask to cool. Pour the remaining water into the measuring cylinder and find its volume as accurately as you can.

Enter the results in the table below.

Time from start/min	Temp. /°C						
I	II	I	II	I	II	I	II
0							
0.5							
1.0							
1.5							
2.0							

Volume of water   240 cm^3

Volume of water remaining   .... cm^3

Volume of water boiled away   .... cm^3
    (Mass of 1 cm^3 water is 1 g)

Mass of water boiled away   .... g

Plot a graph of temperature (vertical axis) against time (horizontal axis).

From the graph calculate the rate of temperature rise (in degrees per minute) from the start of the experiment until the water is boiling.

How many kilojoules of energy did the flame supply per minute?

How long was the water boiling? .... min.

How much energy was supplied by the flame during this time?

What mass of water was changed to steam? .... g

How many moles of molecules (gram-molecules) is this?

How much energy was used to change one mole of molecules of water at boiling point to steam at boiling point?

When Experiment A15.2b has been completed and the results discussed with the class, it is suggested that the teacher repeat the exercise using heptane instead of water. Heptane is suggested because its boiling point is almost the same as that of water (99.6 °C), although it has a much larger molecular mass (100). The principle and procedure are very similar but it is preferable to calibrate the Bunsen burner using water and not heptane, in order to avoid having to explain and use the term specific heat capacity. It is, of course, necessary to do the calibration with the same apparatus as that used for the main experiment.

The film-loop 2–5 'Liquid gas equilibrium' may be used in a discussion of results.

Apparatus

The teacher will need:

Conical flask, minimum size 350 cm³ fitted with an adaptor, condenser and thermometer, −10 to 110 × 1 °C (see figure A15.4 or use Quickfit apparatus)

Bunsen burner and heat resistant mat

Stand, boss, and clamp

Stop-clock (or view of clock with a seconds hand)

Measuring cylinder, 250 cm³

Supply of rubber tubing to connect condenser to cold water supply, etc.)

Heptane

How much energy is required to vaporize 1 mole of heptane molecules?

Procedure
1. Calibration of Bunsen burner
Measure 240 cm³ of water into the flask and adjust the thermometer so that the bulb is in the water. Connect the apparatus (which will also be used in the second part of the experiment) so that the bottom of the flask is 5 cm above the top of the Bunsen burner. Proceed as for A15.2b to obtain a calibration graph. Do not let the water boil. Remove the Bunsen but do not alter the height of the flame. Without altering the height of the clamp, empty the flask, rinse it with a few cm³ of acetone and then a few cm³ of heptane. Dry out any other part of the apparatus which is wet in the same way.

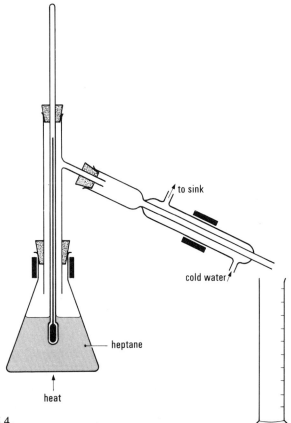

Figure A15.4

2. The heat of vaporization of heptane
Transfer about 250 cm³ of heptane into the flask and place the Bunsen burner beneath the flask. Note the temperature and time at which the heptane starts to boil and time how long it takes for one mole of heptane (146.5 cm³) to be boiled off and condensed into the measuring cylinder. (CAUTION: the rate of delivery can be rather fast!) Remove the Bunsen burner.

In the discussion following these experiments ask pupils where they think any sources of error may arise. Their suggestions may include heat losses caused by draughts, condensation of vapour in the neck of the flask, and so on. Then ask them to look up the accepted value of the heats of vaporization for water and for heptane in the Reference section of the *Handbook for pupils*.

Note. Some teachers may feel that the experiment is unduly complicated and wish to simplify it. This can be done by comparing the times taken to boil off 1 mole of molecules of water and 1 mole of molecules of heptane under identical conditions.

Alternatively an immersion heater of known rating may be used (see suggestions for homework and Topic B23).

Suggestions for homework

1. A Bunsen burner was found to transmit heat energy at the rate of 3530 joules per minute to some water in a flask. After allowing the water to boil for 8 minutes, with the flame unchanged, it was found that 12.5 grams of water had evaporated. What is the heat of vaporization of water per mole of molecules (gram-molecule) in (a) joules and (b) kilojoules?

2. An electric immersion heater, whose output was 310 joules per minute, was used to heat different masses of different liquids. Each liquid was boiled for a certain length of time and weighed. From the difference in the masses, the mass of the liquid that had been converted into its vapour at its boiling point was then deduced. In each of the examples given below calculate the heat needed in kilojoules to vaporize 1 mole of molecules.

Liquid	Formula	Time expressed in minutes	Loss in mass expressed in grams
Tetrachloromethane	CCl_4	5	800
Benzene	C_6H_6	7	550
Cyclohexane	C_6H_{12}	4.5	390
Nitrobenzene	$C_6H_5O_2N$	6	460
Acetone (propanone)	C_3H_6O	4	240

3. 100 grams of trichloromethane ($CHCl_3$) was evaporated in five minutes by an immersion heater. If the energy needed to convert trichloromethane into its vapour at its boiling point is 30 kilojoules per gram-molecule, what was the rate at which the heater supplied the heat energy in kilojoules per minute?

4. Acetone (propanone) (C_3H_6O) required 523 joules of heat to convert 1 gram into its vapour at its boiling point. How many minutes will it take to convert 1 mole of molecules (gram-molecule) of acetone into its vapour at its boiling point if the heater dissipates heat energy at the rate of 2020 joules per minute?

A15.3
Using data on heats of vaporization to provide information about structure

The ideas discussed in the Topic are reviewed and used to provide information about structure.

Objectives for pupils
1. Understanding information about structure provided by heats of vaporization
2. Ability to use tables of data

In the previous section it was shown that the heat of vaporization of water is about 41 kJ mol^{-1} at 100 °C whereas the heat of vaporization of heptane is about 33 kJ mol^{-1}. These results need to be compared with other results given in the *Handbook for pupils*. Pupils will find, for example, that the heats of vaporization of metals suggest that the atoms of metals are held together very strongly indeed. They will also see that water is an exceptional substance. It may be appropriate for the teacher to discuss the energy changes in transforming ice into water, and water into steam. It will be seen that the heat of fusion of water is much lower than the heat of vaporization. Fusion obviously involves some breakdown of the crystal lattice of ice. The heat of fusion is the energy required to do this. The much higher value of the heat of vaporization suggests that liquid water has some regularity of structure and that a considerable amount of energy is needed to destroy this regularity. (Further discussion of this particular point will be found in the *Handbook for teachers* (page 254 and following).)

Ask the pupils whether they think boiling point is significant. Is it a reliable guide to structure? Then refer them to the Reference section of the *Handbook for pupils*. Ask them to plot a graph of boiling point against the heat of vaporization per gram-formula for a representative range of substances.

Suggestions for homework

These questions relate to the following table overleaf. You may assume room temperature to be 15°C.

1. In the table, fill in the values for the heats of vaporization from the *Handbook for pupils*.
2. Which of the substances are liquids at room temperature?
3. Which of the substances are gases at room temperature?
4. From the table, what conditions should be observed in storing bromoethane?
5. Which of the materials do you think have giant structures and which are made up of molecules?

Substance	Formula	Heat of vaporization per gram-formula $kJ\,mol^{-1}$ or $kJ\,g\text{-molecule}^{-1}$	Melting point /°C	Boiling point /°C
Toluene	C_7H_8		-95	111
Bromoethane	C_2H_5Br		-126	38
Silver chloride	$AgCl$		455	1557
Lead	Pb		327	1751
Formic acid	CH_2O_2		9	101
Zinc chloride	$ZnCl_2$		275	765
Copper	Cu		1083	2582
Ethane	C_2H_6		-183	-89
Acetic acid	$C_2H_4O_2$		17	118
Cyclohexane	C_6H_{12}		7	81
Iodine	I_2		114	183

Summary

By the end of this Topic pupils should be able to use the kinetic theory to interpret the three states of matter. They should be aware that energy is required to transform solids into liquids and liquids into gases and be able to measure the quantity of energy required to vaporize a liquid. In addition they should be aware that heats of vaporization can provide information about the bonding in a material, enabling predictions to be made about its structure (for example, whether the material has a molecular structure or a giant structure).

Explaining the behaviour of electrolytes

1. To provide an elementary introduction to electricity and the properties of electrical circuits.
2. To provide examples of the phenomenon of electrolysis.
3. To provide a basis for understanding Faraday's laws.
4. To provide a quantitative basis for electrolysis and to relate this to charges carried by ions.
5. To relate the charges carried by ions to the particulate nature of electricity and to the structure of the atom.
6. To review and distinguish between giant structures and molecular structures.

A16.1 Elementary electricity
A16.2 An investigation of the electrolysis of lead bromide
A16.3 Looking at the migration of ions
A16.4 How much electricity is needed to deposit 1 mole of atoms (gram-atom) of lead, copper, and silver?
A16.5 Investigating the electrolysis of solutions of electrolytes
A16.6 Is the same quantity of electricity always needed to liberate 1 mole of atoms (gram-atom) of copper?
A16.7 The implications of the ionic theory for the structure of the atom
A16.8 Molecules, giant structures, ions, and bonding

Section A16.1 is intended for those classes with no experience of electrostatics and the use of simple electrical circuits. Accordingly, it is difficult to make a recommended time allocation for this part of the Topic. Most sections will need at least a double period each. It is suggested that about six weeks should be allowed for the whole Topic.

We begin by reviewing some elementary ideas about the electrical nature of matter, electrostatics, and the properties of simple electrical circuits. The Topic should include a brief introduction to electricity, stressing those aspects which are unfamiliar to pupils. (Teachers should note that electrostatics is not always studied as part of a physics course, and some coordination with the school physics department is desirable to prevent needless repetition.)

A brief revision of the conductivity of materials, studied in Alternative IA Topic A8 or Alternative IB Topic B6, leads naturally to the electrolysis of fused lead bromide. This experiment confirms and qualifies some of the assumptions which were made during Stage I. A discussion can lead to Faraday's explanation of the phenomenon of electrolysis in terms of ions. Since the movement of ions cannot be observed conveniently in a molten electrolyte, the migration of ions in solution is studied.

The quantitative aspects of electrolysis receive attention next. It is found that definite quantities of electricity are needed to liberate 1 mole of atoms (gram-atom) of various elements. This observation leads to the use of the Faraday or mole of electrons as a unit quantity of electricity and to numerical values of charges on certain ions.

'Is the same quantity of electricity always needed to deposit a mole of atoms (gram-atom) of copper?' The result of an investigation leads to an understanding that the charge on the copper ion may be either one or two. Standard notation is adopted when appropriate from this point on in the course: thus, we refer to copper(II) sulphate $Cu^{2+}(SO_4)^{2-}$; copper(II) chloride, $Cu^{2+}(Cl^-)_2$; etc. There is an opportunity to review the properties of other transition metal ions at this point (see Alternative IIA, Topic A13).

The material presented in this Topic also provides some evidence for the particulate nature of electricity and for the structure of the atom. The Topic ends with a discussion of the nature of ionic bonding and of how this differs from other forms of bonding. The pupils' knowledge of particles and of structure may then be brought together in the following way:

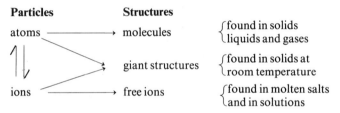

Particles	Structures	
atoms ⟶	molecules	{found in solids / liquids and gases
	giant structures	{found in solids at / room temperature
ions ⟶	free ions	{found in molten salts / and in solutions

Subsequent development

The ideas and concepts developed in this Topic are used in Topic A17 and later Topics. They are also needed in Stage III, especially Options 6, 7, 8, 9, and 10. Option 10 gives pupils an opportunity to follow the development of Davy's and Faraday's contribution to electrochemistry. Electron transfer processes receive further consideration in Option 6.

Alternative approach

The content of this Topic is presented in a modified sequence in Alternative IIB, Topic B19.

Background knowledge

Electrolysis was investigated qualitatively in Stage I, Topic A8 or B6.

Nuffield Physics. Electricity and electrolysis are studied in Year 2. Change, fields and forces, and electron streams are studied during a course on Electrostatics in Year 3. (However, it should be noted that electrolysis is not always studied as part of a physics course.)

Further references

for the teacher

Collected experiments, Chapters 5 and 11, provides additional experiments on electrolysis.
For information on the use of symbols in this Topic teachers are referred to the *Handbook for teachers*, Chapter 4. Chapters 6, 7, and 8 of the *Handbook* provide background information on electrochemistry.
Nuffield Advanced Physical Science, Sections 4, 5, and 7, provides more details and illustrations of the material of this Topic and in particular of the background material introduced in Section A16.1.

Other sources of information
Hockey, S. W. (1972) *Fundamental electrostatics*, Methuen Educational.
CHEMstudy (1963) *Chemistry – an experimental science*, Chapter 5, Freeman.

Supplementary material

Film loops
2–8 'Electrolysis of lead bromide'
2–18 'Electrolysis of potassium iodide solution'

Overhead projection originals
21 Simplified diagram of the electrolysis of lead bromide *(Figure A16.6)*
22 Quantities of electricity required to liberate 1 mole of atoms (gram-atom) of various elements *(Figure A16.11)*

Films
'Electrochemistry' ESSO. Films for Science Teachers No. 24
'Electrostatics' ICI Film Library
'Static electricity in the Chemical Industry' ICI Film Library

Reading material

for the pupil

Chemists in the world, Chapter 4 'Davy and Faraday' and Chapter 1 'Finding out about the atom'.
Handbook for pupils, Chapter 7 'Electricity and matter' and Chapter 9 'Structure of elements and compounds'.

A16.1
Elementary electricity

A knowledge of elementary electrostatics and the properties of simple electrical circuits is fundamental to an understanding of electrolysis. The time required to present this material to a class depends on their background knowledge. The outline suggestions which follow are designed to meet *basic* needs only. If greater detail and information are needed, teachers are referred to the information provided in the various Nuffield Physics guides.

A suggested approach

Objectives for pupils

1. Awareness of the electrical nature of matter
2. Awareness of the electrical basis of electrostatic phenomena
3. Knowledge of the laws of attraction and repulsion for electrically charged bodies
4. An appreciation of the idea that equal quantities of unlike charges can cancel one another
5. An appreciation of the idea that a charged body placed in an electric field experiences a force
6. Knowledge of the properties of simple series circuits and of current indicators (e.g. ammeters, electric lamps)

(Continued)

Pupils should be reminded of a variety of electrical phenomena. When an electric current is passed through a conductor – such as an element in an electric fire or the hot plate of a cooker – it becomes warm. When dry hair is rubbed with a plastic comb, the hair 'stands on end'. On a cold, dry night, sparks fly when we remove clothing made from man-made fibres. Simple demonstrations of rubbed plastic rods attracting pieces of paper and attracting or repelling other rubbed plastic rods can lead to more detailed investigations (Experiments A16.1a, A16.1b, and A16.1c).

The use of a high voltage source for simple electrostatic phenomena is intended to shorten the sequence of demonstrations and to present what may be familiar effects in an entirely new guise. The approach provides a simple link to circuit theory and to electrolysis.

Normally we detect a current flowing in a circuit by means of an indicator, such as an ammeter or an electric lamp. Current is seen to flow only when the circuit is complete. Experiment A16.1d uses this idea and shows the existence of two classes of material: conductors and non-conductors of electricity. This experiment may be used as a substitute for part of the work listed in Topics A8 or B6 in Stage I.

---

7. Knowledge of which types of substance are electrical conductors and which are not

Practical tips when using apparatus to demonstrate electrostatic phenomena

If modern plastics are used, such as polythene and perspex, electrostatic experiments can be made to work under very adverse conditions. Ebonite rods may need cleaning; to remove an oxidized surface from ebonite, rub it with very fine grade sand-paper. Glass rods also need special treatment; they must be dried in an oven before use and, ideally, used while they are still warm.

If possible, arrange to have an electric, reflector-type fire to throw a direct beam of radiant heat along the demonstration bench. Modern plastics acquire and retain electrostatic charges very easily. This can be a handicap – for example when the insulating support of the conductor becomes charged, producing misleading effects. Plastic materials can be discharged quite easily by passing them through the air immediately above a low Bunsen flame. However, it is advisable to keep the Bunsen flame away from a demonstration since the flame is a source of charged particles. It is equally important to see that draughts do not blow ionized air towards the apparatus!

Experiment A16.1a

Apparatus

The teacher will need:

Beaker containing very small pieces of paper

Polythene rod

Perspex rod

Rods of other materials (for example, glass, ebonite, etc.)

Piece of fur

Piece of silk

Duster

Wire jockey

Nylon thread

Burette stand and clamp

Bunsen burner

Simple demonstrations of electrostatic phenomena

Procedure

Rub the polythene rod with the piece of fur or the duster and show the pupils that the rubbed rod attracts small pieces of paper. Repeat the experiment using the perspex rod rubbed with silk or foam plastic. Then ask 'Suppose you were the first to discover this strange effect, what would you try next?' Pupils will probably suggest trying another substance. If pupils have already met some simple ideas from mechanics and have understood something of the nature of force, they may have better suggestions; for example, 'If electrified rods exert forces on bit of paper, does one electrified rod exert a force on another electrified rod?' To test this suggestion, suspend a charged polythene rod in a wire stirrup and bring up a second charged polythene rod. The amount of repulsion will be small – perhaps too small to be noticed. A thin but rigid strip of polythene with a relatively large surface area and low amount of inertia can be used with greater effect. Alternatively, use perspex sheet cut into strips.

It is a good idea to mark one end of the suspended plastic strip and to charge that end only. Next, test rubbed polythene against rubbed perspex and so demonstrate attraction. Show that a finger placed near a charged suspended body also gives rise to attraction, as does an iron stand when placed near the suspended rod. One may therefore conclude 'The only sure test for electrification is repulsion.'

These simple experiments suggest the existence of two different kinds of charge. The real problem which remains is to try and find out if there are more than two kinds of charge. Repeat the experiments using different materials to show 'Any substance which can be charged either repels charged polythene and attracts charged perspex or vice versa.' Therefore, it is necessary to assume the existence of only two kinds of charge. At this point the teacher could introduce the terms positive and negative:

positive charge, a term used for the charge that perspex acquires from silk.

negative charge, a term used for the charge that polythene acquires from fur.

Experiments with an electrometer

The teacher will need:

Perspex box fitted with a copper-plated base, see figure A16.1

2 conducting spheres suspended from terminals inside the box

Connecting leads
(Continued below)

Procedure

Set up the apparatus as shown in figure A16.1. Arrange for the shadows of the two conducting spheres to be visible to the class. Apply a potential difference to the spheres and demonstrate that unlike charges attract one another. Disconnect the power unit and note the effect of joining the spheres together: no charge remains, the spheres no longer attract one another. (See figure A16.2.) Rearrange the wiring as shown in figure A16.3a and demonstrate that like charges repel. Disconnect the power supply and discharge the spheres.

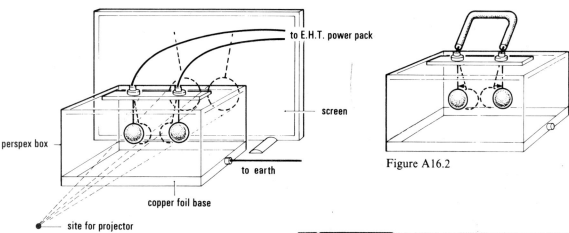

Figure A16.1

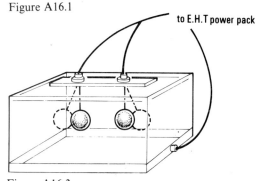

Figure A16.3a

EHT power supply (maximum p.d. about 5000 V d.c.), Nuffield Physics item 14 (see safety note page 180)

Screen

Source of illumination (for example, a point source of light)

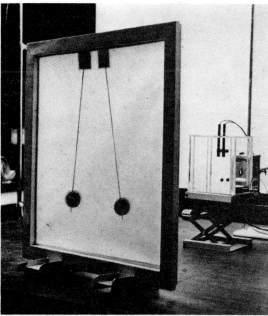

Figure A16.3b

The effect of an electric field on a charged sphere

Apparatus

The teacher will need:

EHT power supply (maximum p.d. about 5000 V d.c.), Nuffield Physics item 14*

2 metal plates fitted with insulating handles

Table tennis ball coated with Aquadag

Reel of nylon thread

3 retort stands and bosses

1 clamp

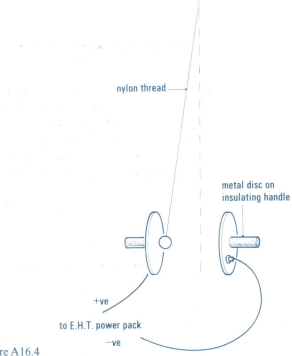

nylon thread

metal disc on insulating handle

+ve

to E.H.T. power pack

−ve

Figure A16.4

Procedure

Arrange the apparatus as in figure A16.4. The handles of the two metal plates are connected to retort stands using the bosses and are set parallel to each other about 10 cm apart. The plates are connected to the positive and negative terminals of the EHT supply using crocodile clips attached to the special lugs on the back of the plates. The table tennis ball coated with Aquadag is suspended by a nylon thread from the third retort stand so that it hangs freely between the two plates. The ball is then allowed to touch the plate connected to the negative terminal of the EHT power pack. The power supply is now switched on **(Caution!)** and a potential difference of between three and four thousand volts is applied to the plates. Note the effect on the ball. Repeat the experiment using a positive charge on the ball and again note the effect.

The demonstration shows the effect of an electric field on a charged object. (It also serves as a model of an ion moving in an electric field.) If the potential difference between the plates and the distance separating the plates are appropriate, the ball may be made to move to and fro between the plates. The ball then represents a positive ion moving one way and a negative ion moving the other way.

*Teachers are reminded of the restrictions relating to power units contained in D.E.S. regulations (see *Safety in Science Laboratories*, D.E.S. Safety Series No. 2, HMSO, 1978). Detailed specifications for such apparatus exist and include those put forward in Nuffield Physics Year 3.

Finding out which substances conduct electricity

The teacher will need:

6 V battery or alternative d.c. supply

Carbon or steel electrodes mounted in a suitable holder

6 V bulb and bulb holder

Connecting wire – with 2 pieces fitted with crocodile clips

Samples of materials such as lead, copper, other metals, naphthalene, sulphur, polythene, wax, etc., all in small dishes

Procedure

Essentially this is the same as section A8.1 and B6.1 in Stage I. The simple series circuit is first demonstrated to the pupils and a lamp is seen to light when the current flows in the circuit. The two probes are applied to pieces of metal and to other substances, in turn. The results may be conveniently summarized in a table. The terms insulator and conductor may then be discussed. If time allows, the tests using fused samples of materials and of lead bromide can be demonstrated by the teacher and used to introduce the next section.

A16.2
An investigation of the electrolysis of lead bromide

The electrolysis of lead bromide is investigated, and the pupils are introduced to ionic theory as an explanation for this effect.

A suggested approach

Objectives for pupils

1. Ability to recall earlier work on conductivity of materials (see A16.1d, A8 or B6)
2. Ability to make deductions from observations
3. Understanding of the ionic theory of electrolytes

The section assumes that pupils are aware of the results of Experiment A16.1d or of the experience gained in Topic A8 or B6. If their only experience is A16.1d, then a rationale for heating lead bromide powder is required. Do powders not conduct because the particles are too far apart? What happens when the powder is heated? Heat is necessary to fuse the particles of the powder. The molten material can then be tested to see whether or not it will conduct electricity. (The fact that decomposition occurs when current is passed through molten lead bromide requires an explanation using the ideas which were developed in A16.1.)

Experiment A16.2

Investigating the electrolysis of lead bromide

Apparatus

The teacher will need:

Safety spectacles

Power supply, 10–12 V d.c.

Rheostat, 12 Ω 5 A

Demonstration ammeter with scale giving a maximum reading of 3 A

Connecting wire

(Continued)

Caution. Lead bromide and the vapours produced during this experiment are toxic; molten lead bromide can cause severe burns. The experiment *must* be performed by the teacher.

Lead bromide. This chemical must be of a good quality otherwise some bromine may be given off on merely melting it. As part of the pre-lesson preparation, it is advisable to check for the absence of ions in the sample to be used.

Procedure

Safety spectacles should be worn during the experiment. Figure A16.5 shows the layout of the apparatus. The experiment may be

2 barrel terminals

Carbon rod (to serve as an anode)

Steel rod (to serve as a cathode)

U-tube, about 15 cm long and made from glass tubing of about 1 cm diameter

Stand, clamp, and boss

Spatula

Bunsen burner

Heat resistant mat

Stop-clock or view of laboratory clock which reads to the nearest minute

Lead bromide

started by warming the bottom of the U-tube with a small flame and adding lead bromide until it reaches the level shown in the diagram. The electrodes are warmed and lowered into the melt. With a current flow of little over 1 A, bromine vapour can be seen almost at once in the anode compartment of the U-tube.

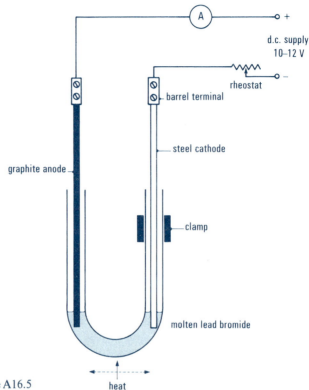

Figure A16.5

The electrolysis can be run for a few minutes in a well ventilated laboratory but thereafter a fume cupboard should be used. Only just enough heat should be used to keep the lead bromide molten. After 4 or 5 minutes, switch off the current and remove the electrodes. Examine the cathode and show the pupils that it marks paper, indicating the presence of a lead coating.

Draw the pupils' attention to the following results:
1. When lead bromide is solid, no current flows and no decomposition takes place.
2. When molten lead bromide is used, an electric current will flow through it and simultaneously bromine is liberated at the anode.
3. At the end of the experiment, the steel cathode is found to mark paper, indicating that lead was liberated at the cathode.

A discussion of these results should lead to the following ideas which were first put forward by Faraday:
1. *If one assumes* that lead bromide consists of charged particles, their movement in the electric field explains the passage of current through molten lead bromide.
2. Since lead is deposited at the cathode (the negatively-charged electrode), the lead atoms in molten lead bromide must be positively

charged; and since bromine is released at the anode (the positively-charged electrode), the bromine atoms in molten lead bromide must be negatively charged.

3. Given the above assumptions, we 'explain' the electrolysis of molten lead bromide. When the charged atoms reach the electrodes, they must be discharged in the following way:

Positively charged lead atoms $\xrightarrow[\text{charge}]{\text{gain}}$ lead atoms

Negatively charged bromine atoms $\xrightarrow[\text{charge}]{\text{lose}}$ bromine molecules

We *need* evidence of ions moving in a specific direction in an electric field to support this view. We also need evidence that neutralization occurs by *gaining* charges in one instance and *losing* charge in the other.

The difference between the properties of charged and uncharged atoms of an element will need to be discussed. Thus, one might talk about the difference between sodium atoms in metallic sodium and sodium 'atoms' in sodium chloride. The two species have quite different properties: sodium atoms in metallic sodium react violently with water whereas sodium 'atoms' in sodium chloride do not. Similarly chlorine 'atoms' in chlorine gas have a nasty smell yet in sodium chloride they do not. So it is with lead atoms in metallic lead and lead 'atoms' in lead bromide, and bromine atoms in liquid bromine and bromine 'atoms' in lead bromide. Where charged 'atoms' are known to exist as the result of experiments such as this one we refer to them as *ions*. Positive ions are called cations and negative ions are called anions. So, when atoms either gain or lose charge, they become new particles with new properties. It is possible to go further and say something about the structure of atoms in simple terms at this point.

The key ideas may now be summarized on the blackboard or by using the overhead projection original which appears in figure A16.6.

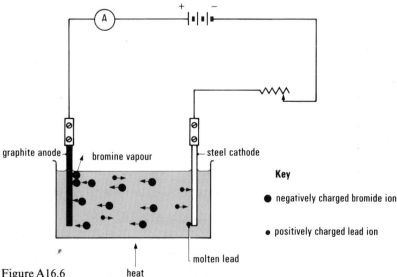

graphite anode
bromine vapour
steel cathode
molten lead
heat

Key

● negatively charged bromide ion

• positively charged lead ion

Figure A16.6
Simplified diagram of the electrolysis of lead bromide. (OHP 21)

It is tempting to show the film loop 'Electrolysis of lead bromide' at this point. However, it is suggested that this loop serves a better purpose if it is left until the end of section A16.4 when the pupils will have been introduced to the following points:
1. The value of the charges on the ions – lead ions Pb^{2+} and bromide ions Br^-.
2. Evidence from electrolysis supporting the particulate nature of electricity.

If the film loop (which shows both the value of the charges on the ions and the electric current in circuit by means of moving electrons) is shown here, it will tend to pre-empt a discussion of these points.

Suggestion for homework

Write a careful description of the deductions you made in this section and the evidence on which they are based.

A16.3
Looking at the migration of ions

In the previous section, it was postulated that during electrolysis the current is carried by the movement of ions in the electrolyte. In this section the movement of ions 'in bulk' is demonstrated visually.

A suggested approach

Objectives for pupils

1. Increasing the familiarity of pupils with the properties of ions and charged particles
2. Awareness of the function of electrodes during the process of electrolysis

Begin by reminding pupils that when a current is passed through an electrolyte it has been assumed to be carried by the ions present. Thus, anions must move towards the anode and cations move towards the cathode. This may be demonstrated using an appropriate source of power and solutions containing coloured ions. Two experiments are described. Additional experiments appear in the book *Collected experiments*, Chapter 5. Alternative IIB, Topic B19, offers an alternative procedure for this section.

Experiment A16.3a

Looking at the migration of ions
This experiment should be done by the teacher.

Apparatus

The teacher will need:

Buchner funnel

Buchner flask

Filter pump

Filter paper

Wide bore U-tube

Stand, boss, and clamp

Power supply about 20 V d.c.

Pipette

(Continued)

Procedure
Part 1 is part of the pre-lesson preparation.

Part 1. Copper(II) chromate is prepared by adding 100 cm^3 of M copper(II) sulphate solution to an equal volume of M potassium chromate solution. The insoluble copper(II) chromate is filtered off using a Buchner funnel, Buchner flask, and pump. The precipitate is then washed thoroughly in distilled water and sucked dry.

The solid copper(II) chromate is dissolved in a *minimum* quantity of 2M hydrochloric acid, and the solution saturated with urea in order to increase its density.

Next, fill one third of the wide bore U-tube with 2M hydrochloric acid and run the copper(II) chromate solution into the U-tube very

Safety suction device for use with the pipette

Connecting wire

2 carbon electrodes

2M hydrochloric acid

M copper (II) sulphate solution

M potassium chromate solution

Urea

Experiment A16.3b

Apparatus

Each pair of pupils will need:

Experiment sheet 62

2 crocodile clips

Microscope slide

2 lengths of connecting wire

2 test-tubes, 150 × 16 mm

2 teat pipettes

Test-tube rack

Source of d.c. (20 V)

Filter paper – cut into strips to fit microscope slide

Small crystal of potassium permanganate

0.1 M silver nitrate

5% potassium chromate

Supply of distilled water

slowly and carefully from a pipette, delivering it at the bottom of the U-tube so that it forms a separate layer and has a clear layer of hydrochloric acid above it on both sides of the U-tube. The pipette will need to be withdrawn carefully to avoid any mixing.

Part 2. Insert a carbon electrode into each arm of the U-tube, so as to dip into the dilute hydrochloric acid, and connect the electrodes to a source of about 20 volts d.c. After about 10 minutes a green colour develops near the cathode and an orange colour near the anode. After about 30 minutes, the blue copper(II) ion and the orange dichromate ion boundaries are clearly seen. Meanwhile pupils may carry out their own experiment as follows.

The migration of ions

Procedure
Experiment sheet 62 provides full details and is reproduced below. Figure A16.7 shows the general layout of the apparatus.

After the d.c. supply has been connected for some minutes, the coloured band due to the permanganate ion in the first experiment will be seen to be moving towards one of the crocodile clips (that is, towards the anode of the cell).

The use of clean teat pipettes enables the transfer of solutions and water to be carried out effectively in the second experiment.

In the second part of the experiment, pupils will see a red colour develop at the centre of the filter paper. This happens because the silver ions moving towards B meet chromate ions moving towards A, and together they form the red insoluble compound, silver nitrate. The simple test-tube reaction mentioned in the *Experiment sheet* is helpful in establishing this explanation.

Experiment sheet 62
Using the apparatus shown in the diagram you can find out whether the movement of coloured ions can be seen when a potential difference is applied to a solution containing them.

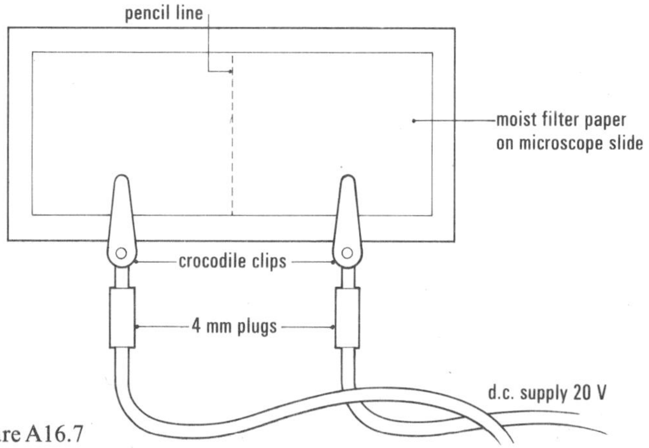

pencil line

moist filter paper on microscope slide

crocodile clips

4 mm plugs

d.c. supply 20 V

Figure A16.7

Draw a faint pencil line across the centre of the filter paper. Moisten the filter paper with tap water and fasten it to the microscope slide with the crocodile clips. Place a small crystal of potassium permanganate in the centre of the filter paper. Connect the crocodile clips to the terminals of a 20 volt d.c. supply, noting which clip is connected to the positive terminal and which to the negative terminal. Observe the filter paper for at least ten minutes. Describe what you see.

What can you deduce about the charge on the permanganate ion? Give your reasons.

Remove filter paper and microscope slide from the clips, discard the filter paper, and clean the microscope slide. Use a fresh strip of filter paper on the slide, arranged in the following way.

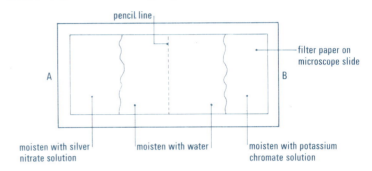

Figure A16.8

Using crocodile clips connect end A to the *positive* side of the 20 volt d.c. supply and end B to the *negative* side. Observe what happens during several minutes. Describe and explain all that you see.

In order to help your explanation, try adding 10 drops of silver nitrate solution to 10 drops of potassium chromate solution in a test-tube.

What is the coloured precipitate?

Suggestion for homework Describe the evidence you have seen for the movement of ions under the influence of an electric potential difference.

A16.4
How much electricity is needed to deposit 1 mole of atoms (gram-atom) of lead, copper, and silver?

The quantity of electricity needed to deposit (or liberate) one mole of atoms (gram-atom) of lead, copper, and silver is measured in the laboratory. The pupils then use tables of data to identify the quantity of electricity required to produce one mole of atoms (gram-atom) of other elements.

A suggested approach

Lead bromide was electrolysed in section A16.2. Discuss how this experiment might be modified to allow a quantitative determination of the quantity of electricity needed to deposit 1 mole of atoms (gram-atom) (i.e. 207 grams) of lead.

Objectives for pupils

1. Understanding of the quantitative aspects of electrolysis and especially that

the number of coulombs needed to deposit 1 mole of atoms (gram-atom) of an element is 96 500 coulombs or a simple multiple thereof

2. Understanding of how these values lead to the values of charges on ions and how they support the particulate nature of electricity itself

Experiment A16.4a

Apparatus

The teacher will need:

Safety spectacles

2 carbon rods mounted in a suitable holder

2 crystallizing dishes

12–14 V d.c. supply (accumulator preferred)

Switch

Connecting wire

Ammeter (demonstration type), 0–5 A

Rheostat, 12 Ω 5 A

Stop-clock

Balance

Spatula

Tripod and gauze

Stand, boss, and clamp

Bunsen burner and heat resistant mat

Tongs

Lead bromide

A quantitative study of the electrolysis of lead bromide

The experiment should be done by the teacher.

Procedure

The experiment must be carried out in a fume cupboard. Safety spectacles should be worn.

A higher voltage is required for this experiment than for Experiment A16.2. As in the earlier experiment, allow the lead bromide to melt before inserting previously warmed electrodes into the electrolyte. Adjust the current to a steady value and maintain the current at this value for the period of electrolysis. After exactly ten minutes of electrolysis, switch off, withdraw the electrodes, and decant the melt carefully into the second dry crystallizing dish. Take care to retain the bead of lead in the first dish. Allow the dish to cool. The bead of lead may now be pressed away cleanly from the glass using a spatula and any adhering lead bromide broken off. Weigh the clean lead bead.

Specimen results

Current passed	1 A
Time for which current passed 6 min.	= 360 s
Mass of lead deposited	0.365 g

Remind pupils that the problem is to find out how much electricity is required to deposit 207 g of lead.

The quantity of electricity is measured in coulombs (C) and 1 coulomb is that quantity of electricity which flows when 1 ampere of electricity is passed for 1 second.

So, if 1 ampere is passed for 360 seconds, the quantity of electricity used is 360 coulombs.

Since 0.365 g lead was deposited by 360 C

then 1 g lead would be deposited by $\dfrac{360}{0.365}$ C

and 207 g lead would be deposited by $\dfrac{207 \times 360}{0.365}$ C

$$= \frac{207 \times 360 \times 1000}{365} \text{C}$$

$$\approx 204\,000 \text{ C}$$

Note on the presentation of the calculation
Some pupils find difficulty in using simple proportion. It may help to avoid decimal fractions as far as possible by using the mass of lead expressed in milligrams. Thus:

As 365 mg lead was deposited by 360 C

So 1 mg lead would be deposited by $\dfrac{360}{365}$ C

Hence 207 grams of lead – i.e. 207 000 milligrams of lead would be deposited by $207\,000 \times \dfrac{360}{365}$ C

$$\approx 204\,000 \text{ C}$$

It is convenient in the calculation to have a current flow of 1 A, but in practice satisfactory results can be obtained with a current flow of up to 5 A.

After the experiment, discuss with pupils the accuracy of the various measurements and the likely sources of error. List the suggestions they make *before* revealing to them that the accepted value of the quantity of electricity required to deposit 207 grams of lead is 193 000 coulombs.

In the additional quantitative experiments (A16.4b) the quantity of electricity needed to deposit 1 mole of copper atoms and 1 mole of silver atoms is found. The determinations may be carried out simultaneously (as described below) or separately. It may be convenient for some pupils to carry out the copper determination while others do the silver determination, or alternatively, pupils could do the copper determination while the teacher does the silver determination.

Note on the electrolysis of silver nitrate
The electrolysis of silver nitrate solution between two silver electrodes, as described in the first edition of this guide, does not usually give a coherent deposit of silver. Consequently it is not easy to obtain accurate results. One way of overcoming this problem is to use a special solution which depresses the concentration of silver ions, and take other precautions, as described in CHEMStudy (1963) *Chemistry – an experimental science:* Teachers' guide, page 443, W. H. Freeman and Co. Alternatively, the electrolysis may be carried out using a steel basin as an anode (as described below). This procedure does not produce a coherent deposit but there is no difficulty in retaining the deposit in the basin itself.

Experiment A16.4b

How much electricity is needed to deposit 1 mole of copper atoms and 1 mole of silver atoms from solution?

The experiment could be introduced by a short discussion on the best way to compare the masses of elements deposited by the same quantity of electricity. By placing two voltameters in series we can ensure that the same quantity of electricity passes through each.

Apparatus

Each pair of pupils will need:

Experiment sheet 63

6 V accumulator or alternative d.c. supply

Rheostat, 10 Ω

Ammeter, 0–1 A

Stop-clock or watch or a view of the laboratory wall clock if it is fitted with a seconds hand

2 beakers, 100 cm³

2 copper foil electrodes, size 5 × 5 cm

Wooden support to hold electrodes

Stainless steel basin, 7 cm diameter

Cork ring to support steel basin

Silver anode (10 cm of silver wire – 2 mm diameter)*

Filter paper and rubber band

Connecting wire

Access to acetone (propanone), steel wool, paper tissues, and to a balance reading to 0.001 g

Stand, boss, and clamp

1 cork

0.5M copper (II) sulphate, 100 cm³

0.1M silver nitrate, 100 cm³

Supply of distilled water

Procedure

Figure A16.9 shows the layout required by each pair of pupils. A suitable holder for the copper foil electrodes is a small piece of hardboard (or wood) with two parallel slits in it 15 mm apart. A suitable current for the determination is 0.2 A but it is quite safe to use solutions of double the stated concentration with a current of 0.4 A in order to complete the experiment in a shorter time. A longer period will be needed if weighing is to be done on a balance reading to 0.01 gram. Practical details are given in *Experiment sheet* 63.

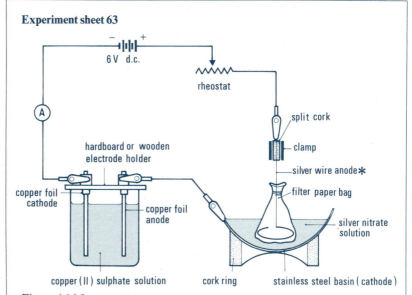

Experiment sheet 63

Figure A16.9

The completed apparatus for this experiment is shown in the diagram above. Assemble it in the following order:

1. Clean carefully the copper foil cathode and anode and the stainless steel basin by rubbing them with steel wool and then with paper tissues. Mark the upper end of the copper foil cathode, in pencil, with a C and the upper end of the copper foil anode with an A.

2. Find the masses of copper cathode, copper anode, and stainless steel basin, and enter them in the table below.

3. Three-quarters fill the beaker with copper sulphate solution and insert the copper electrodes, using the holder provided.

4. Place the stainless steel basin in the cork ring support and add silver nitrate solution to within 1 cm of the rim.

5. Enclose the silver wire anode in a bag by wrapping a filter paper round it and holding the edges together with a rubber band round the vertical portion of the wire. Fix the vertical portion of the wire outside the bag in a split cork held in a clamp attached to a stand. Lower the anode, in its bag, into the silver nitrate solution in the basin, so that the silver ring is just below the centre of the surface.

6. Complete the rest of the circuit, using connecting wires and crocodile clips, but do not attach the crocodile clip on the wire from the ammeter to the copper cathode. Check the rest of the connections, making sure that the copper cathode and anode are in their correct positions.

7. Attach the crocodile clip to the copper cathode, start the stop-clock (or note the time on a clock with a seconds hand), and adjust the rheostat to give a current of 0.20 amp.

*Some teachers may prefer to use stainless steel strip in view of the cost of silver wire.

Allow the current to pass through the solutions for 20 minutes, keeping it at 0.20 amp by adjusting the rheostat when necessary. At the end of 20 minutes disconnect the two copper electrodes and the stainless steel basin. Pour the silver nitrate from the basin into the stock bottle and rinse the basin twice with distilled water and twice with acetone (propanone). *Acetone is highly flammable*; keep it well away from flames. Be careful not to lose any silver during these operations. Allow the remaining acetone to evaporate and warm the basin high above a Bunsen flame to complete the drying. Find the mass of basin + silver and record the mass in the table below. Rinse the copper electrodes with water. The anode (A) will need a strong jet of water from the tap, followed by a firm wipe with a paper tissue to remove the film which collects on the surface. Rinse each electrode twice with acetone, wave in the air for a minute or two to evaporate most of the acetone, and dry by warming high above a Bunsen flame. Find the mass of each electrode separately and record the results in the table.

	Copper cathode /g	Copper anode /g	Steel basin cathode /g
Initial mass			
Final mass			
Change in mass			

Current used .... amp

Time current passed .... sec

Calculation

Quantity of electricity (coulombs)   = current (amp) × time (sec)

Quantity of electricity in experiment   =

.... × .... = ... coulombs

.... g silver deposited by .... coulombs

108 g silver deposited by $\dfrac{\text{................................... } \times 108}{\text{...................................}}$ coulombs

Quantity of electricity needed to deposit 1 mole of atoms (gram-atom) silver is:

.... coulombs.

.... g copper deposited by .... coulombs

64 g copper deposited by $\dfrac{\text{................................... } \times 64}{\text{...................................}}$ coulombs

Quantity of electricity needed to deposit 1 mole of atoms (gram-atom) copper is:

.... coulombs.

Comment on the results obtained.

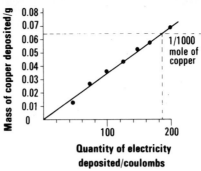

Figure A16.10
Class results for the electrolysis of copper sulphate.

Treatment of results (copper)
The results obtained by various groups for the copper part of the experiment can be presented graphically as shown in figure A16.10. It can then be seen that:

$\frac{1}{1000}$ mole of atoms (0.064 g) of copper is deposited by 193 C.

Therefore, one mole of atoms (64 g) of copper would be deposited by 193 000 coulombs. It is also apparent that:

$$\begin{pmatrix} \text{mass liberated} \\ \text{during electrolysis} \end{pmatrix} \quad \underset{\text{proportional to}}{\text{directly}} \quad \begin{pmatrix} \text{quantity of} \\ \text{electricity passed} \end{pmatrix}$$

This, of course, is the essence of Faraday's first law, which might be stated at this point.

Treatment of results (silver)
The results of the silver experiment may be treated similarly. If only one determination is made, then the results will need to be calculated as follows:

If 0.2 A deposited 0.34 g silver in 25 minutes, then:

$$\text{quantity of electricity passed} = 0.2 \times 25 \times 60 \, \text{C}$$
$$= 300 \, \text{C}$$

0.34 g silver is deposited by 300 C.

So 108 g silver is deposited by $\dfrac{300 \times 108}{0.34}$ C

$$= 95 \ 300 \, \text{C}.$$

Table A16.1 indicates one way in which the teacher might present the data.

A16.4 How much electricity is needed to deposit 1 mole of atoms (gram-atom)   **191**

Element	Number of coulombs required to liberate 1 mole of atoms (gram-atom)
silver	95 500
copper	193 000
hydrogen	96 500
lead	204 000
zinc	193 000
aluminium	289 500

Table A16.1

It may help discussion to plot a bar chart, as in figure A16.11.

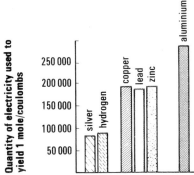

Figure A16.11
Quantities of electricity required to liberate 1 mole of atoms (gram-atom) of various elements (OHP 22)

The pupils should be able to see that the number of coulombs required to liberate 1 mole of atoms (gram-atom) of an element is either 96 500 coulombs or a multiple thereof, within the limits of experimental error. Ask pupils for their criticisms of the procedure used.

At this point, introduce the Faraday (mole of electrons) as a quantity of electricity.

Table A16.2 sums up the findings so far.

Element	Number of Faradays (moles of electrons) required to liberate 1 mole of atoms (gram-atom)
silver	1
hydrogen	1
bromine	1
chlorine	1
copper	2
lead	2
zinc	2
aluminium	3

Table A16.2 Number of Faradays (moles of electrons) required to liberate 1 mole of atoms (gram-atom) of various elements.

These results should now be discussed since they throw light upon the values of the charges carried by ions during electrolysis, and strongly suggest that electricity itself is particulate. Accumulators of about 20 to 30 ampere-hour capacity are useful as a teaching aid. When fully charged, one such accumulator would hold about one Faraday ($26\frac{3}{4}$ ampere hours), the quantity of electricity needed to deposit 1 mole of atoms (gram-atom) of silver from silver nitrate solution. Two such accumulators would contain the quantity of electricity required to deposit 1 mole of atoms (gram-atom) of copper(II) sulphate solution.

Pupils can see from Table A16.2 that the same quantity of electricity is needed to deposit either 1 mole of atoms (gram-atom) of silver or 1 mole of atoms (gram-atom) of hydrogen. They know that both of these quantities of elements contain the same number of atoms (6×10^{23}). Thus, the same quantity of electricity must be associated with 1 atom of either silver or hydrogen. In the case of copper, lead, and zinc, just twice this amount of electricity is needed to deposit a mole of atoms (gram-atom). Since the mole of atoms (gram-atom) of copper, lead, and zinc contain the same number of atoms (6×10^{23}) as the mole of atoms (gram-atom) of silver and hydrogen, twice the quantity of electricity must be associated with each atom of copper, lead, and zinc. In the case of aluminium, three times the quantity of electricity is required for each atom compared with one atom of silver or hydrogen. The conclusion that definite amounts of electricity are associated with ions becomes inescapable.

Other evidence (which they may have met in physics) which supports the idea that electricity is particulate could be mentioned. The particle of electricity is called the electron and has a negative charge.

Next remind the pupils of the electrical circuit used for simple electrolysis experiments – as shown in figure A16.6. An electric current consists of a flow of electrons. In figure A16.6, these electrons flow from the negative terminal of the battery, through the external circuit to the positive terminal of the battery. This electron flow is in the opposite direction to that of the flow of conventional current. The electrons each carry a negative charge which is equal in magnitude and opposite in sign to the charge carried by a hydrogen ion or a silver ion. We may now be more explicit about what happens when ions are discharged at electrodes during electrolysis than we were in section A16.2.

The electrolysis of molten lead bromide provides an illustration (*c.f.* film loop 2–8). Molten lead bromide contains lead ions, Pb^{2+}, and bromide ions, Br^-. During electrolysis these ions migrate to the cathode and anode respectively, where they are discharged.

cathode
$Pb^{2+}(aq) + 2e^- \longrightarrow Pb(s)$

anode
$2Br^- \longrightarrow Br_2(g) + 2e^-$

It may be appropriate to complete this section by summing up the quantitative aspects of electrolysis. Pupils should now know the essence of Faraday's first and second laws of electrolysis: that the more electricity used for electrolysis, the greater the quantity of material deposited, and that the quantity of electricity required to deposit one mole of atoms (gram-atom) of a metal is in simple proportion to that required to deposit one mole of atoms (gram-atom) of silver. When reviewing the evidence for the charge on ions, it may be appropriate to substitute the term mole of electrons for the term faraday.

Suggestions for homework

1. 0.235 grams of copper was deposited on the cathode from a copper anode when a solution of copper sulphate was used as an electrolyte. The current was maintained at 0.25 A and flowed for 48 minutes. Calculate the number of faradays (moles of electrons) needed to deposit 1 mole of atoms (gram-atom) of copper.

2. Humphrey Davy first obtained metallic sodium when he electrolysed fused sodium hydroxide. If you were to repeat Davy's experiment, how long, to the nearest minute, would you need to pass a current of 1.2 A in order to obtain 0.17 grams of sodium?

3. Read the account of Faraday's work in *Chemists in the world*, Chapter 4 'Davy and Faraday'.

A16.5
Investigating the electrolysis of solutions of electrolytes

The Topic now considers the electrolysis of solutions. The part played by water is considered.

A suggested approach

Objectives for pupils

1. Ability to apply earlier knowledge to a new situation
2. Knowledge of the electrolysis of solutions of electrolytes

In Stage I, the pupils electrolysed a number of substances in the fused state and in solution. The present study is an extension of this work. It applies the theoretical ideas developed in this Topic, and leads to an explanation of the electrolysis of solutions in terms of atoms and ions.

Experiment A16.5

The teacher will need

Film-loop projector

Film loop 2–18 'The electrolysis of potassium iodide solution'

Apparatus

Each pair of pupils will need:

Electrolysis cell (see figure A16.12 for details of construction)

2 test-tubes, 75 × 10 mm

(Continued)

Investigating the electrolysis of solutions of electrolytes

Procedure
Pupils investigate the electrolysis of one or more of the solutions provided and test the products of electrolysis. Figure A16.12 shows the apparatus used. The test-tubes should be suspended above the electrodes (as shown in the figure) and not around the electrodes, since this considerably reduces the rate of electrolysis. (This effect can be shown if an ammeter is included in the circuit.)

In the case of the electrolysis of dilute hydrochloric acid, hydrogen is liberated at the cathode and chlorine at the anode. The electrolysis of potassium iodide solution results in the formation of iodine at the anode (identified by colour) and hydrogen at the cathode. The electrolysis of copper(II) chloride results in the formation of a coating of copper on the cathode and the liberation of chlorine at the anode.

2 connecting wires, each fitted with a crocodile clip

Hardboard divider and elastic band

6 V d.c. supply

Stand, boss, and clamp

Indicator paper

Splints

One or more of the following solutions:

0.5M hydrochloric acid

0.5M potassium iodide solution

0.5M copper(II) chloride solution

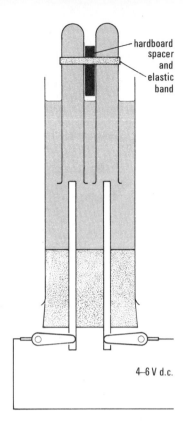

hardboard
spacer
and
elastic
band

4–6 V d.c.

Figure A16.12

Under appropriate conditions, it is possible to obtain oxygen with the halogen in these investigations. Using graphite electrodes, oxygen is difficult to identify. In addition, some hydrogen may be formed at the cathode during the electrolysis of copper(II) chloride.

When the pupils have completed the experiment, tabulate their results. Can the pupils offer a simple explanation for their findings? It looks as though in aqueous solutions, as in fused electrolytes, current is carried by moving ions. The electrode at which the ion is discharged tells us whether the ion is positively or negatively charged. It is clear that hydrogen and copper ions are positively charged and that chlorine and iodine ions are negatively charged.

The teacher will need:

Film-loop projector 2–18 'Electrolysis of potassium iodide solution'

The question may be asked 'Why doesn't potassium appear at the cathode when potassium iodide is electrolysed?' The teacher may remind pupils that potassium reacts vigorously with water. However, they should *not* be allowed to think that potassium is first formed and then reacts with water to form hydrogen. A demonstration of this electrolysis can provide support for this part of the lesson. The film loop 2–18 'Electrolysis of potassium iodide solution' summarizes the ideas used.

Summary

By the end of the section, pupils should be aware that ions carry electric charge through aqueous solutions and that water can play a part in the electrolysis of solutions. The symbol *aq* is used to indicate that ions are in aqueous solution – such as $H^+(aq)$.

A16.6
Is the same quantity of electricity always needed to liberate 1 mole of atoms (gram-atom) of copper?

The pupils investigate whether a copper atom always carries two charges of electricity.

In the previous section Table A16.2 lists the number of Faradays (moles of electrons) required to liberate 1 mole of atoms (gram-atom) of various elements by electrolysis. This information led to the idea that ions possess definite charges: thus, copper ions were found to be Cu^{2+}, lead ions, Pb^{2+}, bromide ions, Br^-, etc. We have assumed that such ions always possess the same quantity of charge without exception. But is this so? In the next experiment we investigate the charge on copper ions.

A suggested approach

Objectives for pupils

1. Understanding the evidence that certain ions (such as copper ions) have variable charge
2. Ability to apply principles and experimental technique to new situations
3. Recapitulation of parts of Topic A13 including the properties of transition metal ions

Experiment A16.6

Is the same quantity of electricity always required to liberate 1 mole of copper atoms?

Apparatus

Each pair of pupils will need:

Experiment sheet 64

6 V battery or alternative d.c. supply

Rheostat (to control the current density of between 10 and 20 mA per cm^2 of the electrode surface)

Ammeter, 0–1.0A

2 beakers, 100 cm^3

Tripod and gauze

Bunsen burner and heat resistant mat

Thermometer, -10 to $+110 \times 1\,^\circ$C

4 copper foil electrodes (about 5×2cm) and 2 supports

Connecting wire

Paper tissues

Fine steel wool

Access to balance – to read to 0.001 g
(Continued)

Procedure

Experiment sheet 64 is reproduced below. It is important to control the current density within the limits suggested. In the procedure suggested, the decrease in mass of the anodes is followed rather than the increase in mass of the cathodes. The temperature control of the alkaline sodium chloride solution is not critical.

Experiment sheet 64

In this experiment you are going to pass a current through two different solutions, using copper electrodes.
The solutions to be used are:
1. Copper sulphate solution at room temperature.
2. Sodium chloride solution made alkaline with sodium hydroxide, at about 80° C.

The object of the experiment is to find the change in mass of the two copper *anodes* when the same current is passed for the same length of time. This can be ensured by using two electrolysis cells connected in series, as in Experiment 63.

The actual value of the current need not be known but it must be kept near the value recommended below. A good deal of care is needed if the experiment is to be successful and accuracy in finding the masses of the anodes before and after the electrolysis is essential.

The circuit used is the same as that used in Experiment 63, except that the stainless steel basin, and its silver electrode, are replaced by a second beaker with two copper foil electrodes dipping into the alkaline sodium chloride solution. This is heated to about 80 °C, by a Bunsen burner, when supported on a tripod and gauze.

Stand, boss, and clamp

0.5M copper(II) sulphate, 100 cm^3

Solution containing 100 g of sodium chloride and 1 g of sodium hydroxide per dm^3, 100 cm^3

First clean the pieces of copper foil that are to be used as anodes, with steel wool, followed by paper tissues. Mark the upper ends '1' and '2' in pencil so that you can identify them later, and find the mass of each.

Connect the circuit as for Experiment 63 (supporting the leads to the alkaline solution by a clamp to keep them away from the flames). Leave the cathode in the first beaker (copper sulphate solution) disconnected. Heat the alkaline sodium chloride to 80 °C and adjust the height of the Bunsen flame to keep it at about this temperature.

Complete the circuit, adjust the rheostat to give a current of 0.1 amp, and allow electrolysis to proceed for 15–20 minutes. The anode in the second beaker (containing the hot sodium chloride solution) must be moved about from time to time, to prevent the orange coloured precipitate which appears from sticking to it.

Disconnect the circuit, remove the two anodes, wash them with a stream of tap water, and rub each firmly with paper tissues to remove any loose film. Rinse each twice with acetone (propanone), partially dry by waving in air for a minute or two, and complete the drying high above a Bunsen flame. Find the mass of each.

	Copper anode in copper sulphate solution /g	Copper anode in alkaline sodium chloride solution /g
Initial mass		
Final mass		
Change in mass		

Take a separate sheet of paper and comment on the results obtained.

From Experiment A16.4b, we know that in the electrolysis of copper sulphate solution two Faradays (moles of electrons) are required to transfer 1 mole of atoms (gram-atom) of copper from the copper anode to the copper cathode.

Experiment A16.6 shows that when the determination is carried out simultaneously using two different sets of conditions, the same quantity of electricity is able to transfer twice as much copper in one electrolysis cell as in the other. This is equivalent to saying that half as much electricity would transfer the same mass of copper in the cell containing the alkaline solution, as for the more usual electrolysis of copper sulphate.

Thus, in the alkaline solution, the following changes must have taken place at the anode:

$$Cu \longrightarrow Cu^+ \quad + \quad e^-$$

these ions pass into solution

these charges return to the battery

Compare this with the copper sulphate cell in which:

$$Cu \longrightarrow Cu^{2+} \quad + \quad 2e^-$$

Copper atoms must be capable of losing either one or two electrons in order to form copper(I) ions, Cu^+, or copper(II) ions, Cu^{2+}, respectively.

The teacher will need;

Chart of Periodic Table
Selection of compounds of
transition elements (see text)

Pupils will need to be reminded that copper is only one example of a special group of elements – the transition elements. It is helpful to refer to both the Periodic Table and a display of compounds of the transition elements. Such a display might include the compounds listed below. Reference might also be made to some of the properties of compounds of transition elements (for example, the formation of coloured ions; the location of the transition element atom in either the cation or the anion of the compounds; the use of transition elements for plating purposes; the fact that the charge on an ion is not constant throughout a range of compounds.)

Transition element	Symbol	Compound	Ion containing the transition metal
chromium	Cr	chromium(III) chloride, $CrCl_3$	Cr^{3+}
		potassium chromate, K_2CrO_4	CrO_4^{2-}
manganese	Mn	manganese(II) chloride, $MnCl_2$	Mn^{2+}
		potassium permanganate, $KMnO_4$	MnO_4^-
iron	Fe	iron(II) sulphate, $FeSO_4$	Fe^{2+}
		iron(III) chloride, $FeCl_3$	Fe^{3+}
		potassium hexacyanoferrate(II), $K_4[Fe(CN)_6]$	$[Fe(CN)_6]^{4-}$
		potassium hexacyanoferrate(III), $K_3[Fe(CN)_6]$	$[Fe(CN)_6]^{3-}$
cobalt	Co	cobalt(II) chloride, $CoCl_2$	Co^{2+}
nickel	Ni	nickel(II) sulphate, $NiSO_4$	Ni^{2+}
copper	Cu	copper(I) chloride,* $CuCl$	Cu^+
		copper(II) chloride, $CuCl_2$	Cu^{2+}
		copper(II) sulphate, $CuSO_4$	Cu^{2+}

A *brief* indication of the relevance of this information to the structure of atoms could be given at the end of this section. (See *Handbook for pupils* Chapter 1, 'Periodicity'.)

Suggestion for homework

Read *Handbook for pupils* Chapter 1, 'Periodicity'.

*Copper(I) chloride does *not* ionize readily and it is only slightly soluble in water.

A16.7
The implications of the ionic theory for the structure of the atom

We have used electrolysis as a means of obtaining evidence for the existence of ions. It is appropriate now to consider a simple explanation of ions in terms of atomic structure.

Faraday's ideas have provided us with explanations for electrolysis, and have enabled us to discover the value of the charges on ions and to infer that electricity itself is particulate. With some pupils it is helpful to revise these points before going on to infer still more about the nature of the atom itself.

It is clear from the work done so far that atoms can gain or lose electrons to form ions and that the converse process may also take place. Atoms must therefore possess electrons. Yet atoms of elements like lead, copper, or hydrogen are electrically neutral. Thus atoms must also contain a zone of positive charge. At this point the teacher may provide the pupils with a simple model of the atom. It seems reasonable to suggest that the atom is made of a positively charged centre of nucleus surrounded by a zone of negative charge containing electrons. Try to avoid creating an image of the electron as a planet rotating around a nucleus since this idea is difficult to eradicate later. It is not essential to develop the structure of the atom in greater detail at this stage, although many teachers may wish to do so.

Suggestions for ways of doing this appear in the Supplement, S1.1– S1.2. The important point to stress is the very big difference in properties between atoms and their related ions. For example, remind pupils how vigorously metallic sodium reacted with water and how unreacting the sodium ions were in sodium chloride. (There is no need to discuss the octet rule at this stage.)

Point out to pupils that a great deal more can be learnt about the structure of the atom, yet this simple picture is adequate for the remainder of the scheme except when discussing radioactivity, when some detailed knowledge of the structure of the nucleus will be needed.

The word 'valency' has not been mentioned in the course. It is inadequate because of the three-dimensional approach to structure introduced in Topic A14. However, it is convenient to be able to arrive at the formulae of ionic compounds, and to see that a pattern is emerging. The problem can be approached by referring to the charge on a metal ion – determined by the number of Faradays required to deposit a mole of atoms (gram-atom) of a metal. Since we know that salts are electrically neutral we can then determine the charge on the negative ion from the formula of the salt. For example, analysis shows that silver sulphate has the formula

Ag_2SO_4. Electrolysis shows that it takes one Faraday to deposit 1 mole of atoms (gram-atom) of silver. The silver ion therefore has one positive charge and is written as Ag^+. The sulphate ion (or 'radical') must in this case have two negative charges so that silver sulphate is electrically neutral; the sulphate ion is written as SO_4^{2-}. In this way, pupils may acquire a knowledge of the charges of the common cations and anions, and appreciate that the charges of these ions are known through experimental work. The list developed at the end of the previous section can now be extended to include a variety of ionic species.

Some common cations and anions
It is useful to list elements whose cations always have the same charge:

Li^+, Na^+, K^+
Mg^{2+}, Ca^{2+}
Al^{3+}

Common anions include:

O^{2-}, S^{2-}
F^-, Cl^-, Br^-, I^-
NO_3^-
CO_3^{2-}, SO_4^{2-}

Pupils will already be aware that certain elements, in particular the transition elements, give rise to more than one ion. It might be appropriate to mention that some elements in Groups V and VI also show this property.

The treatment provided should enable pupils to show how atoms are related to their ions and vice versa, for example:

$$Na^+ + e^- \longrightarrow Na$$
$$Na \longrightarrow Na^+ + e^-$$
$$2O^{2-} \longrightarrow O_2 + 4e^-$$
$$O_2 + 4e^- \longrightarrow 2O^{2-}, \text{etc.}$$

It is sometimes helpful to ask the pupils to guess the formulae of various ionic compounds, for example:
sodium sulphide could be written as $(Na^+)_2 S^{2-}$
potassium chloride as $K^+ Cl^-$
calcium chloride as $Ca^{2+}(Cl^-)_2$
copper(II) sulphate as $Cu^{2+}SO_4^{2-}$, and so on.

Pupils should be clear that an ionic formula should only be written if the compound has been found to be ionized. Some compounds are *not* ionized. (However, teachers will appreciate that most of the compounds likely to be discussed are by no means one hundred per cent ionic in their properties.)

It is important for the teacher not to rush over this section of the work and to allow pupils time to assimilate the ideas and conventions discussed simply.

A16.8
Molecules, giant structures, ions, and bonding

In this section ideas on ions and structures developed in the Topic are brought together and the nature of bonding between elements is discussed.

Objectives for pupils

1. Familiarity with two types of structure: molecular and giant structures
2. Knowledge of ionic and covalent bonding, using a simple model

The particles and structures considered so far make a good starting point for this lesson. Atoms were introduced in Topic A11; molecules and giant structures in Topic A14; ions in this Topic. All matter can be said to be made up of either (neutral) atoms or (charged) ions. Tell the pupils that these building blocks are built up in two essentially different ways. *Atoms* are either arranged in molecules of a specific size or giant structures of an indefinite size. The pupils should have learnt to distinguish between the two possibilities by observing the physical properties of materials and by finding out how much energy is required to pull the atoms apart (Topic A15). *Ions* are found in giant structures in the solid state, and exist as 'free' ions in molten salts or in solutions of salts and other compounds. Pupils should be aware that very few substances exist as single atoms at room temperature and that atoms of most elements are extremely reactive so that they rarely exist for more than a small fraction of a second on their own. When we consider the properties of an element, we usually consider the properties of its giant structure or of its molecules. Only in the case of the inert (noble) gases are atoms stable on their own. Ions, on the other hand, are relatively stable.

In solution, ions possess the ability to transfer charge from one electrode to another during electrolysis.

However, in Topic A23, pupils will see evidence that when a salt dissolves in water a chemical change takes place and so ions are not as unreactive as they first seem.

This is an appropriate place for a discussion of the composition and structure of a number of common substances. Ask the class if they know the structures of such substances as polythene, argon, paraffin, common salt, and copper. 'Are they made up of ions or of atoms?' 'How are the ions or atoms arranged – as molecules or as giant structures?'

It may be helpful to use the *Handbook for pupils*, Chapter 9 'The structures of elements and compounds', as a basis for any detailed discussion which may arise from such questions. It is convenient to summarize the points in the following way:

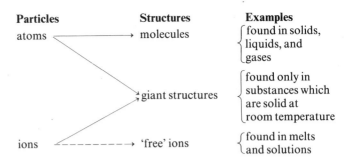

Particles	Structures	Examples

atoms ⟶ molecules — found in solids, liquids, and gases

giant structures — found only in substances which are solid at room temperature

ions ⤏ 'free' ions — found in melts and solutions

The last question to be asked in this Topic is 'How are atoms or ions held together in molecules and in giant structures?' The question is easy to answer in the case of ions. The pupils will know enough electrostatics to understand that unlike charges attract. In the case of atoms, it is not intended that the electronic theory of bonding should be explained, or that electron configuration should be discussed, in detail. The pupils should have a mental picture of the atom as a positively charged nucleus surrounded by negatively charged electrons. We can therefore explain that two positively charged nuclei may be held together by a pair of electrons (one from each atom) with no redistribution of charge. The bond is thus again essentially electrostatic. The theme of atomic structure and bonding is taken up in Chapter 3 of the *Handbook for pupils*, 'Atomic structure and bonding'. The treatment required at this stage is elementary, but includes a simple comparison of the properties of 'model' compounds with reference to the predominant type of chemical bond present. The formulae of the compounds formed between chlorine and the elements in the period sodium to argon can be listed. A definite pattern is apparent. The properties of these compounds show a similar pattern. For example:

Chloride	Properties
sodium chloride	Formula NaCl Crystalline white solid High melting and boiling point
magnesium chloride	Formula $MgCl_2$ Crystalline white solid High melting and boiling point
aluminium chloride	Formula $AlCl_3$ White solid Vaporizes readily on heating
silicon tetrachloride	Formula $SiCl_4$ Colourless liquid Low boiling point
phosphorus trichloride	Formula PCl_3 Colourless liquid Low boiling point
sulphur dichloride	Formula SCl_2 Red liquid Low boiling point

Certain physical properties of a compound provide information about the structure and bonding in that compound, in particular:
1. electrical conductivity of the molten compound
2. melting point of the compound
3. boiling point of the compound
4. energy needed to vaporize 1 mole (gram-formula) of the compound.

The Reference section of the *Handbook for pupils* provides data for a discussion of these points should this seem appropriate.

The chlorides of the elements in the period sodium to argon can now be divided into two types, depending on the physical properties of the compounds.

Ionic chlorides	Non-ionic chlorides
crystalline solids at room temperature – never liquid or gas	frequently liquid or gas at room temperature
conduct electricity when molten	do not conduct electricity when molten
high melting point	low melting point
high boiling point	low boiling point
high heat of vaporization: typical examples, NaCl, $MgCl_2$	low heat of vaporization: typical examples, $SiCl_4$, PCl_3, SCl_2

Explanations for these features might be discussed – for example, ionic chlorides are crystalline due to the regular arrangement of ions in the structure; ionic chlorides conduct electricity when molten because the giant structure of ions is partly broken down on melting, allowing each ion to move towards the electrodes during electrolysis; and so on. The properties of the non-ionic chlorides stem from their molecular structure.

Suggestion for homework

Read the *Handbook for Pupils*, Chapter 3 'Atomic structure and bonding'.

Summary

By the end of this Topic, pupils should have some knowledge of the phenomenon of electrolysis and be able to interpret it in terms of ions. They should know the difference between ionic and non-ionic substances and how this affects structure and properties. They should have some knowledge of the structure of the atom.

Finding the relative number of particles involved in reactions

Purposes of the Topic

1. To demonstrate the interplay of speculation and experimental verification in scientific work.
2. To provide a basis for the determination of reaction stoichiometry in a few simple cases.
3. To provide opportunities for pupils to do calculations associated with reacting quantities of materials and chemical equations.
4. To provide experience of using and writing equations.

Contents

A17.1 A review of terminology: the mole as a unit
A17.2 Reactions involving the formation of a precipitate
A17.3 Reactions involving gases
A17.4 A replacement reaction between a metal and a salt
A17.5 Some uses of chemical equations

Timing

Sections A17.2, A17.3, and A17.4, require between one and two double periods each. Section A17.5 needs only a single period at this point in the course, but may be taken up again subsequently. About 4 to 5 weeks are needed for the whole Topic.

Introduction to the Topic

One of the purposes of the Topic is to show that chemical equations are no more than summaries of investigations on chemical systems. During Topic A11, pupils were encouraged to write simple summaries after finding the formulae of such substances as magnesium oxide, copper(II) oxide, and mercury(II) chloride. In Topic A14, chemical equations were shown to be a way of representing the regrouping of atoms after a chemical change. Pupils were able to represent molecules more precisely and to write equations more fully.

In Topic A16 electrolysis was explained in terms of the behaviour of ions. In this Topic certain chemical reactions can be interpreted in terms of reactions between ions. Ionic equations are then introduced.

The Topic opens with a brief review of terminology concerning the mole. The preparation of a solution containing a mole of solute in a definite volume provides a new means of 'counting particles' and the 'M' notation is introduced as a means of indicating concentration.

Solutions of potassium iodide and lead nitrate are shown to the pupils and are said to contain potassium and iodide ions, and lead and nitrate ions respectively. The solutions should then be mixed together and simple tests performed on both the precipitate, and the residual, or spectator ions. The precipitate is investigated and shown to be lead iodide. An investigation follows to verify that lead iodide has the formula PbI_2. Other precipitation reactions are investigated and the results summarized by writing chemical equations. Reactions involving gases are studied quantitatively, and so are displacement reactions. The Topic concludes with exercises designed to show the value of chemical equations in quantitative studies.

Alternative approach

The material covered in this Topic is treated in Topic B20 in a modified form.

Background knowledge

Pupils studied the composition of simple binary compounds in Topic A11 and learned a simple way of expressing their results by means of word equations. In Topic A14 pupils considered the structure of molecular materials and used equations to represent the regrouping of atoms. Ions were introduced in Topic A16.

Subsequent development

Methods for presenting equations have been introduced in this Topic and are needed for a full appreciation of subsequent Topics.

Further references

for the teacher

Collected experiments, Chapter 12, provides additional experiments for this Topic. *Handbook for teachers*, Chapter 5, has a detailed discussion on an approach to chemical equations.

Supplementary material

Overhead projection original
23 Moles of different substances *(figure A17.1)*

Film
'Gases and how they combine' CHEM study. Distributed by Guild Sound and Vision Ltd.

Reading material

for the pupil

Handbook for pupils, Chapter 4 'Formulae and equations'.

A17.1
A review of terminology: the mole as a unit

In this section, terminology is reviewed and standardized, so that from now on pupils should be fully acquainted with the mole as a unit. The use of solutions of specific concentrations expressed in moles per dm^3 is also introduced.

A suggested approach

Objectives for pupils

1. Knowledge of the mole as a term referring to an amount of substance
2. Knowledge of the mole as a term which may be used to describe the gram-atom, gram-molecule, gram-ion, etc.

(Continued)

The introduction to this section clearly depends on earlier decisions taken by the teacher about the mole concept. Some pupils have met the terms gram-atom, gram-molecule, gram-ion, etc. Each term has been introduced as the need arose. It is suggested that the teacher tells the pupils that although these terms have been useful, most chemists use the term mole, in the manner shown by the following table.

Old usage	New usage
gram-atom	mole of atoms
gram-molecule	mole of molecules
gram-ion	mole of ions
faraday	mole of electrons

This change does not introduce any new concept: it is merely a *change* of terminology. Often the terms *mole of atoms, mole of molecules*, etc. are simply abbreviated to *mole*, and this is acceptable providing the elementary unit is clearly understood.

The following examples illustrate one way in which the teacher might help pupils to 'translate' from the old usage to the new one. Thus, in Topic A11.4 we said:

| 1 gram-atom of copper | *combines with* | 1 gram-atom of oxygen | *to give* | 1 gram-formula of copper(II) oxide |

Now we say:

| 1 mole of copper atoms | *combines with* | 1 mole of oxygen atoms | *to give* | 1 mole of copper(II) oxide |

Or more simply:

| 1 mole Cu | *combines with* | 1 mole O | *to give* | 1 mole CuO |

In Topic A14.8, we said:

| 2 gram-atoms of copper | *combines with* | 1 gram-molecule of oxygen | *to give* | 2 gram-formulae of copper(II) oxide, (CuO) |

Now we say:

| 2 moles of copper atoms | *combines with* | 1 mole of oxygen molecules | *to give* | 2 moles of copper(II) oxide, (CuO) |

Or more simply:

| 2 moles Cu | *combines with* | 1 mole O$_2$ | *to give* | 2 moles CuO |

The use of the formula of the elementary particle after the term mole avoids possible ambiguities. Thus, '1 mole of chlorine' is ambiguous but '1 mole of Cl$_2$' is specific. One anomaly will be apparent: there is no exact 'translation' for the term gram-formula. For a giant structure, such as copper oxide, one must, as in the example above, either refer to 'moles copper(II) oxide' or 'moles CuO'.

The mole is the amount of substance which contains as many elementary units as there are atoms in 0.012 kilograms of carbon-12. The accepted value for the Avogadro constant – the number of atoms of carbon in 0.012 kilograms of carbon-12 – is $(6.022\,52 \pm 0.000\,28) \times 10^{23} \text{mol}^{-1}$. These formal definitions need *not* be given to the pupils but it is a convenient moment to re-emphasize that a mole of a substance always contains the same number of elementary particles – this may be illustrated by using figure A17.1 as an Overhead projection transparency.

In this Topic we investigate the reactions between substances in solution and use solutions containing a definite concentration of solid expressed in mol dm^{-3}. Pupils should be told that a molar

Substance		Number of elementary particles

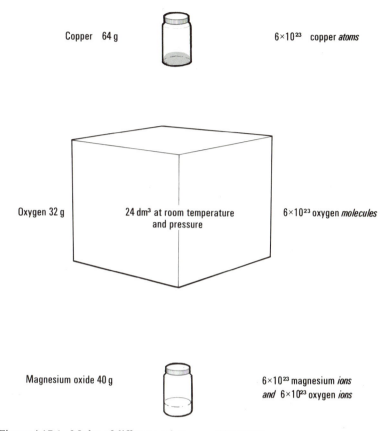

Copper   64 g 6×10^{23} copper *atoms*

Oxygen 32 g   24 dm³ at room temperature and pressure 6×10^{23} oxygen *molecules*

Magnesium oxide 40 g 6×10^{23} magnesium *ions* and 6×10^{23} oxygen *ions*

Figure A17.1   Moles of different substances. (OHP 23)

solution (1.0M) contains one mole of solute dissolved in 1 dm³ solution.

Experiment A17.1

Preparation of 1.0M solution of sodium chloride

Procedure

Sodium chloride is used since it is familiar and relatively cheap. Pupils should be aware that the mass of 1 mole of sodium chloride is 58.5 grams. Show them a labelled jar containing '1 mole of sodium chloride'. Transfer 58.5 grams of sodium chloride to the volumetric flask using a wash bottle and make up the solution in the flask in front of the class. The resulting solution is then labelled as 1.0M solution of sodium chloride.

The point is made that one dm³ of 1.0M solution of sodium chloride contains 1 mole of NaCl.

Apparatus

The teacher will need:

2 volumetric flasks, 1 dm³

1 beaker, 100 cm³

1 glass rod

1 filter funnel

Measuring cylinder, 100 cm³ maximum capacity

Access to balance

Sodium chloride*

Distilled water

*Exactly 58.5 g of sodium chloride should be prepared in advance of the lesson in a screwtop bottle and labelled '1 mole sodium chloride'.

Figure A17.2
A volumetric flask containing 1.0M sodium chloride.

Then transfer $100\,cm^3$ of solution to a second $1\,dm^3$ volumetric flask and ask the pupils how many moles of sodium chloride are in each of the flasks. Make up the solution in the second flask to the $1\,dm^3$ mark with water and repeat the question. The exercise can be repeated until the pupils grasp the point that the quantity of solute and the concentration of solute serve different purposes.

The advantage of working with molar solutions (or convenient multiples or sub-multiples thereof, such as 2.0M or 0.1M) is that reacting volumes are greatly simplified.

It is worth stating:

$1000\,cm^3$ of 1.0M solution contains 1 mole of solute
 $1\,cm^3$ of 1.0M solution contains $\frac{1}{1000}$ mole (i.e. 0.001 mole) of solute
 $50\,cm^3$ of 1.0M solution contains $\frac{50}{1000}$ mole (i.e. 0.050 mole) of solute.

The measurement of volumes of solutions of known strengths can be seen to be 'easier' than the measurement of masses of solids.

A17.2
Reactions involving the formation of a precipitate

In this section one or more precipitation reactions are studied in some detail and the products of reaction are identified. Chemical equations are seen as reaction *summaries.*

A suggested approach

Objectives for pupils

1. Familiarity with precipitation reactions
2. Ability to predict possible products for such reactions given the properties and formulae of reactants
3. Ability to test such predictions qualitatively
4. Ability to formulate a possible equation for the reaction from a knowledge of symbols, formulae, and ionic charges of the species concerned
5. Ability to verify quantitatively a predicted reaction summary (equation)

The lesson might open by reminding pupils of earlier work with precipitation reactions. Then ask pupils to *predict* the reaction between lead nitrate solution and potassium iodide solution. Are these substances ionized? What products might we expect? Which of these products is likely to be soluble and which insoluble? What do we know to support these suggestions? After listening to the views expressed by the pupils, mix samples of the reactants together. The precipitate and the residual liquor may then be investigated. It is unlikely that time will allow pupils to verify every point for themselves.

Whatever is attempted by either the teacher or the pupils, the following points should be considered:
1. The formulae of lead nitrate, $Pb(NO_3)_2$, and potassium iodide, KI, have been determined using techniques similar to those used to determine the formula of copper(II) oxide in section A11.4. (These formulae must be known in order to be able to make up solutions of known molarity.)

2. The charges on the lead and iodide ions are also known and have been determined by finding the number of moles of electrons (Faradays) required to discharge 1 mole of lead ions and 1 mole of iodide ions (see section A16.4).

Thus, after some class discussion, pupils should be able to predict steps leading to the following equations:

$$Pb(NO_3)_2 + 2KI \longrightarrow PbI_2 + 2KNO_3$$

and, in terms of ions:

$$Pb^{2+}(aq) + 2NO_3^-(aq) + 2K^+(aq) + 2I^-(aq) \longrightarrow$$
$$Pb^{2+}(I^-)_2(s) + 2K^+(aq) + 2NO_3^-(aq)$$

It may be necessary to explain the notation used in the second equation and it will be essential to identify 'spectator' ions (i.e. those ions which do not take an active part in the reaction). The equation can then be re-written in the form:

$$Pb^{2+}(aq) + 2I^-(aq) \longrightarrow Pb^{2+}(I^-)_2(s)$$

The purpose of Experiment A17.2a is to see whether this prediction can be verified experimentally.

Experiment A17.2a

To verify the composition of the precipitate formed during the reaction between lead nitrate solution and potassium iodide solution

Apparatus

Each pair of pupils will need:

Experiment sheet 65

2 burettes and stands*

4 test-tubes, 100×16 mm

Test-tube rack

Glass rod

Teat pipette

Watch glass

Beaker, 250 cm^3

Bunsen burner and heat resistant mat

Tripod and gauze

Access to centrifuge

Access to ethanol (I.M.S.), few drops

1.0M lead nitrate, 7 cm^3

1.0M potassium iodide, 10 cm^3

Procedure

Experiment sheet 65 is reproduced below and provides full details. It may be helpful to demonstrate the technique of using a centrifuge before the pupils begin the experiment.

Experiment sheet 65

Using a burette, measure 5.0 cm^3 1.0M potassium iodide solution into a clean, dry 100×16 mm test-tube. Add to this 0.5 cm^3 1.0M lead nitrate solution from another burette. Mix the two solutions with a thin glass rod. Centrifuge the mixture for 30 seconds (do not forget to balance the tube with another containing an equal volume of water). Measure the height of the precipitate, from the bottom of the test-tube, as accurately as you can. Record this in the table described below.

Add another 0.5 cm^3 lead nitrate solution, stir, and again centrifuge for 30 seconds (remember to add another 0.5 cm^3 water to the balancing tube). Measure the new height of the precipitate.

Repeat this process, using successive quantities of 0.5 cm^3 lead nitrate solution, until the height of the precipitate does not change with further additions. Record your results in a table using the headings below.

Volume 1.0M KI(aq) /cm³	Volume 1.0M Pb(NO₃)₂ (aq) /cm³	Height of precipitate /cm

What volume of 1.0M lead nitrate solution [$Pb(NO_3)_2(aq)$] just reacts with 5.0 cm^3 1.0M potassium iodide KI(aq)?

Does this result verify your predicted equation for the reaction? Give your reasons.

---

*Some teachers prefer to use proprietary constant-volume pipettes in place of burettes.

Make up a mixture of 5 cm³ 1.0M potassium iodide solution and exactly the right quantity of 1.0M lead nitrate solution for complete reaction. Add 2–3 drops of alcohol and centrifuge to compact the precipitate. What is the precipitate?

What does the solution above it contain?

Devise experiments to verify your answers. On a separate sheet of paper, describe how you carry them out and state what the results are. (Class discussion will help you with this.)

The experiment shows that, within the limits of experimental error, 5 cm³ of 1.0M potassium iodide reacts with 2.5 cm³ of 1.0M lead nitrate

i.e. 5 cm³ 1.0M KI reacts with 2.5 cm³ 1.0M $Pb(NO_3)_2$

but, 1 dm³ of a 1.0M solution contains 1 mole of solute
therefore, 2 moles KI react with 1 mole $Pb(NO_3)_2$
or 2 moles I^- react with 1 mole Pb^{2+}.

This verifies the equation which was predicted:

$$Pb^{2+}(aq) + 2I^-(aq) \longrightarrow Pb^{2+}(I^-)_2(s)$$

The investigation has depended on a number of assumptions. We should try to obtain evidence to support the one quantitative result obtained. For example, the potassium nitrate solution prepared in this experiment could be crystallized and the crystals compared with those from a solution of potassium nitrate. Also, the lead iodide obtained could be fused and electrolysed to obtain iodine and lead. This part of the investigation must be determined by questions raised during the discussion of the results obtained by pupils.

An alternative approach might use Experiments A17.2a and A17.2b with different groups of pupils at the same time. Confirmatory tests will have to be considered separately, remembering that we are trying to establish an experimental procedure to verify the composition of a precipitate. Some teachers may prefer to use Experiment A17.2b in place of Experiment A17.2a.

Experiment A17.2b

To verify the composition of the precipitate formed during the reaction between barium chloride and sodium carbonate solutions

Apparatus

Each pair of pupils will need:

Experiment sheet 66

2 burettes and stands

4 test-tubes, 100 × 16 mm

Test-tube rack

Glass rod

Teat pipette

Watch glass

Beaker, 250 cm³

(Continued)

Procedure

Details of the procedure are given in the *Experiment sheet* below.

Experiment sheet 66
Your teacher will show you how to make a wire frame to hold several test-tubes in a beaker.

Place a 100 × 16 mm test-tube in a wire frame so that it is immersed to a depth of about 8 cm in a beaker of water kept at a temperature of 80 to 90 °C (small Bunsen flame under gauze on tripod stand). Remove the test-tube, and add to it 5 cm³ 1.0M barium chloride solution and 1 cm³ 1.0M sodium carbonate solution using burettes. Stir the mixture with a glass rod and replace it in the wire frame. Heat the test-tube in hot water for 5 minutes. Centrifuge the mixture in the tube observing the same conditions as in the previous experiment. Measure the height of the precipitate.

Bunsen burner and heat
resistant mat

Tripod and gauze

26 s.w.g. copper wire

1.0M barium chloride

1.0M sodium carbonate

Repeat the process with additional portions of 1 cm^3 sodium carbonate
solution, measuring the height of the precipitate after each addition.
Continue until the height of the precipitate does not change with further
addition. On a separate sheet of paper record the results in a table using the
headings below.

Volume 1.0M $BaCl_2(aq)$ /cm^3	Volume 1.0M $Na_2CO_3(aq)$ /cm^3	Height of precipitate /cm

What volume of 1.0M sodium carbonate solution ($Na_2CO_3(aq)$) just reacts
with 5.0 cm^3 1.0M barium chloride solution ($BaCl_2(aq)$)?

Does this result verify your predicted equation for the reaction? Give your
reasons.

Make up a mixture of 5 cm^3 of 1.0M barium chloride solution and exactly
the right quantity of 1.0M sodium carbonate solution for complete reaction.
Stir and allow the precipitate to settle while the test-tube is immersed in hot
water. What is the precipitate?

What does the solution above it contain?

Devise experiments to verify your answers. Describe how you carry them
out and state what the results are. (Class discussion will help you with this.)

The quantitative studies yield results which confirm the prediction

$$BaCl_2 + Na_2CO_3 \longrightarrow 2NaCl + BaCO_3$$
or
$$Ba^{2+}(aq) + CO_3^{2-}(aq) \longrightarrow Ba^{2+}CO_3^{2-}(s)$$

Pupils should be able to show that the precipitate is a carbonate by
reacting a sample with a dilute acid to produce carbon dioxide.
Similarly, they should be able to show the presence of a chloride in
the supernatant liquid using silver nitrate solution in the presence of
nitric acid. Again, how these qualitative tests are used will depend
on the discussion of results.

Experiment A17.2c extends the range of observable effects when
considering precipitation reactions. It is not seen necessarily as an
alternative to either Experiment A17.2a or A17.2b.

Experiment A17.2c

Apparatus

Each pair of pupils will need:

Experiment sheet 67

2 burettes and stands

6 test-tubes, 100 × 16 mm

Test-tube rack

Glass rod

Teat pipette

Watch glass

(Continued)

**To verify the composition of the precipitate formed during the
reaction between barium chloride and potassium chromate solutions**

Procedure
Experiment sheet 67 is reproduced below. The level and extent of
discussion used when introducing this experiment must necessarily
depend on what has gone before.

Experiment sheet 67
The procedure in this experiment is similar to that described in Experiment
66, except that potassium chromate solution is used instead of sodium
carbonate solution. The precipitate is slow to settle but you will find that
there is a colour change in the reaction mixture when precipitation is
approximately complete. Make up your own table for this experiment and
write notes similar to those for the previous two experiments.

Beaker, 250 cm³

Bunsen burner and heat resistant mat

Tripod and gauze

26 s.w.g. copper wire

0.5M barium chloride solution, 10 cm³

0.5M potassium chromate solution, 12 cm³

Note to teachers
Small variations in the particle size of precipitates affect their volumes. A number of other factors can also influence the volume occupied by a precipitate. By using a sample of one reactant and adding known quantities of the second reactant, mixing, and centrifuging the precipitate, it is possible to see when a further addition fails to achieve precipitation. As will be appreciated, a centrifuge greatly assists the speed with which such investigations can be made.

Suggestion for homework

How would you investigate the reaction between iron(III) chloride and sodium hydroxide solutions? The equation for this reaction has been predicted to be:

$$FeCl_3(aq) + 3NaOH(aq) \longrightarrow Fe(OH)_3(s) + 3NaCl(aq)$$
or
$$Fe^{3+}(aq) + 3OH^-(aq) \longrightarrow Fe^{3+}(OH^-)_3(s)$$

It is known that iron(III) hydroxide decomposes on heating. One of the products of the decomposition is water and the other is a red powder. How would you investigate the composition of this powder?

A17.3
Reactions involving gases

In this section reactions in which there are gaseous reactants and/or products are studied. The volumes of gases are measured and used to verify the overall chemical equation for a reaction. Three systems are suggested for investigation: (A) the reaction between sodium carbonate and dilute hydrochloric acid; (B) the reaction between magnesium and dilute hydrochloric acid; (C) the reaction between gaseous ammonia and gaseous hydrogen chloride.

A suggested approach

Objectives for pupils

1. Familiarity with reactions in which gases are either reactants or products
2. Ability to predict probable equations which summarize such reactions
3. Ability to verify experimentally such predicted equations
4. Knowledge of the use of Avogadro's law

A *The reaction of sodium carbonate with dilute hydrochloric acid*
Reactants and products of a chemical reaction can be solids, liquids, or gases. So far, we have been concerned with systems in which solutions containing ionic species react to form precipitates. Known volumes of solutions were used, each containing a known amount of reactant. This section opens with a study of the reaction between solutions of sodium carbonate and hydrochloric acid. A simple demonstration reveals that a gas is evolved. Pupils should be aware that the gas is carbon dioxide and may be reminded of the lime-water test for the gas.

A discussion of the reaction leads to the following sequence. A simple demonstration of the reaction shows:

sodium carbonate + hydrochloric acid ⟶ carbon dioxide + solution of ?

The reaction between an acid and a carbonate should *not* be totally unfamiliar to the pupils. They will know the formula of hydrochloric acid as HCl from Topic A12 and, since a solution is required for this reaction, we may use the formula HCl(aq) at this point. Carbon dioxide, CO_2, was met in Topic A14, and also appeared in Stage I, Topic A4. It may be necessary to tell the pupils the formula of sodium carbonate, Na_2CO_3. The replacement of words in the reaction summary by chemical formulae leads to the need for simple tests to determine the nature of the unknown product – sodium chloride.

In Experiment A17.3a, pupils are asked to determine the volume of 1.0M hydrochloric acid which reacts completely with $10.0\,cm^3$ 1.0M sodium carbonate. They find that *about* $20\,cm^3$ 1.0M hydrochloric acid is needed. There will be a need to discuss variations in the results obtained and it may be desirable to use an average value. Given the procedure outlined in *Experiment sheet* 68, the end point cannot be more precise than within $1.0\,cm^3$.

Thus, the results may be summarized as:

$$2HCl(aq) + Na_2CO_3 \longrightarrow products$$

The evaporation of a sample of the residual solution produces a white solid which may be shown to have the same chemical properties as purchased sodium chloride; thus:

$$2HCl(aq) + Na_2CO_3(aq) \longrightarrow 2NaCl(aq) + CO_2(g) + \quad ?$$

This leaves 2 atoms of hydrogen and 1 atom of oxygen unaccounted for. These may be assumed to form water, which would be difficult to detect. This point may need to be discussed.

Thus:

$$2HCl(aq) + Na_2CO_3(aq) \longrightarrow 2NaCl(aq) + CO_2(g) + H_2O(l)$$

We could be more confident of this summary if we verified the quantity of carbon dioxide formed. The equation suggests that 1 mole of sodium carbonate, $Na_2CO_3(aq)$, produces 1 mole of carbon dioxide, $CO_2(g)$ – a volume of $24\,dm^3$ at room temperature. Therefore if 0.002 mole of sodium carbonate were used, then 0.002 mole of carbon dioxide would be liberated, i.e. $24\,000 \times 0.002\,cm^3 = 48\,cm^3$. Experiment A17.3b provides a procedure to test this prediction.

If necessary, remind pupils that in Topic A16 we used a simple test for conductivity and developed an explanation of electrolysis in terms of the properties of ions. Solutions of hydrochloric acid, sodium carbonate, and sodium chloride may now be shown to conduct electricity fairly easily and to exhibit the phenomena associated with electrolysis (Experiment A17.3c).

Accordingly, we may now write dilute hydrochloric acid as $H^+(aq) + Cl^-(aq)$, sodium carbonate solution as $2Na^+(aq) + CO_3^{2-}(aq)$, and sodium chloride solution as $Na^+(aq) + Cl^-(aq)$. It follows that the reaction may be represented as:

$$2H^+(aq) + 2Cl^-(aq) + 2Na^+(aq) + CO_3^{2-}(aq) \longrightarrow$$
$$2Na^+(aq) + 2Cl^-(aq) + CO_2(g) + H_2O(l)$$

and we can simplify this by removing 'spectator' ions (i.e. those which do not take an active part in the reaction) from both sides of the equation (see section A17.2).

$$2H^+(aq) + CO_3^{2-}(aq) \longrightarrow CO_2(g) + H_2O(l)$$

Experimental details in the form of three *Experiment sheets* are reproduced below. It is intended that these experiments be carried out either by the pupils or by the teacher as appropriate.

Experiment A17.3a

To determine the volume of 1.0M hydrochloric acid which reacts completely with 10.0 cm³ 1.0M sodium carbonate

Apparatus

The teacher or each pair of pupils will need:

Experiment sheet 68

2 burettes, 50 cm³

2 burette stands

Conical flask, 100 cm³

1.0M hydrochloric acid

1.0M sodium carbonate solution, about 25 cm³

Procedure

Full details appear in *Experiment sheet* 68 which is reproduced below.

Experiment sheet 68

Use a burette to measure 10 cm³ 1.0M sodium carbonate into a 100 cm³ conical flask. From another burette add 1 cm³ 1.0M hydrochloric acid and shake the mixture gently. A gas is given off. What is it?

Continue adding 1 cm³ portions of acid, shaking the flask between each addition, until there is no gas evolved. (Remember that a few air bubbles are trapped when any liquid is shaken. It is the evolution of gas when shaking is finished that you have to look for. If any gas is given off, wait until the evolution is complete before adding the next portion of acid.)

What volume of 1.0M hydrochloric acid just reacts with 10 cm³ 1.0M sodium carbonate solution?

How many moles of hydrochloric acid react with one mole of sodium carbonate?

What is present in the final solution?

Test your answer by experiment and report on this below.

Can you write an equation for the reaction between hydrochloric acid and sodium carbonate?

Experiment A17.3b

To measure the volume of carbon dioxide formed when a known amount of sodium carbonate reacts with dilute hydrochloric acid

Apparatus

The teacher or each pair of pupils will need:

Experiment sheet 69

Glass syringe, 100 cm³

Test-tube, 150 × 25 mm, with bung and right-angle glass tube

Test-tube, 75 × 10 mm, or small specimen tube

(Continued)

Procedure

Before pupils start the experiment, the teacher may need to demonstrate the use of the apparatus, the method of checking gas-tightness, and the procedure for testing whether the product is at atmospheric pressure. *Experiment sheet* 69 provides full details.

Experiment sheet 69

For this experiment you will need to use exactly 0.002 mole sodium carbonate.

What is the mass of this quantity? .... g.
(Check this answer with your teacher.)

Spatula

Stand, boss, and clamp or syringe holder

Rubber connecting tubing

Measuring cylinder, 25 cm³

Access to balance, weighing to 0.001 g

Anhydrous sodium carbonate*

5M hydrochloric acid

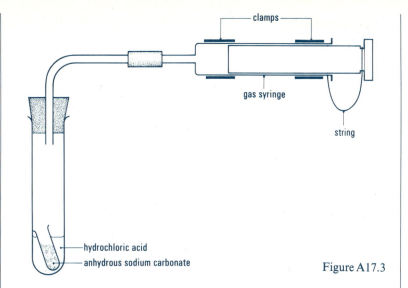

Figure A17.3

Assemble the apparatus shown in the diagram. In the larger test-tube put 5 cm³ 5.0M hydrochloric acid, using a measuring cylinder. Add a measure of sodium carbonate to the acid and wait until all of this has reacted; the purpose of this addition is to saturate the acid with carbon dioxide. Why is this necessary?

Weigh out 0.002 mole sodium carbonate into the smaller test-tube. Slide this tube into the larger tube so that its contents do not come into contact with the acid (see diagram). Connect the larger test-tube to the gas syringe by inserting the rubber bung. Record the reading of the syringe piston. Tilt the larger tube so that the acid makes contact with the sodium carbonate. Carbon dioxide is evolved and pushes out the piston of the syringe. Make sure that all the solid reacts. Record the final reading of the syringe.

Results
Amount of sodium carbonate used .... 0.002 mole

Initial reading of syringe .... cm³

Final reading of syringe .... cm³

Volume of carbon dioxide evolved .... cm³

At room temperature 24 000 cm³ of carbon dioxide contains 1 mole.

Therefore.........................cm³ of carbon dioxide contains.........................mole.

Therefore 1 mole of sodium carbonate gives .... mole carbon dioxide.

You now have all the information required to write the complete equation for the reaction between hydrochloric acid and sodium carbonate.

*It will be found convenient to prepare a number of samples of 0.002 mole (0.212 g) of anhydrous sodium carbonate before the lesson in corked test-tubes (75 × 10 mm).

To find whether hydrochloric acid, sodium carbonate, and sodium chloride solutions are electrolytes

Apparatus

The teacher or each pair of pupils will need:

Experiment sheet 70

4 test-tubes, 150×25mm

Test-tube rack

2 carbon electrodes mounted in a holder

Connecting wires

Ammeter, 0–1 A

6 V d.c. supply

Distilled water

1M hydrochloric acid, 25 cm^3

1M sodium carbonate solution, 25 cm^3

1M sodium chloride solution, 25 cm^3

Procedure

Details are given in *Experiment sheet* 70.

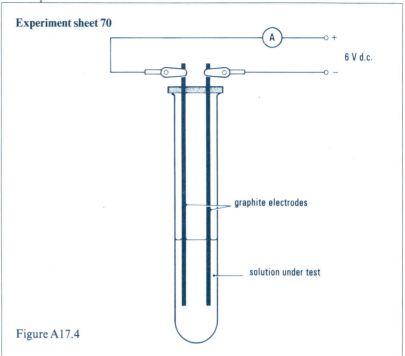

Experiment sheet 70

6 V d.c.

graphite electrodes

solution under test

Figure A17.4

Use the apparatus shown in the diagram. How will you know whether the solution tested is an electrolyte or not?

Try distilled water first, then the three solutions to be tested. Remember to wash the electrodes thoroughly with distilled water between each test. What do you find?

B *The reaction of magnesium with dilute hydrochloric acid*

This reaction is suggested as a follow-up study to that outlined in the previous subsection. A demonstration of the reaction of magnesium with dilute hydrochloric acid to form a gas (which may be identified as hydrogen) provides a convenient opening. The teacher may then review the information available for writing the equation by questioning the class. The argument may be summarized as follows:

a magnesium + hydrochloric acid $\longrightarrow$ product in solution (?)
+ hydrogen

b $Mg(s) + H^+(aq) + Cl^-(aq) \longrightarrow$? $\quad + H_2(g)$

c $Mg(s) + 2H^+(aq) + 2Cl^-(aq) \longrightarrow Mg^{2+}(aq) + 2Cl^-(aq) + \quad + H_2(g)$

d $Mg(s) + 2H^+(aq) \longrightarrow Mg^{2+}(aq) \quad + H_2(g)$

To verify predictions arising from the discussion, we may need to seek experimental evidence. Thus, if equation (**d**) above is correct,

1 mole Mg would liberate 1 mole H_2
that is, 24 g magnesium would liberate 24 000 cm^3 hydrogen at room temperature
that is, 0.024 g magnesium would liberate 24 cm^3 hydrogen at room temperature.

This prediction is tested in Experiment A17.3d. The method of Experiment A17.3c can be used to find out whether magnesium chloride solution is ionized.

Experiment A17.3d

To measure the volume of hydrogen produced when 0.024 g magnesium reacts with dilute hydrochloric acid

Apparatus

The teacher or each pair of pupils will need:

Experiment sheet 71

Glass syringe, 100 cm^3

Test-tube, 150×25 mm, with bung and right-angle delivery tube

Test-tube, 75×10 mm, or small specimen tube

Spatula

Stand, boss, and clamp or syringe holder

Rubber connecting tubing

Measuring cylinder, 10 cm^3

Access to balance

Emery paper

Magnesium ribbon

5M hydrochloric acid, 5 cm^3

Procedure
Magnesium ribbon is usually of a standard thickness which makes measuring it out beforehand easier. Details of the experiment appear in *Experiment sheet* 71.

Experiment sheet 71
The apparatus and procedure are those described in Experiment 69, using magnesium ribbon instead of sodium carbonate. You will not need to saturate the acid with gas beforehand, the solubility of hydrogen in dilute hydrochloric acid is small enough to be neglected. Your only problem is to obtain exactly 0.024 g magnesium ribbon. How will you do this?

Results
Mass of magnesium used is 0.024 g. This is .... mole.

Initial reading of syringe .... cm^3

Final reading of syringe .... cm^3

Volume of hydrogen evolved .... cm^3

This is .... mole.

What is the equation for the reaction between magnesium and hydrochloric acid?

Use the method described in Experiment 70 to find whether the magnesium chloride solution is an electrolyte. (You know already about hydrochloric acid in this respect.) What do you find?

C *The reaction between gaseous ammonia and gaseous hydrogen chloride*
First demonstrate the reaction between hydrogen chloride and ammonia. The production of ammonium chloride as a white smoke from separate gas jars containing colourless gases is quite dramatic! Pupils will need to be told the formula of ammonia, NH_3. The argument then follows the general pattern established for these sub-sections.

1. A statement of the word equation:

ammonia + hydrogen chloride $\longrightarrow$ 'white smoke'

2. The identification of the white smoke as a chloride.
3. An experiment to demonstrate that equal volumes of gaseous

ammonia and hydrogen chloride react together to yield only one product – a white smoke which is a chloride.

4. The formulation of the equation as:

$$NH_3(g) + HCl(g) \longrightarrow (NH_3)HCl$$

5. A demonstration of the fact that ammonium chloride – $(NH_3)HCl$ – dissolves in water and behaves like sodium chloride in that it is also an electrolyte.

6. Finally, the formula for ammonium chloride needs amending to $NH_4^+Cl^-$. Pupils could be shown samples of ammonium chloride and other ammonium salts to familiarize them with the use of $(NH_4)^+$ as the ammonium ion.

Of course, this sequence of ideas can be adjusted to suit a given group of pupils.

One way of finding out the relative number of moles of gases in this system is described in Experiment A17.3e. Pupils have to apply Avogadro's law which was introduced in Topic A14.

Experiment A17.3e

Apparatus

The teacher will need:

2 syringes, 100 cm³ glass

1 three-way stopcock with capillary tubing

Plastic connecting tubing

2 stands and syringe holders

Indicator paper

Apparatus for producing dry hydrogen chloride

Apparatus for producing dry ammonia

To verify the equation for the reaction between gaseous ammonia and gaseous hydrogen chloride
This experiment should be done by the teacher.

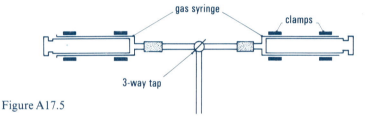

Figure A17.5

Procedure
Set up the apparatus shown in figure A17.5. Fill one syringe with 24 cm³ (0.001 mole) of dry ammonia through the three-way stopcock, flushing the syringe out with the gas two or three times first. Similarly, fill the other syringe with 48 cm³ (0.002 mole) of dry hydrogen chloride. Turn the three-way stopcock so as to connect the two syringes but isolate them from the atmosphere. Push the hydrogen chloride through the stopcock into the ammonia. The gases react to form a white powder, ammonium chloride. There remains 24 cm³ (0.001 mole) of gas that has not reacted. Show that it is hydrogen chloride by passing it over damp indicator paper.

This experiment shows that 24 cm³ (0.001 mole) of ammonia reacts with 24 cm³ (0.001 mole) of hydrogen chloride. Using Avogadro's law, we may now write the lefthand side of this equation as:

$$NH_3(g) + HCl(g) = \ldots .$$

We know that a white solid is formed. Since ammonium chloride is known to be a white solid of formula NH_4Cl, it is reasonable to assume that this is what is formed – although this cannot be regarded as a proof. The equation may then be written as:

$$NH_3(g) + HCl(g) \longrightarrow NH_4Cl(s)$$
or
$$NH_3(g) + HCl(g) \longrightarrow NH_4^+ Cl^-(s)$$

Suggestion for homework *Questions of the type suggested for A17.2 but involving gases may be used.*

A17.4
A replacement reaction between a metal and a salt

In this section, the equation for the reaction of a metal replacing another metal from its solution is determined by a gravimetric procedure.

A suggested approach

Objectives for pupils

1. Ability to determine an equation for a reaction which cannot be predicted in advance
2. Familiarity with replacement reactions involving a solid and a solution
3. Knowledge of gravimetric procedures

Pupils should now understand how chemists are able to predict the equation for a reaction from a knowledge of data, such as the value of charge on an ion. They will have also learnt that it is desirable to *verify* any equations they have written by means of experimental work.

However, it is *not* always possible to predict in advance what the equation for a reaction will be. One example of this is the replacement reaction using iron and a solution of a copper salt. If a known mass of iron is added to a solution of copper(II) sulphate, the iron appears to be replaced by the copper. If pupils are asked to predict the equation for this reaction, they might suggest:

either $Fe(s) + Cu^{2+}(aq) \longrightarrow Fe^{2+}(aq) + Cu(s)$
or $\quad 2Fe(s) + 3Cu^{2+}(aq) \longrightarrow 2Fe^{3+}(aq) + 3Cu(s)$

On the basis of their knowledge at this point in the course, there is no way in which the pupils can decide which equation represents the reaction.

Suppose that 0.01 mole of iron (0.56 g) is used. If the reaction is represented by the first equation, then 0.01 mole of copper (0.64 g) will be produced. If, on the other hand, the reaction is represented by the second equation, then 0.015 mole of copper (0.98 g) will be produced.

The experiment may now be performed to see which of these two equations is the correct one.

Experiment A17.4

An examination of the reaction between iron and copper(II) sulphate solution

Apparatus

Each pair of pupils will need:

Experiment sheet 72

(Continued)

Procedure
Full details are given in *Experiment sheet* 72 which is reproduced below. It is suggested that the teacher explains and/or demonstrates new techniques before pupils start the experiment.

Either 2 test-tubes,
100 × 16 mm or 1 test-tube and
1 beaker 100 cm³

Test-tube rack

Test-tube holder

Glass rod

Teat pipette

Beaker, 100 cm³

Spatula

Bunsen burner and heat
resistant mat

Tripod and gauze

Access to balance

Access to centrifuge

Iron, fine powder, about 1 g

1.0M copper(II) sulphate
solution, 15 cm³

Acetone (propanone)

Distilled water

Experiment sheet 72
Weigh exactly 0.56 g iron powder into a 100 cm³ beaker and add to it at least
15 cm³ 1.0M copper(II) sulphate solution. Heat the mixture just to boiling,
stirring well all the time, and allow it to boil for one minute. Now allow the
contents of the beaker to cool and the precipitate of copper to settle. Pour off
as much of the liquid as you can, being careful not to lose any copper.

Add distilled water (about one-third fill the beaker), and stir the mixture.
Allow the copper to settle and pour off as much liquid as possible. Repeat the
washing with distilled water.

Add about 20 cm³ acetone (propanone) and stir the mixture. Allow the
precipitate to settle and pour off the acetone. *(Be careful, acetone is highly
flammable!)* Allow the beaker to stand for 5 minutes so that most of the
acetone clinging to the precipitate can evaporate. Drive off the remainder by
heating the beaker in an oven at 100 °C for 5 minutes. Allow it to cool and
find the mass of beaker + copper.

Results
Mass of beaker .... g

Mass of beaker + iron powder .... g

Mass of beaker + copper .... g

Mass of iron powder used .... g

This is .... mole.

Mass of copper obtained .... g

This is .... mole.

Which of your two predicted equations is correct?

It will be found that if the copper is properly dried it weighs a little
more than 0.60 g. This may be taken to confirm the first equation.

With a good class, there could be a critical discussion of the assumptions behind the experiment (for example, that the reaction goes to
completion: the very fine grade of iron powder recommended will
react more completely than the coarser grade of filings*).

Suggestions for homework

1. Write a brief account of the assumptions which you think have
been made when examining the reaction between iron and copper
sulphate solution. Indicate how you might investigate your suggestions.
2. Summarize the work you have done so far on chemical equations.
List some of the important uses for equations. Have there been any
assumptions in the investigations made so far?

*See, for example, Grant, J. and Allsop, R. T. (1971) 'Study of the reaction of iron
with copper(II) sulphate'. *School Science Review*, **53**, No. 182, 150.

A17.5
Some uses of chemical equations

It is suggested that some time (perhaps a single period) should now be spent in summing up what the pupils know about chemical equations, and in giving pupils the opportunity to become familiar with their interpretation. Additional lessons are sometimes helpful a little later in the course.

In this Topic the emphasis has been on the use of evidence to support equations (reaction summaries). Pupils should now understand that it is often possible to predict the overall equation for a reaction by using the fact that equations must 'balance' with respect to chemical species, mass, and charge – *but* that *someone* had to carry out quantitative experiments to determine the formulae and value of the charges on the ions and the formulae of the molecules. They will be aware that it is desirable that all equations should be verified experimentally.

Equations may be interpreted in two ways:
1. In terms of the reacting quantities of material expressed in moles, as in A11.4 and in this Topic.
2. In terms of individual atoms or other species, as in A14.10.

Some time could now be spent in reviewing the equations that the pupils have met so far, and in interpreting them in both ways.

Exercises may be set in which the pupil has to work out equations using data given by the teacher or from the Reference section of the *Handbook for pupils*. Alternatively, problems of the kind outlined below may be used.

Finally, the teacher might summarize four relationships involving the mole:
1. One mole of a substance contains 6×10^{23} particles – atoms, molecules, or ions, as may be appropriate.
2. One mole of a substance is equivalent numerically to the formula mass of that substance expressed in grams.
3. One mole of a gas occupies $24\,000\,\mathrm{cm}^3$ at room temperature and normal pressure ($22.4\,\mathrm{dm}^3$ at S.T.P.).
4. The ratio of the substances taking part in the reaction expressed in moles is the same as the ratio of the integers appearing in front of the formulae used in the equation.

Thus, in the example:

$$3CuO(s) + 2NH_3(g) \longrightarrow 3Cu(s) + 3H_2O(l) + N_2(g)$$

3 moles $CuO(s)$ react with 2 moles $NH_3(g)$ to form 3 moles $Cu(s)$, 3 moles $H_2O(l)$, and 1 mole $N_2(g)$.

The visual display of formulae prior to the formal and more usual equation presentation can be helpful. Thus, in this case, one might first write:

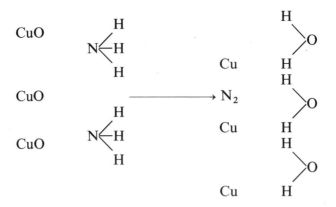

Some teachers find it helpful to provide pupils with a flow diagram summarizing these facts.*

Suggested problems for discussion and evaluation
1. A solution of lead nitrate was treated with dilute sulphuric acid. Experiments showed that the products of reaction were a precipitate of lead sulphate and a solution of nitric acid. A separate investigation showed that 3.31 g of lead nitrate produced a precipitate of 3.03 g of lead sulphate. Is this information consistent with the chemical equation for the reaction?
2. A solution of calcium chloride was treated with silver nitrate solution. The products of reaction were shown by experiment to be a precipitate of silver chloride, and a solution containing calcium nitrate. A separate investigation showed that 1.50 g of anhydrous calcium chloride gave 3.88 g of silver chloride on precipitation with silver nitrate solution. Show that this information is consistent with the chemical equation for the reaction.
3. Each year, the chemical industry in Great Britain manufactures several millions of tonnes of sulphuric acid from sulphur by the so-called 'contact process'. The reactions used in this process are the burning of sulphur in air to form sulphur dioxide; the catalytic oxidation of sulphur dioxide to form sulphur trioxide; and the absorption of sulphur trioxide in sulphuric acid by a special process requiring the addition of water. The process is about 95 per cent efficient. How much sulphur is needed to make 1 tonne of sulphuric acid?
4. In each of the investigations in this Topic, we have assumed that there is no change in total mass when reactants are changed into products. (Historically, this assumption has been found to be valid for a large number of reactions and is referred to as the law of conservation of matter.)

How would you test this idea? Describe the apparatus you might use and state the reaction you would study.

What are the drawbacks of your proposal?

*For example see Head, J. O. (1968) 'Teaching the mole concept in schools'. *School Science Review*, **49**, No. 168, 496.

Summary

When the Topic has been completed, pupils should be able to predict 'balanced' equations to summarize reactions – which they have not personally carried out – given relevant information. After some practice, they should be able to write equations as summaries for the reactions they have met during the course. In addition, they should be able to complete simple calculations associated with a knowledge of chemical equations.

The Topic introduces the use of state symbols; the use of ionic formulae (e.g. $Pb^{2+}(I^-)_2(s)$ rather than $PbI_2(s)$); and the writing of ionic equations. Experience will have been gained in the writing of balanced equations. Pupils should be aware of the significance of 'spectator' ions.

Additional lessons on equations are often helpful during later Topics in the course.

How fast? Rates and catalysts

Purposes of the Topic
1. To introduce the idea that reactions do not necessarily take place instantaneously but at rates which can often be measured.
2. To show that the rate of a reaction can be affected by the concentration or pressure of a reactant; by the temperature of the system; and by the presence or absence of a catalyst.
3. To provide pupils with some simple investigations to illustrate the ideas and concepts used in the Topic.

Contents

A18.1 Measuring the rate of a reaction: how does the concentration of a reactant affect the rate of a reaction?
A18.2 How do changes in temperature affect the rate of a reaction?
A18.3 The effect of particle size on the rate of a reaction.
A18.4 How do catalysts affect the rates of reactions?

Timing

Each of these sections needs at least a double period and section A18.4 may require more. To allow time for discussion and for summarizing the work attempted, about five weeks should be allocated to this Topic.

Introduction to the Topic

The Topic opens with a simple review of the nature of chemical change. Pupils are reminded that most of the reactions they come across are rapid, but that some, for example rusting, are quite slow. In order to study the rate at which a reaction takes place and to find out what factors affect it, it is desirable to choose a reaction which takes place moderately quickly. One such reaction is that between calcium carbonate and hydrochloric acid. The progress of this reaction may be conveniently followed using a balance. The decreasing rate as the reaction proceeds is very evident. The concentration of the acid appears to be a key factor. This factor is investigated by pupils using the reaction between sodium thiosulphate and hydrochloric acid.

The next question is, 'What is the effect of temperature on the rate of the reaction?' Once again, the sodium thiosulphate and hydrochloric acid system is used.

The effect of particle size on a reaction involving a solid reactant is then investigated using the reaction between calcium carbonate and hydrochloric acid.

Finally, the decomposition of hydrogen peroxide is studied to determine the effect of catalysts on the rate of a reaction.

Alternative approach

The central theme for A18 receives a different treatment in the alternative scheme. A partial answer to the question 'How fast?' is provided in Alternative IIB, Topic B13, and leads to an elementary interpretation of a reaction in terms of a collision process, and a qualitative study of catalysis. Topic B22 reviews rate processes in chemistry and seeks to relate and distinguish between radiochemical processes and rates of reaction. Pupils should appreciate that radio-activity is immune to the chemical environment and dependent on a nuclear process whereas reactions are dependent on the chemical environment and on the electronic structure of the species taking part.

Background knowledge

The influence of temperature and the concentration of reactants on the rate of a reaction may be appreciated already by pupils in simple terms. The reactions used in the Topic should be familiar to the pupils – with the possible exception of the precipitation of sulphur from thiosulphate on the addition of hydrogen ions.

Subsequent development

A knowledge of rates of reaction is particularly important in the industrial manufacture of chemicals. References to rates of reaction are made later in the course, for example, the synthesis of ammonia (A22.4).

Further references

for the teacher

See *Collected experiments*, Chapter 14, for additional experiments on this Topic. See *Handbook for teachers*, Chapter 21, for background information on the rates of reaction.

Supplementary material

Film loop
2–13 'Catalysis in industry'

Film
'Catalysis' ICI Film Library

Overhead projection original
24 A chemical reaction considered as a collision process: two possible model processes *(figure A18.5)*

Reading material

for the pupil

Handbook for pupils, Chapter 10 'The chemical industry'.

A18.1
Measuring the rate of a reaction: how does the concentration of a reactant affect the rate of a reaction?

Reaction rates are introduced by reference to reactions already studied and to everyday phenomena. The problem of how to measure the rate of the reaction is then discussed.

A suggested approach

The subject may be introduced by discussing with pupils the rates at which familiar chemical changes take place. Most of these changes, such as burning, are rapid and take only a few seconds to complete. A few, like the rusting of iron or the ripening of an apple, take several days. Some reactions seem to go extremely slowly until the

---

1. Awareness that reactions
proceed at different rates
2. Recognition of the influence
of the concentration of a
reactant on a rate of a reaction
3. Ability to treat experimental
results graphically

temperature is raised: temperature is therefore a factor to be considered. Other reactions require concentrated rather than dilute reagents, and so the concentration of reactants is another factor to be investigated. One or two simple demonstrations may be helpful in identifying these points with a class.

The effects of these factors on the rate of the reaction are most easily studied by choosing situations in which just *one* factor is varied at a time and the rest are left unchanged. In practice this is not always easy to achieve since factors may interact. For example, if a great deal of heat is liberated during the course of a reaction, the temperature of the reactants may be raised and the reaction thereby speeded up. Thus, it may be difficult to estimate the effect of temperature change on the rate of this type of reaction. Most practical problems of this kind can be overcome by a judicious choice of experimental conditions. For example, the rate of reaction of marble chips with hydrochloric acid (Experiments A18.1a and A18.3) is almost temperature-independent and so is an appropriate system to show the effect of particle size on the rate of reaction as well as the effect of acid concentration.

The idea of the rate of a chemical reaction can be acquired at different levels of sophistication. Qualitatively, the effects are very simple and relatively easy to understand. Conditions can be adjusted so that reactions go faster or slower. Pupils can investigate whether a concentrated acid reacts more rapidly with certain substances than a dilute acid or vice versa. Does a hot solution of sodium thiosulphate react with acid more or less rapidly than a cold solution? For many pupils it is best to keep the discussion at a qualitative level. With more able pupils, it will be possible to pursue some of the quantitative aspects of the experiments.

Before any investigations can be made, some techniques for measuring rates of reaction need to be devised. Possible procedures should be discussed with the class, such as colour changes, changes in mass or in volume, and various means of measuring the concentration of a reactant.

The rate of a chemical reaction may be measured by:
1. Estimating the time it takes for a certain amount of reaction to occur.
2. Finding the gradient of a line on a graph showing the change of some quantity, such as mass, volume, etc., with respect to time.

Both methods are useful. However, the teacher should bear in mind that in the first case the rate inferred is always an *average* rate. In the second, although an instantaneous rate can be estimated, it is the rate at a certain point or stage in the course of a reaction. These matters need to be discussed in connection with the individual experiments.

As a first example, the reaction between· calcium carbonate and hydrochloric acid is recommended. The rate can be estimated by measuring losses in mass of the reactants and a typical rate curve obtained. If an excess of marble is used, the change in rate with time must be due to changes in acid concentration.

Experiment A18.1a

The reaction between marble chips and hydrochloric acid
The experiment is most conveniently carried out by the teacher.*

Pupils first encountered the reaction between a carbonate and hydrochloric acid in A17.3 and this experiment extends their knowledge by using calcium carbonate.

$$CaCO_3(s) + 2H^+(aq) \longrightarrow Ca^{2+}(aq) + H_2O(l) + CO_2(g)$$

The course of the reaction is followed by observing the change in mass as carbon dioxide is given off. A direct reading balance is preferable and the experiment will therefore probably have to be demonstrated by the teacher. The pupils may help with weighing, timing, and recording results.

Apparatus

The teacher will need:

3 conical flasks, 100 cm³

Stop-clock

Measuring cylinder, 100 cm³

Direct reading balance reading to 0.1 g

Watch glass

Cotton wool

Marble chips, medium size†

2M hydrochloric acid

The pupils will need:

Access to graph paper

Procedure
Put 40 cm³ of 2.0M hydrochloric acid in a 100 cm³ conical flask and plug the neck of the flask loosely with cotton wool. Place 20 g (an excess) of marble chips on a watch-glass, and place it and the flask on the balance pan. Note the total mass. Add the marble chips to the acid, replace the cotton wool, and start the stop-clock. Note the time taken to lose successive masses of 0.10 g. (Alternatively note the mass of the reaction vessel and its contents every half minute.) Continue the experiment for at least five minutes.

The pupils should tabulate the results in the following way.

Time/minutes	Mass of carbon dioxide given off /g (loss in mass /g)

Some teachers may wish to record the total mass of apparatus and its contents on each occasion. In this case, three columns will be needed for the results.

Ask the pupils to plot the readings obtained on a graph. Curves similar to those shown in figure A18.1 will be obtained.

The curve obtained for the reaction using 2.0M hydrochloric acid is appropriate for discussion. Can the pupils see – by inspection – that the rate of reaction falls as the reaction proceeds? If not, get them to calculate the loss in mass (amount of reaction) during the first minute, second minute, etc. Specimen results are shown in figure A18.2.

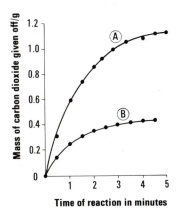

Figure A18.1
Reaction of marble chips (20 g of 10 mm diameter) and dilute hydrochloric acid: (*a*) using 40 cm³ 2M acid; (*b*) using 40 cm³ 1M acid.

*Teachers who prefer to do this as a class experiment involving the measurement of gas evolution with gas syringes will find details in Crawley, D. E. L. (1969) 'Syringes in the Nuffield Chemistry Scheme'. *School Science Review*, **51**, No. 175, 369.

†Marble chips should be washed beforehand in dilute hydrochloric acid and then in water to remove the surface powder.

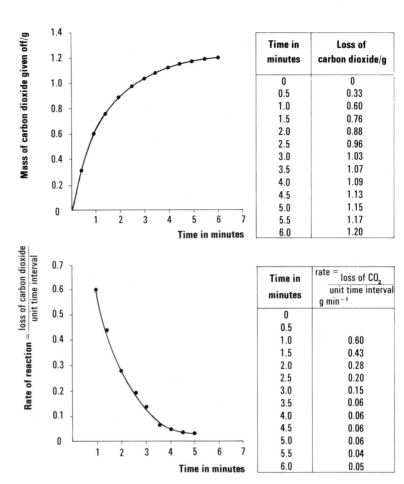

Time in minutes	Loss of carbon dioxide/g
0	0
0.5	0.33
1.0	0.60
1.5	0.76
2.0	0.88
2.5	0.96
3.0	1.03
3.5	1.07
4.0	1.09
4.5	1.13
5.0	1.15
5.5	1.17
6.0	1.20

Time in minutes	rate = $\dfrac{\text{loss of } CO_2}{\text{unit time interval}}$ g min^{-1}
0	
0.5	
1.0	0.60
1.5	0.43
2.0	0.28
2.5	0.20
3.0	0.15
3.5	0.06
4.0	0.06
4.5	0.06
5.0	0.06
5.5	0.04
6.0	0.05

Figure A18.2
Displaying the changing rate of reaction of marble chips and hydrochloric acid.

Ask the pupils why there was no further change in mass after, say, the seventh minute. Was there any marble left? Was there any acid left? They can then deduce that the factor most likely to have affected the rate of reaction, causing it to fall from about 0.6 g of carbon dioxide generated per minute to zero grams of carbon dioxide generated per minute, is the concentration of the acid.

The pupils should now make some measurements for themselves, using sodium thiosulphate and hydrochloric acid. First, demonstrate the reaction using a few cm^3 of a stock solution of sodium thiosulphate and 1 or 2 cm^3 of concentrated hydrochloric acid. Sulphur is precipitated in a recognizable form. Repeat the demonstration using a dilute solution of acid and show that after the reactants are mixed there is a time interval before the precipitate appears. The time required for a fixed amount of sulphur to appear is taken as a measure of the rate of reaction. Pupils should realize (from Experiment A18.1a) that this is an *average* rate during the period of reaction. They can find out how the rate of reaction varies with the initial concentration of the sodium thiosulphate solution, and (in section A18.2) with the temperature. Details of the experiment are given below.

What is the effect of the concentration of the reactants on the rate of reaction between sodium thiosulphate and hydrochloric acid in solution?

Apparatus

Each pair of pupils will need:

Experiment sheet 73

Conical flask, 100 cm³ (labelled 'Hydrochloric acid')

Measuring cylinder, 100 cm³

Measuring cylinder, 10 cm³

Beaker, 100 cm³

Beaker, 250 cm³ (labelled 'Sodium thiosulphate')

Stirring rod

Stop-clock or watch with a seconds hand

Sheet of white paper

Graph paper

2.0M hydrochloric acid, about 40 cm³

Sodium thiosulphate solution, containing 37 g per dm³ (about 0.15M), about 200 cm³

Procedure

The reaction between sodium thiosulphate solution and hydrochloric acid may be represented by the equation:

$$S_2O_3^{2-}(aq) + 2H^+(aq) \longrightarrow H_2O(l) + SO_2(g) + S(s)$$

Details of the investigation are given in *Experiment sheet* 73. The same cross mark must be used throughout the experiment, and pupils need to be warned about not getting the paper wet.

The pupils are asked to use their results to plot graphs of:
1. Concentration of sodium thiosulphate solution against time. The concentration may be measured as the volume of the original solution taken or as a fraction of the original concentration.
2. Concentration of sodium thiosulphate solution against 1/time (since the reciprocal of time is a measure of the average rate of reaction).

Experiment sheet 73

When hydrochloric acid is added to sodium thiosulphate solution, a precipitate of sulphur forms slowly. If we record the time needed for a certain amount of precipitate to form, we can use this as a measure of the rate of the reaction. A simple method of estimating when a fixed amount of sulphur has been formed is to carry out the reaction in a beaker standing on a piece of white paper on which a cross has been marked. On looking down on to the mixture in the beaker the cross will gradually become fainter as the precipitate forms and will no longer be seen when a certain amount of sulphur is present. For this method to be accurate the same depth of liquid must be used in each case. This can be done by having the liquid mixture always of the same volume and using the same beaker and marked piece of paper throughout.

You will be provided with 2.0M hydrochloric acid and 0.15M sodium thiosulphate solution.

Mark a large cross in pencil on a piece of white paper and place it on the bench. Pour 50 cm³ of the sodium thiosulphate solution into a clean dry 100 cm³ beaker and stand it over the cross on the paper. Now add 5 cm³ of the hydrochloric acid to the beaker, stirring the mixture while this is being done, and start a stop-clock (or note the time on a clock with a seconds hand). Look downwards through the solution in the beaker at the cross and note the time when the cross disappears. Enter this in the table below.

Clean and dry the beaker and repeat the above procedure using 40 cm³ sodium thiosulphate solution mixed with 10 cm³ water and add 5 cm³ acid. Again note the time for the cross to disappear.

Repeat the experiment using the other mixtures shown in the table.

Vol. sodium thiosulphate soln/cm^3	Vol. water /cm^3	Vol. hydrochloric acid/cm^3	Time(t) /sec	Concentration of sodium thiosulphate	$1/t$ /sec^{-1}
50	0	5		0.15M	
40	10	5		0.12M	
30	20	5			
20	30	5			
10	40	5			

Plot graphs of:
1. Concentration of sodium thiosulphate (vertical axis) against time (horizontal axis).
2. Concentration of sodium thiosulphate (vertical axis) against $1/t$ (horizontal axis).

What do you learn from these graphs?

The results of the experiment will require interpretation.

For example:
1. How 'far' has the reaction gone?
The time recorded in these experiments is the time it takes for enough sulphur to be produced to obscure the mark on the paper under the beaker. Pupils should realize that, whatever the conditions, this point represents some 'fixed amount of reaction'.
2. How does the rate depend on the concentration of a reactant?
The time taken to obscure the mark gets less as the concentration of thiosulphate used increases. Thus the average rate for that part of the reaction increases with concentration. If the reciprocal of the time to blot out (a measure of the average rate) is plotted against concentration, pupils can see that the mean rate is approximately proportional to the concentration of the thiosulphate.

Discuss with pupils why an increase in the concentration of the thiosulphate should increase the rate of the reaction. A simple picture of an increase in the chance of collision between particles will suffice.

For less able pupils, graphical presentation of these findings may prove unsatisfactory and the approach should be modified. Use 250 cm^3 beakers with 80 cm^3, 40 cm^3, 20 cm^3, and 10 cm^3 of 0.15M sodium thiosulphate (made up to 80 cm^3 with water where appropriate) and 10 cm^3 of 2.0M hydrochloric acid. Halving the concentration of sodium thiosulphate will then be seen to double the time of reaction and so point to the *rate* being halved.

Teachers are reminded *not* to vary the concentration of the acid as a possible additional study: the kinetics of this system are not straightforward!

| Summary | By the end of this section pupils should be aware that the rate of a reaction may be increased by increasing the concentrations of the reactants. |

A18.2
How do changes in temperature affect the rate of a reaction?

The chemical system used in section A18.1b is used again in order to study the effect of temperature on the rate of a reaction.

The effect of change in temperature on chemical reactions is an everyday experience. We put a cake in the oven in order to speed up *desirable* chemical changes and we put milk in the refrigerator in order to slow down *undesirable* chemical changes.

The effect of an increase in temperature on the rates of several chemical reactions was referred to in the previous section and it may be appropriate to revise this by one or two simple demonstrations. The reaction between sodium thiosulphate and hydrochloric acid is easily studied by the pupils themselves.

A suggested approach

Objectives for pupils

1. Measurement of the effect of change in temperature on the rate of a reaction
2. Simple interpretation of the effect of temperature in terms of the kinetic-molecular theory

What is the effect of temperature on the rate of reaction?

Procedure
Practical details are given in *Experiment sheet* 74. It may be appropriate to discuss 'ways and means' before practical work is started.

Some pupils may require assistance with the two graphs mentioned in the *Experiment sheet* – through a confusion of temperature, and times.

Experiment A18.2

Apparatus

Each pair of pupils will need:

Experiment sheet 74

Conical flask, 100 cm^3

Measuring cylinder, 100 cm^3

Measuring cylinder, 25 cm^3

2 beakers, 100 cm^3

Stop-clock or view of laboratory clock with a seconds hand

Thermometer, -10 to $+110 \times 1°C$

Bunsen burner and heat resistant mat

Tripod and gauze

Sheet of white paper

Graph paper

2M hydrochloric acid, about 40 cm^3

Sodium thiosulphate solution containing 37 g per dm^3 (about 0.15M), about 60 cm^3

Experiment sheet 74
The method for this experiment is the same as that for Experiment 73, except that in this case the concentrations of the reacting solutions are kept constant and the temperature of the mixture is varied.

Mix 10 cm^3 sodium thiosulphate solution with 40 cm^3 water in a 100 cm^3 beaker, and stand the beaker over the cross on the sheet of white paper. Add 5 cm^3 hydrochloric acid, with stirring, start the stop-clock, and read the temperature of the mixture. Note the time for the cross to disappear.

Repeat the above procedure but this time heat the diluted thiosulphate solution to just above 30 °C before adding hydrochloric acid. Note temperature and time as before. Then find the times required for temperatures of about 40 °C, 50 °C, and 60 °C.

Results
In each case, use a reaction mixture of 10 cm^3 sodium thiosulphate solution, 40 cm^3 water, and 5 cm^3 hydrochloric acid.

Temperature of mixture /°C	Time (t) /sec	$1/t$ /sec^{-1}

Plot graphs of:
1. Temperature (vertical axis) against time (horizontal axis).
2. Temperature (vertical axis) against $1/t$ (horizontal axis).

What do you learn from these graphs?

Sample results are shown in figures A18.3 and A18.4.

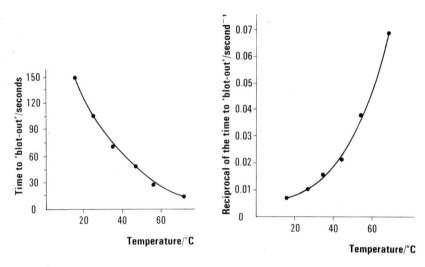

Figure A18.3   Figure A18.4

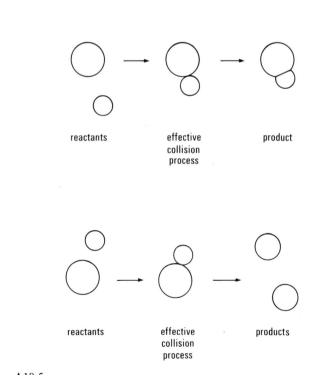

Figure A18.5
A chemical reaction considered as a collision process: two possible model
processes. (OHP 24)

After the experiment, discuss why a reaction might be expected to go faster at higher temperatures. A simple interpretation can be given in terms of the increase in the number of particle collisions per second. Pupils should also be aware that at higher temperatures the energy needed to break bonds (and to make other bonds) is more readily available. Figure A18.5 provides two simple 'models' of chemical change which may form convenient aids for a discussion.

Suggestion for homework

Make a list of as many everyday chemical changes as you can which take place in a measurable time. Have you any evidence that they are slowed down when cooled and speeded up when heated?

A18.3
The effect of particle size on the rate of a reaction

The effect of the size of marble chips (calcium carbonate) on the rate of reaction between calcium carbonate and hydrochloric acid is investigated.

A suggested approach

Objectives for pupils

1. Measurement of the effect of the particle size of a solid reagent on the rate of reaction
2. Appreciation of heterogeneous chemical systems

Experiment A18.3

Apparatus

The teacher will need:

3 conical flasks, 100 cm³

Stop-clock

Marble chips – graded into 2 or 3 sizes, preferably using coarse sieves
(largest size – chips up to about 10 mm maximum dimension; next size – chips up to about 5 mm maximum dimension; smallest size – chips up to about 2 mm maximum dimension.

Note. No powder should be used and any powder present *must* be removed by sieving.)

Experiment A18.3 shows the effect of the particle size of a solid reagent on the rate of reaction. The system used is familiar to pupils (see section A18.1). It may be demonstrated by the teacher assisted by two pupils, other members of the class being required to record the readings, to plot graphs, and to deduce the average rates of reaction during the first minute of reaction. Since the particle size will decrease as the reaction proceeds, the initial rates are taken as an indication of the effect of particle size on the rate of reaction.

What is the effect of particle size on the rate of a reaction?

Procedure
Carry out the experiment as described in Experiment A18.1a. Ask the pupils to plot the loss in mass every minute. They should then compare the loss in mass during the first minute for the three different sizes of marble chips. The differences indicate that the amount of carbon dioxide evolved per minute increases with decreasing particle size – that is, with increasing surface area available for reaction. This is all the experiment is intended to show. However, it may be appropriate to make an estimate of the surface area of the chips in each case and to see whether this is proportional to the amount of reaction during the first minute. Sample results are given in figure A18.6.

(Continued)

Measuring cylinder, 100 cm³

Direct reading balance, reading to 0.1 g

Watch-glass

Cotton wool

2.0M hydrochloric acid

The pupils will need:

Graph paper

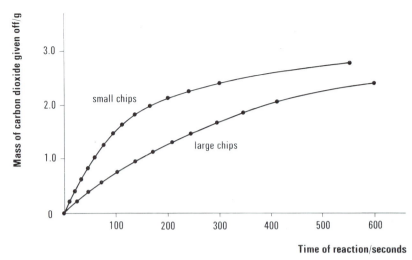

Figure A18.6
Effect of particle size on the rate of reaction of marble chips and dilute hydrochloric acid.

A18.4
How do catalysts affect the rates of reactions?

In this section catalysis is introduced through simple qualitative demonstrations and then pupils investigate the decomposition of hydrogen peroxide using various catalysts.

Objectives for pupils

1. Knowledge of catalysed reactions and the effect of catalysts on rates of reactions
2. Awareness of the industrial importance of catalysts

The teacher will need

Film-loop projector
Film loop 2–13 'Catalysis in industry' (with notes)

The idea that a substance can affect the rate of a chemical change without itself being permanently changed will probably be new to most pupils.

The use of poor quality manganese(IV) oxide in the preparation of oxygen from the thermal decomposition of potassium chlorate is best avoided as a class experiment since it has been known to give rise to an explosion *either* through the presence of impurities in the sample of manganese(IV) oxide *or* through confusing carbon and manganese(IV) oxide. However, many other oxides such as copper(II) oxide, zinc oxide, and even silica, catalyse the reaction and provide a clear demonstration of this phenomenon. Thus, pupils should know that copper(II) oxide will not give off oxygen on being heated yet when mixed with potassium chlorate, oxygen is given off at a strikingly lower temperature than when potassium chlorate is heated alone. Other characteristics of a catalyst may also be illustrated. Details of the experiment are given below.

Experiment A18.4a

A study of the effect of metal oxides on the thermal decomposition of potassium chlorate
This experiment should be done by the teacher with or without assistance by pupils.

Apparatus

The teacher will need:

Safety spectacles – for all participants

8 hard-glass test-tubes, 125 × 16 mm

Test-tube rack

Stand, 2 bosses, and 2 clamps

Splints

Bunsen burner and heat resistant mat

Filter flask, bung, filter funnel, filter paper, and filter pump

Tripod and gauze

Evaporating basin

Distilled water

Potassium chlorate

Copper(II) oxide (dry and stored in desiccator)

Iron(III) oxide (dry and stored in a desiccator)

Manganese(IV) oxide (suitable for the preparation of oxygen)

Apparatus

The teacher will need:

Safety spectacles

Plastic safety screen

2 gas jars and covers

Tongs

Bunsen burner and heat resistant mat

Cylinder of hydrogen, connecting tubing, and a delivery tube

Platinized asbestos

Procedure

Take about 1 cm depth of potassium chlorate in two test-tubes. To one tube add a little *dry* copper(II) oxide and mix well. Clamp the test-tubes side by side and warm them both equally with a non-roaring Bunsen flame. Hold a glowing splint at the mouth of the tubes. The splint held at the mouth of the tube containing the catalyst will ignite some time *before* the other splint.

If time allows, repeat this part of the experiment using a different metal oxide (iron(III) oxide or manganese(IV) oxide of suitable quality).

In this way, the catalytic effect of metal oxides on the decomposition of potassium chlorate can be shown. We need to identify the characteristics of catalysts. How could this be done? Demonstrate the solubility of potassium chlorate and the products of its decomposition in distilled water. Also, show the insolubility of the metal oxide used. Apply this knowledge to recover the metal oxide from the experiment. How could we find out if a loss (or gain) of metal oxide took place in the experiment? Use the discussion to lead the pupils towards a formal definition of the term catalyst.

The following points need to be covered: the fact that catalysts are not changed chemically by a reaction, although their physical state may change, should be mentioned; and the fact that catalysts can change the rate of a reaction. It is not intended that theories to explain their action should be discussed.

Another striking demonstration is the effect of a platinum catalyst on the combination of hydrogen and oxygen. This is described in Experiment A18.4b.

The reaction between hydrogen and oxygen in the presence of a platinum catalyst

This experiment **must** be done by the teacher.

Procedure

The teacher should wear safety spectacles and the safety screen should be placed in an appropriate position in front of the class.

Before the lesson, heat a tuft of platinized asbestos in a Bunsen flame for a few seconds to dry it and then return it to the bottle.

Fill a gas jar with hydrogen. Then, using a pair of tongs, hold the edge of the tuft of platinized asbestos over the mouth of the gas jar. As the hydrogen rises out of the jar, it mixes with the air and the platinized asbestos glows red. This ignites the residual hydrogen and produces the usual gentle explosion. Repeat the experiment without using the platinized asbestos, thereby adding emphasis to the definition of the term catalyst!

Pupils now do their own experiment on another reaction which can be speeded up by a catalyst. The decomposition of hydrogen peroxide in the presence of manganese(IV) oxide is a good example and is described below. It is helpful to allocate particular aspects of the investigation to different pupils before the practical work is started. Pupils can then discuss their problem among themselves and obtain advice as necessary. Problems which could be studied include:

1. Is manganese(IV) oxide speeding up the decomposition of hydrogen peroxide?
2. Is manganese(IV) oxide actually being used up in the reaction?

If time allows, the following supplementary investigations could also be attempted:

3. The effect of other oxides, such as copper(II) oxide, or magnesium oxide, on the rate of this reaction.
4. The effect of the particle size of the catalyst on the rate of reaction.

There is far more work here than can be done by any one pair of pupils during a double period. Nevertheless, if each pair finds out something, the results can be pooled and their significance discussed with *all* members of the class.

Experiment A18.4c

Investigation of the effects of a catalyst on the rate of decomposition of hydrogen peroxide

Apparatus

Each pair of pupils will need:

Conical flask, $100 \, cm^3$ – wide necked and fitted with a bung and delivery tube

Measuring cylinder, $25 \, cm^3$

Stop-clock

Scrap paper

Thermometer, -10 to $+110 \times 1°C$

Spatula

Bunsen burner and heat resistant mat

Tripod and gauze

Teat pipette, with $1 \, cm^3$ graduations

Either (see figure A18.7a) syringe, $100 \, cm^3$ (glass) or $50cm^3$ (plastic)

Syringe holder and stand

Or (see figure A18.7b) burette, $50 \, cm^3$

Stand, boss, and clamp

Trough

Delivery tube

(Continued)

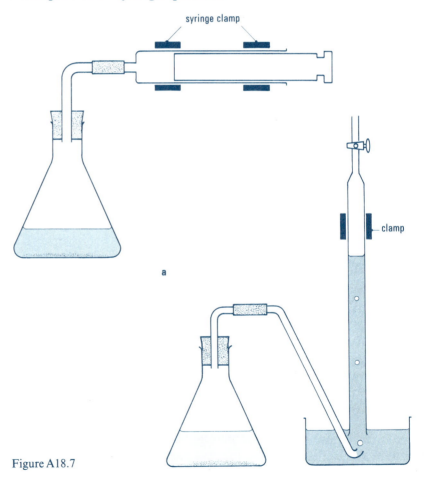

Figure A18.7

20-volume hydrogen peroxide (quantities depend on the range of studies attempted)

Manganese(IV) oxide (quantities depend on range of studies attempted)

Procedure

Tell pupils to add $2\,cm^3$ hydrogen peroxide solution to $48\,cm^3$ water in the flask. Add about half a spatula measure of manganese(IV) oxide and immediately replace the bung. Start the stop-clock and collect the oxygen given off. Record the volume collected at regular intervals so that a graph may be drawn of volume of gas collected against time of reaction.

The slope of this graph at any chosen point is, of course, a measure of the rate of reaction at that point. Attention should be focused on the initial rate of reaction – that is, the slope of the graph at time $t = O$.

Pupils can then investigate the effect on the initial rate of reaction of changing the quantity of manganese(IV) oxide used. After each investigation, a graph showing the volume of gas given off with time should be plotted, as before. Other problems for investigation were suggested before this experiment. Other investigations could include the effect of biological catalysts (enzymes) in the blood and in potatoes on the decomposition of hydrogen peroxide.

After this experiment, some time must be spent in bringing the results together. This is an opportunity to reinforce the principles outlined in the Topic. The film loop 2–3 'Catalysis in industry' provides more information and includes aspects of industrial chemistry.

Suggestions for homework

1. Make a summary of the results of your experiments in this section. Describe clearly the effects of temperature, concentration of reactants, and catalysts, on the rates of the reactions you have studied.
2. Read the *Handbook for pupils*, Chapter 5 'Studying chemical reactions' and Chapter 10 'The chemical industry'.

Summary

By the end of this Topic, pupils should know how to measure the rate of a reaction and should understand the likely effects of concentration of reactants, temperature of the system, and catalysts, on the rate of a reaction.

How far? The idea of dynamic equilibrium

Purposes of the Topic

1. To review evidence for reversible reactions.
2. To provide a basis for the idea of dynamic equilibrium.
3. To demonstrate the effect of concentration change on the position of equilibrium.
4. To review evidence for the dynamic nature of equilibrium.

Contents

A19.1 Looking at reactions which can go both ways.
A19.2 Looking more closely at reversible reactions: the theory of dynamic equilibrium.
A19.3 Evidence for the dynamic nature of equilibria.

Timing

Section A19.1 needs one double period. Section A19.2 needs at least one double and one or two single periods. The experiment in the last section may be demonstrated in a double period. Three weeks should be allowed for all of the Topic.

Introduction to the Topic

In the first section, pupils are reminded that some reactions can go both ways.

Reversible reactions are then investigated more closely. It is found that many reactions do not proceed to completion, and that the point to which they do proceed depends on the concentrations of the reactants and the products.

Discussion of chemical systems in equilibrium leads to a theory of dynamic equilibrium. Some further reactions are then investigated and, in each case, the results are interpreted in the light of the theory.

In the last section the evidence which supports the theory of dynamic equilibrium is discussed and an experiment is carried out to provide direct evidence for the theory.

Background knowledge

Examples of reversible reactions which pupils have met during Stage I are revised briefly at the beginning of the Topic.

Alternative approach

In Alternative IIB, the content of this Topic is divided between Topics B20 and B21. Section B20.5 is concerned with reversible reactions, and the role of the solvent in reactions using solutions of solutes is investigated in section B20.6. Chemical equilibria are considered in section B21.2. The possibility of using radioactive tracer techniques to follow the establishment of equilibrium in a system is discussed in Topic B21 but the experimental details are deferred until Topic B22.

Subsequent development

The ideas developed in this Topic are applied to the study of acids and bases in Topic A20 and to the study of the nitrogen–hydrogen–ammonia equilibrium in Topic A22.

The theory of dynamic equilibrium is used in the study of acid–base and redox reactions in Stage III Option 6. Examples of the relevance of this theory to industrial processes also occur in Stage III Option 8.

Further references
for the teacher

Additional experiments on this Topic occur in *Collected experiments*, Chapter 15. Background information on this Topic will be found in Chapter 22 of the *Handbook for teachers*.
Nuffield Advanced Chemistry (1970) *Teachers' guide I* (Topic 12) and *Teachers' guide II* (Topic 15), and Nuffield Advanced Physical Science (1972) *Teachers' guide I* (Sections 4 and 6) provide advanced treatments of the concepts used.

Supplementary materials

Film loops
2–5 'Liquid–gas equilibrium'
2–6 'Solid/liquid equilibrium'
2–19 'Dynamic equilibrium'

Films
'Chemical equilibrium', ESSO. Films for Science Teachers. No. 23
'Equilibrium' CHEMstudy. Distributed by Guild Sound and Vision Ltd.

Overhead projection original
25 Traffic-flow model *(figure A19.1)*

Reading material
for the pupil

Handbook for pupils, Chapter 5 'Studying chemical reactions'.

A19.1
Looking at reactions which can go both ways

In this section some examples of reversible reactions which pupils have already met are discussed and the idea of an equilibrium between reactants and products is introduced.

A suggested approach

Objectives for pupils

1. Extending pupils' knowledge of the nature of chemical change
2. Familiarity with the idea that certain reactions can be reversed
3. Ability to explain observations

A survey of chemical reactions and of their common characteristics reveals that several chemical systems are able to go both ways. Pupils can be shown again the effect of heat on copper sulphate crystals and the addition of water to anhydrous copper sulphate. The decomposition of mercury(II) oxide by heat and the synthesis of mercury(II) oxide by heating mercury in air provides a second system which they have met before. They should also be able to discuss the synthesis and decomposition by heat of calcium carbonate and calcium hydroxide.

In each instance, pupils should be able to write chemical equations for the changes described. Thus, the reactions between hydrated copper(II) sulphate, anhydrous copper sulphate, and water may be summarized as follows:

$$CuSO_4 5H_2O(s) \underset{\text{add water to anhydrous salt}}{\overset{\text{heat crystals}}{\rightleftarrows}} CuSO_4(s) + 5H_2O(g)$$

The equilibrium sign ($\rightleftharpoons$) should be avoided at this stage, since it is used to describe a reaction in equilibrium. The concept of chemical equilibria is developed in the next section, A19.2.

The following experiment may be used to provide plenty of material for discussion.

A study of the reaction between iodine and chlorine
This experiment should be done by the teacher.

Procedure

The experiment should be performed in a fume cupboard. Place a few small crystals of iodine in the bottom of the U-tube and then pass chlorine through the tube. Tell pupils to watch carefully. Ask one or two pupils to touch the U-tube.

The pupils will be able to see the iodine change into a brown liquid. They will also find that the bottom of U-tube gets hot. On passing more chlorine through the tube yellow crystals appear on the walls of the U-tube.

The question now arises, 'What are the brown liquid and the yellow solid?' The brown liquid is only produced when chlorine meets the iodine. It must therefore be a compound of chlorine and iodine. This may surprise the pupils at first since they have not been conditioned to expect compounds to be formed between elements within the same group. However, they should soon realize that if chlorine can combine with itself to form Cl_2, then it may also combine with an atom of iodine to form ICl, iodine monochloride. From their knowledge of the Periodic Table, this type of behaviour should not be totally unexpected. The matter can be pursued a little further with the class, by discussing expected changes in properties on going down a group within the Periodic Table. 'What can we say about the yellow solid?' Being a solid, we would expect it to have a higher molecular mass than the volatile liquid which was formed first. The pupils should appreciate that the yellow solid is a compound of chlorine and iodine monochloride. It is, in fact, iodine trichloride.

When the chlorine supply is detached and the U-tube removed from the fume-cupboard, tip the U-tube slowly until it is almost upside down. Then turn it upright again. The yellow crystals disappear and the brown liquid with its associated vapour are seen again.

Now pass more chlorine through the U-tube until the crystals reappear. The experiment may be repeated several times. A large sheet of white card held behind the U-tube will enable the pupils to see the change clearly.

This effect is unexpected! After a little while, someone will see that the comparatively dense chlorine gas is being tipped out of the tube during the pouring process. Pupils will also see that the yellow solid is produced again when chlorine is passed over the brown liquid. Accordingly, pupils may be able to deduce:

chlorine + brown liquid/vapour $\longrightarrow$ yellow solid

and   yellow solid $\longrightarrow$ chlorine + brown liquid/vapour

The teacher may then sum up the situation:

$$Cl_2 + ICl \underset{\text{remove Cl}_2 \text{ from ICl}_3}{\overset{\text{pass Cl}_2 \text{ over ICl}}{\rightleftarrows}} ICl_3$$

It is suggested that the *reversibility* of the reaction be stressed and that all reference to the concept of chemical equilibrium be deferred.

Suggestion for homework

What experiments could you devise to confirm the formulae of the brown liquid (iodine monochloride) and the yellow solid (iodine trichloride) in the reaction between iodine and chlorine? Outline some possibilities. If possible, give equations for the reactions you describe.

A19.2
Looking more closely at reversible reactions: the theory of dynamic equilibrium

In this section, the reversibility of reactions is used to lead to the idea of an equilibrium condition. A theoretical interpretation of this situation leads to a theory of dynamic equilibrium which may then be expressed in terms of a simple collision mechanism.

A suggested approach

Objectives for pupils

1. To show that certain reactions do not proceed to completion, but reach a 'point of balance'
2. To show that the point to which such reactions proceed depends on the concentrations of the reactants and products
3. To provide a theory of chemical equilibrium and to show that it gives a satisfying explanation of the systems investigated.

The teacher will need

Film loop projector
2–5 'Liquid–gas equilibrium'

2–6 'Solid/liquid equilibrium'

2–19 'Dynamic equilibrium'

In the previous section, pupils studied several chemical reactions which could be classed as 'reversible' reactions. They found that the addition of an excess of one reagent could move the reaction either backwards or forwards. In this section, we introduce the idea of dynamic equilibrium by means of an experiment using iodine, chloroform (trichloromethane), and potassium iodide solution.

The distribution of iodine between two immiscible solvents

Apparatus

The teacher or each pair of pupils will need:

Experiment sheet 75

2 test-tubes, 100 × 16 mm, with corks

Teat pipette

Spatula

Chloroform (trichloromethane), 20 cm³

1.0M potassium iodide solution, 20 cm³

Iodine crystals

Procedure

The principle of dynamic equilibrium is illustrated in this experiment by the distribution of iodine between two immiscible solvents, potassium iodide solution and chloroform (trichloromethane). There is no need to explain that the iodine in the potassium iodide is largely represented by I^{3-}(aq) ions. Iodine may be considered to be dissolved in potassium iodide solution.

Details of the procedure appear in *Experiment sheet* 75. The experiment could be demonstrated by the teacher or used as a class experiment.

Experiment sheet 75

1. Put potassium iodide solution into a test-tube to a depth of about 2 cm, add a *small* crystal of iodine, and shake the tube until the iodine has dissolved.

What is the colour of the solution?

Now add about an equal volume of chloroform (trichloromethane) and shake the test-tube gently. What do you see?

2. Put chloroform (trichloromethane) into a test-tube to a depth of about 2 cm. Add a small crystal of iodine as nearly the same size as that used in (1) as you can manage. Shake the tube until the iodine has dissolved.

What is the colour of the solution?

Add about an equal volume of potassium iodide solution and shake the test-tube gently. What happens?

3. Cork the test-tubes used in (1) and (2) and shake both vigorously for one minute. Allow them to stand until the two layers of liquid in each have separated. Compare their appearances.

4. What do you think will happen if you remove the potassium iodide layer from one test-tube, replace it by a fresh solution, and shake the tube vigorously again?

Now try it, using a teat pipette to remove the potassium iodide layer. Were you right?

Account as fully as you can for what you have seen during this experiment. (Class discussion should help here.)

The significance of this experiment should be discussed and an explanation given in terms of dynamic equilibrium. (A static picture of equilibrium – as given by a see-saw analogy – is unsatisfactory.) The experiment can only be explained in terms of a dynamic equilibrium in which the two processes are moving in opposite directions at the same rate.

Other instances of dynamic equilibria should now be considered. The film loops 2–6 'Solid/liquid equilibrium' and 2–5 'Liquid–gas equilibrium' are examples of physical equilibria.

Chemical equilibria may also be investigated and the results interpreted in the light of the theory of dynamic equilibrium. Thus pupils could try out the effect of acid and alkali on:

1. Bromine water
2. Potassium chromate solution

Both of these investigations are qualitative and can be carried out using test-tubes. Each system can be seen to be reversible.

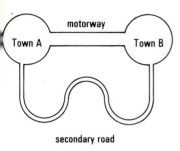

Figure A19.1
Traffic-flow model. (OHP 25)

Since both reactants and products are particulate, the situation may be likened to traffic flow between two towns joined by roads of different load-carrying capacities. The traffic flows more readily, and at greater speeds, on a motorway than via a secondary road (see figure A19.1). If the traffic flow from Town A to Town B is on the motorway and traffic flow from Town B to Town A is via the secondary road, then the situation would be similar to that which obtains when a reversible reaction takes place with a greater tendency under a given set of conditions to go from A to B rather than from B to A.

During these experiments, it is suggested that the teacher moves around the class, discussing ways in which the pupils can find a 'balanced position' for a system. How could they identify components other than by colour? Might there be traces of all components present in a given system even when the reaction is apparently completely over to one side?

Experiment A19.2b

Two examples of reversible reactions

Apparatus

Each pair of pupils will need:

Experiment sheet 76

White tile or sheet of white paper

Beaker, 100 cm³

2 teat pipettes

Stirring rod

Bromine water (10 g dm⁻³), 10 cm³

2M sulphuric acid

2M sodium hydroxide

0.1 M potassium chromate, 10 cm³

Distilled water

Procedure

In this experiment, pupils are asked to examine two different chemical systems. Full details appear in *Experiment sheet* 76.

> **Experiment sheet 76**
> In this experiment you are going to investigate two reactions that can be made to go 'either way'.
>
> **1.** Put 10 cm³ bromine solution into a 100 cm³ beaker and stand this on a white surface. Add sodium hydroxide solution, five drops at a time, from a teat pipette stirring between each addition, until no further change takes place. What do you observe?
>
> Now add dilute sulphuric acid similarly; what happens now?
>
> By adding alkali or acid, in smaller portions, try to establish a position between the two extremes of excess acid and excess alkali. What is the colour of the mixture now?
>
> **2.** Repeat the above procedure, using potassium chromate solution instead of bromine solution, but *adding the acid first.*
>
> Effect of acid:
>
> Effect of alkali:
>
> Colour of intermediate stage:
>
> Write equations for the two equilibria which you have investigated. (Your teacher will help you with these.)

When the pupils have completed their experiments, gather the class together to discuss their findings, and give the equations for the different reactions. For example:

$$Br_2(aq) + H_2O(l) \xrightarrow[\text{gain of products}]{\text{loss of reactants}} H^+(aq) + Br^-(aq) + HOBr(aq)$$

reactants   products

As this reaction proceeds, there is a loss of the reactants from the system and a simultaneous formation of products. This process may be termed the *forward reaction*. However, as soon as the products have formed, they begin to react:

$$Br_2(aq) + H_2O(l) \xleftarrow[\text{loss of products}]{\text{gain of reactants}} H^+(aq) + Br^-(aq) + HOBr(aq)$$

reactants   products

This process is called the *back reaction*. As the concentrations of the reactants decrease, so does the rate of the forward reaction. When the concentrations of the products increase, so does the rate of the back reaction. Eventually, we reach a state of dynamic equilibrium and the reaction *appears* to stop. In reality, both the forward and the back reactions continue.

$$Br_2(aq) + H_2O(l) \rightleftharpoons H^+(aq) + Br^-(aq) + HOBr(aq)$$

The symbol $\rightleftharpoons$ is used to show that a dynamic equilibrium exists. The theory of dynamic equilibrium convincingly explains the effect of concentration of reactants on the position of an equilibrium.

The second system can be explained in a similar way. The equilibrium here is:

$$2CrO_4^{2-}(aq) + 2H^+(aq) \rightleftharpoons Cr_2O_7^{2-}(aq) + H_2O(l)$$

Two more examples are now suggested for investigation by pupils and used to give emphasis to the widespread application of dynamic equilibria, before summing up the Topic using the film loop 2–19.

Experiment A19.2c

Two other examples of systems in chemical equilibrium

Apparatus

Each pair of pupils will need:

Experiment sheet 77

6 test-tubes, 100×16 mm

4 test-tubes, 150×25 mm

3 corks to fit 100×16 mm test-tubes

Test-tube rack

Glass rod

Procedure
Details are given in *Experiment sheet* 77.

Experiment sheet 77
Caution. Care is needed when handling the chemicals mentioned below.

1. *Hydrolysis of bismuth chloride*
Put a spatula measure of bismuth chloride in a test-tube and, using a teat pipette, add concentrated hydrochloric acid until the solid has *just* dissolved. You will need about 2 cm^3 of the concentrated acid.

a. Two-thirds fill a 150×25 mm test-tube with water and add a few drops of the bismuth chloride solution. What happens?

(Continued)

Spatula

3 teat pipettes

Access to a centrifuge

Bismuth(III) chloride

Concentrated hydrochloric acid, $2\,cm^3$

0.1M silver nitrate, $10\,cm^3$

0.1M iron(II) sulphate, $10\,cm^3$, freshly prepared

1.0M iron(III) nitrate, $10\,cm^3$

Freshly prepared potassium hexacyanoferrate(III) solution

0.1M potassium thiocyanate solution

b. To study this equilibrium more carefully, two-thirds fill three other 150×25 mm test-tubes with water. To the first add 5 drops concentrated hydrochloric acid, to the second 10 drops, and to the third 15 drops. Mix the contents of each tube thoroughly with a glass rod. Add 5 drops of bismuth chloride solution to each tube. Note any differences between the three precipitates and between the speed with which they are formed.

Tube 1 (5 drops conc. acid).

Tube 2 (10 drops conc. acid).

Tube 3 (15 drops conc. acid).

Write an equation for this equilibrium.

2. *Reaction of silver nitrate and iron(II) sulphate in solution*
To study the second reaction (between iron(II) ions, Fe^{2+}(aq), and silver ions, Ag^+(aq)), you will need to know how to test for iron(II) ions and iron(III) ions, Fe^{3+}(aq).
 Iron(II) ions, Fe^{2+}(aq), give a dark blue colour or precipitate with potassium hexacyanoferrate(III) solution.
 Iron(III) ions, Fe^{3+}(aq), give a deep red colour with potassium thiocyanate solution.

Put iron(II) sulphate solution into a 100×16 mm test-tube to a depth of 2 cm. Add an equal volume of silver nitrate solution. Cork the tube and shake well. What happens?

Transfer a few drops of the solution with a teat pipette to another test-tube and test for the presence of Fe^{3+}(aq) ions. What is the result?

Transfer a few drops to another test-tube and test for Fe^{2+}(aq) ions. What is the result?

Do you think a state of equilibrium has been reached? If so, write an equation to represent it.

Now try to reverse the reaction. Centrifuge the mixture, remove the upper liquid layer, wash the precipitate (add water, centrifuge, remove liquid), and add iron(III) nitrate solution to a depth of about 4 cm. Cork the tube and shake well for 2 minutes. Allow any precipitate to settle, remove a few drops of liquid, and test for Fe^{2+}(aq) ions. What is the result?

After the pupils have completed these investigations, their findings should be discussed.

The hydrolysis of bismuth chloride may be represented by:

$$BiCl_3(aq) + H_2O(l) \rightleftharpoons BiOCl(s) + 2H^+(aq) + 2Cl^-(aq)$$

The equation for the reaction between silver nitrate and iron(II) sulphate in solution may be written as:

$$Ag^+(aq) + Fe^{2+}(aq) \rightleftharpoons Ag(s) + Fe^{3+}(aq)$$

The film loop 2–19 'Dynamic equilibrium' reviews the reaction of iron(II) ions with silver ions and forms a satisfactory summary of the theory of dynamic equilibrium.

Suggestion for homework Make a list of all the reactions you have met which go both ways. Describe in each case what conditions are necessary to make the reaction go in each direction.

A19.3
Evidence for the dynamic nature of equilibria

In this section evidence for the dynamic nature of equilibrium processes is summarized and assessed. (Alternatively, the viewpoint put forward in B21.2 can be adopted and this section, in part, deferred until A24.3.)

A suggested approach

Objectives for pupils

1. Assessment of evidence which supports the idea of the dynamic nature of equilibria
2. Direct evidence for the dynamic nature of equilibria between lead chloride and its saturated solution

After reviewing the previous section, the teacher might ask what evidence there is to support a theory of dynamic equilibrium.

Pupils should now realize that the theory provides a satisfying explanation for the fact that the position of equilibrium may be altered by changing the concentration of one or more of the reacting species. It is this evidence which should be stressed here. Refer to 'physical systems' and point out that equilibrium between the liquid and the gaseous states is entirely consistent with the kinetic theory (see the film loop 'Liquid–gas equilibrium'). It is tempting to develop a kinetic theory for use with chemical systems (for example, consider a mental picture of bromine molecules colliding and reacting with water molecules for Experiment A19.1a) but this picture must be used with care since it can so easily give an incorrect impression of the nature of the interacting species.

One experiment which provides more direct evidence for the dynamic nature of equilibria uses a radioactive tracer to investigate the equilibrium between lead chloride and its saturated solution.

$$PbCl_2(s) \rightleftharpoons Pb^{2+}(aq) + 2Cl^-(aq)$$

We have to show that both the forward and the backward reactions take place when such a system is in equilibrium. The existence of radioactive isotopes enables us to 'label' some of the reacting particles. Lead chloride and its saturated solution are suitable because the lead can be 'labelled' with a radioisotope, lead-212. Some lead chloride is precipitated in the presence of a thorium salt which contains lead-212 (so that the precipitate of lead chloride is radioactive). The washed precipitate is then shaken with saturated lead chloride solution and centrifuged. The solution is shown to be radioactive, thereby proving that there is a two-way movement of lead ions between the solid and the saturated solution.

Note: reference should be made to the precautions to be taken when handling radioactive substances (see Alternative IIA, section A24.1).

Experiment A19.3

Investigating the equilibrium between solid lead chloride and its saturated solution using a radioactive tracer technique

This experiment **must** be done by the teacher.

Procedure
This is an opportunity to demonstrate the procedures adopted when using a radioactive isotope. The experiment is carried out on a

The teacher will need:

Beaker, 100 cm^3

Beaker, 500 cm^3

Stand, boss, and clamp

2 test-tubes, 100 × 16 mm

Teat pipette

Mechanical stirrer or shaker

Bunsen burner and heat resistant mat

Tripod and gauze

Scaler

GM liquid counter

Centrifuge and test-tubes, 100 × 16 mm

Tray lined with absorbent paper

Polythene gloves

Talcum powder

Thorium nitrate solution (5%)

Lead nitrate solution (5%)

2M hydrochloric acid

Acetone (propanone)

Distilled water

'disposable' surface (a prepared tray lined with absorbent paper) and disposable polythene gloves are used in a sequence of handling operations which avoid contaminating apparatus, etc. Pupils should be made aware of the need for caution when handling radioactive material. Given the conditions for this demonstration, a large beaker may be used in place of the more usual liquid waste disposal bottle.

The equilibrium between the solid lead chloride and its solution is followed using a radioactive tracer. First take a count of the background radiation. Then make a saturated solution of lead chloride by adding just enough 2M hydrochloric acid to about 5 cm^3 of 0.2M lead nitrate solution to precipitate most of the lead. Shake well, centrifuge the mixture, and separate the liquid from the solid using a teat pipette. Transfer 5 cm^3 of the solution to the GM tube and take a count over a period of a few minutes. It should give no more than the background count.

To about 10 cm^3 of 0.1M thorium solution in a beaker add 5 cm^3 of the lead nitrate solution, and precipitate lead chloride by the addition of hydrochloric acid. This precipitate will contain the radioactive isotope lead-212. Centrifuge and separate the components of the mixture. Add the precipitate to the saturated lead chloride solution prepared above, and shake or stir mechanically for about twenty minutes. Centrifuge and separate the mixture. Use the 'clear' liquid for a count over a period of a few minutes. The liquid will now be active: this is not due to lead chloride dissolving because the solution was already saturated, but to the interaction of lead ions between the solution and the solid. The count rate will be several times greater than for the original saturated lead chloride solution.

Since this will be the first occasion when a radioactive technique is demonstrated to the pupils, it may be necessary to emphasize the principles which we are trying to establish–namely, those relating to dynamic equilibrium.

Suggestion for homework

Make a summary of the principles you have learnt in this Topic.

Summary

By the end of this Topic, pupils should understand the concept of dynamic equilibrium as applied to physical and chemical systems. They should be aware that, in general terms, it is possible to shift the equilibrium in a chemical system by changing the concentrations of the reactants in that system. The effect of temperature and pressure on equilibria are *not* studied in this Topic.

Topic **A20**

Acids and bases

Purposes of the Topic

1. To review evidence for operational and conceptual definitions of the terms acid, base, alkali, and salt.
2. To review the evidence for the Brønsted-Lowry theory of acids and bases.
3. To provide pupils with practical experience of the preparation of salts.

Contents

A20.1 Investigating the acidic properties of some substances.
A20.2 Concerning acids and bases.
A20.3 Assessing the role of water.
A20.4 Investigating the reaction between an acidic solution and an alkaline solution.
A20.5 Preparing salts.

Timing

A total of six weeks for the Topic allows adequate time for discussion and for practical work.

Introduction to the Topic

In this Topic, pupils are led to a review of operational definitions of acid and base (that is, how such substances can be recognized in the laboratory) and of conceptual definitions for these substances (that is, definitions which explain their behaviour.) The acceptance of the conceptual definition of an acid as a proton donor and of a base as a proton acceptor leads to the somewhat surprising idea that water can behave either as an acid or as a base.

So far we have classified chemicals in a variety of ways. Materials have been shown to be pure or impure; compounds or elements; compounds with a common characteristic, such as chlorides or oxides,.etc. This Topic opens with a simple survey of the properties of oxides and leads directly to an operational definition of an acid.

An experimental investigation of some common acids and of some other substances, which are not listed as acids, leads to a greater appreciation of this simple definition. The pupils discover by experiment that perfectly dry 'acids' do *not* exhibit acidity, whereas some other substances in aqueous solution do exhibit acidity or basicity. The part played by water forms a separate section of this Topic. The properties of hydrogen chloride are reviewed and evidence is put forward to suggest that a chemical reaction occurs when the gas is dissolved in water:

$$HCl(g) + H_2O(l) \longrightarrow H_3O^+(aq) + Cl^-(aq)$$

The reaction between an acidic and an alkaline solution is then investigated conductimetrically, leading to another method for verifying the overall equation for a reaction. The acidity or alkalinity exhibited by certain salts in aqueous solution may also be discussed.

The final section of the Topic is concerned with salts and their preparation.

Simple operational definitions of acids and bases appear in Topic B12. The Brønsted-Lowry theory of acids and bases is considered as an extension of the properties of ions in solution, in Topics B20 and B21.

The subject of acidity–alkalinity was broached in the Stage I Topics, A1 or B2. The treatment provided in Topic A20 depends to some extent on an appreciation of the nature of chemical equilibrium (Topic A19). The Topic may be regarded (alongside other Topics, such as A18 and A19) as illustrating yet another general characteristic of chemical change. Section A20.4 assumes a knowledge of the properties of ions (Topic A16).

Acidity and alkalinity are studied further in Stage III Option 6.

Additional experiments for this Topic are given in *Collected experiments*, Chapter 8. Nuffield Advanced Chemistry *Teachers' guide I* (part of Topic 6) and *Teachers' guide II* (part of Topic 15), and Nuffield Advanced Physical Science *Teachers' guide I* (parts of sections 4 and 6) provide a more advanced treatment.

Overhead projection original
26 Formation of ions *(figure A20.2)*

A20.1
Investigating the acidic properties of some substances

Pupils discuss the knowledge of acidity–alkalinity gained from their experiments in Stage I and perform additional experiments to find out more about acidity and substances labelled 'acids'.

Objectives for pupils

1. Knowledge of the properties of acids
2. Awareness that certain substances not usually listed as acids can exhibit acidity
3. Ability to summarize an extensive range of experimental results
4. Knowledge of the operational definition of the term acid
5. Awareness that 'acids' require water in order to exhibit acidity

Open the lesson with a simple review of the ways in which compounds have been classified (for example, oxides, chlorides, etc.) The properties of oxides may be reviewed briefly. Some oxides are water soluble: their solutions affect the colours of indicators and possess other distinctive properties of acids. During Stage I (A1.2 and B2) pupils explored some of these characteristics under the general heading of acidity–alkalinity. They found that acidity could be 'cured' or neutralized by another group of substances, called alkalis. By using Full-range Indicator, they found that solutions of a variety of substances possessed different degrees of acidity or alkalinity, and used the pH scale as a means of describing how acidic or how alkaline such solutions were. Much of this earlier material will need to be revised before asking 'How can we explain these properties of acids and alkalis?', 'What causes acidity?', and so on. An obvious starting point is with those substances which are labelled as acids. Reactions of acids with indicators, metals, carbonates, metal oxides, etc., provide a series of possible investigations. By dividing the tasks between the pupils and asking them to report their findings to the class, a fairly detailed study of a number of acids can be carried out in a relatively short time.

The effect of the presence of water should also be shown after careful preparation of apparatus and chemicals. For example, dry glacial acetic acid (ethanoic acid) does not show the usual acid reaction with dry indicator paper whereas dilute acetic acid does give a reaction. Pupils will find out that the acids only show the 'common' characteristics of acids if the compound is dissolved in water. An explanation of this finding is left until section A20.3.

The properties of compounds such as ammonium chloride, bismuth nitrate, aluminium sulphate, sodium hydrogen sulphate, sodium hydrogen carbonate, sodium dihydrogen phosphate, etc. in aqueous solution could also be studied.

Note on nomenclature
Several of the acids used (citric, tartaric, etc.) have systematic names which are given in the Apparatus list for the teacher's information. These are not in common use and should not be given to the pupils as they are difficult to remember and their significance will not be grasped.

Experiment A20.1a

Investigating the acidic properties of some compounds

Procedure
This experiment is a preliminary or supplementary demonstration to accompany Experiment A20.1b. (The teacher is not expected to demonstrate all possible reactions but should complement those experiments to be attempted by the pupils.) For example, the dried glacial acetic acid and the benzoic acid in acetone solution with dry indicator paper, magnesium ribbon, etc., can be used to show the importance of the presence of water. Other experiments can illustrate the variety and the range of the properties to be investigated.

Apparatus

The teacher may need:

12 test-tubes, 150×16 mm

Test-tube rack

Spatula

Teat pipette

2 beakers, $250 \mathrm{~cm}^3$

Bunsen burner and heat resistant mat

Splints

Distilled water

Indicator papers (very carefully dried and stored in a desiccator)

Dry glacial acetic acid (prepared by adding a few cm^3 of acetic anhydride to the acid the previous day)

Boiled out distilled water in a covered beaker

Benzoic acid (benzene carboxylic acid) in acetone (propanone) solution

Tartaric acid crystals (2,3-dihydroxybutanedioic acid)

Citric acid crystals (2-hydroxypropane-1,2,3-tricarboxylic acid)

2M hydrochloric acid

M sulphuric acid

2M acetic acid (ethanoic acid)

Magnesium ribbon

Marble chips (calcium carbonate)

Sodium carbonate crystals

Limewater

Copper(II) oxide

Bismuth nitrate

Ammonium chloride

Sodium hydrogen sulphate

Sodium hydrogen carbonate

Sodium dihydrogen phosphate

Full-range Indicator solution

Investigating the acidic properties of some compounds

Apparatus

Each pair of pupils will need:

Experiment sheet 78

5 test-tubes, 100 × 16 mm

Test-tube rack

Test-tube holder

Spatula

Teat pipette

Bunsen burner and heat resistant mat

Splints

Distilled water

A solution of a mineral acid in water (for example, 2M hydrochloric acid or M sulphuric acid), 7 cm³

Crystals of a carboxylic acid (for example oxalic acid [ethanedioic acid], citric acid [2-hydroxypropane-1,2,3-tricarboxylic acid], or tartaric acid [2,3,dihydroxybutane-dioic acid]), 1 spatula measure

Sample of a salt which has acidic properties (for example ammonium chloride, aluminium sulphate, or sodium hydrogen sulphate), 1 spatula measure

Magnesium ribbon, 3 pieces 2 cm long

Dried indicator paper (litmus and Full-range Indicator)

Marble chips (calcium carbonate), 3 chips

Copper(II) oxide, 3 spatula measures

Sodium carbonate crystals, 3 crystals

Limewater (calcium hydroxide solution), 10 cm³

Procedure

It is not intended that each pair of pupils should attempt every test, but all should compare the properties of a solution of a mineral acid with those of a solution of an organic acid. If time allows, a substance which is not usually labelled as an acid should also be investigated. If possible, at least one group should be provided with a substance which does not behave like an acid (for example, sodium hydrogen carbonate) so that during the summing up the teacher may stress the need for experimental investigations and the value of an operational definition of an acid.

Experiment sheet 78 is reproduced below.

Experiment sheet 78

You will be provided with a range of substances to investigate. Test each substance used with *dry* indicator paper. Then make a solution of each solid substance: use one measure of solid in half a test-tube (100 × 16 mm) of water.

Test separate small portions of each solution with:

a. Indicator paper (state which one was used and use the same type throughout).

b. Magnesium ribbon (about 2 cm); try to identify any gas evolved.

c. A crystal of sodium carbonate; try to identify any gas evolved.

d. A small piece of marble; try to identify any gas evolved.

e. A small amount of copper(II) oxide; warm the mixture and allow to stand.

Record observations and conclusions in the table below.

Test	Substance used		
Solid with dry indicator paper (　　　　　　　)			
Solution with indicator paper (　　　　　　　)			
Magnesium			
Sodium carbonate			
Marble			
Copper(II) oxide			

---

1. Use the results of your investigations to write a detailed statement which could be used to define the term acid.

2. *(If pupils have access to a range of books in the school library)*
Tabulate the acids you have met and give their formulae. In a separate column, list their everyday uses.

A20.2
Concerning acids and bases

In this section, pupils investigate the properties of bases, and the terms base, alkali, and acid are defined operationally.

The homework set for the previous section could be reviewed and used to derive acceptable definitions of the terms acid and maybe the term base. This leads naturally to Experiment A20.2a.

Alternatively, the properties of bases could be demonstrated and used to lead to a more formal statement of operational definitions of the terms acid and base using the experience given by Experiment A20.2a.

In section A20.1, pupils found that substances which are usually termed acids do not always exhibit 'acid' properties. They also found that certain substances which are not usually labelled as acids can exhibit 'acid' properties. 'What do we mean by the term "acid"?' One may merely state how we recognize an acid. The experimental work in section A20.1 showed that all the substances tested which are normally listed as acids were soluble in water and that the solutions had the following properties:
1. They changed the colour of litmus paper to red.
2. They reacted with metals such as magnesium or zinc giving off hydrogen.
3. They reacted with metal carbonates such as sodium carbonate or calcium carbonate to give off carbon dioxide.

A simple demonstration (part of Experiment A20.2a) shows that solutions of acids in water conduct electricity and exhibit electrolysis. We can then extend the operational definition of an acid by stating:
4. They conduct electricity and exhibit electrolysis (that is, the formation of products at the surfaces of the electrodes), and so must contain ions.

This definition does *not* exclude substances such as bismuth nitrate or ammonium chloride and so it is necessary to state as well that:
5. An acid is a substance containing hydrogen which may be replaced directly or indirectly by a metal to form a salt.

This may be illustrated by showing that when magnesium ceases to react with dilute sulphuric acid and to liberate hydrogen, the residual solution must contain some product of reaction: a white powder, magnesium sulphate. The reaction need not take a great deal of time. It can be summarized using both word and chemical equations.

Solutions of the oxides of magnesium or calcium can then be prepared and shown to turn litmus solution blue, an alkaline property. During Stage 1, pupils saw that alkaline solutions possess some form of 'anti-acid' property. The fact that solutions of the hydroxides of potassium, sodium, lithium, barium, and calcium also show this property may be noted. Other metal oxides are insoluble and therefore do not react with indicators.

All metal oxides react with hot solutions of dilute acids and the reaction of copper(II) oxide with dilute sulphuric acid may be used to supplement the pupils' knowledge of bases (Experiment A20.2b).

It will be necessary to clarify and emphasize the differences between oxides and hydroxides; bases and alkalis.

It is suggested that pupils are also asked to list examples of acids, bases, and salts as indicated in the table below.

Acid	Formula of acid	Example of the corresponding salt	Formula
Hydrochloric acid	HCl	sodium chloride	NaCl
Nitric acid	HNO_3	potassium nitrate	KNO_3
Sulphuric acid	H_2SO_4	sodium sulphate	Na_2SO_4
		copper(II) sulphate	$CuSO_4$

Base	Formula of base	Example of the corresponding salt	Formula
Barium oxide	BaO	barium chloride	$BaCl_2$
Calcium oxide	CaO	calcium nitrate	$Ca(NO_3)_2$
Copper(II) oxide	CuO	copper(II) sulphate	$CuSO_4$
Sodium hydroxide	NaOH	sodium chloride	NaCl

Apparatus

The teacher will need:

Power supply, 2–6 V d.c.

Current indicator (a lamp or an ammeter)

1 pair of carbon or platinum electrodes suitably mounted

Connecting wire

Evaporating dish

Bunsen burner and heat resistant mat

Tripod and gauze

6 test-tubes, 100×16 mm

Test-tube rack

Filter paper

Solutions of various acids and alkalis in beakers

(Continued)

Properties of acids and alkalis

Procedure

The experiment is intended to illustrate the introduction to this section. Precise details are not given since these necessarily depend on circumstances. However, the following matters should be considered.

It is intended that the electrolysis of various acids and alkalis be demonstrated using the d.c. power supply, current indicator, and electrodes in a series circuit with a variety of aqueous solutions.

It is suggested that the reaction of an excess of magnesium with about 4 or $5 \, cm^3$ 1M sulphuric acid be demonstrated. After the removal of the unused magnesium metal, the presence of magnesium sulphate (a white powder) in solution may also be shown, thereby linking operational definitions of the terms acid and salt.

Finally, solutions of various oxides in water should be shown to possess alkaline properties.

Supply of distilled water

Magnesium

M sulphuric acid

Metal oxides (for example, calcium, magnesium, copper)

Full-range Indicator solution

To examine the reaction between dilute sulphuric acid and copper(II) oxide

Apparatus

Each pair of pupils will need:

Experiment sheet 79

2 test-tubes, 100 × 25 mm

Test-tube rack

Test-tube holder

Filter funnel

Filter paper

Filter stand

Spatula

Bunsen burner and heat resistant mat

M sulphuric acid, 10 cm^3

Copper(II) oxide, 3 spatula measures

Procedure

The object of the experiment is to add small quantities of copper(II) oxide to warm dilute sulphuric acid until no further change takes place. The reaction mixture is then filtered and shown to consist of a blue filtrate, copper(II) sulphate solution, and a residue on the filter paper of unused copper(II) oxide. After the experiment, ask the pupils to summarize the reaction by writing an equation:

$$CuO(s) + H_2SO_4(aq) \longrightarrow CuSO_4(aq) + H_2O(l)$$

An operational definition of a base can then be written.

Experiment sheet 79 is reproduced below.

Experiment sheet 79
One-third fill a test-tube with dilute sulphuric acid. Hold the test-tube in a holder and heat it with a Bunsen flame until the acid just boils. Place the test-tube in a stand and add copper(II) oxide to the hot acid, a measure at a time, until no more dissolves. Stir the mixture after each addition of solid. Filter the mixture while still hot into another test-tube.

What is the residue on the filter paper?

What colour is the filtrate?

What do you think the filtrate contains?

Write an equation for the reaction between solid copper(II) oxide and dilute sulphuric acid.

Make a list of all the names and formulae of acids that you have met so far.

A20.3
Assessing the role of water

Objectives for pupils

1. Understanding of the role of water in acidity and alkalinity

(Continued)

In this section the effects of solvents on the properties of a solution of hydrogen chloride are investigated and pupils are introduced to conceptual definitions of the terms acid and base.

The role of water in the reactions of acids and bases should be familiar to pupils from section A20.1 and A20.2. The acidic nature of an aqueous solution of hydrogen chloride has also been mentioned in Topic A12. In this section pupils compare the properties of hydrogen chloride in water and in toluene solution, and formulate and

2. Recognition of the fact that terms such as acid and base may be defined both operationally and conceptually
3. Understanding of the terms hydrogen ion, $H^+(aq)$, oxonium ion, H_3O^+, and hydroxide ion, OH^-

test hypotheses to account for differences in the properties of these two solutions.

Experiment A20.3a

A comparison of the properties of hydrogen chloride in water and in toluene solution

Apparatus

The teacher will need:

Gas generators for supplying dry hydrogen chloride

Corked flask for drying toluene

Access to fume-cupboard

6 dry test-tubes, 100 × 16 mm

2 test-tubes, 150 × 25 mm, fitted with corks

2 beakers, 250 cm³

1 beaker, 100 cm³

Test-tube rack

2 teat pipettes

6 V battery or d.c. supply

2 steel electrodes mounted in a suitable holder

Connecting wire

Paper tissues

Indicator paper stored in a desiccator

Limewater

Marble chips

Magnesium ribbon

Distilled water

Concentrated sulphuric acid

Sodium chloride

Anhydrous calcium chloride

Toluene (methylbenzene)

2M hydrochloric acid labelled 'solution of hydrogen chloride in water'

Procedure
Part 1. *Before the lesson*
Prepare a saturated solution of dry hydrogen chloride in dry toluene. (To ensure that the toluene is absolutely dry, it should stand at least overnight in a corked flask in contact with anhydrous calcium chloride.) The hydrogen chloride from the gas generator should be dried by bubbling it through concentrated sulphuric acid before passing it through the dry toluene. This solution should then be kept corked until required for use. It is suggested that 2M hydrochloric acid is used as the 'solution of hydrogen chloride in water'.

Part 2. *The demonstration*
The tests *a*, *b*, *c*, and *d* should be carried out with both solutions of hydrogen chloride. It is helpful to build up a summary of the results on the blackboard as the demonstration proceeds.

a. The effect of the two solutions of hydrogen chloride on dry indicator paper.
b. Reaction of the solutions of hydrogen chloride with marble chips and the detection of any gas given off.
c. The reaction of the solutions of hydrogen chloride with magnesium ribbon and the detection of any gas given off.
d. The effect of immersing two steel electrodes in samples of each solution when the electrodes are connected to the d.c. supply and an electric bulb. (*Note*. The electrodes should be placed in the toluene solution first to avoid transporting any water to the toluene.)

e. The teacher should then transfer about 10 cm³ of the toluene solution to a corked 150 × 25 mm test-tube and shake it with an equal volume of distilled water. After removing the upper layer of toluene, each of the tests *a*, *b*, *c*, and *d* should be applied to the aqueous layer.

From these tests it should be clear that some at least of the hydrogen chloride has passed into the aqueous layer and that in doing so, it has developed different properties.

The evidence obtained suggests that:
1. Ions are produced when hydrogen chloride dissolves in water.
2. Acidic properties are associated with the solution containing ions.

The electrolysis of a strong solution of hydrogen chloride shows that hydrogen is evolved at the cathode and chlorine at the anode.

Thus, one might predict the existence of hydrogen cations (H^+) and chloride anions (Cl^-). (*Note.* Such a demonstration can be very effective when it is used to complement the qualitative investigation outlined in Experiment A20.3a.)

Let us now consider the formation of hydrochloric acid from hydrogen chloride and water. If a reaction occurs between these chemicals, then one would expect to find heat being given out as the gas dissolves. If there is only the breaking of the hydrogen–chlorine bond to form H^+ ions and Cl^- ions, then heat would be absorbed and the temperature of the system would drop. The same test should also be made using toluene in place of water.

Experiment A20.3b

Apparatus

The teacher will need:

2 gas generators for dry hydrogen chloride

2 gas jars and lids

4 filter tubes, 150 × 25 mm

Corks

Delivery tubes (see figure A20.1)

Rubber connectors

2 thermometers

2 stands, 4 bosses, and clamps

Access to a fume-cupboard

Toluene (methylbenzene) (dry)

Distilled water

The preparation of solutions of hydrogen chloride in toluene and in water

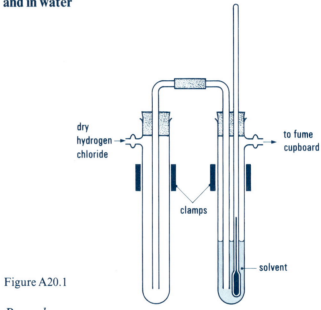

Figure A20.1

Procedure

To show the effect of solvent, a thermometer bulb could be dipped into toluene and then into a jar of hydrogen chloride. The thermometer shows no rise in temperature. Now take the thermometer out of the jar, dry it, and dip the bulb into water before placing it in a jar of hydrogen chloride. This time the temperature rise will be about 10 °C. This suggests that a reaction has taken place between hydrogen chloride and water, but not between hydrogen chloride and toluene. A more detailed and alternative study of this effect may now follow.

Assemble the two gas generators and trains of absorption tubes as shown in figure A20.1. Note the initial temperatures of the solvents. On passing dry hydrogen chloride into dry toluene very little temperature change takes place. The absorption of hydrogen chloride in water, however, results in a marked rise in the temperature of the solution.

We may therefore conclude that reaction (*a*) is unlikely but that reaction (*b*) is likely:

(a) $HCl(g) \xrightarrow{\times} H^+ + Cl^-$

(b) $HCl(g) + H_2O(l) \longrightarrow H_3O^+(aq) + Cl^-(aq)$

Further, it is likely that the hydrogen ion, $H^+(aq)$, is surrounded by a number of water molecules rather than being associated with a single molecule of water and so we write H_3O^+ (aq). No real proof can be offered here for the existence of oxonium ions, $H_3O^+(aq)$. Since pupils have a simple picture of atoms from A16.7, the following point can be made. The proton, H^+, is extremely small, much smaller than a hydrogen atom. Its positive charge is likely to attract electrons associated with the oxygen atom in a water molecule. (This can be shown using scale models or scale drawings on an overhead projection transparency.)

The fact that reaction (b) is reversible can be shown by heating a little concentrated hydrochloric acid and identifying the hydrogen chloride gas evolved. Indeed, concentrated hydrochloric acid is another example of dynamic equilibrium:

$$HCl(g) + H_2O(l) \rightleftharpoons H_3O^+(aq) + Cl^-(aq)$$

It might be appropriate to state that although acidic solutions contain oxonium ions, H_3O^+, it is quite usual to refer to them as hydrogen ions. Chemists accept that such ions are normally hydrated in solution and that the symbol $H^+(aq)$ also implies the interaction of H^+ with water.

Concerning the definition of the terms we use
We can now define the terms acid and alkali by explaining the way in which chemicals react: that is, conceptual definitions. An *acid* is a substance containing hydrogen which in solution yields hydrogen ions. An *alkali* is a substance containing hydroxide ions or hydroxyl groups which dissolves in water to yield hydroxide ions.

The reaction of hydrogen chloride and water has already been investigated and pupils know that heat is generated.

$$HCl(g) + H_2O(l) \rightleftharpoons H_3O^+(aq) + Cl^-(aq): \text{heat lost by the system}$$

Concentrated nitric acid may be shown to behave in a similar manner:

$$HNO_3(l) + H_2O(l) \rightleftharpoons H_3O^+(aq) + NO_3^-(aq): \text{heat lost by the system}$$

Likewise, when concentrated sulphuric acid is added to water, the reaction generates heat. (If $50\,cm^3$ of concentrated sulphuric acid is added to $1000\,cm^3$ of cold water, *taking the usual precautions,** a temperature rise of some $16\,°C$ will be obtained.)

$$H_2SO_4(l) + H_2O(l) \rightleftharpoons H_3O^+(aq) + HSO_4^-(aq): \text{heat lost by the}$$
$$\text{system}$$
$$HSO_4^-(aq) + H_2O(l) \rightleftharpoons H_3O^+(aq) + SO_4^{2-}(aq): \text{heat lost by the}$$
$$\text{system}$$

*When concentrated sulphuric acid is added to water, pour a small volume of the acid down a glass rod into the water and stir well before repeating the process.

Pupils will note that in each case, the acids produce hydrogen ions and that these ions can be written as either $H_3O^+(aq)$ or $H^+(aq)$.

For the equilibrium,

$$HCl(g) + H_2O(l) \rightleftharpoons H_3O^+(aq) + Cl^-(aq)$$

we may think of the hydrogen ions being pulled in two different directions.

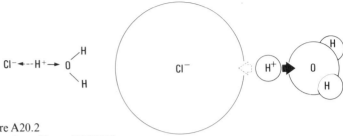

Figure A20.2
Formation of ions. (OHP 26)

Representations can be made using models or scale drawings on an overhead projection transparency. The water molecules 'win' the major share of the hydrogen ions in the solution and the solution is strongly acidic. The teacher may point out an even more general definition of the terms *acid* and *base*: an acid is a proton donor and a base is a proton acceptor. The acceptance of this definition leads to somewhat surprising conclusions.

$$HCl(g) + H_2O(l) \rightleftharpoons H_3O^+(aq) + Cl^-(aq)$$
acid base   acid   base

On the basis of this definition, water must be regarded as a base since it is acting as a proton acceptor in the reaction.

One may also show how alkalis give rise to the presence of hydroxide ions: thus,

$$Na^+OH^-(s) \xrightarrow{\text{solution in water}} Na^+(aq) + OH^-(aq)$$

$$Ba^{2+}(OH^-)_2(s) \xrightarrow{\text{solution in water}} Ba^{2+}(aq) + 2OH^-(aq)$$

$$NH_3(g) + H_2O(l) \rightleftharpoons NH_4^+(aq) + OH^-(aq)$$

The teacher should indicate that hydroxides such as sodium or barium hydroxide are largely ionized in the solid state, whereas ammonia requires water for the formation of ammonium ions and hydroxide ions and the establishment of the equilibrium represented in the equation above. It is best to avoid the use of the misleading term ammonium hydroxide.

The example of ammonia solution may be used to extend the definition of the terms acid and base. Water may be regarded as an acid (a proton donor) *in this instance.*

$$NH_3(g) + H_2O(l) \rightleftharpoons NH_4^+(aq) + OH^-(aq)$$
base acid   acid   base

The acceptance of the (conceptual) definition of the term acid leads to the idea that substances such as water, which are not acids or bases, according to operational definitions, can nevertheless exhibit both acidic and basic properties. All of these interpretations and definitions have their uses. In particular, the relationship:

$$acid \rightleftharpoons base + proton$$

helps to clarify our thinking about acids and bases. However, it would be confusing for chemists to interpret this idea rigidly and to say, for example, that water is an acid. Even chemists cannot altogether dispense with operational definitions!

Suggestion for homework

Write a short account to show the various meanings given to the terms 'acid' and 'base'.

A20.4
Investigating the reaction between an acidic solution and an alkaline solution

The problem is studied through the case of the reaction between sulphuric acid and barium hydroxide.

A suggested approach

Objectives for pupils

1. Application of previous knowledge of the properties of acids and bases
2. Ability to use both conceptual and operational definitions of the terms acid, alkali, and base
3. Understanding of the reaction between an acid and a base in terms of hydrogen and hydroxide ions
4. Understanding of the term neutral solution (pH 7) as one possessing an equal number of hydrogen and hydroxide ions

The previous section considered a variety of characteristics of both acids and bases, and the homework set will show whether pupils understand the distinction between operational and conceptual definitions of these classes of compound. It may be necessary for the teacher to review the key points before starting any new work.

Pupils should be aware that the hydrogen ion (hydrated proton) plays a vital role in acid–base reactions. For example, the teacher may ask 'What happens to an acid when it loses its acidic properties through the addition of a base?'

It would be helpful to study just one system in detail and then to apply the findings to other systems to see whether our ideas are of general application.

When an acid is dissolved in water, ions exist in solution. A similar statement has been made for alkalis. We know that acids react with metal oxides (and metal hydroxides) to form salts (see section A20.2). Previous experience of the properties of aqueous solutions of salts (see Stage I and Topic A16) shows that these solutions also contain ions.

So, if we use a conductivity test to follow the course of the reaction between an acid and a base, it would help if we chose a chemical reaction in which the products did not ionize or were insoluble. Such products would not interfere with the investigation. One such reaction is that between dilute sulphuric acid and barium hydroxide solution. Two possibilities now arise: the teacher may follow this discussion with either a class experiment (Experiment A20.4a) or a demonstration experiment (Experiment A20.4b). Either experiment can show the pupils that at the end point the solution is nonconducting, thereby providing evidence to support the equation for the reaction. Whichever approach is used, the apparatus, concentration of the solutions, and specification of electrical equipment, need to be adhered to closely if satisfactory results are to be obtained.

Experiment A20.4a

Apparatus

Each pair of pupils will need:

Experiment sheet 80

Test-tube, 100×16 mm

Teat pipette

Beaker, 100 cm^3

2 carbon electrodes in a holder

3 lengths of connecting wire

Bulb holder and 6.5 V 0.06 A bulb or a.c. ammeter (0–100 mA)

Crocodile clips

Source of 12 V a.c.

Burette, $50 \times 0.1 \text{ cm}^3$

Burette stand

Measuring cylinder, 25 cm^3

Glass rod

Phenolphthalein solution

0.05M barium hydroxide, standardized, 60 cm^3

0.5M sulphuric acid, standardized, 20 cm^3

Distilled water

To find the ratio in which barium and sulphate ions combine

Procedure
Details are given in *Experiment sheet* 80.

Note. Barium hydroxide solution is poisonous and so should *not* be transferred from one vessel to another using a pipette. It reacts rapidly with the carbon dioxide in the air to form barium carbonate and 'cloudiness' is seen in the solution.

Precision is only possible when pupils know how to use an a.c. milliammeter rather than a simple bulb current indicator.

Experiment sheet 80

You are provided with 0.05M barium hydroxide solution and 0.5M sulphuric acid. Using an electric bulb and an alternating current you are going to investigate the changes in conductivity when the sulphuric acid solution is added to the barium hydroxide solution.

As a preliminary experiment, put about a 1 cm depth of the barium hydroxide in a test-tube and add a few drops of the dilute sulphuric acid. What do you see? What do you think it is?

Use your beaker and test each solution for conductivity. Wash out the beaker *and* the electrodes with distilled water before testing the other reactant. Record your results.

Dilute sulphuric acid:

Barium hydroxide solution:

For the main part of the investigation use the apparatus shown in the diagram.

Use a measuring cylinder to put 50 cm^3 of the barium hydroxide solution into the 100 cm^3 beaker and add 5 drops of phenolphthalein solution. Put the sulphuric acid solution into the burette, see that the jet is filled with solution, and fix the burette in a stand so that it is above the beaker. Put the two electrodes in the solution vertically and connect them in series with the a.c. supply and bulb.

Run the sulphuric acid solution into the beaker 1 cm^3 at a time, stirring between each addition. Note the appearance of the bulb and the mixture in the beaker after each addition. When the bulb starts to become dim add the acid one drop at a time. When there is enough sulphuric acid to react with all the barium hydroxide, the bulb will go out and the phenolphthalein will lose its pink colour.

How much sulphuric acid has been added when this happens? .... cm^3.

Continue to add acid; what happens to the lamp now?

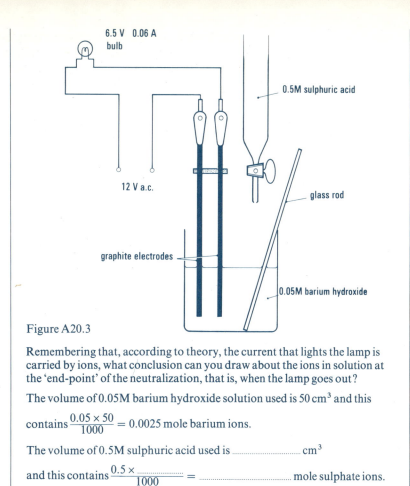

6.5 V 0.06 A bulb

0.5M sulphuric acid

12 V a.c.

glass rod

graphite electrodes

0.05M barium hydroxide

Figure A20.3

Remembering that, according to theory, the current that lights the lamp is carried by ions, what conclusion can you draw about the ions in solution at the 'end-point' of the neutralization, that is, when the lamp goes out?

The volume of 0.05M barium hydroxide solution used is 50 cm^3 and this

contains $\dfrac{0.05 \times 50}{1000} = 0.0025$ mole barium ions.

The volume of 0.5M sulphuric acid used is .... cm^3

and this contains $\dfrac{0.5 \times \text{.....................}}{1000} =$ .... mole sulphate ions.

How many moles of sulphate ions react with one mole barium ions?

.... mole.

Write the equation for the reaction between barium ions and sulphate ions.

What has happened to the hydrogen and hydroxide ions?

Experiment A20.4b

Apparatus

The teacher will need:

Test-tube, 100×16 mm

Beaker, 100 cm^3

Teat pipette

2 carbon electrodes in a holder

5 lengths of connecting wire

Bulb holder

6.5 V 0.06 A bulb

(Continued)

To find the ratio in which barium and sulphate ions combine

Procedure

Demonstrate the reaction between barium hydroxide and sulphuric acid. Ask the pupils to 'explain' their observations. After discussing their ideas, demonstrate the main experiment.

Transfer 50 cm^3 of barium hydroxide solution to a 100 cm^3 beaker. Place the two electrodes in the solution and connect them to a rheostat, bulb, milliammeter, and the 4 V a.c. supply (figure A20.4). Adjust the rheostat so that the milliammeter reads 10 mA at the start. A few drops of phenolphthalein can now be added.

Now set up the burette containing the sulphuric acid (10 times the strength of the barium hydroxide solution to speed up the experiment and to prevent a large volume change in the beaker). Add the

a.c. milliammeter reading to 0–10 mA (demonstration model preferred)

Crocodile clips

Source of 4 V a.c.

Rheostat, 20 Ω

Magnetic stirrer

Phenolphthalein solution

0.01M barium hydroxide solution, made from solid and standardized

0.1M sulphuric acid, standardized to match the barium hydroxide solution

sulphuric acid 1 cm³ at a time at first. Stir the solution and record the volume added and the milliammeter reading. When the reading is less than 2 mA, add the acid in 0.5 cm³ portions, and as the reading approaches zero use even smaller portions.

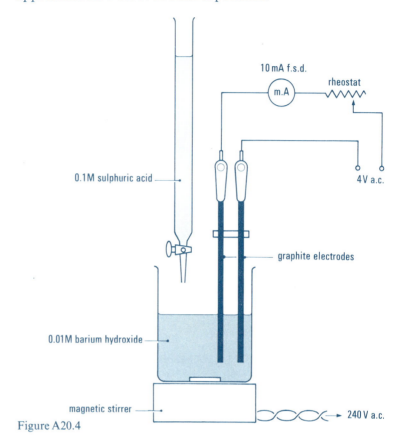

Figure A20.4

The indicator turns colourless when the current is almost zero, due to a lack of ions in solution. The meter reading increases again when an excess of acid has been added. Some specimen results are shown in figure A20.5.

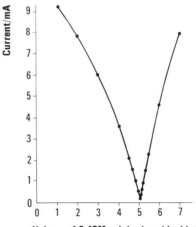

Figure A20.5

Topic A20 Acids and bases

The results of this experiment confirm our expectations. We can now write the lefthand side of the equation as:

$$Ba(OH)_2(aq) + H_2SO_4(aq) = \ldots$$

Since the reaction results in the removal of hydrogen ions and hydroxide ions, we may now write:

$$H^+(aq) + OH^-(aq) \longrightarrow H_2O(l)$$

and

$$Ba^{2+}(aq) + SO_4^{2-}(aq) \longrightarrow BaSO_4(s)$$

A neutral solution may now be defined as one of pH7, i.e. as one containing an equal number of hydrogen and hydroxide ions. We may mention that the reactions studied in this experiment are both equilibria.

$$H^+(aq) + OH^-(aq) \rightleftharpoons H_2O(l)$$

and

$$Ba^{2+}(aq) + SO_4^{2-}(aq) \rightleftharpoons BaSO_4(s)$$

Optional material (non-examinable at Stage II)
One problem has been omitted from the discussion so far: certain salts which contain neither hydroxide ions nor hydrogen ions exhibit some alkaline or acid properties. To interpret such findings requires the distinction to be made between strong and weak acids, and strong and weak bases.

A solution of sodium carbonate in water could give rise to the following chemical equilibria:

$$2Na^+(aq) + CO_3^{2-}(aq) + 2H_2O(l) \rightleftharpoons$$
$$H_2CO_3(aq) + 2Na^+(aq) + 2OH^-(aq)$$

$$H_2CO_3(aq) \rightleftharpoons H^+(aq) + HCO_3^-(aq)$$

$$HCO_3^-(aq) \rightleftharpoons H^+(aq) + CO_3^{2-}(aq)$$

Since carbonic acid is a weak acid, very few hydrogen ions are formed and the solution contains an excess of hydroxide ions. Hence sodium carbonate in aqueous solution is alkaline.

In the case of copper(II) sulphate solution, the following equilibria are feasible:

$$Cu^{2+}(aq) + SO_4^{2-}(aq) + 2H_2O(l) \rightleftharpoons$$
$$Cu(OH)_2(s) + SO_4^{2-}(aq) + 2H^+(aq)$$

$$Cu(OH)_2(s) \rightleftharpoons Cu^{2+}(aq) + 2OH^-(aq)$$

Since copper hydroxide is a weak base, very little is ionized so there is an excess of hydrogen ions, which explains this unexpected property of copper sulphate solution.

Suggestion for homework   Write ionic equations (including state symbols) for a variety of reactions between acids and bases.

A20.5
Preparing salts

The meaning of the term salt is revised (see section A20.1). Pupils then spend some time preparing a variety of salts.

First revise the meaning of the term salt and remind pupils of their earlier work on salts..Details of some well-known preparations of salts are given below.

Objectives for pupils

1. Understanding the term salt and knowledge of the methods by which salts may be made in the laboratory
2. Ability to design experiments for the preparation of salts

Experiment A20.5a

Preparation of copper sulphate from (1) copper(II) oxide; (2) copper(II) carbonate

Apparatus

Each pair of pupils will need:

Experiment sheet 81

Beaker, 100 cm³

Measuring cylinder, 25 cm³

Tripod and gauze

Bunsen burner and heat resistant mat

Filter funnel, filter paper, and filter stand

Crystallizing dish

Cloth

Spatula

2M sulphuric acid, 25 cm³

Copper(II) oxide (powder), 4 g

Copper carbonate (powder), 6 g

Procedure
In this preparation, the experience gained in Experiment A20.1a is extended and crystals of copper sulphate are obtained by one of two procedures. Full details appear in *Experiment sheet* 81.

Experiment sheet 81

1. Put 25 cm³ 2M sulphuric acid into a 100 cm³ beaker and heat it, on a gauze over a tripod, until it is nearly boiling. Add copper(II) oxide, in small quantities, until no more will dissolve (about 4 g is required). Filter the solution while hot (hold the beaker with a cloth) into a clean crystallizing dish. Add 2 or 3 drops sulphuric acid to the filtrate and allow it to cool. Crystals of copper sulphate will form. Pour off the liquid and dry the crystals on blotting paper or filter paper.

2. The method using copper(II) carbonate is the same as in (1) above except that cold acid is used. With copper(II) carbonate there will be much frothing owing to the formation of carbon dioxide gas, so more care is needed when adding the solid to the acid. Use very small quantities each time and stir the mixture after each addition. A total of about 6 g copper(II) carbonate is required.

Write the equation for the reaction you use.

Make a drawing of a well-formed copper sulphate crystal. The formula for crystalline copper sulphate is $CuSO_4 5H_2O$.

Experiment A20.5b

Preparation of magnesium sulphate from (1) magnesium; (2) magnesium carbonate

Apparatus

Each pair of pupils will need:

Experiment sheet 82

Beaker, 100 cm³

Procedure
This experiment may be used to extend earlier experiences of the preparation of salts. Full details appear in *Experiment sheet* 82 reproduced below.

(Continued)

Experiment sheet 82

1. Put 25 cm^3 2M sulphuric acid in a 100 cm^3 beaker and add magnesium, in small quantities, until no more will react (about 1.5 g is required). Stir well after each addition. Heat the liquid to boiling and filter while hot into a clean evaporating basin, add a few drops of sulphuric acid, and evaporate the filtrate to about half its volume. Allow to cool when crystals will form. Pour off the liquid and dry the crystals with blotting paper or filter paper.

2. The same procedure is used if you start with magnesium carbonate, but about 5 g of this will be needed. Care is needed when adding the solid to the acid – use very small quantities each time and stir the mixture after each addition.

Write the equation for the reaction you use.

Make a drawing of a well-formed magnesium sulphate crystal. The formula for crystalline magnesium sulphate is $MgSO_4 7H_2O$.

Salts may also be prepared by other methods. A sparingly soluble salt such as lead chloride may be conveniently made from two soluble salts, as described below.

Experiment A20.5c

Preparation of lead chloride from lead nitrate solution

Apparatus

Each pair of pupils will need:

Experiment sheet 83

2 test-tubes, 100 × 16mm–to fit the centrifuge

Teat pipette

Glass rod

Bunsen burner and heat resistant mat

Beaker, 100 cm^3

Measuring cylinder, 25 cm^3

Access to centrifuge

M lead nitrate solution, 5 cm^3

2M sodium chloride solution, 5 cm^3

Distilled water

Procedure
The addition of lead nitrate solution to sodium chloride solution should be demonstrated and the technique of using a centrifuge should be discussed and demonstrated by the teacher before the experiment is attempted by the class. Full details appear in *Experiment sheet* 83 which is reproduced below.

Experiment sheet 83

Put 5 cm^3 M lead nitrate solution into a 100 × 16 mm test-tube and add 5 cm^3 2M sodium chloride solution. Stir the mixture well. What happens?

Write the equation for the reaction.

Centrifuge the mixture and pour off the clear liquid. Add a little distilled water (to a depth of about 2 cm), stir the mixture, and again centrifuge. Pour off the clear liquid, scrape the solid on to filter paper, and allow it to dry.

Why is a different method used in this preparation from those for copper sulphate and magnesium sulphate?

Finally, a little time could be spent on the occurrence of salts in nature and this could be related to the periodic classification (see Topic A13). Depending on the class, this could be as short as making the point that carbonates, sulphates, chlorides, phosphates, etc., all occur in nature. A more extended approach is described in B12.3.

Suggestions for homework

Summarize the procedures available for the preparation of salts in the laboratory. Give examples of each of the procedures you describe and include equations as summaries.

Summary

By the end of the Topic pupils should know operational and conceptual definitions of acids, bases, and salts. They should be able to distinguish between alkalis and bases and be acquainted with the evidence for the Brønsted–Lowry theory of acids and bases. They should also be able to describe and perform the various procedures available for the preparation of salts in the laboratory.

Optional material for this Topic includes the hydrolysis of salts and a brief account of the occurrence of salts in nature related to the periodic classification.

Breaking down and building up large molecules

Timing

Most of the sections may be covered in a double period, but section A21.2 may require two double periods and section A21.8 needs only one single period. Six weeks should give plenty of time for the whole Topic.

Introduction to the Topic

This Topic is primarily concerned with large molecules, both natural and synthetic, and their importance in biochemistry and the chemical industry. It opens with an elementary review of the range of carbon compounds and their relevance to daily life. Starch is taken as an example of a naturally occurring large molecule and is studied through a stage-by-stage breakdown into maltose, glucose, ethanol, and ethene (ethylene), which enables the biological significance of large molecules to be considered. The term hydrolysis is introduced during this sequence. The end product of this series, ethene, is an extremely important raw material for the chemical industry. The pupils are told of the production of ethene from a crude oil fraction. The process is imitated by the pupils who crack medicinal paraffin and obtain a mixture of ethene and other gases. This experiment is followed by another in which naturally occurring substances, this time fats and oils, are broken down by alkali to produce a soap, or by an acid to produce a soapless detergent.

The last two sections of the Topic are directed to the building-up of large molecules from smaller ones. Two polymers, perspex and nylon, are prepared and the importance of chemistry in the manufacture of entirely new materials is again stressed.

Alternative IIB offers the material discussed in this Topic as part of Topic B15, 'Everyday materials: large molecules and metals'. Different contexts are suggested, one of which depends on a formal introduction to structure.

Background knowledge

This Topic assumes that pupils are familiar with concepts which have been used in earlier Topics of Alternative IIA, especially those relating to structure.

Subsequent development

The themes begun in this Topic are developed in Stage III, Option 5 'Plastics' and Option 3 'Drugs and medicines'.

Further references

for the teacher

Additional experiments are provided in Chapter 10 of *Collected experiments*. References to the biological importance of carbohydrates as foodstuffs and to the digestion of starch are given in Revised Nuffield Biology (1975) Text 2 *Living things in action*, Chapters 6, 7 and 9.
Nuffield Advanced Chemistry (1970) *Teachers' guide II*, Topic 18, provides an advanced treatment of the material used in this Topic. *The chemist in action*, Chapter 4, also relates to this Topic.
Nuffield Advanced Physical Science (1974) *Teachers' guide III*, Options G7 and M2, provides an alternative advanced treatment, and the *Sourcebook*, Part 10, relates to this Topic.

Supplementary materials

Film loops
2–9 'Cracking hydrocarbons'
2–10 'Plastics'
3–4 'Giant molecules – proteins'

Films
'Polymers' ICI Film Library
'What is soap?' Unilever Film Library
'Chemistry of soapless detergents' Unilever Film Library
'Outline of detergency' Unilever Film Library
'Hard water' Unilever Film Library

Overhead projection original
27 Facts about the organic chemical industry *(figure A21.1)*

A21.1
Molecular materials: carbon chemistry

This section introduces the chemistry of carbon compounds.

A suggested approach

Objectives for pupils

1. An indication of the range of carbon compounds
2. Knowledge of the detection of carbon in carbon compounds
3. Familiarity with the use of models for molecular species

The approach adopted must depend very much on the ability and interest of the pupils. One approach is to use a simple display of materials, chemicals, and molecular models and to direct a discussion towards the relevance of carbon compounds in daily life, in medicine, and in industry. The point needs to be made that the majority of known chemical compounds contain carbon. Factual and topical information about the oil industry or about the pharmaceutical industry may be used as illustrations. A convenient way of summarizing facts about the organic chemical industry is shown in figure A21.1.

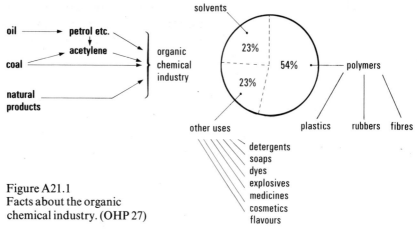

Figure A21.1
Facts about the organic
chemical industry. (OHP 27)

How can we really be sure that a material is made of or contains carbon? Refer back to experience gained during Topic A13.5. With some pupils, it may be appropriate to refer to such methods of analysis as the oxygen flask technique.* The effect of heating mixtures of carbon – and subsequently carbon compounds – with dry copper(II) oxide to produce carbon dioxide (and, in the case of many carbon compounds, water) needs to be fully tested. The range of chemicals used should include starch. With some classes, it will be unnecessary to attempt all aspects of Experiment A21.1.

Experiment A21.1

Apparatus

Each pair of pupils will need:

Experiment sheet 84

6 test-tubes, 150×16 mm

Test-tube rack

Bung fitted with a right-angled delivery tube

Teat pipette

Test-tube holder

Bunsen burner and heat resistant mat

Spatula

Glass rod

Powdered charcoal ⎤

Sucrose ⎪
            ⎬ *
Starch ⎪

Naphthalene ⎦

*about 1–2 g of each substance to be tested

(Continued)

The detection of carbon and hydrogen in organic chemicals

Procedure
The purpose of this experiment is to provide pupils with an opportunity to perform simple qualitative tests for the carbon dioxide and water produced when organic chemicals are heated with copper(II) oxide.

Full details are given in *Experiment sheet* 84 which is reproduced below.

Experiment sheet 84
The method you will use is to heat the substance to be tested with dry copper(II) oxide. This will convert any carbon in the substance to carbon dioxide, which will be given off as a gas, and any hydrogen present to steam, which will condense to water. You should know tests for carbon dioxide and water; describe them below.

Test for carbon dioxide:

Test for water:

Put a measure of dry copper(II) oxide powder into a dry 100×16 mm test-tube. Add half a measure of the substance to be tested and mix the two thoroughly, with a glass rod. Add another measure of copper(II) oxide and do *not* mix it with the other contents of the tube. Heat the test-tube for about

*Some recent references to the applications of this method of analysis include *School Science Review* (1973), **54**, No. 189, 778; *Journal of Chemical Education* (1965), **42**, 270.

Dried copper(II) oxide, prepared and stored in a desiccator before the lesson, 2 g

Anhydrous copper(II) sulphate, prepared before the lesson and stored in a corked test-tube, 1 g

Limewater, 10 cm³

a minute with a small flame. Remove a sample of gas from the tube with a teat pipette and test it for carbon dioxide (prepare for this beforehand). Look for drops of liquid on the cooler parts of the tube and test these for water. Enter the results in the table below, use a tick to indicate the presence of carbon or hydrogen.

Substance tested	Contains carbon	Contains hydrogen

A21.2
The breakdown of starch

The study of starch is used as an introduction to the chemistry of large molecules. Starch is hydrolysed by saliva and by dilute hydrochloric acid.

A suggested approach

Objectives for pupils

1. Awareness of large organic molecules
2. Knowledge of the relationship of starch to maltose, and of maltose to glucose
3. Skill in the technique of chromatographic analysis
4. Understanding of the process of hydrolysis

The teacher will need:

Models of glucose and maltose molecules

The structures which the pupils have studied up to now have either been giant structures – covalent or ionic – or small molecules. It is the object of this Topic to extend the pupils' knowledge to large molecules. This is done first through a substance which is familiar to them – starch. The lesson may start with a simple discussion on the structure of molecular materials, particularly of those materials used in the previous lesson.

Reference should then be made to the presence of starch in foodstuffs, such as bread (wheat) and potatoes, and of its importance as a source of energy for the body. Tell the pupils that chemists have discovered that the molecules of starch are extremely large and have a molecular mass of the order of 100 000 relative to an atom of hydrogen.

Refer to the simple tests carried out during Experiment A21.1, in which pupils were able to show that starch contains carbon and hydrogen. A more detailed analysis – carried out quantitatively – shows that starch is made of carbon, hydrogen, and oxygen. The large molecule is broken into a number of smaller molecules which can be recognized fairly easily. There are several ways of doing this. We might try to imitate the process of digestion in which the enzyme present in saliva breaks the starch down (hydrolyses it) into sugar, or we might use dilute hydrochloric acid to hydrolyse starch to form glucose. Discuss these two possibilities with the pupils. They may already know that if you chew bread for a long time it begins to taste sweet.

Experiment A21.2

Apparatus

Each pair of pupils will need:

Experiment sheet 85

(Continued)

Identification of the breakdown products of starch using a chromatographic method of analysis

Procedure
Make sure that the pupils are familiar with the starch test and the Fehling's test before allowing them to begin the main section of this experimental work. *Experiment sheet* 85 is reproduced below.

Microscope slide

Test-tube, 150 × 25 mm

Test-tube, 125 × 16 mm, with
side arm and fitted with a bung
and delivery tube as
'a cold-finger condenser'

2 lengths of rubber tubing

4 test-tubes, 100 × 16 mm

Test-tube rack

Test-tube holder

Measuring cylinder, 25 cm³

Glass rod

Stand, boss, and clamp

Tripod and gauze

Bunsen burner and heat
resistant mat

4 teat pipettes

Microscope slide

Rectangle of Whatman No. 1
paper to form a cylinder inside
a gas jar

Thermometer, − 10 to
+ 110 × 1°C

2 paper clips – or a needle and
thread

Gas jar and cover slip

Capillary tubes

Litmus or Full-range Indicator
paper

In addition the pupils will
require access to:

2 or more sets of apparatus for
concentrating solutions under
reduced pressure using filter
pumps

1 or more plastic troughs (or
shallow beakers) to contain
the solution of locating agent
which has been selected for
them

Towards the end of the lesson, the teacher can make the point that
the procedure using the enzyme is 'better' (quicker) under the conditions used than the purely chemical procedure.

Experiment sheet 85
This is a long experiment, but an interesting one. You will have to work
carefully if it is to be successful.

You will need to be able to test for the presence (or absence) of starch and
certain kinds of sugar. For these you will use iodine solution and Fehling's
solution (a rather complicated mixture containing copper compounds).
Try these two reagents out on solutions of starch, glucose, and maltose,
as follows.

1. Put separate drops of starch, glucose, and maltose solutions on a
microscope slide (make it slightly greasy first by wiping it with your finger)
and add one drop of iodine solution to each. What colour is:

the mixture containing starch? ....

the mixture containing glucose? ....

the mixture containing maltose? ....

2. Put five-drop portions of starch, glucose, and maltose solutions into
separate test-tubes. Add five drops of each of the two Fehling's solutions to
each and boil the mixtures over a small flame (use a test-tube holder). What
happens:

with starch?

with glucose?

with maltose?

You should now know how to distinguish between starch and the two
sugars, glucose, and maltose.

The action of dilute acid on starch
Have ready several separate drops of iodine solution on a microscope slide.

Put 10 cm³ starch solution into a 150 × 25 mm test-tube and add 10 drops
2M hydrochloric acid. Fit the test-tube with a 'cold-finger' condenser (see
diagram) and heat the mixture to boiling over a small flame. Regulate the
size of the flame so that the mixture is just kept boiling and no more. The
condenser reduces water loss by evaporation. After 5 minutes boiling,
remove the Bunsen burner, lift the condenser, and with a glass rod transfer
one drop of the mixture from the boiling tube to a drop of iodine on the
microscope slide.

Does the mixture still contain starch? ....

If so, continue the boiling for 2 minutes, stop the heating, and again test for
starch. Continue doing this until no starch can be detected in the mixture.
Then stop the heating and remove the condenser.

Using a teat pipette, transfer 5 drops of the final solution to a test-tube, add
5 drops 2M sodium hydroxide to make it alkaline, and then 5 drops of each
of the two Fehling's solutions. Boil the mixture for one minute (test-tube
holder). What do you see?

What has happened to the starch?

The unused solution has now to be neutralized with sodium hydroxide
solution and evaporated to one-twentieth of its volume for further
investigation. Your teacher will arrange for this to be done.

(Continued)

2M hydrochloric acid, $2\,cm^3$

2M sodium hydroxide, $2\,cm^3$

1% starch solution, $20\,cm^3$ per pair of pupils (freshly prepared)

Fehling's solution No. 1 and No. 2 (see Experiment B15.1), $2\,cm^3$ of each

0.01M iodine in 0.1M potassium iodide solution, $2\,cm^3$

10% glucose solution, $1\,cm^3$

10% maltose solution, $1\,cm^3$

Mixture of propan-2-ol, acetic acid, and water in ratio: 3 :1 :1, $25\,cm^3$

Locating agents

Either aniline (phenylamine), acetone (propanone), diphenylamine, phosphoric acid

or p-anisidine hydrochloride, butan-1-ol

or benzene-1,3,diamine, tin(II) chloride, ethanol

or saturated aqueous aniline oxalate solution

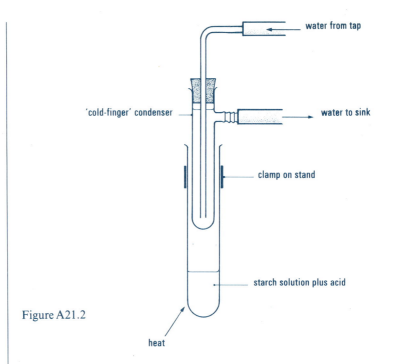

Figure A21.2

The action of saliva on starch

Have ready several separate drops of iodine solution on a microscope slide as before. Half fill a $250\,cm^3$ beaker with water and warm it to $40\,°C$ (no higher). Mix $10\,cm^3$ starch solution with about $1\,cm^3$ saliva in a $150 \times 25\,mm$ test-tube and place it in the warm water. After 5 minutes test for starch as before and continue this at 2-minute intervals until no more starch is present in the mixture. Boil 5 drops of the mixture with 5 drops of each of the two Fehling's solutions for one minute. What do you see?

What has happened to the starch?

The unused remainder of this mixture must also be handed to your teacher to be concentrated (under reduced pressure) for further investigation.

You should now suspect that when starch is mixed with acid or saliva a sugar is produced. You will not know whether the sugar is glucose $(C_6H_{12}O_6)$ or maltose $(C_{12}H_{22}O_{11})$. You can find this out by using paper chromatography. Your teacher will explain this in more detail. The following notes are for general guidance.

Use the rectangle of filter paper provided. Rule a pencil line 2 cm from the bottom of the paper and make four small vertical marks at 2 cm intervals along this line, two on either side of the middle point. Label the marks, *in pencil*, A, B, C, D, underneath the line. At mark A put one small drop of glucose solution, from a narrow glass tube, and *allow it to dry*. Continue this until five small drops have been added, allowing each one to dry before the next is added.
At mark B, repeat this process with maltose solution.
At mark C, with the concentrated solution from the action of acid on starch.
At mark D, with the concentrated solution from the action of saliva on starch.

Roll the paper into a cylinder which will stand in a gas jar without touching the sides and fasten the edges with clips. Put a mixture of propan-2-ol, acetic acid (ethanoic acid), and water (3 : 1 : 1 by volume) in the gas jar to a depth of 1 cm and stand the cylinder of paper in it, with the pencil line and spots at the bottom. Put a cover on the gas jar and leave it for 6–12 hours.

Remove the paper cylinder, open it out, and hang it up to dry in a stream of warm air. While the cylinder was in the gas jar, the solvent mixture will have moved upwards through it, taking the substances in the spots with it. You have now to find the new positions of the spots. Your teacher will explain which method you will use for this. Describe it briefly on a separate sheet of paper. Draw a diagram to show the final appearance of the paper.

Complete the following statements:

starch $\xrightarrow{\text{acid}}$ ....

starch $\xrightarrow{\text{saliva}}$ ....

Locating agents
Note. The first agent is the most suitable one. Tell pupils to take care not to get any of the reagents on their hands.

1. Roll the paper and dip it into a measuring cylinder containing a freshly prepared mixture of the following solutions:
2% aniline (phenylamine) in acetone (propanone), 5 volumes
2% diphenylamine in acetone (propanone), 5 volumes
85% phosphoric acid, 1 volume
and air dry before heating in an oven at 100 °C for 2–3 minutes or warm cautiously above a gauze heated by a Bunsen flame. Different colours appear for the various sugars:
Glucose – greenish, fading to greenish-brown after an hour or so
Maltose – bluish-grey.

The positions on the paper of the glucose and maltose spots should be compared with those of the two hydrolysates.

2. Roll up the paper and dip it into a measuring cylinder containing a freshly prepared solution made up as follows:
Benzene-1, 3 diamine, 0.5 g
Tin(II) chloride, 1.2 g
Acetic acid (ethanoic acid), 20 cm³
Ethanol, 80 cm³
Air dry the dipped paper and heat in an oven at 100 °C for 5 minutes or above a gauze heated by a Bunsen flame, taking care to avoid scorching the paper.

3. Alternatively, the procedure in (2) above may be followed using *p*-anisidine hydrochloride in butan-1-ol as the locating agent.

4. Dip the dried paper in a saturated aqueous solution of aniline oxalate and heat cautiously some distance above a gauze heated by a Bunsen burner or before an electric fire until brown spots appear.

After the experiment bring the class together for a discussion of the results. Models of glucose and maltose molecules are useful when the chemistry of the process is being discussed.

Suggestion for homework   Write an account of all of the ways in which men get starch from plants for food. Why do you think different plants are used in different parts of the world?

Summary

By the end of this section pupils should have some idea of the size of large organic molecules and know that they can be broken down (hydrolysed) in various ways. They will have used the important technique of chromatography to identify breakdown products of starch, and have found that glucose is one of those products.

A21.3
Can glucose be broken down further?

Glucose is broken down to ethanol and carbon dioxide by fermentation.

A suggested approach

Objectives for pupils

1. Knowledge of the fermentation of glucose to form ethanol
2. Identification of the products of fermentation of glucose as ethanol and carbon dioxide
3. Awareness of the industrial application of fermentation
4. Knowledge of glucose as one of the energy providers in living systems

If the suggested homework for the last section was given to the class, the lesson could start with a discussion of the sources of starch in the world (mainly grain, rice, and potatoes) and their subsequent processing and uses. One of these uses is the manufacture of beer and spirits by fermentation involving a further breaking down of the starch molecule through glucose to ethanol. Discussion may lead to ways in which the pupils themselves can perform a fermentation experiment.

This is an appropriate time to discuss the importance of glucose to man. Tell pupils that glucose is made in plants from carbon dioxide, water, and energy from the Sun. Man does not usually use the energy from the Sun directly, but he can make use of the Sun's energy by eating plants, and by using the glucose as a provider of energy.

Experiment A21.3

Apparatus

Each pair of pupils will need:

Experiment sheet 86

Conical flask, 100 cm³, fitted with cork or bung and delivery tube

2 test-tubes, 100 × 16 mm

Spatula

Stand, boss, and clamp

Glucose, 5 g

Baker's yeast, granulated yeast or, better, fresh brewers' yeast, 1 g

Limewater

(Continued)

To break down the glucose molecule by fermentation

Procedure
Experimental details are given in *Experiment sheet* 86.

To get a reasonable sample of ethanol, it is best for the teacher to pool the results of the class experiments and distil all the fermented products in a large flask with a fractionating column and condenser. It is advisable to decant or filter the solution from any solid residue, to prevent excessive frothing during the distillation. The first few cm³ to come over will contain enough ethanol to burn.

Experiment sheet 86

Put 5 g glucose in the flask and dissolve it in 50 cm³ water. (See figure A21.3.) Mix two measures of baker's yeast with a little water to form a thin paste and add this to the glucose solution. Half fill the test-tube with limewater and stand the apparatus in a warm place (near a radiator in winter). Observe it at intervals for a few days. What do you see?

Name one product of the reaction.

To investigate the presence of any other product, the mixture remaining in the flask has to be distilled. This is best done on a larger scale by combining the mixtures from the whole class, so your teacher will take over here.

For the distillation, the teacher will need:

Anti-bumping granules

Round bottomed flask, 500 cm³

Fractionating column, 20 cm long, packed with short pieces of glass rod or tubing

Thermometer, −10 to +110 × 1 °C

Bunsen burner and heat resistant mat

Tripod and gauze

Liebig condenser

2 lengths rubber tubing

Receiving flask or beaker

2 stands, bosses, and clamps

Test-tube rack

Test-tubes, 100 × 16 mm

Corks to fit flasks and fractionating column

Cork borers

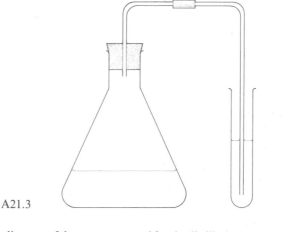

Figure A21.3

Draw a diagram of the apparatus used for the distillation on a separate sheet of paper.

Describe briefly what is done, giving the results on the distillate.

Complete the statement:

glucose $\xrightarrow[\text{in yeast}]{\text{zymase}}$ .... + ...

Write an equation for the reaction.

The chemical processes involved in the fermentation of glucose to ethanol may now be discussed with the class. Once more the use of models may help to explain what is happening during the reaction. Alternatively, an overhead projection original showing representations of such models may be used for this purpose. The products of fermentation can be identified as carbon dioxide and ethanol and the uses of ethanol should be mentioned during the discussion.

Suggestions for homework Describe the way in which man gets energy from the Sun through glucose. Could glucose be used as a fuel for engines? How?

A21.4
Can ethanol also be broken down further?

The degradation of starch is taken to its final stage in which ethanol is decomposed into ethene (ethylene) and water by passing it through a heated tube.

A suggested approach

Objectives for pupils

1. Ability to design an experiment

Various means have been used to break down starch to ethanol. Ask the class how they would try to break ethanol down yet further. 'Biochemical' processes will not help. Heating in air produces carbon dioxide and water. What happens if ethanol is heated in the absence of air? Refer to a molecular model of ethanol during the discussion with pupils and then let them try the method described below.

(Continued)

2. Knowledge of the production of ethene from ethanol and its properties
3. Understanding the use of molecular models to explain the properties of ethene

The teacher will need:

Molecular models of ethanol, ethene, ethane, and water

Experiment A21.4

Can ethanol be broken down further?

Apparatus

Each pair of pupils will need:

Experiment sheet 87

3 test-tubes, 125×16 mm, and corks

Hard-glass test-tube, 150×16 mm, fitted with cork and a delivery tube (see diagram)

Broken pieces of porous pot – to fill a 125×16 mm test-tube

Teat pipette, 1 cm³ capacity

Mineral wool

Stand, boss, and clamp

Trough

Bunsen burner and heat resistant mat

Ethanol, I.M.S. grade, 2 cm³

Bromine water, 2 cm³

0.01M potassium permanganate, 2 cm³

2M sulphuric acid, 1 cm³

Procedure
Experiment sheet 87 is reproduced below.

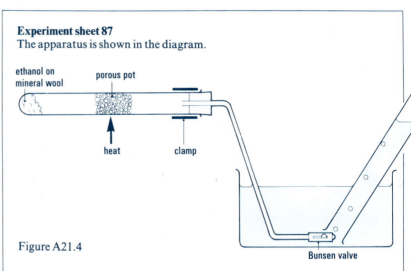

Experiment sheet 87
The apparatus is shown in the diagram.

Figure A21.4

Pack the end of a 150×16 mm hard glass test-tube with mineral wool to a depth of 2 cm and saturate it with 2 cm³ ethanol from a teat pipette. Nearly fill the rest of the tube with pieces of porous pot. Clamp the tube horizontally, insert the cork with delivery tube, and arrange for the collection of a gas over water. Heat the porous pot strongly but do not heat the mineral wool, enough heat is conducted along the test-tube to vaporize the ethanol. Allow a minute or two for air to be expelled. Collect three tubes of gas, cork them, and then remove the delivery tube from the trough before the apparatus cools.

Perform the following tests on the gas samples in order to identify the gas.

1. Remove the cork and apply a lighted taper to the mouth of the tube. Describe what happens.

2. Remove the cork, add about 2 cm³ bromine solution, replace the cork, and shake the tube. What happens and what is formed?

What was the gas collected in the test-tubes? ....

Write the equation for the reaction which took place when ethanol vapour was heated.

3. Repeat test 2 using dilute potassium permanganate solution acidified with dilute sulphuric acid instead of bromine solution. What happens and what is formed?

After the experiment tell the pupils that the gas ethene (ethylene) has the formula C_2H_4. 'Someone had to show that 1 mole of ethene contains 2 moles of atoms of carbon (C) and 4 of hydrogen (H).' Show the pupils models of ethanol, ethene, and water, and relate them to the equation which summarizes the reaction. Alternatively, show such representations on an overhead projection transparency.

Pupils may not have come across the tests which are performed during the experiment and it is worth explaining to them that they are all tests for 'ethene-like' compounds.

Using models of ethene and ethane it is possible to show that ethene has fewer hydrogen atoms per carbon atom than ethane. It is therefor said to be 'unsaturated'. This makes ethene a more reactive compound than ethane and hence an extremely important industrial chemical from which many other materials are made.

Finally, summarize the series of reactions by which starch has been broken down to glucose and then to ethanol and finally to ethene.

Suggestion for homework

Summarize the work of the last three sections in which starch has been broken down by stages to ethene.

A21.5
Cracking petroleum to obtain new products

One of the products of oil fractionation, medicinal paraffin, is broken down to ethene (ethylene). This provides a link with the breaking down of starch, and demonstrates an important industrial process.

A suggested approach

Objectives for pupils

1. Understanding the process of cracking
2. Knowledge of the dehydrogenation of saturated compounds to form unsaturated compounds
3. Awareness of the relevance of cracking to the petroleum industry
4. Knowledge of the process of catalysis
5. Knowledge of the process of polymerization

Pupils learnt something of the chemical importance of petroleum during Stage I. One of the refined products of crude oil is medicinal paraffin, and in this section the pupils perform a reaction comparable to one of the processes carried out in the petroleum industry. Explain to the pupils that there is a general problem in the industry – the light hydrocarbon molecules are the most useful ones but the heavier molecules make up the largest part of the crude oil. Chemists have solved this problem by a process called 'cracking' in which the large molecules are passed through a heated tube and broken into smaller molecules. Related to this is the process of dehydrogenation, in which the saturated hydrocarbon molecules lose hydrogen and become 'unsaturated'.

Ethene, which pupils came across in section A21.4, is one of the most important 'building blocks' in the whole chemical industry. It can be reacted directly with water in the presence of a catalyst to form ethanol (the opposite process to that which the pupils carried out) and can be polymerized to make polythene, a substance which should be very familiar to them.

The experiment which the pupils are to do involves both cracking and dehydrogenation of a mixture of molecules containing 20 or more carbon atoms each.

Cracking petroleum products

Procedure
The experiment uses a technique similar to that of Experiment A21.4. The gas that collects after the air has been expelled contains alkenes and will be found to have a characteristic smell, to burn with a yellow flame, and to decolorize weak bromine water. Full details appear in *Experiment sheet* 88 which is reproduced below.

Each pair of pupils will need:

Experiment sheet 88

4 test-tubes, 150×25 mm

Hard-glass test-tube, 125×16 mm, fitted with cork and delivery tube and Bunsen valve (i.e. piece of rubber tubing containing a slit and fitted with a piece of glass rod)

Bunsen burner and heat resistant mat

Mineral wool

Small pieces of porous pot, sufficient to fill a 125×16 mm test-tube

Stand, boss, and clamp

Trough

Teat pipette

Medicinal paraffin, 2 cm^3

Dilute bromine water, 2 cm^3

Experiment sheet 88
Arrange the apparatus as shown in the diagram.

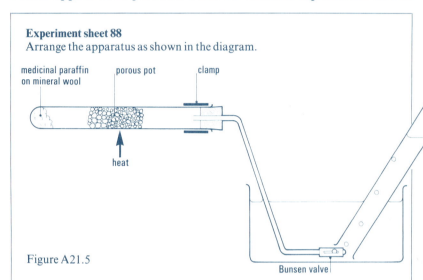

Figure A21.5

Pack the end of a 150×16 mm hard glass test-tube with mineral wool to a depth of 2 cm, and saturate it with medicinal paraffin. Nearly fill the rest of the tube with pieces of porous pot. Clamp the test-tube in position and fit the delivery tube – as shown above.

Heat the porous pot strongly but do not heat the mineral wool until the tube is quite hot.

Allow for air to be expelled from the apparatus before collecting several test-tubes of gas.

1. Note the smell of the product.

2. Try to burn the gas.

3. Examine the reaction between the gas and bromine solution.

What do you think the gas is?

After the experiment discuss the findings, differentiating these results from those obtained in Experiment A21.4. The film loop 2–9 'Cracking hydrocarbons' could be shown here.

Find out as much as you can about the usefulness of ethene as a 'raw material' for the chemical industry. (*Chemists in the world* Chapter 6 'Polymers from petroleum,' contains much useful information.)

A21.6
Preparing soaps and soapless detergents from castor oil

A soapless detergent is formed by reacting castor oil with sulphuric acid and a soap is formed by hydrolysing castor oil with an alkali.

The lesson may be started by reminding pupils that they were able to find out a great deal about starch by breaking the molecule down. Castor oil is another naturally occurring substance which could be investigated by this process. The fact that castor oil is a liquid should suggest that its molecules are smaller than those of starch. Ask pupils how they might try breaking down castor oil. Their answers are likely to be either hydrolysis or cracking since these are the two main processes used in the earlier parts of this Topic. In this case, the cracking would be too drastic to be useful, but hydrolysis yields very interesting results. Now tell pupils they are going to study two reactions using castor oil – with acid and with alkali. Warn pupils about the reagents and conditions to be used before allowing them to try the experiment described below.

Experiment A21.6

Apparatus

Each pair of pupils will need:

Experiment sheet 89

Safety spectacles

4 test-tubes, 100×16 mm

Test-tube, 150×25 mm

Test-tube rack and holder

Teat pipette

Measuring cylinder, 25 cm^3

2 beakers, 100 cm^3

Evaporating basin

Tripod and gauze

Bunsen burner and heat resistant mat

Glass rod

Filter flasks, funnel, filter papers, and filter pump

Spatula

Castor oil, 3 cm^3

5M sodium hydroxide, 10 cm^3

Sodium chloride, 10 g

Concentrated sulphuric acid, 2 cm^3

The reaction of castor oil with (1) sodium hydroxide; (2) sulphuric acid

Procedure

This is described in *Experiment sheet* 89 reproduced below.

Experiment sheet 89
You should wear safety spectacles for this experiment

1. *Reaction with sodium hydroxide*
Put 2 cm^3 castor oil in an evaporating basin and add 10 cm^3 5M sodium hydroxide solution (this solution is **very corrosive**, so be careful not to get any of it on your skin or on the bench). Place the basin on top of the beaker in which water is boiling and heat it in the steam for 10 minutes, stirring well.

Add to the mixture in the basin 40 cm^3 water in which 10 g sodium chloride has been dissolved, stir and heat for another 5 minutes. Allow the mixture to cool after removing the dish from the steam. A solid will separate on the top of the liquid; skim off with a spatula and wash it with a little cold water in a small beaker. Pour off the water. Transfer a small quantity of the solid to a 100×16 mm test-tube, half fill the test-tube with water, and shake well. What happens?

What do you suppose the solid is?

2. *Reaction with sulphuric acid*
Put 1 cm^3 castor oil in a 100×16 mm test-tube and fasten this in a test-tube holder. Hold the test-tube over a beaker and add carefully 2 cm^3 concentrated sulphuric acid, keeping your face well away from the mouth of the tube. Stir the mixture with a glass rod. What happens?

When reaction appears to be finished pour the contents of the tube into a 3 cm depth of water in a 150×25 mm test-tube. Stir the mixture and carefully pour off the water into the sink. Wash the solid twice more with about the same amount of water, pouring off the liquid each time.

Shake a small quantity of the solid with half a test-tube of water. What happens?

What do you suppose the solid is?

After the experiment discuss the nature of the products formed in each case. In the case of the reaction with sodium hydroxide, a *soap* was formed; and in the case of the acid treatment, a *soapless detergent* was formed. Once again this is an opportunity to emphasize the importance of chemistry in daily life. (The history and chemistry of detergents has been well described in a Unilever educational booklet, *Detergents* (Revised Ordinary Series No. 1). This account is a little too advanced for most pupils who will be studying Topic A21 but provides useful background reading for the teacher.)

The question 'How much of the chemistry of these processes do we expect the pupils to understand?' must be faced at this point. There is no reason why they should not appreciate that naturally occurring oils – such as castor oil – are composed of mixtures of closely related compounds, each of which is a compound of an acid and an alcohol. The facts that the predominant acid is a fatty acid, ricinoleic acid, and the alcohol is glycerol (propane-1,2,3-triol) are not important at this stage. Nevertheless, if pupils are inquisitive give them the formula of glycerol, $CH_2OH.CHOH.CH_2OH$, and of ricinoleic acid, $CH_3(CH_2)_5CHOHCH_2CH=CH(CH_2)_7CO_2H$

These formulae will show them the relative sizes of the molecules. They are not, of course, expected to remember the formulae. Show the pupils models of these molecules. They should understand that the alkaline hydrolysis produces the sodium salt of the acid – which is a soap (one type of detergent) – and the action of the acid on the oil produces a soapless detergent.

Note for teachers

The reactions of castor oil (ricinoleic acid) with sodium hydroxide and with concentrated sulphuric acid are atypical: sodium hydroxide yields octan-2-ol and sodium sebacate (disodium decanedioate); and concentrated sulphuric acid yields a complex mixture of the hydrogen sulphate of ricinoleic acid in which the hydroxyl group is esterified and a compound in which the sulphuric acid has added to the double bond. Esterification and addition do not occur together in the same molecule of ricinoleic acid. This information is supplied so that no over-simplification will be made when interpreting these two reactions. It is helpful to tabulate the similarities and differences of soaps and soapless detergents at this point, especially their reactions with hard water and their cleansing actions (see Alternative IIB, Topic B15.4).

Suggestion for homework

Write an account of the cleansing action of detergents. (See *Handbook for pupils*, Chapter 12 'Man, chemistry and society'.)

A21.7
Breaking down and building up perspex

The study of large molecules is extended to man-made polymers. A polymer, perspex, is depolymerized and then repolymerized.

Objectives for pupils

1. Knowledge of the meaning of the terms monomer and polymer, and of the process of polymerization
2. Knowledge of the industrial applications of polymerization

One of the most striking developments of the last twenty years has been the increasing use of man-made polymers. This might be made the theme of the first part of this section. Ask the class how many they can name. They are probably familiar with PVC (polyvinylchloride), polythene, and perspex. Explain that all of these substances are made by taking a relatively simple molecule, the 'monomer', and making it join up with itself to form very long chains of particles. In this respect, man is imitating nature in much the same way as the pupils found in section A21.2 when they discovered that starch was a compound containing long chains of 'glucose-like' molecules. An example of an everyday use of polymerization is to be found in modern glues in which a 'resin' is formed when a monomer is mixed with a catalyst (the 'hardener').

The experiment below is designed to show how monomers polymerize.

Experiment A21.7

Depolymerizing perspex and polymerizing the monomer

Procedure

This experiment *must* be done in a fume cupboard by the teacher since *poisonous* fumes are given off during the heating of perspex.

5 g of perspex chips are placed in the test-tube fitted with a cork and delivery tube. The test-tube is heated with a non-roaring Bunsen flame and the distillate is collected in a second test-tube which is kept cool by standing it in a beaker of water.

Apparatus

The teacher will need:

2 hard-glass test-tubes, 125 × 16 mm, one fitted with a cork and delivery tube and the other with a cork, delivery tube, and thermometer, 0–360 × 1 °C

6 test-tubes, 75 × 10 mm

2 beakers, 100 cm³

Tripod and gauze

Bunsen burner and heat resistant mat

2 stands, bosses, and clamps

Access to oven

Chips of perspex (acrylic granules), about 5 g

Lauroyl peroxide, about 0.10 g

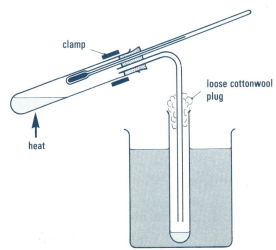

Figure A21.6

The liquid distillate contains impure monomer which cannot be polymerized without purification. This is done by fractional distillation using the apparatus shown in figure A21.6. The fraction collected is that boiling between 100 and 105 °C. This is the pure monomer. Repolymerization is brought about by heating the monomer in the small test-tube standing in a beaker of hot water at about 80 °C. A small quantity (about 0.1 g) of lauroyl peroxide is added to the monomer to catalyse the polymerization. The monomer begins to thicken after about five minutes, and goes solid in about 10 minutes.

After the experiment ask the pupils some questions about it. How can they tell the distillate contains smaller molecules than the polymers? The evidence, which is not conclusive, is that it is a liquid rather than a solid at room temperature.

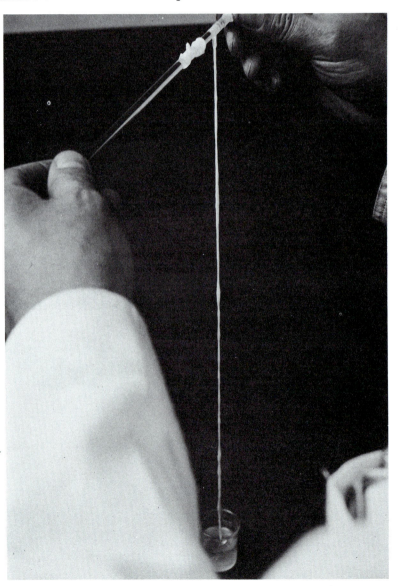

Figure A21.7
Making nylon.

A21.8
Making nylon

This section is designed to give more time and opportunity to develop the theme of polymers and their uses. The pupils perform an experiment in which nylon is produced.

One of the most useful polymers yet made by man is nylon. This section could be started by explaining that it was one of the first polymers to be discovered by Carothers in 1935. Its remarkable tensile strength should be mentioned. Polymers such as terylene and nylon are made of long chains containing two different units.

The production of nylon from hexamethylenediamine and adipyl chloride can be performed by the pupils. The teacher can make the point that the type of linkage formed when nylon is made is a 'man-made' modification of the type of linkage found in proteins. The film loop 3–4 'Giant molecules – proteins' *may* be appropriate to use with some pupils.

A suggested approach

Objectives for pupils

1. To extend pupils' knowledge of polymers and their uses
2. To provide further opportunity for using molecular models in reaction summaries

Experiment A21.8

Apparatus

Each pair of pupils will need:

Experiment sheet 90

Crucible or beaker, 5 cm^3

Forceps

Glass rod

Hexamethylenediamine (hexane-1,6-diamine), 5% solution in water

Adipyl chloride (hexanedioyl dichloride), 5% solution in carbon tetrachloride (tetrachloromethane)

To prepare some nylon

Procedure
Details are provided in *Experiment sheet* 90. See also figure A21.7.

Experiment sheet 90
You will be provided with a solution of adipyl chloride in carbon tetrachloride (tetrachloromethane) (solution A) and a solution of hexamethylenediamine (hexane-1,6-diamine) in water (solution B).

Put about 2 cm^3 solution A into a crucible or a very small beaker and carefully add 2 cm^3 of solution B *so that the two solutions do not mix*. A film will form where the two liquids meet; pull out a thread from this with a pair of tweezers and attach it to a glass rod. This is a thread of one kind of nylon. Rotate the glass rod above the liquid mixture so that a continuous thread is wound on to the rod. When you have collected about six feet of thread break it off with the forceps and wash it in a gentle stream of water. You can now handle it safely and test its strength. The process can be repeated until one of the solutions is used up.

After the experiment some time should be spent in discussing the reaction. It is not intended that pupils should be taught the equation for the reaction although some classes may find this helpful. The key point is that polymerization is possible because each molecule possesses two reactive groups, one on each end. Thus, the amine group on each end of the hexamethylenediamine molecule reacts with an adipyl chloride molecule. Each adipyl chloride molecule, so linked, then has a free acyl chloride group to link with another hexamethylenediamine molecule, and so on. Models of the two molecules may be used to clarify the explanation, and reference may be made to the structure of natural fibres which can be regarded as polyamides.

It may be appropriate to conclude the Topic by reviewing the various terms introduced. Some teachers find it helpful to spend a little time on the nature of the chemical bond, using the models of molecules which have been mentioned in the Topic.

Summarize the work you have done during this Topic.

Summary

By the time they have completed this Topic, pupils should be aware of the chemical significance of large molecules. They should know that large molecules play a most important part in all living systems. Cracking and hydrolysis are shown as processes which have important industrial applications. The pupils will have been introduced to the world of polymer chemistry and will have made at least two polymers in the laboratory, perspex and nylon.

Ammonia, fertilizers, and food production

Purposes of the Topic

1. To provide a study of the chemistry of ammonia and its industrial synthesis.
2. To provide a simple investigation of some of the properties of proteins and of their relationship to the use of ammonia as a fertilizer.
3. To indicate the social significance of applied chemistry.
4. To apply various chemical concepts to new situations.

Contents

A22.1 Getting ammonia from natural substances
A22.2 What elements are present in ammonia?
A22.3 What is the formula of ammonia?
A22.4 Making ammonia from nitrogen and hydrogen
A22.5 Ammonia as an alkali
A22.6 Fertilizers and food production

Timing

Each section will probably require one double period. To allow adequate time for discussion, some six or seven weeks are needed for the Topic.

Introduction to the Topic

The underlying theme of this Topic is the usefulness of chemistry, illustrated by the part which chemists are playing in solving the world food problem. The first step is to find out what elements are present in living substances. The pupils discover that it is possible to obtain ammonia from proteins. They then find out more about ammonia, showing that it contains nitrogen and hydrogen and seeing an experiment in which ammonia is 'cracked' in order to determine its formula. Pupils learn that ammonia and its compounds play an important part in the chemistry of living things and consequently in the manufacture of fertilizers. The key problem is how to synthesize very large quantities of ammonia using nitrogen from the air and hydrogen from a crude oil fraction. The Haber process is demonstrated and the reaction is compared with the 'cracking' of ammonia. These two reactions are seen to be different sides of the same equilibrium. The idea of dynamic equilibrium developed in Topic A19 and ideas on catalysis from Topic A18 are used in studying these reactions.

The Topic then leads to the need to convert ammonia into a 'solid' form for general agricultural use. Pupils investigate the properties of ammonia and prepare a sample of ammonium sulphate which they can then use in an experiment with plants. The Topic concludes with a summary of the world food problem, some information about fertilizers in common use, and a discussion on the nitrogen cycle.

Topic B14 adopts the wider theme 'Chemistry and the world food problem'. It considers the problem of food supply, food storage, use of fertilizers, etc. and then investigates the production and properties of ammonia.

Sections A22.3 and A22.4 use the concepts of equilibrium and catalysis which were considered in Topics A19 and A18 respectively. The properties of ammonia solution were mentioned in Topic A20 and these are revised and extended in the section dealing with the preparation of ammonium salts (A22.6).

Nuffield Advanced Chemistry (1970) *Teachers' guide I* (Topic 12), (1970) *Teachers' guide II* (Topic 19); and Nuffield Advanced Physical Science (1972) *Teachers' guide I* (Section 6) provide a more detailed account.

Film loops and strips
2–11 'Ammonia manufacture'
2–12 'Ammonia – uses'
2–13 'Catalysis in industry'
3–4 'Giant molecules – proteins'

Films
'Ammonia' ICI Film Library
'Cereals in the 70s' ICI Film Library
'Weeding by spraying' ICI Film Library
'Nothing to eat but food' Unilever Film Library

Chemists in the world, Chapter 7 'Fertilizers'.
Handbook for pupils, Chapter 11 'The world food problem', and Chapter 12 'Man, chemistry, and society'.

A22.1
Getting ammonia from natural substances

In this section the world food problem is introduced. The first step is to find out about the composition of living matter. It is found that ammonia can be obtained from a number of natural substances.

Objectives for pupils

1. Awareness of the world food problem
2. Recognition that ammonia can be obtained from natural substances
3. Knowledge of the use of activated charcoal for absorbing gases

Pupils now consider some of the relationships between food production and chemistry. The rise in world population has led to an ever-increasing need for food, and therefore to more and more intensive farming. Supplies of natural fertilizers have become insufficient to replenish the soil. During the late nineteenth century it became vital to supplement these natural fertilizers with artificially produced ones. Before this could be done, it was necessary to find out what substances plants needed in order to grow. Pupils know already that carbon and hydrogen are found in living materials but they may not realize that these elements are found in all living things. 'What else do living things contain?' One of the answers to this question may be obtained by heating substances like meat, fish, corn, and hair with alkali. In each case a rather foul-smelling substance which turns red litmus blue comes off. It may remind some pupils of one property of ammonia. However, the smell is not very much like ammonia.

Investigating foodstuffs (1)

Apparatus

Each pair of pupils will need:

2 hard-glass test-tubes, 100 × 16 mm

Test-tube holder

Test-tube rack

Bunsen burner and heat resistant mat

Spatula

Indicator papers (litmus and Full-range Indicator)

Access to a fume cupboard

Samples of materials which contain protein: e.g. meat, fish, corn, hair, cheese

Soda lime

Procedure

It is suggested that pupils heat a small quantity of a protein-containing foodstuff with twice its bulk of soda lime. The fumes evolved are tested with indicator paper and found to be alkaline. The teacher should note that the fumes are particularly unpleasant and it is suggested that the experiment be performed in a fume cupboard or a well ventilated laboratory.

When the pupils have completed this experiment, discuss their results. An alkaline gas or gases are given off when a protein-containing foodstuff is heated with soda lime. The most common alkaline gas is ammonia, with which they may be familiar through its use in the kitchen or through Topic A20. Ammonia has a characteristic smell. We are faced with the practical problem of detecting ammonia in the presence of other strongly smelling substances. Is it possible to filter gases in the same way that we filter solid particles from a liquid? Tell the pupils that during the Second World War gas masks were produced which contained a type of carbon called 'activated carbon'. Industrial gas masks also make use of it. The 'activated carbon' absorbs dense poisonous gases and allows less dense gases like oxygen, nitrogen, and ammonia to pass through it. Ask for suggestions as to how the experiment could be modified to allow the gases to pass through activated carbon and then let pupils try Experiment A22.1b.

Experiment A22.1b

Investigating foodstuffs (2)

Apparatus

Each pair of pupils will need:

Experiment sheet 91

Hard-glass test-tube, 100 × 16 mm

Test-tube holder

'Calcium chloride' drying tube filled with activated carbon (charcoal) and plugged with cotton wool

Cork or bung to fit test-tube and drying tube

Bunsen burner and heat resistant mat

Indicator paper

(Continued)

Procedure

Details are given in *Experiment sheet* 91.

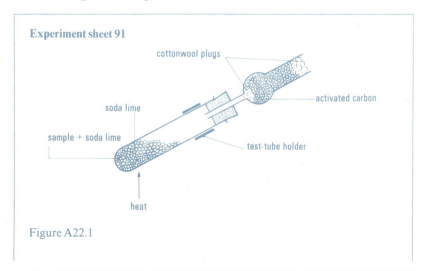

Experiment sheet 91

cottonwool plugs

activated carbon

soda lime

sample + soda lime

test-tube holder

heat

Figure A22.1

Glass rod

Spatula

Concentrated hydrochloric acid

Soda lime granules, 6 spatula measures

Gelatine, 3 spatula measures

Milk powder, 3 spatula measures

In the test-tube put 2 measures of the sample mixed with 2 measures of soda lime and cover this with a further 2 measures of soda lime. Heat the mixture *until* a gas comes out of the activated carbon tube (smell cautiously). Test the gas with (a) wet indicator paper; (b) hydrogen chloride gas, from a glass rod dipped in concentrated hydrochloric acid.

Record your results and conclusions below.

After the pupils have tried the second experiment and have detected at least a trace of ammonia in the evolved gases by smell, an experiment may be demonstrated in which the ammonia produced is unmistakable. The decision whether or not to do this experiment *must* depend on the time available and on the relative success of the last two experiments.

Experiment A22.1c

The detection of ammonia as a product of the decomposition of a protein
This experiment **must** be done by the teacher.

Apparatus

The teacher will need:

Kjeldahl flask

'Calcium chloride' drying tube, containing activated carbon (charcoal) and plugged with cotton wool, carrying cork (or bung) to fit flask

Beaker, 100 cm^3

Bunsen burner and heat resistant mat

Tripod and gauze

Stand, boss, and clamp

Measuring cylinder, 25 cm^3

Anti-bumping granules

Indicator papers

Spatula

Access to a fume-cupboard

Egg albumin (white of egg)

Potassium hydrogen sulphate

Concentrated sulphuric acid

8M sodium hydroxide

Procedure
The teacher should do this experiment in a fume cupboard.

Put 5 grams of egg albumin into the Kjeldahl flask and cautiously add 15 cm^3 of a solution containing a spatula measure of potassium hydrogen sulphate in concentrated sulphuric acid. Boil the contents of the flask in a fume cupboard for about 10 minutes. Allow to cool and then add the residue very carefully to a beaker containing 10 cm^3 of water. Return the solution to the flask and add sufficient 8M sodium hydroxide (or solid pellets) to make the solution strongly alkaline. Put some anti-bumping granules into the flask and fit the drying tube on the top. Place a wad of damp cotton wool on the open end of the drying tube and heat the flask gently. After boiling for five minutes the cotton wool will be saturated with ammonia. Pass it around the class for the pupils to smell.

Ask pupils what they know about proteins. They should be aware that proteins are an important constituent of foods, and should be told that ammonia can be formed from the decomposition of proteins.

Suggestion for homework

What is the world food problem? Suggest some ways which would increase the world's supply of food.

A22.2
What elements are present in ammonia?

An experiment is devised to find out what elements are present in ammonia. Ammonia is passed over heated copper oxide, and water, nitrogen, and copper are produced.

What elements are present in ammonia? Pupils may already have the answer from general knowledge or from labels on ammonia bottles. The question then becomes 'How can we show that nitrogen and hydrogen are present in ammonia?' The clue lies in the pupils' own experiments with carbohydrates in section A13.6 where they used copper oxide to convert hydrogen into water and carbon into carbon dioxide. The pupils may be reminded of this and asked what they think would happen if a similar experiment were performed with ammonia. Discuss the apparatus to be used and then let the pupils try the experiment.

To remove hydrogen from ammonia

Procedure
Details are given in *Experiment sheet 92*.

Experiment sheet 92

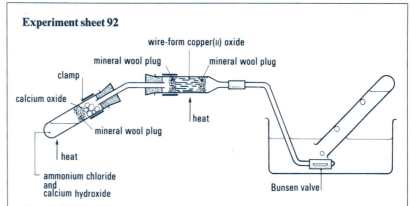

Figure A22.2

The apparatus is shown above. Ammonia, prepared by heating a mixture of ammonium chloride and calcium hydroxide and dried with calcium oxide, is passed over heated copper(II) oxide.

Mix 3 measures of ammonium chloride with an equal bulk of calcium hydroxide in the 125 × 16 mm test-tube. Add a plug of mineral wool (see diagram), and then add small lumps of calcium oxide. Put 3 measures of copper(II) oxide in the combustion tube.

Heat the bottom of the test-tube and the combustion tube. What happens to the copper oxide?

What does this tell you?

What do you notice on the cooler part of the combustion tube?

What does this tell you?

The pupils have now shown that ammonia contains hydrogen (they know that water is an oxide of hydrogen from Topic A7.5) and an inert gas which we must assume to be nitrogen without additional evidence. We now need to find out more about ammonia: its formula and how to make it from the nitrogen of the air and hydrogen. This is discussed in the following sections.

Suggestion for homework

Read *Chemists in the world*, Chapter 7 'Fertilizers': sections on 'Food for plants' and 'The nitrogen problem'.

A22.3
What is the formula of ammonia?

The volume composition of ammonia is determined by the teacher, using $100 \, cm^3$ syringes.

A suggested approach

Objectives for pupils

1. Understanding of 'volume composition' experiments
2. Knowledge of Gay-Lussac's and Avogadro's laws
3. Knowledge of the formula of ammonia

The problem to be solved is, 'How can we find the formula of ammonia?' If pupils know the formula of ammonia to be NH_3, it becomes a matter of confirming the formula.

Pupils should know that a mole of any gas occupies the same volume under given conditions of temperature and pressure (see Topic A14). To put it another way, a given volume of a particular gas at a particular temperature and pressure contains the same number of molecules as an equal volume of any other gas at the same temperature and pressure. If we can find the volumes of the gases which react together to produce a known volume of the gaseous product, and we also know the formulae of the reacting gases, we can deduce the formula of the product. The problem then becomes a technical one of finding ways in which to measure the reacting volumes of gases.

In the case of ammonia we have yet another problem. Under 'normal' conditions, hydrogen and nitrogen do not react together to any appreciable extent to form ammonia. Fortunately, this problem is readily solved by considering the reverse process. Ammonia is easily decomposed into hydrogen and nitrogen. The gas may be first decomposed, the increase in volume noted, and then the hydrogen removed by reaction with copper(II) oxide. An experiment in which this is done, using syringes, is described below.

Experiment A22.3

The volume composition of ammonia
This experiment should be done by the teacher.

Procedure
Set up the apparatus as shown in figures A22.3 and A22.4.

Apparatus

The teacher will need:

Apparatus for producing dry ammonia:

Hard-glass test-tube, 150×25 mm

Drying tube containing glass wool and fresh pellets of potassium hydroxide

Rubber bung and a delivery tube

Bunsen burner and heat resistant mat

The teacher will also need:

Either Syringe bench arranged as shown in figure A22.4

Or 3 syringes, 100 cm³, in glass

3 syringe holders with stands, or a special syringe bench

2 three-way stopcocks with capillary tubing

2 combustion tubes, 15 cm long, 7 mm in diameter (ordinary hard-glass tubing is not satisfactory: transparent tubing made of silica or a high silicon glass is essential)

4 pieces of hard-glass rod about 2 cm long, with a diameter slightly smaller than the internal diameter of the tubes

Thick walled rubber connecting tubing

Indicator paper

Cylinders of compressed gases: hydrogen and nitrogen

Iron wool (clean and free from grease)

Copper(II) oxide, wire form

0.880 ammonia solution

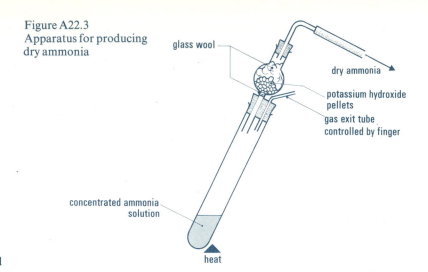

Figure A22.3
Apparatus for producing dry ammonia

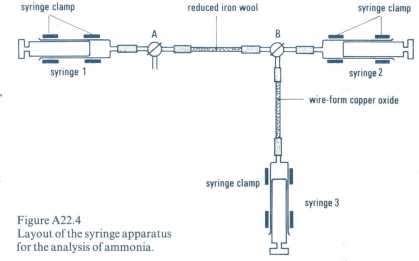

Figure A22.4
Layout of the syringe apparatus for the analysis of ammonia.

Make sure that the iron wool is freshly reduced by heating it in a stream of hydrogen from the cylinder to remove any trace of oxide.

Flush the whole apparatus with nitrogen from a cylinder in order to drive out all air. Fill syringe 1 with 48 cm³ (0.002 mole) of dry ammonia through stopcock A by warming a sample of 0.880 ammonia solution in the test-tube, using the drying tube suggested earlier. Heat the combustion tube containing the iron wool very strongly with a roaring Bunsen burner (the iron wool must be white hot). Have stopcock B set so that the syringes 1 and 2 are connected, and pass the ammonia backwards and forwards over the heated iron wool from syringe 1 to syringe 2. Do this several times until no further increase in volume takes place. This may require several passes. Now cool the combustion tube with a damp cloth, pass all the gas into syringe 1 and read the increase in volume. The volume of gases (now a mixture of hydrogen and nitrogen) should have nearly doubled. Pass the gases into syringe 2 and turn the stopcock B so that syringes 2 and 3 are connected. Heat the combustion tube containing copper(II) oxide and pass the gas from syringe 2 over the copper oxide into syringe 3 and back to syringe 2 until there is no further decrease in volume. The hydrogen will reduce the copper oxide to form water, and the volume should now be just over a quarter of the volume of the mixture of gases. Hence, the result may be expressed as:

$$\underset{\text{ammonia}}{2\text{ volumes of}} \longrightarrow \underset{\text{nitrogen}}{1\text{ volume of}} + \underset{\text{hydrogen}}{3\text{ volumes of}}$$

Sources of experimental error

1. It is *most important* that the ammonia should be *dry*. Otherwise some ammonia will dissolve in the water and the volume will be more than doubled when the gas is passed over heated iron.

2. If there is (*a*) residual air in the apparatus or (*b*) iron oxide on the iron wool, some of the hydrogen will react with this and the result will be low.

3. In the second part of the experiment water is produced. This means that the nitrogen remaining is saturated with water vapour whereas the other two volumes were measured dry. The result is rather *high*.

Specimen results

48 cm^3 (0.002 mole) of ammonia
96 cm^3 (0.004 mole) of nitrogen and hydrogen
25 cm^3 (0.001 mole) of nitrogen

The experiment shows that one volume of nitrogen and three volumes of hydrogen produce two volumes of ammonia. Using Gay-Lussac's law and Avogadro's law and assuming the formulae of nitrogen and hydrogen, we can now write:

$$2\text{N}_x\text{H}_y(g) \longrightarrow \text{N}_2(g) + 3\text{H}_2(g)$$

The formula of ammonia must therefore be N$_1$H$_3$ – that is, NH$_3$.

We can now ask the pupils if they think that the reaction went to completion. 'Was all of the ammonia converted into nitrogen and hydrogen?' On the evidence of volumes of gases produced this may appear to be the case, but to confirm it we need to see if there is any ammonia remaining. This can be done by repeating the first part of Experiment A22.3 and expelling the gases through the stopcock onto a small piece of damp indicator paper. The indicator paper will show the presence of an alkaline gas and it is usually possible to smell the ammonia. Thus, the reaction does not go to completion. 'Does this show us that the ammonia, nitrogen, and hydrogen are in equilibrium? If so, is it a clue to how ammonia can be synthesized?' The synthesis of ammonia is the subject of the next section.

Suggestion for homework

Describe the evidence to support the view that ammonia has the formula NH$_3$.

A22.4
Making ammonia from nitrogen and hydrogen

The world food problem is reviewed. Pupils should understand that nitrogen is a key element in life. They now consider methods of synthesizing ammonia and replacing the 'fixed' nitrogen in the soil.

1. Understanding the synthesis of ammonia and its social significance
2. Knowledge of the range and composition of common fertilizers

The first three sections of this Topic have shown nitrogen to be a key element in living things. If the suggestions for homework have been used, pupils will be familiar with some of the points raised in this section.

Discuss the importance of nitrogen and the nitrogen cycle. The problem which chemists had to solve was how to replenish the nitrogen of the soil. Gaseous nitrogen is of no direct value as plants are unable to use it. However plants can use nitrogen in the form of ammonium compounds or as nitrates. Naturally occurring nitrates from Chile were used at first but the great advance came when Haber and Bosch succeeded in synthesizing ammonia from nitrogen and hydrogen.

The experiment in the previous section led to the idea that nitrogen, hydrogen, and ammonia are in equilibrium. If this is so, it should be possible to produce ammonia from a mixture of nitrogen and hydrogen. It is suggested that the teacher repeats the experiment using nitrogen and hydrogen as starting materials instead of ammonia. Care must be taken that no residual ammonia is in the apparatus since this would invalidate the results.

Experiment A22.4

The synthesis of ammonia
This experiment should be performed by the teacher.

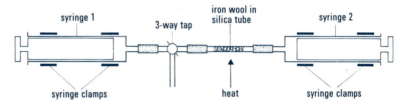

Figure A22.5
Apparatus for the synthesis of ammonia.

Apparatus

The teacher will need:

2 syringes, 100 cm³

2 syringe holders, bosses, and stands, or a syringe bench

Combustion tube, 15 cm long, 7 mm diameter

2 pieces of hard-glass rod, about 2 cm long, diameter slightly smaller than the internal diameter of the combustion tube

1 three-way stopcock with capillary tubing

Bunsen burner and heat resistant mat

Thick-walled rubber tubing

Indicator paper (preferably litmus)

Iron wool

Cylinders of compressed gases: nitrogen and hydrogen, with connecting tubing

Procedure
Set up the apparatus as shown in figure A22.5. Flush out the apparatus with nitrogen and then with hydrogen. Transfer 100 cm³ of hydrogen to syringe 1, close the stopcock, and then heat the iron wool strongly using the Bunsen burner. Pass the gas to and fro over the hot iron wool to reduce any iron oxide which may be present. Continue this process until no further reduction in gaseous volume takes place. Then remove the Bunsen burner and discharge the residual gas from syringe 1. Flush out the entire apparatus with nitrogen and then transfer 20 cm³ of nitrogen and 60 cm³ of hydrogen from the gas cylinders to syringe 1 through the stopcock. (It is very easy to break the syringe plunger by blowing it out of the syringe end with gas from a gas cylinder. This may be prevented by attaching a short length of string to the plunger and the syringe barrel as a safety device.) The following sequence of operations is advisable.
1. Open the cylinder valve before connecting the cylinder with the stopcock and set the fine adjustment control on the cylinder to produce a very slow rate of flow. This ensures that all air is swept out of the connecting tubing.

---

2. Hold the rubber tube from the cylinder lightly against the stopcock when passing gas into the syringe, so that it can be removed as soon as enough gas is in the syringe.
3. When the gases are in syringe 1, turn the stopcock to connect syringe 1 with syringe 2.
4. Heat the combustion tube containing the iron wool strongly with a Bunsen burner and pass the gases two or three times over the iron wool.

When the gas has been treated, turn off the Bunsen, open the stopcock, and eject the gases onto a piece of damp red litmus paper. It will turn blue showing the presence of ammonia.

A great deal can be learnt from this experiment. Some ammonia is synthesized under these conditions but not enough for the process to be a commercial proposition. Haber had to think of ways to increase the rate of production of ammonia from nitrogen and hydrogen. Pupils can probably understand that compressing the gases will make them more likely to form ammonia as this has a smaller volume than the reactants, hydrogen and nitrogen. (The fact that temperature also affects this equilibrium may be mentioned, but it is not necessary to discuss this in detail.) Chapter 7 'Fertilizers' of *Chemists in the world* includes a section on this Topic.

Suggestions for homework

Read about Haber's solution to the nitrogen problem – see *Chemists in the world*, Chapter 7, 'Fertilizers'.

A22.5
Ammonia as an alkali

In this section the properties of ammonia are discussed in terms of its ability to accept protons. An ammonium salt is prepared.

A suggested approach

Objectives for pupils

1. Knowledge of ammonia as a proton acceptor
2. Recapitulation of knowledge of acid–base theory

The pupils should know from their work in Stage I and Topic A20 that a solution of ammonia in water is alkaline and will neutralize acidic solutions. Remind pupils of the concept of an acid as a proton donor (see Topic A20). 'What happens when ammonia is added to an acid?' The pupils know that an acidic solution is one which contains oxonium ions, $H_3O^+(aq)$ (hydrogen ions, $H^+(aq)$), in a state of equilibrium. Explain that the addition of ammonia molecules upsets this equilibrium because the ammonia reacts with the oxonium ion accepting the proton to form ammonium ions and water:

$$NH_3(aq) + H_3O^+(aq) \rightleftharpoons NH_4^+(aq) + H_2O(l)$$

The equilibrium position is well over to the right, and the fact that the ammonia 'wins' most of the protons from the water shows that ammonia is a stronger base than water. Emphasize that in the above equilibrium both ammonia and water are bases (proton acceptors) and the oxonium ion and the ammonium ion are both acids (proton donors).

Then turn to a practical problem concerned with the chemistry of ammonia. Ammonia is a good fertilizer but, being a gas, cannot easily be applied directly to the soil. Its alkaline nature is also a disadvantage. Direct injection of ammonia into the soil has been tried in the United States of America, but it has not been widely adopted. The problem is to get the ammonia into a neutral, solid, and soluble form. This problem will be discussed more fully in the next section, but pupils will quickly see that a salt such as ammonium sulphate provides at least a partial answer. Ammonium sulphate or some other ammonium salt can now be prepared.

Preparation of ammonium sulphate

Apparatus
Each pair of pupils will need:

Experiment sheet 93

Evaporating basin or a beaker, 100 cm³

Glass rod

Bunsen burner, heat resistant mat, tripod, and gauze

Measuring cylinder, 25 cm³

Microscope slide

Filter flask, funnel, filter paper, and filter pump

M sulphuric acid, 20 cm³

2M ammonia solution 25 cm³

Procedure
Details of the procedure are given in *Experiment sheet* 93.

Pupils may wish to retain a sample of the product in a specimen tube.

Experiment sheet 93
Put about 20 cm³ M sulphuric acid into an evaporating basin. Add 2M ammonia solution, a little at a time, with stirring, until the mixture has a definite smell of ammonia. Evaporate the solution to about one-fifth of its original volume and set aside to cool. How will you tell when the solution will crystallize?

When crystallization is complete, filter off the crystals and dry them on blotting paper or filter paper. Keep them for the next experiment. Write the equation for the reaction.

Why is it not necessary to add exactly the correct amount of ammonia solution?

Suggestions for homework

Read the rest of *Chemist in the world*, Chapter 7 'Fertilizers'.

A22.6
Fertilizers and food production

In this section the Topic is concluded with a discussion on fertilizers.

A suggested approach

Objectives for pupils

1. Knowledge of the use of ammonium compounds as fertilizers
2. Knowledge of the use of phosphates and potassium salts in fertilizers

In the last section, chemical fertilizers were mentioned briefly and pupils prepared ammonium sulphate. Begin this section by discussing the usefulness of ammonium sulphate. Pupils could try using it on plants, with a control, to see if it really does encourage growth. For example, it may be possible to mark out some areas of grass and to apply different quantities of ammonium sulphate to each area. (Co-operation with the Biology department in the school is necessary here to avoid duplication of work.)

Two questions can be discussed before the results of the experiment are seen. 'Is the salt neutral?' and 'What proportion of nitrogen is there in the salt?' The first question is important since the fertilizer will affect the pH of the soil to which it is added. The pupils can dissolve some ammonium sulphate in distilled water and find that it has distinctly acidic properties. This can be related to an earlier discussion on acidic properties of chemicals which are not labelled 'acid' (see Topic A20.1). Ask them how they think farmers solve this problem (lime).

The second question is important as it affects both transport costs and the cost of putting it on the land. Ask the pupils to work out the percentage of nitrogen in ammonium sulphate $(NH_4)_2SO_4$ (21.2%); ammonium nitrate, NH_4NO_3 (35%); and in urea, CON_2H_4 (46.7%). The answer accounts for the growing popularity of ammonium nitrate and urea as fertilizers. Some attempt should be made to review the current range of nitrogenous fertilizers. The 'nitrogen cycle' provides a convenient summary of this part of the section (see section A22.4).

Although this Topic is primarily concerned with the role of ammonia in solving the world food problem, the importance of phosphates and of potassium salts should also be mentioned (for example, with respect to nitrogen, phosphorus, and potassium content). The film loop 2–12 'Ammonia – uses' could also be shown. Finally, the teacher should pick out the most important aspects of the Topic, thereby providing a framework for a summary.

Suggestion for homework

Make a summary of the work done during the Topic.

Summary

By the end of this Topic, pupils should be aware of the way chemists have helped to solve the world food problem through their work on ammonia. Pupils should be familar with the simple chemistry of ammonia and have used their knowledge of equilibria and of rates on the nitrogen–hydrogen–ammonia equilibrium. They will also have used and extended their knowledge of formulae and equations.

Topic A23

Energy changes in chemical systems

Purposes of the Topic

1. To provide some techniques for measuring the energy changes associated with chemical changes.
2. To introduce and use the ΔH (heat of reaction) notation.
3. To introduce the idea that the energy associated with a reaction can be converted into useful work.
4. To demonstrate that energy can be converted from one form to another.
5. To reveal the origin of the energy liberated or absorbed during a chemical change – namely, the energy difference between that needed to break bonds and that transferred when other bonds are made.
6. To discuss the importance of fuels.

Contents

A23.1 Energy changes in chemical systems
A23.2 What is the source of the energy liberated in a chemical change?
A23.3 Chemicals as fuels
A23.4 Energy transformation and related topics
A23.5 Ways of obtaining energy for use in industry

Timing

About 5 weeks are required for this Topic.

Introduction to the Topic

So far pupils have encountered examples of a wide range of chemical reactions including:
a. Reactions between molecular species, for example, the reaction between hydrogen and oxygen.
b. Electron transfer processes (redox reactions), for example, the reaction between iron and copper(II) ions in solution.
c. Ion-combination processes resulting in the formation of a precipitate, for example, the reaction between silver ions and chloride ions to form silver chloride.

Pupils also know that many reactions take place with little or no apparent energy change but that others occur with explosive violence, or quite slowly. They know that it may be possible to show dynamic equilibrium between reactants and products.

The purpose of this Topic is to explore and interpret energy changes in chemical systems in elementary terms.

The Topic opens with a review of various types of chemical reaction and a series of demonstrations of reactions which result in (1) a release of energy and (2) the absorption of energy. This requires the introduction – but not the theoretical interpretation – of such terms as exothermic and endothermic reaction. Pupils should be aware of examples of both kinds of reaction.

The interpretation of such effects requires a semi-quantitative measurement of a heat of reaction and the recognition of the principle of conservation of energy. The ΔH notation for the heat of reaction (enthalpy of reaction) is introduced and applied to simple examples using the conservation principle. The heat of reaction of an ion combination process and of a displacement (electron-transfer) reaction is used to introduce a discussion on the source of the energy liberated in a chemical change. This is illustrated by the energy changes when hydrogen chloride is formed from its elements.

An elementary study of 'chemicals as fuels' follows. A discussion of possible criteria for a comparison of fuels leads to another use of the mole concept. The quantity of heat evolved when different substances are burned in air is worked out and this is related to the measurement of energy change. Experimentally determined values of heats of combustion are discussed and compared with published values for a series of related fuels. The origin of heats of combustion arises naturally from this study.

The fourth section considers the production of other forms of energy from chemicals. A simple voltaic cell is made and the heat of reaction is shown to be made up of two components, energy which can be used to produce work and energy which only appears as thermal energy. The Topic is concluded by discussing ways of obtaining energy for use in industry, including fuel cells.

This approach to 'energy changes in chemical systems' differs in several respects from that used in the Sample Scheme. Where a more detailed treatment is required it is hoped that teachers will use local resources to extend this presentation.

Background knowledge

During Stage I, energy changes were noted but their significance was not discussed. Pupils will be familiar with the use of thermal energy for bringing about changes of state and for chemical decomposition. They will also be aware of the liberation or absorption of energy for decomposition.

Topic A15 examines changes from an energy viewpoint and correlates heat of vaporization with structure.

The Nuffield Physics course includes a detailed study of energy: it relates food with manual work and considers energy transfer by 'machines', such as levers and pulleys. This is followed by the measurement of force and work, using the joule as the principal unit.

In the Nuffield Biology course, energy changes associated with respiratory processes are studied, and the direct conversion of 'chemical' to 'mechanical' energy is considered. The calorific value of food materials is also measured.

Further references

for the teacher

The *Handbook for teachers*, Chapters 15 to 19, provides a detailed discussion on energy changes, and teachers who are unfamiliar with the presentation of ideas on this Topic at this level should consider the material in these chapters carefully. *Collected experiments*, Chapters 16 and 17, offers some additional experiments for this Topic.

Topic A23 Energy changes in chemical systems

Nuffield Advanced Chemistry uses the general theme of energy changes and their interpretation as an important part of that course – see *Teachers' guide I* Topics 4, 7, 10 and 11; *Teachers' guide II*, Topic 17.

Nuffield Advanced Physical Science attaches similar importance to the theme of energy changes and their interpretation – see *Teachers' guide I*, Sections 1, 3, 4, 6 and 7, *Teachers' guide II*, Section 8, and *Teachers' guide III*, Options G1 and G2. A concise account of the underlying principles occurs in Dawson, B. E. (1971), *Energy and chemistry*, Methuen.

Supplementary materials

Film loop
2–14 'Energy changes in HCl formation'

Films
'Change of state' ESSO
'Energy in chemistry (1): energy levels, latent heats and heats of reaction' ESSO. Films for Science Teachers No. 21
'Energy in chemistry (2): chemical energy, electrical energy, and work' ESSO. Films for Science Teachers No. 22

Reading material

for the pupil

Handbook for pupils, Chapter 6 'Energy changes and material changes' reviews energy changes and material changes in reactions and offers a data section for use with this Topic.
Chemists in the world, Chapter 5 'Energy and chemicals' reviews the production of energy and chemicals from coal, gas, and oil.

A23.1
Energy changes in chemical systems

The theme is introduced qualitatively before energy-level diagrams are used as a means of representing energy changes in chemical reactions. Calorimetric procedures are then used in quantitative determinations of the heat of reaction, assuming the general applicability of the principle of conservation of energy.

A suggested approach

Objectives for pupils

1. Familiarity with examples of chemical reactions in which measurable energy changes occur
2. Understanding energy-level diagrams and the ΔH convention
3. Knowledge of the terms exothermic reaction and endothermic reaction
4. Familiarity with calorimetric procedures for measuring energy changes
5. Ability to calculate the heat of reaction from data obtained by a calorimetric procedure

When a chemical reaction takes place, heat energy is released to the surroundings, *or* is absorbed from the surroundings. This idea has been mentioned from time to time throughout the course but this Topic explores and interprets such energy changes in chemical systems using simple terms.

The teacher should show a few qualitative demonstrations of energy changes resulting from chemical reactions. Reactions which release energy to the surroundings include:

1. The reaction of an acid with a base, such as the reaction of dilute hydrochloric acid with sodium hydroxide solution, or of dilute acetic acid with calcium hydroxide.
2. The addition of concentrated sulphuric acid to water, using a beaker of water and stirring the mixture during the addition of the acid.
3. The addition of a limited quantity of water to a sample of freshly prepared anhydrous copper(II) sulphate.

Examples of systems which must absorb energy from the surroundings if they are to remain at a constant temperature should also be shown. For example:

2 test-tubes, 150 × 25 mm

4 beakers, 1000 cm³

6 thermometers

4 measuring cylinders, 100 cm³

Swabs

Anhydrous copper(II) sulphate, freshly made and kept in a corked test-tube

Concentrated sulphuric acid

Ammonium nitrate, 80 g, on a sheet of paper

M calcium chloride solution

M sodium carbonate solution

Barium hydroxide octahydrate, 32 g, on a sheet of paper

Ammonium thiocyanate, 8 g, on a sheet of paper

2M hydrochloric acid

2M sodium hydroxide

Distilled water

4. The solution of ammonium nitrate in water. (Stir the mixture throughout the addition of the solid to the water in a beaker.)
5. The addition of calcium chloride solution to sodium carbonate solution.
6. The mixing of molar proportions of barium hydroxide octahydrate and ammonium thiocyanate, in the solid state.

Although the pupils' experience of chemical reactions may suggest that most reactions release energy to the surroundings, both types of energy change do occur. This raises the need for a convention. The convention '$-q$' and '$-$heat' is an obvious addition to the list of products of reaction *when* the energy change results in a lowering of the total energy content of the system and a consequent release of energy to the surroundings. The convention '$+q$' and '$+$heat' on the products side of an equation may be less clear – *unless* the teacher makes the point that to restore the reaction system to room temperature, heat needs to be added to the system and so the overall energy content of the products *when compared with that of the reactants* is higher. A simple form of energy-level diagram at this point enables such points to be emphasized. For example, the reaction of dilute hydrochloric acid with dilute sodium hydroxide might be represented as shown in figure A23.1.

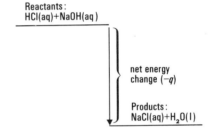

Figure A23.1
Simple energy-level diagram for the reaction between dilute hydrochloric acid and sodium hydroxide solution.

It remains for the teacher to stress the need for a quantitative measurement of such effects so that an even better theoretical interpretation can be attempted. It will be found more appropriate to apply the idea to each system studied rather than to adopt a simple generalized presentation immediately and without preliminary discussion. The entire section can then be summarized as in figure A23.2.

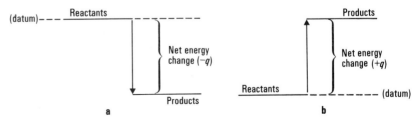

Figure A23.2
Energy-level diagrams for (*a*) exothermic reactions, and (*b*) endothermic reactions.

Calorimetric procedures provide a means whereby the quantity of energy absorbed or evolved may be measured. Pupils should recognize:

a. the difference between quantity of heat and heat level (i.e. between heat and temperature, respectively); and

b. the need to adopt the principle of energy conservation – 'energy can be neither created nor destroyed'.

Where pupils have some experience of calorimetry from physics, many of these ideas, together with those raised during Experiment A23.1a, will merely required revision. In other instances, greater detail may be necessary.

Heat transfer experiments

Procedure

Heat some water in a beaker to between 80–90 °C and then transfer 200 g to a 400 cm³ plastic beaker. Record the temperature and the mass of water transferred. Add 50 g cold water (of known initial temperature – between 15 °C and 20 °C) to the water in the plastic beaker. Stir the mixture and record its final temperature.

Repeat the experiment using 200 g water at the same high temperature but only 25 g of cold water.

Tabulate the data from these experiments and compare them. Suitable headings for a table are given below.

Expt. no.	Mass of hot water used (M_h/g)	Initial temperature of hot water (T_h/°C)	Mass of sample added (M_c/g)	Initial temperature of sample (T_c/°C)	Final temperature of water and sample (T_m/°C)	Change in temperature of hot water ($-\Delta T_h$/°C)	Change in temperature of cold water (ΔT_c/°C)	Ratio $\dfrac{-\Delta T_h \times M_h}{\Delta T_c \times M_c}$

The experiment shows that the addition of cold water to hot water results in an exchange of heat energy: the hot water becomes colder and the cold water becomes hotter. This should not be unexpected! A four-fold decrease in the mass of cold water yields a corresponding temperature change. Thus, the change in temperature of the 'hot water' is only $\frac{1}{4}$ of that of the 'cold water'.

Using the symbols given above, it follows that:

$$-\frac{\Delta T_h \times M_h}{\Delta T_c \times M_c} = 1.0 \text{ (approximately)}$$

This ratio is given a negative sign since ΔT_h – the change in temperature of the hot water – is negative, *and* the overall heat energy gained by the cold water is that heat energy lost by the hot water. It will be found that the values for this ratio are not exactly constant. We have assumed in our calculation that the beaker neither absorbed nor lost heat. If a thermos flask is used for this demonstration, it can be shown to be an ideal vessel in which to conduct a heat exchange experiment since *very* little heat energy is lost to the vessel over a short period of time.

If care is taken, the point can be made fairly easily that heat is neither gained nor lost during such demonstrations.

If dilute solutions of chemicals are used in place of water in this calorimetric experiment, the heat of reaction can be related to the temperature change, the quantity of liquid used, and the quantity of chemicals.

Originally the *calorie* was defined as the quantity of heat required to raise the temperature of 1 g of water through 1 °C. Subsequent work showed that this unit depended on the temperature interval selected, but the variation only became significant in really precise determinations. The heat unit now adopted is the *joule* and all that needs to be remembered at this stage is:

1 calorie = 4.2 joules

However, some teachers may wish to omit this brief reference to the calorie on the grounds that it could be confusing to pupils. The basic unit may then be defined as that quantity of heat needed to raise 1 g of water through 1 °C – namely, 4.2 joules. So heat transferred will be given by:

4.2 × (temperature change) × (mass of water in g)   joules

It is suggested that the teacher now demonstrate on a large scale the experiment on measuring a heat of reaction which pupils carry out on a small scale in Experiment A23.1b. Use 500 cm^3 2M sodium hydroxide and 500 cm^3 2M hydrochloric acid. Both solutions should be at the same initial temperature, room temperature. On mixing these solutions, the temperature rise at the centre of the mixture will be about 13.3 °C. This addition results in the formation of a solution which is 1.0M with respect to *all* species present and the volume will be about 1000 cm^3. Since the density and heat capacity of this solution are approximately the same as those for water, the heat of reaction will be given by:

$4.2 \times 13.3 \times 1000 \, \text{J mol}^{-1}$

$= 55.9 \, \text{kJ mol}^{-1}$

This heat must be lost if the system is to return to the initial (room) temperature and so the products formed during the reaction will contain *less* energy than the reactants. The symbol used for heat content (at constant pressure) is H, and so in going from reactants to products in this case:

Topic A23 Energy changes in chemical systems

$$\Delta H = H_{\text{reactants}} - H_{\text{products}} = -55.9 \, \text{kJ mol}^{-1}$$

and we may now write:

$$HCl(1.0M \, aq) + NaOH(1.0M \, aq) \longrightarrow NaCl(1.0M \, aq) + H_2O(1)$$
$$: \Delta H = -55.9 \, \text{kJ mol}^{-1}$$

This same result will be obtained when 50 cm³ or even 5 cm³ of each of these solutions are mixed together. Pupils should be asked to explain this fact and then to verify it.

Experiment A23.1b

To measure the heat of reaction of a neutralization reaction

Apparatus

Each pair of pupils will need:

Experiment sheet 94

2 beakers, 100 cm³

Polythene bottle, 60–70 cm³, or expanded polystyrene cup (see figure A23.3)

Thermometer, −10 to +110 × 1 °C, long form

Measuring cylinder, 25 cm³

2M sodium hydroxide, 25 cm³

2M hydrochloric acid, 25 cm³

Procedure
Tell each pair of pupils the volume they are to use (say, 10, 15, 20, or 25 cm³). Experimental details are given in *Experiment sheet* 94.

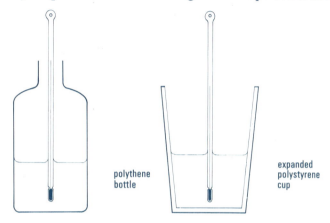

polythene
bottle

expanded
polystyrene
cup

Figure A23.3

Experiment sheet 94
In this experiment you are going to mix equal volumes of 2.0M solutions of sodium hydroxide and hydrochloric acid and find the temperature change during the reaction. Your teacher will tell you the actual volume of each that you are to use.

Measure the appropriate volumes of acid and alkali into separate beakers, and take the temperature of each solution.

Pour the alkali into the calorimeter (a polythene bottle) and add the acid, stir the mixture well, and record the highest temperature reached.

Results

Volume of each solution used	.... cm³
Temperature of acid	.... °C
Temperature of alkali	.... °C
Average temperature	.... °C
Highest temperature of mixture	.... °C

Calculate the heat of reaction per mole of reactants as follows.

Since equal volumes of solutions are used, and since each solution is 2M, the concentration of each reactant in the mixture will be ....M.

Hence if the heat of reaction (ΔH) is measured in kJ,

$$\Delta H = -\text{(temperature change)} \times 4.2\,\text{kJ}$$

$$= -(\dotfill - \dotfill) \times 4.2\,\text{kJ}$$

$$= \dotfill \text{kJ}$$

Write the equation for the reaction, and include the ΔH value that you have calculated.

Pupils should see from this experiment that the same heat of reaction in kJ mol^{-1} is found regardless of the starting volumes of stock solution.

Two additional reactions are suggested for class work: the precipitation of silver chloride in Experiment A23.1c, and a displacement reaction between zinc or iron and copper(II) sulphate solution in Experiment A23.1d.

In Experiment A23.1c, interest is added if different groups use different chloride solutions to form silver chloride. The results can be collected and displayed as shown in figure A23.4. Pupils may then appreciate the significance of the equation:

$$\text{Ag}^+(\text{aq}) + \text{Cl}^-(\text{aq}) \longrightarrow \text{AgCl}(\text{s})$$

as a summary for such studies. They can also compare the accuracy obtained by other pupils using the same procedure.

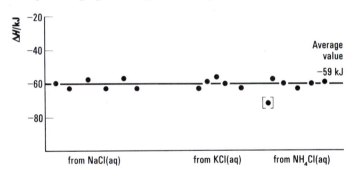

Figure A23.4
Heat of reaction measurements: precipitation of silver chloride.

Experiment A23.1c	**To measure the heat of reaction of an ion combination reaction**

Apparatus

Each pair of pupils will need:

Experiment sheet 95

Polythene bottle, 60–70 cm³ fitted with a bung and thermometer, -10 to $+110 \times 1\,°C$ (and large enough to have 20 °C mark visible)

Measuring cylinder, 25 cm³

(Continued)

Procedure

The method of using the polythene bottle depends on the type of thermometer used. The short type is more robust but the thread is then inside the bottle when it is upright and readings have to be taken with the bottle inverted, and the bulb just inside the rubber stopper. Longer thermometers can be read upright, as long as the bulb is properly covered by the liquid.

Experiment sheet 95
The reaction whose heat of reaction you are going to measure is

$$\text{Ag}^+(\text{aq}) + \text{Cl}^-(\text{aq}) \longrightarrow \text{AgCl}(\text{s})$$

Topic A23 Energy changes in chemical systems

0.5M silver nitrate solution, 25 cm^3

Either 0.5M sodium chloride solution, 25 cm^3

or 0.5M potassium chloride solution, 25 cm^3

or 0.5M ammonium chloride solution, 25 cm^3

You will be provided with 0.5M solutions of silver nitrate (source of Ag$^+$(aq) ions) and of a soluble chloride (source of Cl$^-$(aq) ions).

Write the name of the chloride solution you use here: ....

Put 25 cm^3 (measuring cylinder) of the silver nitrate solution into a polythene bottle (the calorimeter). Insert the rubber bung and the thermometer and take the temperature of the solution; you may have to invert the bottle to make sure that the thermometer bulb is immersed in the solution. Wash the measuring cylinder with distilled water and use it to add 25 cm^3 of the chloride solution to the silver nitrate solution. Stopper the bottle and shake it well. Watch the thermometer and record the highest temperature reached.

Results

Volume of 0.5M silver nitrate used   25 cm^3

Volume of 0.5M .... used   25 cm^3

Initial temperature   .... °C

Final temperature   .... °C

Temperature rise   .... °C

Total volume of solution used   50 cm^3

Assuming that this has the same specific heat capacity as water, 4.2 joule g^{-1} °C^{-1}, and a mass of 50 g:

Heat evolved during reaction $50 \times 4.2 \times$ .... joules

Number of moles of silver chloride precipitated

$$\frac{25 \times 0.5}{1000} = 0.0125 \text{ mole}$$

Heat evolved if 1 mole of silver chloride was precipitated:

$$= \frac{\text{........................} \times \text{........................}}{\text{........................}} \text{joules}$$

$$= \text{........................} \text{ kilojoules}$$

Therefore,

Ag$^+$(aq) + Cl$^-$(aq) $\longrightarrow$ AgCl(s): ΔH = ... kJ

With the second system, a powdered metal (zinc or iron) can be added to copper(II) sulphate solution.

Experiment A23.1d

Apparatus

Each pair of pupils will need:

Experiment sheet 96

Polythene bottle, 60–70 cm^3, fitted with a bung and thermometer, −10 to +110 × 1 °C

Measuring cylinder, 25 cm^3

(Continued)

To measure the heat of reaction for a displacement reaction

Procedure

Practical details are given in *Experiment sheet* 96. Pupils should be able to determine the heat of reaction and show that:

$$\Delta H = -4.2 \times 25 \times 0.2 \times (t_{\text{final}} - t_{\text{initial}}) \text{ kJ}$$

Experiment sheet 96

You will measure the heat of reaction for either:

     Cu^{2+}(aq) + Fe(s) $\longrightarrow$ Cu(s) + Fe^{2+}(aq)

or   Cu^{2+}(aq) + Zn(s) $\longrightarrow$ Cu(s) + Zn^{2+}(aq)

0.2M copper(II) sulphate
solution, 25 cm³

Either zinc powder, about 0.5 g
in a specimen tube

or iron powder, about 0.5 g in
a specimen tube

The procedure is, in general, the same as that outlined in Experiment 95,
except that the second reagent to be added is a solid. Use 25 cm³ 0.2M
copper(II) sulphate solution, measure its temperature, add about 0.5 g of the
metal used (iron or zinc powder), shake the mixture, and record the highest
temperature reached.

Make up your own table of results and calculate the heat of reaction.
Assume that the specific heat capacity of the copper(II) sulphate solution is
4.2 joule g^{-1} $°C^{-1}$.

At the end of the section raise the question, 'Where does the energy
evolved come from?' (This forms the theme for A23.2.)

Suggest improvements to the experimental procedure you have used
to measure the heat of a reaction.

Summary

By the end of this section pupils should have a reasonable knowledge of simple calorimetric procedures for estimating heats of
reaction. They should understand the energy-level diagrams for
each of the reactions investigated. For the convenience of the
teacher, the latter are reproduced in figure A23.5.

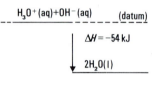

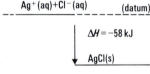

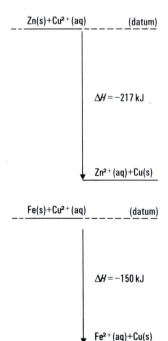

Figure A23.5

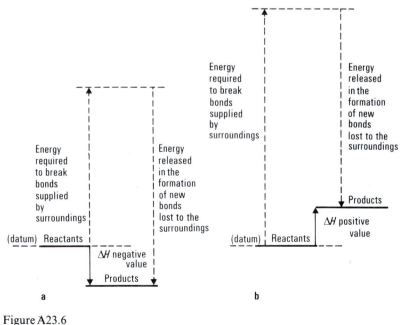

Figure A23.6
A simple interpretation of the origins of energy changes for (*a*) exothermic reactions,
and (*b*) endothermic reactions.

Topic A23 Energy changes in chemical systems

A23.2
What is the source of the energy liberated in a chemical change?

This section is intended to be treated very simply in order to answer the question which it is hoped pupils have considered in the last section.

Objectives for pupils

1. Application of energy-level diagrams to solve problems
2. Appreciation of the principle of conservation of energy

The lesson could open with a discussion of the following points:
1. When chemicals rearrange themselves during a reaction, they have to lose or gain energy so as to remain at the same temperature.
2. The need to gain or lose energy must be due to the 'new' ions or molecules – the products – requiring more or less energy than the 'old' ions or molecules – the reactants.
3. What energies do the particles – ions, molecules, etc. – need?

Pupils should be able to appreciate that:

$$\text{the energy of a chemical} = \left[\begin{array}{l} \text{energy used to} \\ \text{hold the atoms} \\ \text{together in} \\ \text{particles} \end{array} \right] + \left[\begin{array}{l} \text{energy of movement} \\ \text{of the particles,} \\ \text{giving them kinetic} \\ \text{energy which is} \\ \text{proportional to} \\ \text{temperature} \end{array} \right]$$

We need to examine this idea closely using a simple example of a chemical change. The reaction of hydrogen and chlorine has been studied very carefully over a range of conditions:

$$H_2(g) + Cl_2(g) \longrightarrow 2HCl(g)$$

The heat of reaction has been determined over a wide range of temperatures, and is as follows:

Temperature/°C	0	100	200	300
ΔH/kJ	-184.4	-184.8	-185.6	-186.1

Temperature/°C	400	500	600	700
ΔH/kJ	-186.9	-187.7	-188.2	-188.6

The energy change is more or less the same at all temperatures. This is a common pattern: *in most cases the heat of a reaction varies very little with temperature.* What can we conclude? The simplest explanation is that the heat of the reaction is chiefly due to differences in overall energy of bond breaking and bond making.

We can refer to an energy-level diagram – such as that shown in figure A23.6 (opposite).

An exothermic reaction is therefore evidence of overall stronger bonds being formed. If ΔH is positive (an endothermic reaction), then the energy for making bonds is greater than the energy for breaking them; if ΔH is negative (an exothermic reaction) the reverse is true.

In the case of the reaction between hydrogen and chlorine, we know that hydrogen chloride is formed. We do *not* know *how* this happens – but whichever way it happens, two things *must* occur:
1. The hydrogen molecules and the chlorine molecules must be broken into atoms – these processes require energy.
2. The atoms must recombine to form hydrogen chloride – this process gives out energy.

The energies required to pull molecules apart have been measured.

Energy required to break 1 mole of hydrogen molecules into atoms = +437 kJ

Energy required to break 1 mole of chlorine molecules into atoms = +244 kJ

Energy required to break 1 mole of hydrogen chloride molecules into atoms = +433 kJ

Ask pupils to use this information. Can they 'explain' the loss of 185 kJ when 1 mole of hydrogen and 1 mole of chlorine combine to form 2 moles of hydrogen chloride?

It is important for pupils to appreciate that not all of the hydrogen molecules have first to be dissociated into atoms and then all of the chlorine molecules into atoms and lastly all of the atoms recombined. Rather the calculations show the 'balance sheet' of energy: energy cannot be created neither can it be destroyed – but it can be accounted for!

The teacher will need:

Film-loop projector
2–14 'Energy changes in HCl formation'

To reinforce this argument, the film loop 2–14 'Energy changes in HCl formation' could now be shown. The information given by energy-level diagrams *is* limited. It does *not* include information about the *actual* process of reaction. In fact, a study of the rate at which this reaction takes place suggests that a few chlorine molecules are dissociated into atoms and that these then react with a few hydrogen molecules. The energy liberated enables a few other processes of these kinds to occur and so on. *Energy-level diagrams only summarize the energy changes!*

It is important *NOT* to take the matter too far at this stage and only to provide some indication of the answer to the question 'Where does the energy come from?'

Suggestion for homework

Summarize the work covered in this section.

A23.3
Chemicals as fuels

After a brief discussion of reactions in which energy changes occur, attention is focused on combustion processes. A method for measuring the heat evolved when different substances burn in air is worked out and the results are expressed so that useful comparisons may be made.

Topic A23 Energy changes in chemical systems

1. Familiarity with examples of chemical changes which are accompanied by energy changes
2. Knowledge of the concept of heats of combustion
4. Ability to analyse an experimental procedure
5. Ability to interpret heats of combustion in simple terms

The lesson could be opened by a review of chemical reactions which are accompanied by a noticeable loss of heat energy. Some examples, such as the following, can be put on the blackboard or on an Overhead projector transparency:

$$H^+(1.0M\,aq) + OH^-(1.0M\,aq) \rightarrow H_2O(l); \Delta H = -55.6\,kJ\,mol^{-1}$$

$$C(graphite) + O_2(g) \qquad \rightarrow CO_2(g); \Delta H = -393.1\,kJ\,mol^{-1}$$

$$Mg(s) + 2H^+(aq) \qquad \rightarrow Mg^{2+}(aq) + H_2(g);$$
$$\Delta H = -461.4\,kJ\,mol^{-1}$$

$$Mg(s) + \tfrac{1}{2}O_2(g) \qquad \rightarrow Mg^{2+}O^{2-}(s);$$
$$\Delta H = -602\,kJ\,mol^{-1}$$

Other reactions may be added to illustrate the absorption of heat from the surroundings to maintain a steady temperature, such as:

$$Ca^{2+}(aq) + SO_4^{2-}(aq) \longrightarrow Ca^{2+}SO_4^{2-}(s); \Delta H = +7.1\,kJ\,mol^{-1}$$

$$Mg^{2+}(aq) + CO_3^{2-}(aq) \longrightarrow MgCO_3(s); \quad \Delta H = +25\,kJ\,mol^{-1}$$

The most common way of liberating energy as heat is to burn a substance in air, such as oil in an oil-fired boiler, or coal or coke in an appropriate fire-grate. The substance burned is referred to as a fuel, the second reactant for this process being oxygen from the air.

Ask the class to list some common fuels and give reasons for their use in particular instances. Points for discussion should include the cost, the amount of heat liberated, storage problems, and possible effects on the environment. If possible, matters of topical or local interest should be mentioned in discussion. For example, when considering fuel for use in a space craft, we need to discuss oxidants, costs, mass of reactants, bulk of reactants, and so on.

How can the heat produced when a fuel is burned be measured? It is suggested here that the heat produced can be measured by allowing it to heat a known mass of water and measuring the temperature rise (remind the pupils of section A15.2). Other methods are, of course, available.

What fuel should be used? Coal? Gas? Lighter fuel? A discussion of likely difficulties leads to the choice of liquid fuels that burn easily and 'cleanly'. Introduce alcohol as a special sort of fuel. Four alcohols are chosen. What are the products of combustion in each case? Since the fuel is a liquid in each of these cases, the mass burned can be found by weighing a suitable 'spirit lamp' burner before and after the experiment. To enable a comparison to be made, the same type of burner and water container must be used by different groups of pupils. Does all of the liberated heat warm the water? Are there other matters that need to be considered?

Pupils may now do Experiment A23.3a, with different groups burning different alcohols. Afterwards, results for the different liquids can be collected and errors which are inherent in the procedure discussed. This activity can have a real value for most pupils

---

and be used to introduce a more specialized form of heat of combustion apparatus (see Experiment A23.3b).

Experiment A23.3a

To measure the heat of combustion of a liquid fuel

Apparatus

Each pair of pupils will need:

Experiment sheet 97

Metal can of capacity about 300–500 cm³ (*Note:* all groups should use the same sort of can)

Spirit lamp (improvised – see diagram)

Stand, boss, and clamp – with wide jaws

Measuring cylinder, 500 cm³

Thermometer, −10 to +110 × 1 °C

Draught shield – 2 heat resistant mats are suitable

Teat pipette

Access to balance

Methanol

Ethanol

Propan-1-ol

Butan-1-ol

} *

*Quantities depend on the type of spirit lamp used

Procedure
Full details and a diagram of the apparatus appear in *Experiment sheet* 97 which is reproduced below. To enable comparisons to be made, pupils should use the same type of burner and water container.

Experiment sheet 97
You can get an approximate value for the heat evolved when a known mass of a liquid burns in air by burning it in a spirit lamp which is used to heat a known amount of water. The class will investigate a number of different liquids in this way; your teacher will tell you which you are to use.

Write its name here: ....

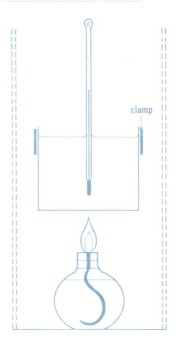

clamp

Figure A23.7

Support a metal can containing 300 cm³ water in a clamp on a retort stand and place a thermometer in the water. Put some of the liquid in a spirit lamp, place it under the metal can, and adjust the height of the can so that there is a space of 3 cm between the top of the spirit lamp and the bottom of the can. Arrange to shield the flame of the lamp from draughts by using asbestos sheets.

Find the mass of the spirit lamp and liquid as accurately as you can. Replace it below the metal can. Take the temperature of the water in the can. Light the spirit lamp. Stir the water with the thermometer and watch the temperature; when it has risen about 30 °C, extinguish the lamp and record the highest temperature reached. Find the mass of the spirit lamp and remaining liquid as quickly as possible. Why?

This procedure yields consistent results which are about two-thirds of the accepted value. It is helpful to compare experimental values with accepted values of the various heats of combustion (see Table A23.1) and to use this comparison as a basis for indicating sources of experimental error. (For example, does all the liberated heat warm the water?) A more efficient procedure can then be devised (see Experiment A23.3b). This apparatus allows solids such as carbon, sulphur, solid foods, and coal as well as liquids to be burned. Several versions of the apparatus are available from suppliers although home-produced versions made of metal and fitted with an appropriate form of heat insulation shield can be more effective. Figure A23.8 shows one form of home-produced apparatus.

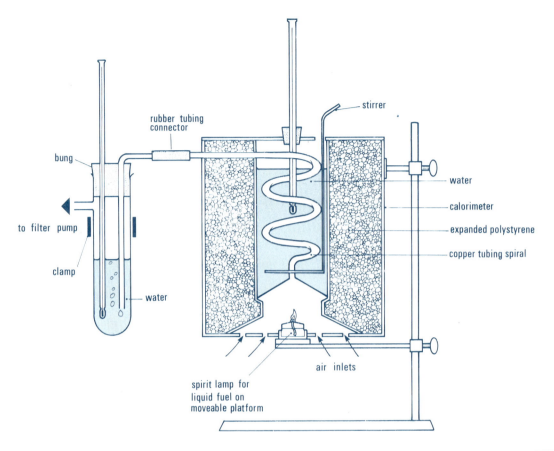

Figure A23.8
Heat-of-combustion
apparatus: (a) diagram;
(b) photograph; of 'home-
made' version.
Note. Since naked flames are
used, apparatus should be
encased in non-flammable
material (e.g. aluminium sheet).

To measure the heat of combustion of a fuel

This experiment should be performed by the teacher.

Figure A 23.8 shows an apparatus with an arrangement to check air flow and whether heat losses occur as a result of too rapid an air flow through the apparatus. (Note: the temperature of the water in the filter tube should remain constant.)

Apparatus

The teacher will need:
Heat-of-combustion apparatus (see figure A23.8)

Hard-glass test-tube, 150 × 25 mm, fitted with bung + drying tube containing silica gel or anhydrous calcium chloride

Stand, 2 clamps, and bosses

Filter pump and connecting tubing

Measuring cylinder, 500 cm^3

Beaker, 600 cm^3

Thermometer (−5 to +50 × 0.1 °C, or −10 to +110 × 1 °C)

Bunsen burner and heat resistant mat

Access to balance

Carbon, in lumps of about 0.5 g

Oxygen cylinder

Alcohols as in Experiment A23.3a

Procedure

1. *For carbon.* Select lumps of carbon, for example wood charcoal, weighing about half a gram each and dry them by heating them in a test-tube. While cooling, the test-tube should be fitted with a drying tube and the carbon then kept in the test-tube until required.

Note (or determine) the water equivalent of the calorimeter and then fill it to within 1 cm of the top with a known volume of water and clamp it securely. Weigh the crucible with about half a gram of the carbon and support the crucible on the asbestos platform a few centimetres below the calorimeter.

Apply moderate suction by connecting the filter pump to the spiral. Stir the calorimeter contents and take the temperature when it becomes constant. Connect the oxygen supply to the asbestos platform and carefully regulate it to give a gentle stream of gas. (About 3 p.s.i. on the outlet pressure gauge.) Ignite the carbon lump with a small Bunsen flame and when the carbon is just glowing raise the asbestos platform to fit securely below the base of the calorimeter. The carbon will now glow brightly in the air enriched with oxygen. It should burn quietly with no spluttering (and no loss of mass). Should the carbon splutter, the oxygen supply must be reduced until quieter combustion is obtained. The water is stirred throughout the combustion; this should last about five minutes for the mass of carbon suggested.

When combustion has ceased, turn off the oxygen supply and the filter pump; stir until the maximum temperature is observed and record this temperature.

When cool, remove the crucible and reweigh it to find the amount of carbon consumed.

The heat of combustion of 12 g of this form of carbon can then be calculated in kilojoules per mole.

2. The apparatus may also be used to compare (more precisely than in Experiment A23.3a) the heats of combustion of a series of alcohols. A special 'spirit' burner is used and the loss in weight during combustion of a few cm^3 of an alcohol measured.

An adequate air supply should be maintained by the filter pump, supplemented, if necessary, by leaving a small space between the asbestos platform and the calorimeter. Do not use an oxygen supply in this experiment.

A temperature rise of about 5 °C is required. A close fitting cap should be placed on the burner to prevent evaporation of alcohol as it cools.

Compound	Molecular formula	ΔH° /kJ mol^{-1}	Increment /kJ
methanol	CH_3OH	-726	
ethanol	CH_3CH_2OH	-1367	641
propan-1-ol	$CH_3CH_2CH_2OH$	-2017	650
butan-1-ol	$CH_3CH_2CH_2CH_2OH$	-2675	658
pentan-1-ol	$CH_3CH_2CH_2CH_2CH_2OH$	-3323	648
hexan-1-ol	$CH_3CH_2CH_2CH_2CH_2CH_2OH$	-3976	653

Table A23.1
Heats of combustion of alcohols

The data shown in Table A23.1 support the view that increments in the heats of combustion for a series of related compounds – such as these alcohols – are related to the increments in composition. It would seem that the methylene group,—CH_2—, contributes a definite quantity of energy to each heat of combustion value. In discussion *suggest* that each C—H, C—C, C—O, and O—H bond present must also require a specific quantity of energy to break it. The presentation of these ideas can be assisted by the use of molecular models.

A simple interpretation of enthalpy changes in both endothermic and exothermic reactions may now be put forward (see figure A23.6). It *suggests* that these changes could be calculated from the energies required to make or break covalent bonds in reactions such as those studied in Experiments A23.3a and A23.3b.

The determination of bond energies need *not* concern pupils at this point. (The teacher may note in passing that the bond dissociation energy of diatomic molecules can be found by spectroscopic means but the spectra from polyatomic molecules are too complicated for this purpose. Electron impact techniques may be used to determine the enthalpy of atomization of methane and to estimate the strength of the C—H bond.) The significant point to stress is *not* the detail but rather the idea conveyed in figure A23.6. These ideas entail no more than an application of the principle of conservation of energy.

Suggestion for homework

Read *Chemists in the world*, Chapter 5 'Energy and chemicals'.

A23.4
Energy transformation and related topics

This section considers energy transformation and the construction and use of a simple fuel cell.

A suggested approach

Objectives for pupils

1. Awareness that energy released during a chemical reaction can give rise to a variety of effects

Pupils have met ideas about energy in chemistry, physics, and biology without necessarily having an opportunity to relate these ideas and to see general applications. It is worth spending a little time demonstrating some interconversions of energy, and in particular bringing out a link between substances undergoing change and the production of movement.

2. Awareness that energy can be converted from one form to another
3. Appreciation of the importance of the principle of conservation of energy
4. Appreciation of the unifying theme of energy in science and technology

Apparatus

The teacher will need

3 or 4 SP2 batteries

Small electric motor fitted with a 'lay' shaft (operating voltage 4 V or less)

Cord and small weight

The interconversion of different forms of energy

One example of interconversion of energy is given here. Others are needed (see, for example, Nuffield Physics) and all should be demonstrated by the teacher.

Procedure

The object of this experiment is to show pupils that chemical energy (in a battery) can be converted into other forms of energy. The battery drives a motor and the motor lifts a load. Assemble the cells and motor in a series circuit. Attach a cord to the motor shaft with a small weight on the floor. On completing the circuit, the cord will wind round the motor shaft and lift the weight.

It is helpful to bring the discussion down to earth. For example, 4.2 kJ of energy is needed to warm 1000 g of water by 1 °C or to raise a one-hundredweight (about 50 kg) sack of potatoes nearly 28 feet (about 8.5 m). (Actually, about four times that quantity of energy will be needed to lift the sack since the human body loses about 3 kJ of thermal energy for every 1 kJ energy it uses to do work.)

A boy of 10 stone (63.5 kg) needs to convert about 197 kJ of chemical energy to mechanical energy in order to climb a hill 100 feet (about 30 metres) high. He uses 25 kJ each time he goes upstairs in a house. His muscles convert about 3 times as much energy to thermal energy as to mechanical energy, so he will need about as much energy as is given out every time 1000 cm^3 of 1.0M sulphuric acid is made from concentrated sulphuric acid and water. This corresponds to 4–5 g of glucose (a teaspoon of sugar or a 'square' of chocolate).

In Topic A15 we referred to the energy required to pull particles of water apart at 100 °C and give them the necessary energy to become a gas.

$$H_2O(1) \longrightarrow H_2O(g): \Delta H_{373} = -40.7 \text{ kJ mol}^{-1}$$

This amount of energy – i.e., that needed to boil away 18 g of water at 100 °C – *if* it could be converted to mechanical energy, could be used to lift a 1 ton girder 13 feet from the ground (or 1 tonne about 4 m)!

However, not all energy transfers involve thermal energy. Consider the possibility of obtaining electrical energy from chemicals. The reaction of zinc dust with copper(II) sulphate solution was studied

---

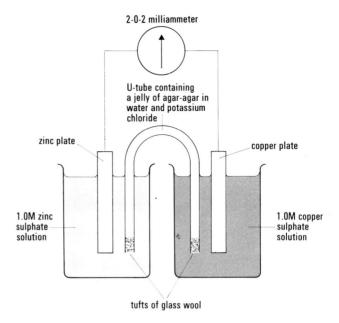

Figure A23.9

in Experiment A23.1d. This reaction can be regarded as an electron transfer process. Thus, the reaction:

$$Zn(s) + Cu^{2+}(aq) \longrightarrow Zn^{2+}(aq) + Cu(s)$$

may be *imagined* to be the summation of two processes:

$$Zn(s) \longrightarrow Zn^{2+}(aq) + 2e^-$$

and $Cu^{2+}(aq) + 2e^- \longrightarrow Cu(s)$

We can demonstrate this idea placing each metal in contact with a solution containing its ions, connecting the two pieces of metal through a milliammeter, and connecting the two solutions electrically with a conducting bridge of agar-agar jelly and potassium chloride *or* by some other means.

A current is seen to flow in the circuit, corresponding to a transfer of electrons from metallic zinc to metallic copper through the milliammeter.

If the milliammeter is replaced by a high resistance voltmeter, we can estimate the e.m.f. of the cell. Pupils may now make some voltaic cells and relate their findings to earlier studies on the reactivity series.

To measure the e.m.f.s of some voltaic cells

Apparatus

Each pair of pupils will need:

Experiment sheet 98

Simple voltaic cell (see figure A23.10)

Metal foil electrodes (copper, zinc, silver, lead, magnesium, etc.)

2 beakers, 100 cm³

Access to a high resistance voltmeter (0–3 V)

Saturated potassium nitrate solution, 10 cm³

Filter paper

Glass rod

Emery paper

M Copper(II) sulphate solution, 25 cm³

M Zinc sulphate solution, 25 cm³

M Silver nitrate solution, 25 cm³

M Lead nitrate solution, 25 cm³

M Magnesium sulphate solution, 25 cm³

Procedure

The arrangement of the cell and the practical details are shown in *Experiment sheet* 98 which is reproduced below.

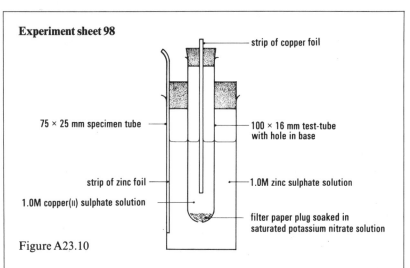

Experiment sheet 98

- strip of copper foil
- 75 × 25 mm specimen tube
- 100 × 16 mm test-tube with hole in base
- strip of zinc foil
- 1.0M zinc sulphate solution
- 1.0M copper(II) sulphate solution
- filter paper plug soaked in saturated potassium nitrate solution

Figure A23.10

The method of preparing the cell shown in the diagram is described here. You may be asked to prepare one containing different metals and solutions, but the procedure is the same in all cases. Clean the surface of the metal strips with emery paper.

The solutions which form part of the two electrode systems in the cell must be kept from mixing and for this purpose a plug of filter paper soaked in saturated potassium nitrate solution is used. Tear a filter paper into small pieces and immerse them in saturated potassium nitrate solution. Remove some of the pulp thus obtained, squeeze it with your fingers, and place it in the bottom of the inner vessel of the cell (remove this inner from the outer vessel to do this). Compress the filter paper into a layer about 1 cm thick with a stout glass rod.

Two-thirds fill the outer vessel with 1.0M zinc sulphate solution, add a strip of cleaned zinc foil (about 10 cm long) and insert the cork holding the inner tube, so that the foil is held between the cork and the wall of the outer tube. Pour 1.0M copper(II) sulphate solution into the inner tube, so that the levels of liquid in both tubes are the same. Add a strip of cleaned copper foil (about 10 cm long) to the inner tube and hold it between the two halves of a split cork.

To measure the e.m.f. of the cell connect the two pieces of foil to a high resistance voltmeter using wire and crocodile clips. You will have to find by trial and error which foil has to be connected to the positive terminal of the meter and which to the negative terminal.

Repeat the measurements using cells prepared by other members of your class. Record the results in the table below.

Negative electrode system	Positive electrode system	e.m.f. of cell /V
Zn(s)|1.0M Zn²⁺(aq)	1.0M Cu²⁺(aq)|Cu(s)	

Reaction	$E^{\ominus}_{298}$
$Mg^{2+}(aq) \mid Mg(s)$	-2.37
$Al^{3+}(aq) \mid Al(s)$	-1.66
$Zn^{2+}(aq) \mid Zn(s)$	-0.76
$Fe^{2+}(aq) \mid Fe(s)$	-0.44
$Ni^{2+}(aq) \mid Ni(s)$	-0.25
$Pb^{2+}(aq) \mid Pb(s)$	-0.13
$H^{+}(aq) \mid [H_2(g)]Pt$	0.00
$Cu^{2+}(aq) \mid Cu(s)$	$+0.34$
$I_2(aq), 2I^{-}(aq) \mid Pt$	$+0.54$
$Fe^{3+}(aq), Fe^{2+}(aq) \mid Pt$	$+0.77$
$Ag^{+}(aq) \mid Ag(s)$	$+0.80$
$Br_2(aq), 2Br^{-}(aq) \mid Pt$	$+1.09$
$[MnO_2(s) + 4H^{+}(aq)], [Mn^{2+}(aq) + 2H_2O(l)] \mid Pt$	$+1.23$
$Cl_2(aq), 2Cl^{-}(aq) \mid Pt$	$+1.36$
$[MnO_4^{-}(aq) + 8H^{+}(aq)], [Mn^{2+}(aq) + 4H_2O(l)] \mid Pt$	$+1.51$
$[H_2O_2(aq) + 2H^{+}(aq)], 2H_2O(l) \mid Pt$	$+1.77$

$Pt[H_2(g)] \mid H^{+}(aq)$

Table A23.2 Standard electrode potentials ($E^{\ominus}_{298}$)

After collecting the results from the class, it is helpful to relate the findings to standard electrode potentials (see Table A23.2) and to previous experience of the reactivity series, before considering cell notation. It is customary to represent the interface between solid terminal materials and a solution by a vertical line. Any porous partition or salt bridge is represented by a broken line. The sign preceding the e.m.f. value for the cell indicates the polarity of the right hand terminal of the cell. For example, the cell represented in figures A23.9 and A23.10 – a Daniell cell – is represented by:

$Zn(s) \mid Zn^{2+}(1.0M) \mid Cu^{2+}(1.0M) \mid Cu(s); \ E^{\ominus} = 1.10 \text{ V}$

Table A23.2 lists the potentials of half cells such as $Zn(s) \mid Zn^{2+}$ (1.0M) and $Cu^{2+}(1.0M) \mid Cu(s)$. These refer to the potential differences for the cells:

$Pt[H_2(g)] \mid H^{+}(aq, 1.0M) \mid Zn^{2+}(1.0M) \mid Zn(s)$

which implies the reaction:

$\frac{1}{2}H_2(g) + \frac{1}{2}Zn^{2+}(aq) = H^{+}(aq) + \frac{1}{2}Zn(s)$

and

$Pt[H_2(g)] \mid H^{+}(aq, 1.0M) \mid Cu^{2+}(1.0M) \, Cu(s)$

which implies the reaction:

$\frac{1}{2}H_2(g) + \frac{1}{2}Cu^{2+}(aq) = H^{+}(aq) + \frac{1}{2}Cu(s)$

respectively. In both examples the electrode on the left is the standard hydrogen electrode and so these potential differences are *relative* electrode potentials or, more briefly, *electrode potentials*.

These electrode processes produce the current in the external circuit and the voltmeter records the e.m.f. of the cell. How is the e.m.f. of the cell related to the energy that can be got out? The voltmeter is calibrated in volts but, as may have been discussed by pupils during a physics class, it could equally well be calibrated in joules per coulomb without changing the scale.

1 volt $\equiv$ 1 joule per coulomb

In Topic A16 a mole of electrons was found to be 96 500 coulombs or 1 faraday.

1 volt $\equiv$ 96 500 joules per faraday.

Hence, the energy transferable from a voltaic cell as electrical energy can be calculated from the e.m.f. of the cell. However, if we pass a current from a cell through a resistance and measure both the current and the potential difference of the cell under various conditions, the energy transferable as electrical energy from the cell to the circuit will depend on how the cell is used in that circuit. What will be the maximum electrical energy that can be transferred? A demonstration will show that the maximum voltage is obtained when the cell is on open circuit – i.e. the e.m.f. of the cell.

Pupils now do an experiment to find the maximum amount of energy available as work which can be obtained from a cell.

Experiment A23.4c

Apparatus

Each pair of pupils will need:

Experiment sheet 99

Daniell cell

d.c. supply, approximately 2 V

Ammeter, 500–0–500 mA

Voltmeter, 0–3 V or 0–5 V

Rheostat, 5–10 Ω

Switch

Connecting wires

Graph paper

To determine the maximum amount of energy available as work from a voltaic cell

Procedure

Note. If a centre-reading ammeter is not available, some rewording of this *Experiment sheet* will be needed.

Pupils may require help with wiring their circuits.

Pupils are asked to calculate the maximum amount of energy in kJ per mole of copper deposited, available as work from the cell, after discussion with their teacher.

Experiment sheet 99

Assemble the circuit shown in the diagram. A Daniell cell is a convenient type to use. Start with the sliding contact of the rheostat at end A. Read the ammeter and voltmeter. Move the sliding contact towards B until the voltage has increased by 0.1 to 0.2 volts; read ammeter and voltmeter. Continue in this way until the sliding contact is at B. At some stage the current in the circuit will reverse; when this happens give the current reading a negative sign. Make a table on a separate sheet of paper with the headings below.

Current /amp	Potential difference /volt

Plot a graph of current/amp (vertical axis) against potential difference/volt (horizontal axis).

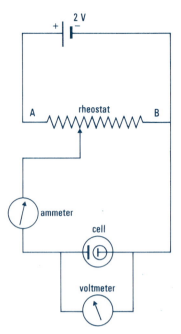

Figure A23.11

What is the voltage when the current is zero? .... volt

This is the point where the cell voltage is just balanced by the voltage from the d.c. supply and gives the maximum e.m.f. of the cell. What happens in the cell when the voltage from the d.c. supply exceeds this value?

From the maximum e.m.f. of the cell, calculate the maximum amount of energy, in kilojoules per mole of copper (64 g) deposited, available as work from the cell. Your teacher will discuss with you how to do this.

The calculation of this energy could be presented as follows:

When 1 coulomb of electricity moves through a potential difference of 1 volt, 1 joule of work is done.

To liberate 64 g (1 mole) of copper in the reaction:

$$Cu^{2+}(aq) + Zn(s) \longrightarrow Cu(s) + Zn^{2+}(aq)$$

the total quantity of electricity used is $2 \times 96\,500$ coulombs.

The maximum e.m.f. for this system is found to be 1.1 volts and so:

$$
\begin{aligned}
\text{work done} &= -2 \times 96\,500 \times 1.1 \text{ joules per mole of copper} \\
&= -212.3 \text{ kJ mol}^{-1} \\
&= \text{maximum energy available to do work}
\end{aligned}
$$

Pupils should appreciate, that, when a chemical change occurs, a definite amount of energy has to be transferred if the final products of the reaction are to be at the same temperature and pressure as the reactants. However, it is very seldom that all of the energy can – even in principle – be transferred as mechanical energy. Nevertheless, the guiding principle is always the same: the principle of conservation of energy.

The energy transferred as thermal energy and the energy transferred as mechanical or electrical energy must *always* add up to the same value – the heat of reaction or enthalpy change for the reaction, ΔH.

We have already calculated that the energy available to do work in the Daniell cell is $212.3\,kJ\,mol^{-1}$. The heat of reaction for this system:

$$Zn(s) + Cu^{2+}(aq) \longrightarrow Zn^{2+}(aq) + Cu(s)$$

has been measured (Experiment A23.1d). (From tables it is $-218.4\,kJ\,mol^{-1}$.)

We can therefore summarize as follows for this reaction:

$$Zn(s) + Cu^{2+}(aq) \longrightarrow Zn^{2+}(aq) + Cu(s):$$

Heat of reaction $= \Delta H = -218.4\,kJ\,mol^{-1}$

Maximum amount of energy available as work (ΔG):
$$= nFE = -212.3\,kJ\,mol^{-1}$$

Heat energy to particles $= -218.4 - (-212.3) = -6.1\,kJ\,mol^{-1}$

It will be seen that for the Daniell cell (displacement of copper by zinc system) the values of the heat of reaction and the maximum amount of energy available as work (ΔG) are very close indeed.

If the cell is made by the system in which silver is displaced by copper, a greater difference is found.

$$Cu(s) + 2Ag^{+}(aq) \longrightarrow Cu^{2+}(aq) + 2Ag(s):$$

Heat of reaction $= \Delta H = -147\,kJ\,mol^{-1}$

Maximum amount of energy available as work (ΔG):

$$= nFE = -89\,kJ\,mol^{-1}$$

Heat energy to particles $= -147 - (-89) = -58\,kJ\,mol^{-1}$

So, we can make the point:

$$\text{heat of reaction} = \begin{bmatrix} \text{net energy} \\ \text{change in going} \\ \text{from reactants} \\ \text{to products} \end{bmatrix} = \begin{bmatrix} \text{energy which} \\ \text{can be turned} \\ \text{to mechanical} \\ \text{work} \end{bmatrix} + \begin{bmatrix} \text{energy which} \\ \textit{cannot } \text{be} \\ \text{turned into} \\ \text{mechanical} \\ \text{work and} \\ \text{appears as} \\ \text{thermal} \\ \text{energy} \end{bmatrix}$$

since the conservation of energy principle holds. Some systems yield a very high conversion for ΔH values into nFE or ΔG values. Others do not.

A note to teachers
Teachers should be aware that nFE is related to ΔG (*see Handbook for pupils*, Chapter 6). However, it is *not* intended to set examination questions requiring pupils to use this idea or other approaches to free energy change, ΔG.

A23.5
Ways of obtaining energy for use in industry

This section considers the relative effectiveness of different methods of transferring energy for specific processes. The fuel cell is considered as a basis for future development.

Objectives for pupils

1. Awareness of the industrial need to get mechanical work as efficiently as possible from a chemical change
2. Knowledge of the inefficiency of getting mechanical work from the heating effects of the energy of a reaction
3. Knowledge of the fuel cell as a technique for getting mechanical work from the electrical effects of the energy of a reaction

The discussion must be kept simple. One could ask the class to suggest fuels commonly used by industry – coal, coke (both of which are mainly carbon), oil, North Sea gas (mostly methane). 'Atomic energy' may be mentioned and the opportunity can be taken to explain briefly how uranium and plutonium differ from conventional fuels. Solar energy and energy from hydroelectric sources may also be mentioned.

Pupils should be aware of how to find the maximum energy from a given process which can be used to do work. A brief list of ΔH and nFE (ΔG) values for some reactions occurring when conventional fuels are burned can now be introduced, for example:

Energy changes of the system kJ g-equation^{-1}

Fuel used	Reaction	a. If *all* as thermal energy	b. If maximum work demanded: as mechanical energy	as thermal energy
hydrogen	$2H_2(g) + O_2(g) \rightarrow 2H_2O(g)$	-485	-460	-25
	(If $2H_2(g) + O_2(g) \rightarrow 2H_2O(l)$)	-573	-477	-96
carbon (in some form)	$C(s) + O_2(g) \rightarrow CO_2(g)$	-393	-393	0
carbon monoxide	$2CO(g) + O_2(g) \rightarrow 2CO_2(g)$	-569	-519	-50
methane	$CH_4(g) + 2O_2(g) \rightarrow CO_2(g) + 2H_2O(g)$	-803	-803	0
propane	$C_3H_8(g) + 5O_2(g) \rightarrow 3CO_2(g) + 4H_2O(g)$	-2046	-2075	$+29$
diesel fuel	$C_9H_{20}(l) + 14O_2(g) \rightarrow 9CO_2(g) + 10H_2O(g)$	-5732	-5858	$+126$

These examples show that most of the energy released by burning conventional fuels should be available for useful work. However, much of the energy from the reaction in fact appears as heat instead of work, resulting in low efficiency. The sequence fuel grate – steam boiler – steam engine or turbine – dynamo – electric motor must be inefficient: no one step is 100 per cent efficient, each step wastes some energy.

Even in the internal combustion engine, 70–80 per cent of the energy available for work appears as heat. Research into the problem has led to the development of the fuel cell in which energy from the oxidation of fuel is converted directly into electricity. Reactions such as:

$$2H_2(g) + O_2(g) \longrightarrow 2H_2O(g)$$

are made to take place at nearly ordinary temperatures so that most of the energy is transferred as electrical energy, and used to run an electric motor or some other electrical appliance. One solution to the problem is to pass hydrogen and oxygen through separate tubes to porous nickel electrodes in a tank through which a concentrated (30%) solution of potassium hydroxide is circulating. The operating temperature of this early type of cell is about 200 °C. The reactions which take place at the electrodes when current flows may be represented as follows (see figure A23.12):

1. At the hydrogen-saturated nickel electrode (i.e. negative electrode):

$$2H_2(g) + 4OH^-(aq) \longrightarrow 4H_2O(l) + 4e^-$$

2. At the oxygen-saturated nickel electrode (i.e. positive electrode):

$$O_2(g) + 2H_2O(l) + 4e^- \longrightarrow 4OH^-(aq)$$

and these equations add up to: $2H_2(g) + O_2(g) \longrightarrow 2H_2O(l)$

This use of the hydrogen–oxygen reaction to obtain electrical energy has been known since the experiments of Grove in the nineteenth century. Others worked on this system and the cell made by F. T. Bacon in the U.K. led to the development in America of the Pratt and Whitney fuel cell used in the Apollo space programme.

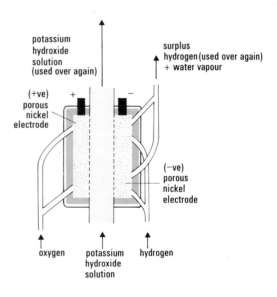

Figure A23.12
Diagram of the Bacon fuel cell.

Other reactions have been examined as potential sources of energy in fuel cells; for example:

$$2CH_3OH(l) + 3O_2(g) \longrightarrow 2CO_2(g) + 4H_2O(l); \Delta G = -1406 \, kJ$$

$$C_3H_8(g) + 5O_2(g) \longrightarrow 3CO_2(g) + 4H_2O(l); \Delta G = -2109 \, kJ$$
(propane)

An article by Williams, K. R. (1964) *School Science Review*, **45**, No. 157, 521, is helpful, both in general terms and in suggesting experiments. Further experimental details are also given in *Collected experiments*, Chapter 16 and 17.

The *Chemical Society Special Publication No. 26* (1974) 'Chemistry and the needs of society' contains much useful background information on energy resources (and other topics).

There are several demonstration fuel cells available from laboratory suppliers or other sources. Where a fuel cell is demonstrated, *pupils should see it being used to do something*. In this way they can appreciate that here is a technique for getting energy from chemicals.

Experiment A23.5

A very simple hydrogen–oxygen fuel cell

Apparatus

Each pair of pupils will need:

Experiment sheet 100

Electrolysis cell, with carbon electrodes (see figure A23.13)

2 test-tubes, 75 × 10 mm

d.c. supply, 4–6 V

2 leads with crocodile clips on one end

Access to a high resistance voltmeter (0–3 V) fitted with leads and crocodile clips

M sodium hydroxide, 50 cm^3

Procedure

Details are given in *Experiment sheet* 100. The gases produced are absorbed on the carbon electrodes and provide the reactants for the cell reaction. Some carbon dioxide is bound to be produced during electrolysis but the alkali removes this.

Experiment sheet 100

We are now going to try to 'burn' hydrogen in a cell, that is to make hydrogen and oxygen combine in an arrangement that will produce a voltage, or a source of electrical energy which will do work. This is known as a 'fuel cell'.

Set up the apparatus as shown in the diagram. Separate the two tubes with a wooden splint resting on the top of the wide glass tube, as shown.

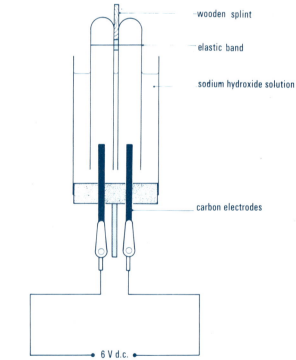

- wooden splint
- elastic band
- sodium hydroxide solution
- carbon electrodes

6 V d.c.

Figure A23.13

Electrolyse the solution until *both* tubes are filled with gas. Now disconnect the d.c. supply and put a high-resistance voltmeter across the carbon electrodes.

What does it read? .... volt

Write equations:

1. For the reactions at the electrodes during electrolysis:

anode ....

cathode ....

2. For the reactions when the system is acting as a fuel cell:

positive electrode ....

negative electrode ....

One way in which this section could be concluded is by relating experimental results to the general theme of ways of obtaining energy for use in industry. Why bother with fuel cells? Discussion should bring out three points.

1. If coal or oil is burned and the net energy change is liberated as heat first, this thermal energy must then bring about movement or be changed into electrical energy in order to get work done. The best machines cannot convert more than about 30 per cent of this into work; the rest still stays as thermal energy. If the same fuel could be changed in a fuel cell about 50 per cent of the maximum energy available as work (nFE [ΔG]) could perhaps be obtained. The choice is between 30 per cent ΔH and 50 per cent nFE (ΔG). Simple calculations based on the information given above help here. However, there are many engineering difficulties yet to be overcome.

2. All industry is concerned with:

either changing raw materials in order to get energy from them. (This energy is used for heating or for doing work. The new materials formed are usually unimportant except that they may be a nuisance, for example in the fuel industry.)

or using energy which has been obtained by someone else to change raw materials into new materials that we do want, for example in the chemical industry.

3. It is found that the sign of maximum energy available as work for a reaction (ΔG) is a certain indication of whether the change is possible, that is, whether it could happen spontaneously as written.

A loose piece of rock may be found on a hillside. If given an initial push it could fall downwards but it could not fall upwards. Only those changes take place spontaneously from which energy might be transferred to do work.

Whether a system has to lose or gain energy in order to stay at the same temperature when it changes gives no indication whether it is likely to happen: endothermic and exothermic reactions are known. Whether the system can be made to supply energy capable of performing work determines if it is to take place spontaneously.

This Topic can profitably be concluded by asking the class to find details of reactions in which energy is used to produce wanted new materials and then discussing how and for what purpose the energy is used (or how the development of excess energy is prevented). Some good examples are: the Haber process, any method of extracting metals, electrolysis of brine, manufacture of polythene, and 'lime burning'. Much fundamental chemistry can be brought out in such discussions.

Some consideration may also be given to energy transfer in biological systems: living organisms are highly successful manipulators of ΔG.

1. An oxygen–hydrogen fuel cell has already been used in space capsules. Can you think of any other uses for fuel cells? What advantages would fuel cells have in a car?
2. Write a summary of the principal ideas you have met in this Topic.

Summary

By the end of this Topic pupils should be aware that when a chemical change occurs, there is a net energy change in the system. If the net energy change results in the liberation of energy to the surroundings, the reaction is said to be exothermic; whereas if thermal energy has to be absorbed from the surroundings to maintain the system at room temperature the reaction is said to be endothermic. In either case, the macro-effect can be interpreted in terms of the behaviour of atoms and molecules. Pupils should be aware of the value and use of fuels and of the possibility of transferring the heat of a reaction into electrical energy capable of doing work. The Topic is concluded by referring to ways of obtaining energy for use in industry.

Radiochemistry

1. To provide some simple experiences of radiochemical phenomena.
2. To introduce the notion of growth and decay, and of half-life.
3. To discuss the uses of radioactive isotopes as tracers.
4. To make pupils aware of the hazards associated with radiation.

Contents

A24.1 Introduction to radiochemistry
A24.2 The study of the decay and growth of some radioactive chemicals
A24.3 The use of radioactive isotopes as detectives

Timing

About three weeks (a double period and a single period in each week) should be enough for this Topic.

Introduction to the Topic

The general features of the Periodic Table are used as a setting for this Topic. The pupil will learn something of the occurrence and detection of radioactive materials, the way in which they decay and grow, and some of their uses. The approach is historic, starting with Becquerel's discovery and followed by detailed attention to the work of Madame Curie, Crookes, Rutherford, Soddy, and others. The experimental work deals mainly with the thorium decay series, from which two of the three isotope separation experiments are taken – the other experiment uses protactinium. The only application of radioactive materials discussed is their use as chemical tracers.

Further references

for the teacher

Further information on radiochemistry is given in the *Handbook for teachers,* Chapter 24.
Nuffield Advanced Chemistry, *Teachers' guides I* and *II*, provide brief references to radiochemical techniques in Topics 12 and 14.
Nuffield Advanced Physical Science, *Teachers' guides I, II,* and *III,* refer to radiochemical techniques in Sections 3, 12, and Option G2.
Nuffield Physics *Teachers' guide V*, reviews radioactivity extensively.
Nuffield Advanced Physics provides additional comment in Unit 5 *Atomic structure*.
The following book contains more detailed information about experimental work: Faires, R. A. (1970) *Experiments in radioactivity*, Methuen Educational.

Supplementary materials

Film loop
2–15 'Radioactive materials – uses'

Film
'Chemical equilibrium' ESSO. Films for Science Teachers No. 23

Overhead projection original
28 The thorium decay series *(figure A24.1)*

Chemists in the world, Chapter 1 'Finding out about the atom' and Chapter 2 'Using ideas about the atom'.
Handbook for pupils, Chapter 8 'Radioactivity'.

A24.1
Introduction to radiochemistry

A brief reference is made to the Periodic Table. The fogging of a photographic plate introduces the idea that there is something special about substances that cause this effect. Other ways of detecting these 'invisible rays', such as a Geiger-Müller tube and a scaler, are mentioned and this apparatus may be used to show that the 'radioactivity' of an element is unaffected by the state of chemical combination, the temperature, and several other factors.

A suggested approach

Objectives for pupils

1. Awareness of the contributions of Becquerel, Madam Curie, and Rutherford to our knowledge of radioactivity
2. Knowledge of the types of radiation emitted by radioactive substances
3. Awareness of the techniques needed for measuring radioactive radiations
4. Knowledge of the general patterns portrayed by the Periodic Table

The pupils should have some elementary knowledge of the Periodic Table and of atomic masses (see Topics A11 and A13). The main features of the Periodic Table can be reviewed and used as a setting for the Topic.

The study of radioactivity is desirable for its own sake and because it can be used to clarify other aspects of chemistry. The teacher may need to restrict the work attempted since certain aspects of radioactivity may be more conveniently covered in a physics course. These include experiments with electroscopes, the characteristics of a Geiger-Müller tube, deflection and absorption experiments, an introduction to the statistics of counting, and an analysis of decay curves (see Nuffield Physics).

The Topic enables the role of chance in scientific discovery and the nature of scientific inquiry to be discussed. The discovery by Becquerel of the fogging of photographic plates provides an appropriate opening. This experiment can be demonstrated simply (see Experiment A24.1) and accompanied by a brief account of the discovery and isolation of radium by Madame Curie, who was given this task by Becquerel for her doctoral thesis.

The types of radiation emitted by radioactive substances, alpha-, beta-, and gamma-rays, should then be discussed. The original methods of detection such as the discharge of an electroscope and the fogging of photographic plates can be shown (see Experiment A24.1a). The scintillation-counter technique employed by Rutherford may be mentioned and scintillations viewed in a darkened room. (Experiment A24.1b).

A more modern method for detecting radioactive change should then be introduced: the Geiger-Müller tube (GM tube) with a scaler and, if one is available, a scintillation counter. The Geiger-Müller tube can be used to demonstrate the random nature of radioactivity by counting a low-powered source (Experiment A24.1c).

Apparatus

Several suitable types of scaler and rate meter are available. For experiments in radiochemistry using low activity sources a scaler is usually more suitable than a rate meter although it takes longer to do the experimental work.

Experimental work

Before carrying out any experimental work, it is necessary to explain and demonstrate some radiochemical techniques, such as the preparation of samples for counting; the exclusion of all mouth operations by using rubber bulbs attached to pipettes or plunger devices for withdrawing and transferring liquids; the use of rubber or polythene gloves and tissues for handling radiochemical specimens (although it should be explained that this is only to prevent contamination of the hands and does *not* protect the wearer from radiation); the importance of conducting all radiochemical operations over a tray lined with absorbent paper, and of placing all contaminated solid waste in a bin labelled 'radioactive waste', and liquid waste in a separate waste bottle suitably labelled.

It is important to stress that 'carriers' have always been used in radiochemical experiments. The radioactive isotopes cannot be used in a pure condition owing to the very small quantity employed, for example, 10^{-10} to 10^{-15} g. The carrier consists of inactive atoms of the same element in the same oxidation state as the radioactive tracer. If a carrier is not employed, most of the radioactive atoms will be lost on the sides of the glass vessels.

Safety precautions

Detailed references to safety measures to be taken by teachers in schools or other educational establishments appear in Department of Education and Science *Administrative Memorandum* 2/76 (17 May 1976), obtainable from the Department of Education and Science, Elizabeth House, 39 York Road, London SE1 7PH. It should be noted that except for work (*a*) with the usual laboratory compounds of potassium, uranium, and thorium, (*b*) with some closed radioactive sources of very low activity equal to that, for example, of a luminous watch and (*c*) with equipment in which electrons are accelerated by a potential difference of less than 5 kV, a radioactive substance and all devices in which electrons are accelerated must only be used, and in the case of radioactive substances must only be obtained, in accordance with the conditions set out in the Administrative Memorandum and in the associated explanatory booklet issued by the DES.

Experiment A24.1a

Apparatus

The teacher will need:

Box of photographic plates or

(Continued)

Becquerel's chance discovery

This experiment should be done by the teacher.

Procedure

The fogging on a photographic plate by radiation from a radioactive mineral was first observed by Becquerel. In a dark room, place a fast photographic plate (the type does not greatly matter) on a sheet

X-ray film (for example,
Kodirex general purpose
X-ray film from Kodak)

Access to dark room

Piece of black paper, 30 cm^2

Key or coin

Uranium nitrate or thorium
nitrate

Developing solution

of black photographic paper about 30 cm^2. Wrap it in the black paper to form a parcel, sensitive side up. Now bring this into the daylight and place a coin or key on the plate. Sprinkle the top with some powdered uranium or thorium salt (use two or three spatula measures) and put the plate in a cupboard for several days. Develop the plate. It should now be sufficiently fogged to show the shadow of the coin or the key.

Next pupils do their own experiment on radioactivity.

Experiment A24.1b

Apparatus

Each pair of pupils will need:

Experiment sheet 101

Scintillation plate (taken from the packet and kept in the dark for *at least* 24 hours)

Magnifying lens

Use of a darkened room

Specimens of non-radioactive luminous paint

An examination of scintillations

Procedure
Experimental instructions are given in *Experiment sheet* 101. A darkened room is necessary for this experiment, so pupils should read the instructions and make preparations for the experiment beforehand. About five minutes is needed for dark adaptation.

Alternatively, use an optical microscope, with slide platform and base, inside a dark box covered with a dark cloth. Place a scintillation plate and a slide with non-radioactive luminous paint on the covered slide platform under a low power lens. Focus onto the specimen. Allow the class to view the specimens side by side.

Experiment sheet 101
Note: Most of this experiment must be done in the dark when you will not be able to read these instructions, so you must try to remember what you have to do.

When the lights are on, put a specimen of non-radioactive luminous paint in front of you and observe it through a hand lens. Continue your observation when the lights have been turned off.

Does the light from the paint fade as time goes on?

Is the light given off all over the surface or from some parts only?

Next, you will be given, *in the dark*, a scintillation plate, which contains radioactive material. Put it on the bench next to the non-radioactive paint and look at both through the lens.

Is light given off from the scintillation plate?

Is the light given off all over the surface or from some parts only?

If you have a watch with luminous figures, look at them through the hand-lens. Does the paint on them contain radioactive or non-radioactive material?

If your watch dial behaves like a scintillation plate, let your teacher know: it may be useful in a later experiment.

After the experiment explain the function of the radiation source or the scintillation plate as well as the role played by the zinc sulphide or barium cyanoplatinate phosphor.

Then introduce the GM tube as a method for detecting radioactive change.

A first use of a Geiger-Müller tube and scaler

The teacher will need:

Scaler and GM tube – for use with solutions

Stopclock

Measuring cylinder, 25 cm³

Beaker, 100 cm³

0.1M uranyl nitrate solution

4M potassium hydroxide solution

8M potassium hydroxide solution

8M hydrochloric acid

Distilled water

This experiment **must** be done by the teacher. **Standard radio-chemical precautions should be observed** (see details given on page 329).

Procedure

The object of this experiment is to give elementary experience in the use of a scaler and a GM tube, and to compare the radioactivity of potassium salts with that of uranium salts. The scaler is first switched on and adjusted to the recommended working voltage for the GM tube.

A count of the 'background' radiation should be taken over two or more minutes with the liquid counter filled with distilled water. The water is then replaced by 0.1M uranyl nitrate solution and a count made. The same solution is then diluted to double its original volume and a count of about half the first count should be obtained. The solution is again diluted and a further count taken. The results should show that the radioactivity is proportional to the dilution.

The GM tube is then cleaned, filled with 4M potassium chloride solution and a count taken. Calculate the result you might expect by diluting this solution to 0.1M and compare it with that for the 0.1M uranyl nitrate solution.

To investigate the effect of chemical change on radioactivity, mix equal volumes of 8M hydrochloric acid and 8M potassium hydroxide in a beaker. Stir thoroughly and transfer a sample of a neutral solution to the GM tube. Make a count and compare the result obtained with that from a 4M potassium chloride or 4M potassium hydroxide solution.

Find out what you can about the discovery of radioactivity.

A24.2
The study of the decay and growth of some radioactive chemicals

The idea is introduced that a radioactive element, while emitting rays, is slowly changing into another element. This process is studied by plotting both decay and growth curves, and the concept of half-life is introduced. The concept of isotopes arises from a discussion of the nature of radioactive decay.

The next section should include the discovery of thorium-234 (uranium-X) by Sir William Crookes. He added a carbonate solution to a uranium salt until the precipitate first formed was almost completely re-dissolved. The residue, thorium-234, can be shown to contain almost all of the activity of the uranium. This resulted in

---

1. Awareness of the
contributions of Crookes and
of Soddy
2. Knowledge that decay and
growth of a radioactive
element involve conversion
to another element
3. Knowledge of the concept of
isotopes
4.Knowledge and use of the
concept of half-life
5. Familiarity with techniques
for measuring radiation

Rutherford's discovery of growth and decay. The work of Soddy should also be mentioned and used to introduce the concept of isotopes through his measurement of the variation of the atomic mass of lead according to its origin. He found that natural lead, lead from pitch-blende, and lead from thorianite had different relative atomic masses. From this discovery, the disintegration hypothesis can be established and explained, and the idea of a radioactive series introduced. Growth and decay are probably best understood by reference to a decay series, such as that of thorium, figure A24.1.

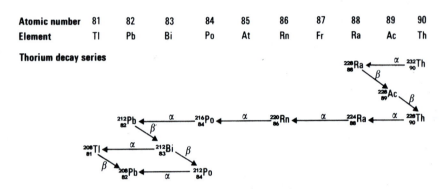

Figure A24.1
The thorium decay series. (OHP 28)

Parts of the series can be studied experimentally at this level. The decay of lead-212 to bismuth-212 is demonstrated in Experiment A24.2a. Experiment A24.2b is about the decay of bismuth-212 to lead-208. Experiment A24.2c relates to the decay of protactinium-234 and Experiment A24.2d extends the notion of half-life.

Part 1 of Experiment A24.2a takes a long time. Nevertheless, the experiment is included here for those teachers who wish to use it. Experiment A24.2b is similar. Good results can be obtained and a set of specimen results is included in the account. Experiment A24.2c involves a very short half-life isotope and is relatively easy to complete.

Experiment A24.2a

The decay and growth curves of lead-212
This experiment should be done by the teacher.

Apparatus

The teacher will need:

Scaler and GM liquid counter

Centrifuge and 4 test-tubes,
100 × 16 mm

Watch-glass and beaker of
boiling water

(Continued)

Procedure
Standard radiochemical precautions should be observed (see page 329).
The two parts of this experiment can be run concurrently.

Part 1. To about $10 \, cm^3$ of the thorium nitrate solution add three drops each of 0.2M solutions of lead nitrate and barium nitrate. Add 0.880 ammonia solution in drops until the precipitation of hydroxides is complete. This precipitate consists of thorium and lead hydroxides, the latter carrying the radioactive isotope lead-212,

Sellotape

0.1M thorium nitrate

0.2M lead nitrate

0.2M barium nitrate

0.880 ammonia solution

Concentrated nitric acid

2M nitric acid

2M sulphuric acid

from the original thorium nitrate. The radium-224 isotope which is also a decay product of the thorium remains in the filtrate. The barium is acting as a 'holdback' carrier. The precipitate of thorium and lead hydroxides should be warmed to coagulate it and centrifuged. Decant and retain the liquid for Part 2. Dissolve the precipitate in the minimum quantity of nitric acid and re-precipitate with 0.880 ammonia, rejecting the washings. The object of this is to free the precipitate from all absorbed ions which may have been carried down when it was first precipitated. Now wash the precipitate with hot water, centrifuge, and dissolve it in 2–3 cm^3 of 2M nitric acid. Place this in the GM liquid counter and use the scaler to take a count over a period of, say, two minutes, repeating this every hour or at convenient intervals during the next twenty-four hours. The count is due to particles emitted by the lead-212, the particles from the radium-228 being of too low an energy to affect the GM tube. Plot the decay curve for the lead-212. The results should be discussed and the term half-life introduced. The half-life of lead-212 is about 10.6 hours.

Part 2. This follows the growth of the same isotope, lead-212 from radium-224, itself a decay product of thorium. The radium was not precipitated by the ammonia solution, but remained in the filtrate which was retained. Treat this filtrate (which contains the barium ions added to the original solution) with sulphuric acid until no further precipitation occurs. The barium sulphate will carry down with it the radium sulphate. (It will also bring down radium-228, if present.) The growth curve will then include actinium-228 produced from the radium-228. Whether or not this happens to an appreciable extent depends on the age of the thorium salt.

Centrifuge the precipitate, and decant the liquid. Transfer the precipitate to a watch-glass and dry it over a beaker of boiling water. Stick a piece of Sellotape over the precipitate to secure it on the watch-glass and to retain the radon which is a decay product of the radium and in turn decays to lead-212, the growth curve of which is being plotted. Invert the watch-glass over the GM tube and count at convenient intervals over a period of twenty-four hours. The GM liquid counter has a glass shield which cuts out alpha particles but is transparent to beta particles, so that it is possible to measure the growth of the lead-212, which is a beta emitter, in spite of the presence of radium-224 and radon-220 which are alpha emitters.

Experiment A24.2b

Apparatus

The teacher will need:

Scaler and GM liquid counter

Centrifuge and 4 test-tubes,
100 × 16 mm

(Continued)

The decay of bismuth-212
This experiment should be done by the teacher.

Procedure
Standard radiochemical precautions should be observed (see page 329). In this experiment we measure the decay of the next isotope in the series, namely bismuth-212, which is formed as a result of beta emission from the lead-212 studied in the previous experiment. This is a slightly easier decay curve to plot because the half-life of the bismuth-212 is about one hour rather than ten hours.

Hydrogen sulphide generator

Stopclock

Solutions of thorium nitrate (5% w/v), lead nitrate (5% w/v), and bismuth nitrate (5% w/v)

2M sulphuric acid

Concentrated nitric acid

Distilled water

To about 20 cm³ of the thorium nitrate solution add 1–2 cm³ of the lead nitrate solution and a few drops of the bismuth nitrate solution. The latter acts as a 'holdback' carrier for the bismuth-212 at this stage. Add 2M sulphuric acid in drops to precipitate the lead and centrifuge. To ensure that all the lead is removed from the solution, repeat the precipitation by adding a few more drops of lead nitrate to the filtrate and more 2M sulphuric acid. Centrifuge off the precipitate. Pass hydrogen sulphide into the clear solution to precipitate the bismuth as sulphide, centrifuge, and reject the clear liquid.

Dissolve all the bismuth sulphide, which includes the bismuth-212, in a few drops of concentrated nitric acid, dilute with 2 or 3 cm³ of distilled water, and pour into the GM tube. Count immediately for a minute and repeat the count every five minutes for about half an hour and then every ten minutes for a further half hour. If possible continue to count at intervals for another three hours or so. Plot the decay curve of the bismuth-212. The pupils can then be given other decay curves to examine and led to the idea of half-life empirically. The half-life for bismuth-212 is 60 minutes. A set of specimen results is shown in figure A24.2.

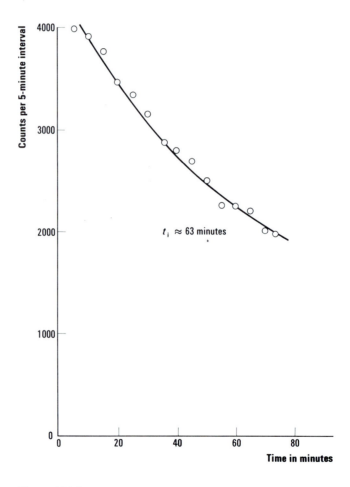

Figure A24.2
Sample results for the decay of bismuth-212.

Topic A24 Radiochemistry

The decay of protactinium-234

This experiment should be done by the teacher.

Apparatus

The teacher will need:

Scaler and GM liquid counter

Stopclock

Separating funnel, 50 cm³

Measuring cylinder, 25 cm³

Beaker, 100 cm³

Uranyl nitrate

Amyl acetate (pentyl ethanoate)

Concentrated hydrochloric acid

Distilled water

Procedure

Standard radiochemical precautions should be observed (see page 329). Dissolve 1 g of uranyl nitrate in 3 cm³ distilled water. Add 7 cm³ concentrated hydrochloric acid. This changes the molarity of the solution to 7M with respect to hydrochloric acid. Transfer this solution to a separating funnel and shake it with 10 cm³ of amyl acetate for half a minute. Remove the lower aqueous layer. Transfer 5 cm³ of the amyl acetate layer to the liquid counter and start counting immediately. It takes a finite time to make and record an observation, so it is best to count for 15 seconds and then to record for 5 seconds before continuing to count for a further 15 seconds and recording for a further 5 seconds, and so on. Thus, the first count occurs at 15 seconds and is recorded as being the value for half-way during the period – that is, at 7.5 seconds. The second count is at 35 seconds and is recorded as being for 17.5 seconds, and so on. A graph showing counts per second against time is plotted to give a decay curve. The characteristic of this curve can be discussed and it can be shown that the half-life for protactinium-234 is about 1.2 minutes.

A growth curve for protactinium-234 can be obtained by counting the aqueous layer.

The layers may then be shaken together again and the experiment repeated. A sample set of results for the first part of this experiment are shown in figure A24.3.

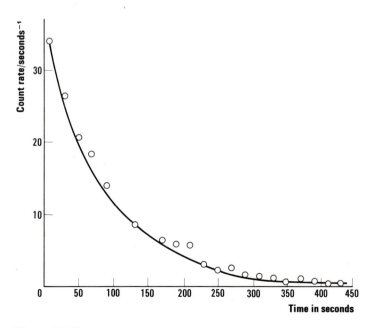

Figure A24.3
Sample results for the decay of protactinium-234.

A24.3
The use of radioactive isotopes as detectives

This section is concerned with the use of radioactive tracers.

Objectives for pupils

1. Knowledge of the practical applications of radioactive tracers
2. Awareness of the safety precautions required when radioactive tracer techniques are employed

The essential factor in the use of radioactive isotopes as 'detectives' is the ease with which relatively small numbers of these atoms can be detected. A small proportion of radioactive atoms in a compound enables those atoms to be followed through very many stages of a chemical synthesis or of a biological process. For example, plants may be subjected to an atmosphere containing carbon dioxide labelled with carbon-14 atoms, and the uptake of radioactive material studied by analysis of the materials in the plant to see what has happened to the carbon-14. This type of study has been carried out on many chemical processes to find out more about the reactions involved.

Radioactive isotopes can also be of use in solubility measurements. If the specific activity (that is, activity per gram) of a solid is known, then, by shaking an excess of the solid with the solvent and measuring the activity transferred to the solution, the solubility can be found. Additional experiments can be found in Faires, R. A. (1970) *Experiments in radioactivity*, Methuen Educational, page 92. The partition of a solid between two liquids can also be measured by this procedure. Further applications of radioactive isotopes are discussed in Alternative IIB, Topic B22.

If the pupils have already seen Experiment A19.3 (A24.3a) they could now be reminded of it and the use of a radioactive isotope in the study of equilibria discussed. If they did not see it then, do it now. This experiment requires some care. The activity obtained may be as low as ten counts per minute above background. Nevertheless, it is included because it makes an important point. Similar experiments with other isotopes, for example silver-110, are very effective but *cannot* be used in school (see Faires, *op cit*).

Evidence for the dynamic nature of equilibria

This experiment is described in Topic A19, Experiment A19.3.

1. Make a summary of the work you have done in this Topic.
2. *Chemists in the world*, Chapters 1 'Periodicity' and Chapter 2 'Finding out about matter' relate to the atom. Revise your ideas by reading these chapters.

By the end of this Topic, pupils should have some knowledge of radiochemistry including the idea of growth and decay and the concept of half-life. They should be aware of the use of radioactive isotopes as tracers and of the hazards associated with radioactivity. They should have some knowledge of the history of the phenomena discussed.

Part 2 Alternative IIB

Safety in science laboratories
Teachers are reminded of the need to observe those safety regulations which apply to schools in their area. General advice on safety in school science laboratories in England and Wales is conveyed through Department of Education and Science publications (see page 650 for the list relating to the period immediately before this guide was prepared for publication). Teachers are advised that it *may* be necessary for them to adapt some of the suggestions in this guide in order to conform to their local safety regulations.

Elements and compounds

Purposes of the Topic

1. To revise knowledge of the chemistry of elements and compounds.
2. To recall that most elements react with oxygen to form oxides.
3. To recall the properties of the oxides of non-metallic elements and of metallic elements, and from these to formulate operational definitions of the terms acid and alkali.
4. To give pupils an opportunity to do some qualitative investigations.
5. To introduce pupils to the properties of hydrogen chloride and to determine its composition.

Contents

B11.1 Elements, compounds, and mixtures
B11.2 Properties of oxides
B11.3 The 'salt-gas' problem
B11.4 The composition of 'salt gas'

Timing

This Topic revises a number of ideas which appear in Stage I of Nuffield Chemistry or Nuffield Combined Science and which should be familiar to pupils before they attempt subsequent Topics in Alternative IIB. The time needed for this will therefore depend on the experience and ability of the pupils. It is suggested that between one and two weeks will be needed for the revision and that a further two weeks should be spent investigating the 'salt-gas' problem. A total of about three or four weeks will enable all aspects of the Topic to be covered given that pupils have appropriate background knowledge.

Introduction to the Topic

The purpose of the first section of the Topic is to provide pupils with a brief review of much of the work attempted during Stage I or an equivalent introductory course. Two important themes are considered: the distinction between mixtures and pure substances; and the classification of pure substances as either compounds or elements. Mention is made of the use of atomic symbols.

The properties of oxides form the second section and the opportunity is taken to distinguish between the oxides of metallic and non-metallic elements. The formation of water by burning hydrogen in air or by passing hydrogen over a hot metal oxide may also be considered. The use of word equations as reaction summaries is recommended and, for some pupils, it may be appropriate for the teacher to supplement this with chemical equations in a few instances.

Finally, the 'salt-gas' problem provides an opportunity for pupils to investigate the properties and composition of an 'unknown' substance. Simple investigations are used to lead to the fact that 'salt gas' contains both hydrogen and chlorine. The study enables the terms 'analysis' and 'synthesis' to be revised.

The Topic is extended by Topics B12, B13, and B14. Quantitative aspects of the theme occur in B18 and the theoretical aspects of acidity and basicity feature in B20. This delay in the introduction of quantitative measurements is deliberate and is compatible with the content of Alternative IIA.

Alternative approach

The development of Topic A11 does not require the revision of the properties of oxides. Topic A12 is devoted to the 'salt-gas' problem and Topic A20 deals with acids and bases as a separate theme.

Supplementary materials

Film loops
1–1 'Salt production'
1–2 'Chlorophyll extraction'
1–5 'Petroleum fractionation'
1–6 'Liquid air fractionation'
1–7 'Gold mining'

Overhead projection original
29 Common elements and their abundance in the Earth's crust and in the human body *(Table B11.1)*

Reading material

for the pupil

Study sheets:
The words chemists use
Where chemicals come from
The chemical elements
Analysis
Chemicals and rocks

B11.1
Elements, compounds, and mixtures

The pupils review their knowledge of the techniques used for purifying chemicals and revise their understanding of the terms compound and element.

A suggested approach

Objectives for pupils

1. Knowledge of the common processes used to separate mixtures of chemicals
2. Appreciation of the terms element, compound, and mixture
3. Knowledge of the formation of compounds from elements
4. Awareness of the percentage composition of the Earth's crust and of the human body

This section revises work done during Stage I or during Nuffield Combined Science. Remind pupils of the sequence of experiments on identifying and subsequently separating mixtures of chemicals. These included the preparation of pure salt from a sample of crude rock salt; the separation of certain inks into their component dyes; and the fractionation of crude oil. Then establish the common characteristics of all such separations, namely the use of a difference in the properties of the materials which make up a mixture. Refer to other examples of mixtures and invite pupils to apply this principle to mixtures of, say, salt and water, powdered iron and powdered sulphur, or to a sample of air. A discussion of this last example might include some description of the separation processes and subsequent storage of pure gases from the atmosphere. (It might even include a demonstration of the properties of liquefied gases!) Where appropriate, Stage I film loops may be shown. The purpose of this work is to ensure that pupils can distinguish between mixtures and single substances.

A classification of single substances as either compounds or elements may then follow. Pupils should know that water is a compound of hydrogen and oxygen and that it may be formed either when hydrogen is burned in oxygen, or when a mixture of hydrogen and oxygen is exploded, and the products of reaction condensed. Compare the properties of the compound and its constituent elements and make the point that the compound possesses quite distinctive properties from those of its constituent elements. Elements are described as substances which cannot be broken down into simpler substances by chemical means. Draw some attention to the fact that when elements combine to form a compound, there is a net energy change. Although no formal proof is given at this stage, it could be added that compounds possess a definite composition by mass whereas the composition of mixtures is quite arbitrary.

Finally, refer to chemical analysis which has enabled chemists and geologists to find out about the abundance of the elements in nature. Table B11.1 lists the abundance of elements in the Earth's crust and in the human body. Only nine elements are needed to account for some 98 per cent of the Earth's crust and only six are needed to account for some 99 per cent of the composition of the human body. Oxygen is the most abundant element in the Earth's crust and occurs in combined form as water and in many rocks, as well as being in the free (or elemental) state in the atmosphere. The table also lists the symbols needed to represent one atom of each of the elements used, and something could be said about the origins of these symbols. The *Study sheets* listed at the beginning of the Topic provide much useful information and can be used to supplement this section.

Element	Symbol used to represent 1 atom of the element	Abundance in the Earth's crust (% by mass)	Abundance in the average human body (% by mass)
Aluminium	Al	7	—
Calcium	Ca	3	2
Carbon	C	very small*	18
Chlorine	Cl	very small*	0.15
Hydrogen	H	1	10
Iron	Fe	4	very small
Magnesium	Mg	2	—
Nitrogen	N	very small*	3
Oxygen	O	50	65
Phosphorus	P	very small*	1
Potassium	K	$2\frac{1}{2}$	0.4
Silicon	Si	26	—
Sodium	Na	$2\frac{1}{2}$	0.15
Sulphur	S	very small*	0.3

Table B11.1 Common elements and their abundance in the Earth's crust and in the human body [OHP 29]

*The remaining 2 per cent of the Earth's crust is made up of these and 78 other elements.

B11.2
Properties of oxides

Objectives for pupils

1. Knowledge of the formation of oxides of some metallic and some non-metallic elements
2. Knowledge of the properties of these oxides
3. Recall that water is hydrogen oxide
4. Recall of the reducing properties of hydrogen
5. Knowledge of the terms element, atom, symbol, compound, and formula

Experiment B11.2

Apparatus

The teacher will need:

Cylinder of compressed oxygen

Connecting tubing

Bunsen burner and heat resistant mat

6 gas jars and cover slips

6 deflagrating spoons

Knife

Filter paper

Crystallizing dish

Emery paper

Access to water

Powdered roll sulphur

Small pieces of charcoal

White phosphorus*

Magnesium ribbon

Iron wool (grease free)

Indicators (litmus and Full-range Indicator)

In this section, various oxides are made and their properties investigated. This information is then used to formulate operational definitions of the terms acid and alkali.

Introduce the theme of this section by asking pupils about the meaning of the terms element and compound. Remind pupils about the property of oxygen to support combustion and then demonstrate the formation and properties of various oxides.

The formation of oxides

Procedure
This experiment is a revision of Stage I Experiment A4.1 or B4.1. Heat small samples of each element in turn in a Bunsen flame on a clean deflagrating spoon. Then lower the hot sample into a gas jar of oxygen. Draw attention to the various reactions which occur. Investigate the solubility of the products of reaction in water and note the effect of these solutions on an indicator.

When the results of these experiments are discussed, bring out the differences in the properties of metallic and non-metallic oxides. Lead the discussion towards descriptive definitions of the terms acid and alkali. Thus, an acidic solution will change the colour of litmus to red: an alkaline solution will change the colour of litmus to blue. Word equations may be introduced at appropriate points in the lesson as summaries of chemical reactions.

Many elements react with oxygen to form oxides. Water is hydrogen oxide and the formation of water by burning hydrogen in air and condensing the product should be revised. (Compare Experiment A7.4b or Experiment B8.1b.)

The explosion of a mixture of hydrogen and oxygen in a rubber balloon suspended on a long pole behind a properly mounted safety screen, could be demonstrated. (**Caution!** Compressed gases should be used and precautions taken to avoid contaminating each source of gas!) The effect of passing hydrogen over heated metal oxides to yield water should also be revised (compare Experiment

---

Safety measure. Fill vessels which are suspected of being contaminated with unburnt phosphorus with M copper(II) sulphate solution, and allow to stand for one week before rinsing with water.

A7.5 or B10.1b). The significance of these studies can be established by asking the pupils to indicate differences in properties between elements and their compounds (compare Section B8.2).

Throughout this section, pupils will have had opportunities to show their knowledge and understanding of the terms element, compound, and atom. Symbols for the elements were mentioned in Section B11.1, and with some classes it may be practicable to introduce a few formulae. With an able class, it may even be appropriate to supplement the word equations. If this is attempted, it is important to restrict the examples to the simplest cases. For example, the combustion of elements in oxygen may be summarized:

$$sulphur + oxygen \longrightarrow sulphur\ dioxide$$
$$S\ +\ O_2\ \longrightarrow\ SO_2$$

Or the reduction of a metal oxide by hydrogen might be quoted:

$$copper\ oxide + hydrogen \longrightarrow copper + water$$
$$CuO\ +\ H_2\ \longrightarrow\ Cu\ +\ H_2O$$

Or the solution of sulphur dioxide in water to make sulphurous acid:

$$sulphur\ dioxide + water \longrightarrow sulphurous\ acid$$
$$SO_2\ +\ H_2O \longrightarrow\ H_2SO_3$$

It is *not* recommended that a full and detailed treatment of formulae and equations be attempted.

Suggestions for homework

1. Complete, as far as you are able, the following table on a separate sheet of paper.

Element	Symbol	State whether the properties of the oxide of the element are (a) acidic; (b) alkaline; (c) neutral
Hydrogen		
Sulphur		
Sodium		
Magnesium		
Carbon		
Phosphorus		

2. Make a list of ten elements and their symbols. Find out about the derivation of the words from which the symbols are made.

B11.3
The 'salt-gas' problem

In this section, pupils investigate the gas given off when concentrated sulphuric acid is added to some common salt. This gas is called 'salt gas'. They determine some of the properties of the gas.

A suggested approach

The starting point for this investigation is a simple demonstration of the effect of adding concentrated sulphuric acid to some common salt. The pupils then repeat this demonstration and investigate some

1. Knowledge of the preparation and properties of an important compound
2. Ability to frame a simple hypothesis and devise experiments to test it

of the properties of the fuming gas evolved: its reaction towards indicators; its flammability; and whether it is soluble in water.

Pupils are unlikely to get good results with the solubility test and will need to have this demonstrated to them. The ensuing discussion can lead to a demonstration of the fountain experiment and subsequently to the preparation of a solution of the gas. However, the technicalities of the fountain experiment can produce a barrier to those pupils who fail to appreciate the point of it.

Experiment B11.3a

Investigating some properties of 'salt gas'

Apparatus

Each pair of pupils will need:

Experiment sheet 46

4 test-tubes, 100 × 16 mm

Cork with right-angled delivery tube to fit test-tube

Small trough or basin to contain water

Spatula

Splint

Full-range Indicator paper

Sodium chloride, one spatula measure

Concentrated sulphuric acid*, few drops

Procedure
Instructions are given in *Experiment sheet* 46. It will be found necessary to give some help to pupils in the planning of the various tests suggested. This is most easily done by discussing possibilities before practical work is attempted. The solubility test is likely to prove inconclusive and may need to be followed by a demonstration. It can be used to lead to a demonstration of the fountain experiment (Experiment B11.3b) and to a more effective procedure for making a solution of a soluble gas in water (Experiment B11.3c).

Experiment sheet 46
Place a spatula measure of sodium chloride in a test-tube and add a few drops (about 5 or 6) of concentrated sulphuric acid to it. **(Concentrated sulphuric acid is a very corrosive liquid, so handle it carefully. If any of it is accidentally spilled on your skin, wash it off IMMEDIATELY with a stream of water from a tap, and then report the mishap to your teacher.)** Describe all that happens, paying special attention to what you see inside the test-tube and what you see just outside the open end.

Devise and carry out tests on the gas to find out:
1. whether it is acidic, neutral, or alkaline;
2. whether it burns or allows a lighted splint to burn in it;
3. whether it is soluble in water.
(Your teacher will probably give some help in planning these tests.)
Describe below what you do, what happens, and the conclusions that you reach.

You may be shown another way of finding whether 'salt gas' is soluble in water. If so, draw a diagram of the apparatus used.

Experiment B11.3b

The fountain experiment
It is best for this to be a teacher demonstration since a spectacular result can only be achieved if the flask is very thoroughly filled with 'sat gas'. The filling must be done in a fume cupboard and inevitably involves the escape of a large quantity of gas.

Apparatus
The teacher will need:

Gas generator for 'salt gas':

Conical flask or filter flask, 500 cm³

Tap funnel

Delivery tube and rubber connector

(Continued)

Procedure
Fill the flask with salt gas by downward delivery as shown in figure B11.1a. The difficulty lies in displacing all the air from the flask, and a separate delivery tube leading to the bottom of the flask is therefore preferred to the use of the central jet with the cork loosened.

Safety note. The use of concentrated sulphuric acid by pupils requires close supervision.

Bung
Stand, boss, and clamp

Dry round-bottom flask, preferably 1000 cm³, fitted as shown in figure B11.1 *or* as in the special arrangement shown in figure B11.2

Glass trough

Cloth

Bunsen burners and heat resistant mats

Litmus solution

Rock salt (lumps, not powder – to prevent frothing)

Concentrated sulphuric acid

Allow a great deal of gas to escape before carefully withdrawing the delivery tube and inserting the cork fitted with the jet. Invert the flask in a trough of water which has been coloured blue with litmus solution and one or two drops of alkali (figure B11.1b). In order to start the 'fountain' release the clip and warm the flask carefully with an almost luminous Bunsen flame until a few bubbles of gas (mainly air from the tube) have escaped as a result of the expansion; then cool the flask with a wet cloth and, as the gas contracts, the water will rise in the central tube. Once water enters the flask through the jet, the 'fountain' will operate without further assistance. The acidic solution which forms in the flask will be coloured red by the litmus. This heightens the effect of the demonstration.

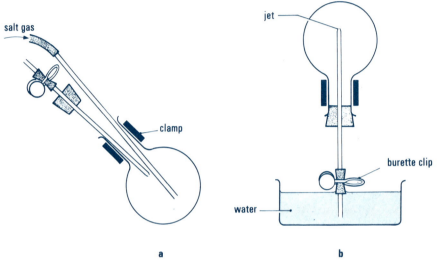

Figure B11.1
Fountain experiment: (*a*) arrangement for filling the flask; (*b*) apparatus ready for the demonstration.

Alternatively, set up the apparatus shown in figure B11.2. Use a dry 1000 cm³ round-bottom flask for the fountain effect. Connect this flask to a salt-gas generator and fill it with salt gas. Close clips A and B. On standing, the fountain usually starts spontaneously. If necessary, the flask may be cooled with a little water or diethyl ether (ethoxyethane) *or* one can blow into tube C, to start the effect.

Discuss how to make a solution of salt gas and then show the pupils a convenient laboratory method for making this solution.

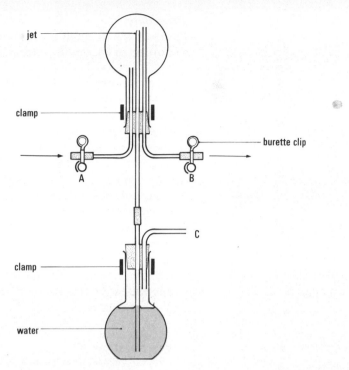

jet

clamp

burette clip

A

B

clamp

C

water

Figure B11.2
Fountain experiment: to operate remove clips A and B, and connect A to gas
generator; when the flask is full of salt gas, close A and B. On standing (or on
blowing into C), a fountain is formed.

Experiment B11.3c

Apparatus

The teacher will need:

Gas generator used in B11.3b

Either 2 filter tubes,
150 × 25 mm
Or 2 Drechsel bottles, fitted as
shown in figure B11.3

Preparing a solution of 'salt gas'

This experiment should be done by the teacher.

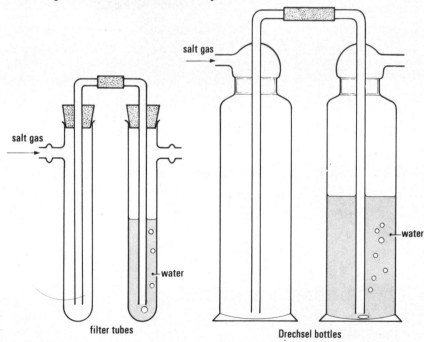

salt gas

salt gas

water

water

filter tubes

Drechsel bottles

Figure B11.3   a   b

Procedure
Pass salt gas through the apparatus shown in figure B11.3 to obtain a solution of the gas in water.

After these demonstrations, ask pupils to devise their own experiments to decide what substance causes the steamy fumes made when salt gas comes into contact with the air. In answer to the question in *Experiment sheet* 47, they will probably suggest carbon dioxide and water as possible substances to test. The apparatus listed for the experiment is based on this assumption.

Finding out what causes the steamy fumes when 'salt gas' is exposed to air

Apparatus

Each pair of pupils will need:

Experiment sheet 47

6 test-tubes, 100 × 16 mm

4 corks to fit the test-tubes

Cork with right-angled delivery tube to fit test-tubes

Teat pipette

Spatula

Sodium chloride, one spatula measure

Concentrated sulphuric acid, few drops

Small marble chips, about 6

Dilute hydrochloric acid, 10 cm^3

Cotton wool (previously dried in oven)

Procedure
Experiment sheet 47 is reproduced below and gives full details. Pupils may require assistance with these investigations and a demonstration of the effect of dry carbon dioxide and of steam on salt gas may be necessary.

Experiment sheet 47

You will have noticed that 'salt gas' appears clear and colourless until it comes out into the air, when it forms steamy fumes. These fumes increase greatly if the gas comes into contact with your breath. What substances are there in your breath which also occur, but in smaller proportions, in ordinary air? You can probably think of two such substances and can now try to devise simple experiments which will enable you to decide whether one or both of them are responsible for the fuming.

You will have to decide how to produce separate specimens of the two substances that you wish to test. It may help you to know that a plug of oven-dried cotton wool, placed in the mouth of a test-tube in which a gas is being produced, will dry the gas as it passes through.

Write down the names of the substances, in the air and in your breath, which you think may be responsible for the fuming of 'salt gas'.

Describe the experiments you try and the results you obtain.

What are your conclusions from these experiments?

In the discussion following the experimental work, pupils should understand that when concentrated sulphuric acid is added to common salt, a gas (provisionally called salt gas) is obtained which gives an acid reaction towards indicators; does not burn; has a high solubility in water; and fumes in air due to the presence of water vapour.

B11.4
The composition of 'salt gas'

'What is the composition of salt gas?' 'Where should we start?' A hint may be taken from the acidity of the gas. What properties of acids have they met before? The reactions of acids with metals to yield hydrogen may require revision. Perhaps salt gas contains hydrogen since it has acidic properties? Discuss ways of finding out about this. Make sure that pupils appreciate that no test for hydrogen is valid if water is also present. Now demonstrate the effect of passing salt gas over heated iron. Remind pupils of the fountain experiment and point out the danger of 'sucking back' which is prevented in this experiment by the use of a simple valve (a Bunsen valve).

Experiment B11.4a

Apparatus

The teacher will need:

Buchner flask fitted with cork and tap funnel as an apparatus for generating 'salt gas'

Combustion tube, 125 × 16 mm – with end delivery tube

Delivery tube fitted with a Bunsen valve (see figure B11.4 – a Bunsen valve can be made from a piece of rubber connecting tubing with a short slit in it and a piece of rod)

Small trough or basin

Test-tube, 100 × 16 mm

Bunsen burner

Stand, boss, and clamp

Spatula

Rock salt

Concentrated sulphuric acid

Iron powder

Finding one of the elements in 'salt gas'
This experiment should be done by the teacher.

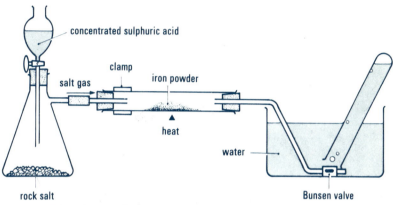

Figure B11.4

Procedure
Place three or more spatula measures of iron powder in a combustion tube and assemble the apparatus, as shown in figure B11.4, with the Bunsen valve under water in the basin. Drive the air out of the apparatus with a steady stream of salt gas, so that the bubbles are seen to escape through the valve. As soon as most of the gas appears to be dissolving, heat the iron powder with a moderate flame and collect the bubbles of gas which then begin to escape through the valve. The way in which this gas burns, when the tube is held to a flame, shows that it is hydrogen. Near colourless crystals of iron(II) chloride form in the combustion tube.

From the results of this experiment, it should be clear to pupils that the hydrogen must have come from the salt gas. 'What else is there in salt gas?' 'If we remove the hydrogen, what will we find?' 'How else can hydrogen be removed from salt gas?' (By combining it with oxygen to form water.) 'Perhaps the oxygen can be supplied by a compound?' Pupils will have come across a number of oxygen-containing compounds in Stage I: copper(II) oxide, manganese(IV) oxide, and so on. Pupils now try some of these substances in a further set of investigations.

Experiment B11.4b

To find out what other substance is present in 'salt gas'

Warning. On no account should the following experiment be carried out on an open bench, in a badly ventilated laboratory, or in an inefficient fume cupboard.

Each pair of pupils will need:

Experiment sheet 48

Test-tube, 100×16 mm

Full-range Indicator paper

Splint

Bunsen burner and heat resistant mat

Teat pipette

Manganese(IV) oxide, half a spatula measure

Copper(II) oxide, half a spatula measure

Potassium permanganate, half a spatula measure

Supply of concentrated hydrochloric acid labelled 'Solution of salt gas in water'

It is essential that the quantities of manganese(IV) oxide, copper(II) oxide, and potassium permanganate taken by pupils should be very strictly supervised.

Procedure
Throughout this experiment, pupils should be warned *not* to use larger quantities of materials than stated in the instructions and *not* to breathe the gas formed. If anyone accidentally inhales some chlorine, he should *gently* sniff the bottle of dilute ammonia solution. *Experiment sheet* 48 is reproduced below.

Experiment sheet 48

You have seen an experiment which shows that one constituent of salt gas is hydrogen. If you can find a way of removing hydrogen from salt gas you may be able to release the other substance present. You will have discussed this in class and prepared a list of substances that might be needed for this purpose. Test each of these substances as follows:

Add not more than half a measure of the substance to about one cm³ of a solution of salt gas in water in a test-tube. If there is no sign of anything happening when the mixture is cold warm it gently. Test any gas evolved with damp indicator paper – remember that unchanged salt gas may be escaping from the mixture.

(Do not use larger quantities than those stated and keep the test-tube well away from your nose and eyes.)

Record your results and conclusions below.

Which of the substances releases a gas from salt gas which behaves differently from salt gas towards indicator paper?

What is the name of this gas?

As the result of these investigations, pupils should know that salt gas contains hydrogen and chlorine. They will have been able to detect chlorine by its colour (greenish-yellow) and by its bleaching action on moist Full-range Indicator paper. Chlorine is formed when strong salt-gas solution is used with either manganese(IV) oxide or potassium permanganate. However, pupils may be confused by smells and by the greenish-yellow colour of copper(II) chloride solution formed when copper(II) oxide is used. Pupils may report that chlorine is formed in this last case *unless* the need for a positive bleaching test has been insisted on.

The next question is 'Does salt gas contain *only* hydrogen and chlorine?' Discuss any suggestions. The term synthesis should be introduced and used when introducing Experiment B11.4c.

Experiment B11.4c

Apparatus

The teacher will need:

Full-range Indicator paper

Splints

(*Continued*)

Burning hydrogen in chlorine
This experiment must be done by the teacher.

Procedure
Prepare one or two gas jars of chlorine.

Chlorine may be most conveniently prepared by adding concentrated hydrochloric acid in drops to potassium permanganate and collecting the gas in a gas jar by the displacement of air upwards.

Delivery tube for burning
hydrogen in a gas jar

Rubber connection tubing

Cylinder of hydrogen

Gas jars full of chlorine

Access to chlorine generator

Use a steady stream of hydrogen from a cylinder. Light the hydrogen at the jet, adjust the pressure to give a flame 2–3 cm high, and then lower the burning hydrogen jet into a gas jar of chlorine.

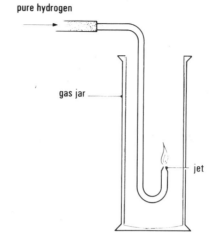

pure hydrogen

gas jar

jet

Figure B11.5

The hydrogen continues to burn, but with a white flame. The colour of the chlorine disappears, and its place is taken by steamy fumes of hydrogen chloride, salt gas. This may be tested using damp Full-range Indicator paper, and a lighted splint.

This experiment shows that salt gas is a compound of hydrogen and chlorine only and may therefore be given the more explicit name of 'hydrogen chloride'.

Suggestions for homework

1. What do you understand by the terms *analysis* and *synthesis*? Write a short explanation of these terms. Provide and explain two examples of each.
2. Read the Study sheets *Analysis* and *The chemical elements*.

Summary

During this Topic pupils have an opportunity to revise a great deal of work from Stage I. They will have revised the meanings of the terms element and compound, and know that most elements react with oxygen to form an oxide. In addition, they should be aware of differences in the properties of the oxides of non-metallic elements and of metallic elements, and have formulated simple definitions of the term acid and alkali. They will have investigated the 'salt-gas' problem and shown that this gas consists of hydrogen and chlorine only, and this investigation will have led to a discussion of the terms analysis and synthesis. Pupils should also be aware of the value of synthesis as a proof of the composition of a compound. The value of word equations as reaction summaries has been stressed throughout the Topic.

Note. The quantitative determination of the composition of hydrogen chloride does not appear in Topic B11 but is left until Topic B18, in contrast with the treatment of the salt-gas problem in Topic A12.

Topic B12

Acids, bases, and salts

Introduction to the Topic

In this Topic, substances are classified as acids, bases, or salts by using simple operational (descriptive) definitions. In the first section, the reaction of 'salt gas' with a metal is compared with the reaction of magnesium with dilute sulphuric acid and other similar reactions. Pupils can then identify acids as being compounds which:
1. dissolve in water to give solutions which change the colour of indicators;
2. react with metals, such as magnesium or zinc, to form hydrogen and a salt.

Subsequent experimental work reveals that not all materials labelled as acids give colour changes with indicators. Neither do they necessarily always react with metals to yield hydrogen, unless they are first dissolved in water.

Solutions of acids in water also possess the following general properties:
3. they react with a metal oxide to form a salt and water;
4. they react with a metal carbonate to form a salt, water, and carbon dioxide.

The characteristic properties of metal oxides (bases) are considered next. It will be recalled that they can be prepared by synthesis (see Topic B11), and also by the action of heat on metal carbonates (see Stage I, Topic A9). However, it may be necessary to revise the effect of heat on metal carbonates and to identify the products of reaction. The solubility of calcium hydroxide and barium hydroxide in water needs to be demonstrated, as does the effect of heat on these compounds to produce water and the corresponding oxide.

The final section of the Topic is directed to the preparation of various salts. The occurrence of salts as minerals is also considered. The Topic concludes with a short review of inorganic nomenclature together with operational definitions of the terms which have been used in the Topic.

The properties of acids are used in the approach to reaction rates adopted in Topic B13, and the subject of alkalinity occurs again in Topic B14 when the chemistry of ammonia is considered. The properties of ions in solutions are studied in Topic B20 and this includes the interpretation of the reactions of an acid with an alkali in terms of ions. Acidity and basicity also feature in Topic B21 when the Brønsted-Lowry theory is considered. An extended treatment of the Brønsted-Lowry theory occurs in Stage III Option 6, 'Change and decay'.

Alternative approach

The general theme of acidity and basicity and the preparation of salts are featured in Topic A20. The chemistry of ammonia is considered in Topic A22.

Background knowledge

This Topic follows from Topic B11, which revised some of the material introduced in Stage I. Simple descriptive definitions of acids and alkalis were developed during Stage I, Topics A1.4 and B2.

Further references

for the teacher

Collected experiments, Chapter 8, provides further experiments on this Topic. Nuffield Advanced Chemistry, Topics 6 and 15 *(Teachers' guides I and II)*, and Physical Science, Sections 4 and 6 *(Teachers' guide I)*, provide advanced treatments of this Topic.

Supplementary materials

Film loops
1–7 'Gold mining'
1–8 'Iron extraction'

Overhead projection originals
30 Composition of some minerals *(Table B12.1)*
31 Inorganic nomenclature: acids and corresponding salts; bases and corresponding salts *(Table B12.2)*

Reading material

for the pupil

Study sheet:
Chemicals and rocks

B12.1
Investigating acids

In this section, the properties of acids are investigated and a simple operational definition of the term acid is formulated.

A suggested approach

Objectives for pupils
1. Knowledge of the properties of acids

An obvious place to start this investigation in a school laboratory is with those substances labelled 'acids'. Do acids possess the same properties as the solutions of non-metallic oxides (Topic B11)? What other properties do these compounds show? A fairly detailed comparative study of a number of acids can be carried out by dividing the tasks between pupils and pooling the results. Pupils should appreciate the importance of the presence or absence of water. Thus, the teacher can demonstrate the difference in properties

(Continued)

2. Awareness that certain substances not usually listed as acids can exhibit acidity
3. Ability to summarize an extensive range of experimental results
4. Knowledge of an operational definition of the term acid
5. Awareness that 'acids' require water in order to exhibit acidity

between *dry* glacial acetic acid and dilute (*wet*) acetic acid. However, no explanation of this finding is offered until later in the course (Topic B20).

Experiment B12.1a

Investigating the acidic properties of some compounds

Apparatus

The teacher may need:

12 test-tubes, 150 × 16 mm

Test-tube rack

Spatula

Teat pipette

2 beakers, 250 cm³

Bunsen burner and heat resistant mat

Splints

Distilled water

Indicator paper (very carefully dried and stored in a desiccator)

Dry glacial acetic acid (Make sure that the acid is really free from water by adding a few cm³ of acetic anhydride the day before the acid is required)

Boiled-out distilled water in a covered beaker

Procedure

This experiment is a preliminary or supplementary demonstration to Experiment B12.1b. (It is *not* intended that the teacher should demonstrate all the possible reactions but that he should complement those experiments attempted by the pupils.) For example, the reactions of glacial acetic acid, and benzoic acid in acetone solution, with dry indicator paper, magnesium ribbon, etc. can be used to show the importance of the absence of water.

Benzoic acid (benzene carboxylic acid) in acetone (propanone) solution	Sodium carbonate crystals
	Limewater
Tartaric acid crystals	Copper(II) oxide
Citric acid crystals	Bismuth nitrate
2M hydrochloric acid	Ammonium chloride
M sulphuric acid	Sodium hydrogen sulphate
2M acetic acid (ethanoic acid)	Sodium hydrogen carbonate
Magnesium ribbon	Sodium dihydrogen phosphate
Marble chips (calcium carbonate)	Dried Full-range indicator paper

Experiment B12.1b

Investigating the acidic properties of some compounds

Apparatus

Each pair of pupils will need:

Experiment sheet 78

5 test-tubes, 100 × 16 mm

Test-tube rack

Test-tube holder

Spatula

Teat pipette

Procedure

Divide the tests between pairs of pupils so that each pair can compare the properties of a mineral acid with those of a solution of an 'organic' acid. If time allows, each pair could investigate a further substance, such as a substance which is not usually labelled as an 'acid'. These investigations may be supplemented by demonstrations (see Experiment B12.1a). Stress the need for experimental studies of this kind and point to the value of a descriptive (operational) definition of the term acid. *Experiment sheet* 78 is reproduced below.

(*Continued*)

Topic B12 Acids, bases, and salts

Bunsen burner and heat resistant mat

Splints

Distilled water

A solution of a mineral acid in water, for example, 2M hydrochloric acid or M sulphuric acid, 7 cm³

Crystals of a carboxylic acid, for example, oxalic acid (ethanedioic acid), citric acid, or tartaric acid, 1 spatula measure

A sample of a salt which has acidic properties, for example, ammonium chloride, aluminium sulphate, or sodium hydrogen sulphate, 1 spatula measure

Magnesium ribbon, 3 pieces 2 cm long

Dried indicator paper, litmus and/or Full-range Indicator

Marble chips (calcium carbonate), 3 chips

Copper(II) oxide, 3 spatula measures

Sodium carbonate crystals, 3 crystals

Limewater, 10 cm³

Experiment sheet 78

You will be provided with a range of substances to investigate. Test each substance used with *dry* indicator paper. Then make a solution of each solid substance: use one measure of solid in half a test-tube (100×16 mm) of water.

Test separate small portions of each solution with:
a. Indicator paper (state which one was used and use the same type throughout).
b. Magnesium ribbon (about 2 cm); try to identify any gas evolved.
c. A crystal of sodium carbonate; try to identify any gas evolved.
d. A small piece of marble; try to identify any gas evolved.
e. A small amount of copper(II) oxide; warm the mixture and allow to stand.

Record observations and conclusions in the table below.

Test	Substance used		
Solid with dry indicator paper ( )			
Solution with indicator paper ( )			
Magnesium			
Sodium carbonate			
Marble			
Copper(II) oxide			

The points to be established in the discussion are listed below and can be used to formulate an operational definition of the term acid, that is, a definition based on the range of properties which enables us to distinguish acids from other classes of compound. Thus, pupils should appreciate that substances which are said to be acids possess the following properties:

1. They are soluble in water.
2. Their solutions change the colour of litmus or Full-range Indicator to red.
3. Their solutions react with metals, such as magnesium or zinc, and one of the products of reaction is hydrogen.
4. Their solutions react with sodium carbonate and one of the products of these reactions is carbon dioxide.
5. Their solutions react with metal oxides.

These properties taken together define a substance as an acid.

One of the demonstrations accompanying either the introduction or (preferably) the review of the lesson might well be the reaction of magnesium with dilute sulphuric acid. Magnesium is added to a small quantity of dilute sulphuric acid and hydrogen is given off and detected in the usual way. More magnesium is added to the residual acid until no further change takes place. After removing the excess magnesium, the residual solution is evaporated to leave a white powder. The discussion which follows should relate the terms acid and salt: a salt is formed when the displaceable hydrogen of an acid is replaced by a metal. The reaction of a metal oxide with a solution of an acid can be explained in terms of the formation of a salt and water.

Word equations should be used and the reactions are sufficiently elementary to be used as a means of introducing chemical equations.

Suggestion for homework

Write a summary of the main points raised in this section.

B12.2
Concerning bases and alkalis

The preparation of metal oxides is considered briefly and used to introduce the properties of bases and alkalis. The terms base and alkali are then defined.

A suggested approach

Objectives for pupils

1. Knowledge of the preparation and properties of bases
2. Knowledge of the operational definitions of the terms base and alkali

In Topic B11, pupils found that some metal oxides were soluble in water but others were insoluble. Metal oxides which were soluble in water were said to be alkaline and their solutions changed the colour of indicators. The formation of metal oxides by heating metals in oxygen was also demonstrated. Other methods of preparation include heating metal carbonates and metal hydroxides. (The effect of heat on copper carbonate (malachite) was studied in A9.2 and could be revised. The need to continue heating the carbonate until no further change takes place and also how to decide when to stop heating, could be discussed.) The identification of the products of such reactions should also be mentioned. The fact that calcium hydroxide and barium hydroxide are both soluble in water and that both can be used to prepare the corresponding metal oxides can be demonstrated. This information can then be used to point to general methods for making a metal oxide. Word equations (and chemical equations) can be used as reaction summaries, as in the previous section. Finally, descriptive (operational) definitions of the terms base and alkali can be established after a brief investigation of the properties of these two classes of compound.

Experiment B12.2

The preparation and properties of some bases

Procedure
The *Experiment sheet* is in two parts:

Apparatus

The teacher or each pair of pupils will need:

Experiment sheet 102

4 test-tubes, 100 × 16 mm

Test-tube rack

Test-tube holder

Bunsen burner and heat resistant mat

Teat pipette

2 beakers, 250 cm³

Access to:

2M hydrochloric acid

2M nitric acid

2M sodium hydroxide

Copper(II) oxide

Copper carbonate

Copper(II) hydroxide (prepared before the lesson)

Lead carbonate

Lead monoxide

Magnesium

Calcium oxide

Calcium hydroxide

Barium oxide

Barium hydroxide

Zinc

Litmus

Full-range Indicator

Limewater or barium hydroxide solution

Distilled water

Anhydrous copper(II) sulphate (freshly made)

Suggestion for homework

1. *Preparing metal oxides.* Discuss this with pupils before they begin practical work. They should heat a small quantity of the corresponding metal carbonate or hydroxide until no further change takes place. Remind pupils of tests to detect the production of carbon dioxide and water during these reactions. Details of such tests do not appear in the *Experiment sheet.*

2. *Properties of metal oxides.* Details are given in *Experiment sheet* 102 and need to be discussed before practical work is attempted. Pupils may be able to write word equations as reaction summaries but will probably require help in writing chemical equations.

Experiment sheet 102
1. In the first part of this experiment you will investigate the effect of heat on a metallic hydroxide or carbonate. Your teacher will discuss this before you begin. On a separate sheet of paper, write an account of what you do, stating your conclusions and writing equations for all the chemical changes that you observe.

2. For the second part of the experiment you will be provided with samples of metallic oxides and hydroxides. For each of these perform tests to find out whether:
a. it is soluble or insoluble in water;
b. its solution (if soluble) changes the colour of an indicator;
c. it reacts with dilute nitric acid (warming the test-tube if necessary);
d. it reacts with sodium hydroxide solution.
Enter your observations in the table.

Substance used	(a) Soluble /insoluble in water	(b) Colour with	(c) Reaction with dilute nitric acid	(d) Reaction with sodium hydroxide solution

Write equations for the reactions which take place.

After this experiment, pupils should be able to define a base as a substance which reacts with an acid to give a salt and water only, and should know that an alkali is the name given to a base which dissolves in water and, in consequence, changes the colours of indicators.

Write a summary of the main points raised in this section.

B12.3
Salts and their occurrence as minerals

In this section, salts are prepared in a number of ways, and their occurrence as minerals is mentioned briefly. In addition, an opportunity is taken to systematize the inorganic nomenclature which has been used so far and to review the work of the Topic.

Objectives for pupils

1. Understanding of the term salt and knowledge of the methods by which salts may be made in the laboratory
2. Ability to design experiments for the preparation of salts
3. Awareness of the composition of some minerals
4. Knowledge of simple inorganic nomenclature

The lesson may be opened with a simple discussion and, if necessary, a demonstration of the preparation of a salt by the reaction of a metal with an aqueous solution of an acid (for example, the reaction of magnesium or zinc with dilute sulphuric acid). The effect of using the corresponding metal oxide or some other compound in place of the metal may then follow (see section B12.1). Pupils are then asked to prepare one or more of the following salts: copper(II) sulphate; magnesium sulphate; lead chloride.

Experiment B12.3a

Apparatus

Each pair of pupils will need:

Experiment sheet 81

Beaker, 100 cm^3

Measuring cylinder, 25 cm^3

Tripod and gauze

Bunsen burner and heat resistant mat

Filter funnel

Filter paper

Filter stand

Crystallizing dish

Spatula

Cloth

2M sulphuric acid, 25 cm^3

Copper(II) oxide, powder, 4 g

Copper carbonate, powder, 6 g

Preparations of copper sulphate from (1) copper(II) oxide; (2) copper(II) carbonate

Procedure

Pupils can do one or both preparations. Full details are given in *Experiment sheet* 81 which is reproduced below.

Experiment sheet 81

1. Put 25 cm^3 2M sulphuric acid into a 100 cm^3 beaker and heat it, on a gauze over a tripod, until it is nearly boiling. Add copper(II) oxide, in small quantities, until no more will dissolve (about 4 g is required). Filter the solution while hot (hold the beaker with a cloth) into a clean crystallizing dish. Add 2 or 3 drops sulphuric acid to the filtrate and allow it to cool. Crystals of copper sulphate will form. Pour off the liquid and dry the crystals on blotting paper or filter paper.

2. The method using copper(II) carbonate is the same as in (1) above except that cold acid is used. With copper(II) carbonate there will be much frothing owing to the formation of carbon dioxide gas, so more care is needed when adding the solid to the acid. Use very small quantities each time and stir the mixture after each addition. A total of about 6 g copper(II) carbonate is required.

Write the equation for the reaction you use.

Make a drawing of a well-formed copper sulphate crystal. The formula for crystalline copper sulphate is $CuSO_45H_2O$.

Experiment B12.3b

Apparatus

Each pair of pupils will need:

Experiment sheet 82

Beaker, 100 cm^3

Glass rod

Measuring cylinder, 50 cm^3

Tripod and gauze

Bunsen burner and heat resistant mat

(*Continued*)

Preparation of magnesium sulphate from (1) magnesium; (2) magnesium carbonate

Procedure

Details are given in *Experiment sheet* 82 below.

Experiment sheet 82

1. Put 25 cm^3 2M sulphuric acid in a 100 cm^3 beaker and add magnesium, in small quantities, until no more will react (about 1.5 g is required). Stir well after each addition. Heat the liquid to boiling and filter while hot into a clean evaporating basin, add a few drops of sulphuric acid, and evaporate the filtrate to about half its volume. Allow to cool when crystals will form. Pour off the liquid and dry the crystals with blotting paper or filter paper.

2. The same procedure is used if you start with magnesium carbonate, but about 5 g of this will be needed. Care is needed when adding the solid to the

Filter funnel, paper, and stand

Evaporating basin

Spatula

2M sulphuric acid, 25 cm^3

Magnesium carbonate, powder, 5 g

Magnesium ribbon, 1.5 g

acid – use very small quantities each time and stir the mixture after each addition.

Write the equation for the reaction you use.

Make a drawing of a well-formed magnesium sulphate crystal. The formula for crystalline magnesium sulphate is $MgSO_47H_2O$.

Salts may also be prepared by other methods. A sparingly soluble salt, such as lead chloride, may be conveniently made from two other soluble salts as described in the next experiment.

Preparation of lead chloride from lead nitrate solution

Apparatus

Each pair of pupils will need:

Experiment sheet 83

2 test-tubes, 100×16 mm (to fit the centrifuge available)

Teat pipette

Glass rod

Bunsen burner and heat resistant mat

Beaker, 100 cm^3

Measuring cylinder, 25 cm^3

Access to centrifuge

M lead nitrate solution, 5 cm^3

2M sodium chloride solution, 5 cm^3

Distilled water

Procedure

The addition of lead nitrate solution to sodium chloride solution and the technique of using a centrifuge should be demonstrated beforehand.

Experiment sheet 83

Put 5 cm^3 M lead nitrate solution into a 100×16 mm test-tube and add 5 cm^3 2M sodium chloride solution. Stir the mixture well. What happens?

Write the equation for the reaction.

Centrifuge the mixture and pour off the clear liquid. Add a little distilled water (to a depth of about 2 cm), stir the mixture, and again centrifuge. Pour off the clear liquid, scrape the solid on to filter paper, and allow it to dry.

Why is a different method used in this preparation from those for copper sulphate and magnesium sulphate?

At the end of this section, spend a little time on the occurrence of salts as minerals and then sum up this Topic through a simple review of inorganic nomenclature.

Pupils will be aware that those parts of the earth accessible to man and from which chemicals can be extracted include:
1. The atmosphere (see Topic A3 or B4, when the industrial extraction of oxygen, nitrogen, and inert gases was mentioned).
2. The seas (see Topic A10) and other waters.
3. The Earth's crust to a depth of a few miles (see Topic A9 which mentions the use of iron ore, malachite, and limestone). Some pupils may be aware that the thickness of the Earth's crust (lithosphere) is about 20 miles. Almost all of its composition can be accounted for in terms of only a few elements (see B11.2).

The properties of the halogens were used as a basis for studying the composition of the salts from the sea in Topic A10. The waters of the oceans contain 34 parts per thousand of dissolved salts, of which about 55 per cent by weight is chlorine and 31 per cent by weight is sodium. Other important metallic elements present in these dissolved salts are magnesium, calcium, potassium, and strontium. In addition to salts of hydrochloric acid in sea water, salts of the following acids can be detected: sulphuric acid, carbonic acid, hydrobromic acid, and hydrofluoric acid. The salinity of sea water varies only slightly, except near land masses where it may be greatly reduced by the influx of river water. Point out here that the salts originating from river waters are quite different from the average salt content of the oceans and that much of the river-borne material is abstracted by plants and animals and involved in biological cycles.

The extraction of metals from their ores was dealt with briefly in Stage I (Topic A9). It may be convenient to group ores in order of economic importance, pointing out that several of them are classed as salts. At present, the chief ores from which metals are extracted are oxides, followed by sulphides, chlorides, and carbonates. Sulphides and carbonates are often converted to oxides first before extracting the metal. Table B12.1 gives the names of some minerals, their chemical names, and formulae. Pupils should be familiar with this information, rather than being asked to remember every detail.

Class of compound	Chemical name	Mineral	Composition of mineral
oxide	iron(III) oxide	haematite	Fe_2O_3
	tin(IV) oxide	cassiterite	SnO_2
	aluminium oxide	bauxite	$Al_2O_3 2H_2O$
	silicon dioxide	{ quartz	SiO_2
		silver sand	
sulphide	lead(II) sulphide	galena	PbS
	zinc sulphide	zinc blende	ZnS
chloride (salts of hydrochloric acid)	sodium chloride	rock salt	$NaCl$
	potassium magnesium chloride	carnallite	$KClMgCl_2 6H_2O$
carbonate (salts of carbonic acid)	calcium carbonate	{ limestone chalk marble	$CaCO_3$
	sodium sesquicarbonate	trona	$Na_2CO_3 NaHCO_3 2H_2O$
	magnesium carbonate	magnesite	$MgCO_3$
nitrate (salts of nitric acid)	sodium nitrate	Chile saltpetre	$NaNO_3$
sulphate (salts of sulphuric acid)	magnesium sulphate	kieserite	$MgSO_4 H_2O$
	calcium sulphate	anhydrite	$CaSO_4$
		gypsum	$CaSO_4 2H_2O$
	barium sulphate	barytes	$BaSO_4$
phosphate (salts of phosphoric acid)	calcium phosphate	phosphorite	$Ca_3(PO_4)_2$

Table B12.1 Composition of some minerals [OHP 30]

The teacher could show samples of some of the minerals to accompany this table.

Alternatively, the reactivity series discussed in Stage I (Topic A6 or B10) may be used to relate compositions of ores and the method used for the extraction of the metal. The film loops 1–7 'Gold mining' and 1–8 'Iron extraction' may also be used.

There are two ways of illustrating simple inorganic nomenclature and these are given in Table B12.2. The intention is to offer pupils a basis for classifying the names of chemicals which have been used throughout this Topic and earlier.

Clearly it is useful to construct such summaries of salts, acids, and bases by way of a simple question and answer session before using a completed table.

First it is suggested that salts be related directly to their parent acid:

Acid	Formula of acid	Examples of corresponding salts	Formula of salt
hydrochloric acid	HCl	sodium chloride	NaCl
		potassium chloride	KCl
		magnesium chloride	$MgCl_2$
		calcium chloride	$CaCl_2$
		barium chloride	$BaCl_2$
		aluminium chloride	$AlCl_3$
nitric acid	HNO_3	sodium nitrate	$NaNO_3$
		potassium nitrate	KNO_3
		magnesium nitrate	$Mg(NO_3)_2$
		calcium nitrate	$Ca(NO_3)_2$
		barium nitrate	$Ba(NO_3)_2$
sulphuric acid	H_2SO_4	sodium sulphate	Na_2SO_4
		potassium sulphate	K_2SO_4
		copper(II) sulphate	$CuSO_4$
		aluminium sulphate	$Al_2(SO_4)_3$
		sodium hydrogen sulphate	$NaHSO_4$
phosphoric acid	H_3PO_4	sodium phosphate	Na_3PO_4
		disodium hydrogen phosphate	Na_2HPO_4
		sodium dihydrogen phosphate	NaH_2PO_4

Table B12.2a Inorganic nomenclature: acids and corresponding salts [part of OHP 31]

Secondly, it is helpful to relate the composition of salts to their parent base:

Base	Formula of base	Examples of corresponding salts	Formula of salt
barium oxide	BaO	barium chloride	$BaCl_2$
		barium nitrate	$Ba(NO_3)_2$
		barium sulphate	$BaSO_4$
calcium oxide	CaO	calcium chloride	$CaCl_2$
		calcium nitrate	$Ca(NO_3)_2$
		calcium sulphate	$CaSO_4$

copper(II) oxide	CuO	copper(II) chloride	$CuCl_2$
		copper(II) nitrate	$Cu(NO_3)_2$
		copper(II) sulphate	$CuSO_4$
sodium hydroxide	NaOH	sodium chloride	NaCl
		sodium hydrogen sulphate	$NaHSO_4$
		sodium sulphate	Na_2SO_4
potassium hydroxide	KOH	potassium nitrate	KNO_3

Table B12.2b Inorganic nomenclature: bases and corresponding salts [part of OHP 31]

Suggestions for homework

1. Summarize the ways in which you can prepare salts in the laboratory.
2. How would you convert a sample of powdered malachite into a sample of pure copper(II) sulphate crystals? Explain each of the steps required in the process and indicate what additional steps are necessary to obtain a good yield of crystals.

Summary

By the end of this Topic pupils should be able to state simple, descriptive (operational) definitions of the terms acid, base, alkali, and salt. They should be able to use simple qualitative tests to identify acids, bases, and alkalis, and be capable of making and isolating simple salts by a variety of experimental procedures. In addition, they should be familiar with the use of word equations as reaction summaries and know something about simple inorganic nomenclature. The formulae of some simple chemicals and chemical equations have been introduced and used without formal proof. Pupils should also be aware of the occurrence of salts as minerals.

How fast? A study of reaction rates

<table>
<tr><td>

Purposes of the Topic

</td><td>

1. To introduce the idea that reactions do not occur instantaneously but at rates which can often be measured.
2. To show that the rate of a reaction can be affected by the concentration of the reactants, by the temperature of the system, and by catalysts.
3. To provide opportunities for some investigations to illustrate the ideas used in the Topic.

</td></tr>
<tr><td>

Contents

</td><td>

B13.1 Measuring the rate of a reaction: how does the concentration of a reactant affect the rate of a reaction?
B13.2 Investigating the influence of temperature on the rate of a reaction
B13.3 What is a catalyst?

</td></tr>
<tr><td>

Timing

</td><td>

About three weeks are needed for this Topic.

</td></tr>
<tr><td>

Introduction to the Topic

</td><td>

Much basic information about chemical reactions will have been revised in the previous Topics. Pupils should know that when a chemical reaction takes place, new substances are formed; and that these new substances possess different properties from those of the reactants. They will also know that heat is given out or taken in during the reaction. The range of examples used in earlier Topics demonstrates that reactions do not all proceed at the same rate. Chemists need to know how to control the rate of a reaction and in this Topic, pupils investigate a number of factors. The influences of temperature and of the concentration of reactants are studied and subsequently interpreted by a simple collision theory. The effect of catalysts on the rate of a reaction is also investigated.

</td></tr>
<tr><td>

Subsequent development

</td><td>

The theme is extended and related to other rate processes in Topic B22 (rather later than in Alternative IIA) by which time pupils have acquired experimental skills and a greater awareness of the ideas that chemists use.

</td></tr>
<tr><td>

Alternative approach

</td><td>

Topic A18 covers much the same material as Topic B13 and parts of Topic B22. Differences from the treatment in Alternative IIA are that a semi-formal study of the nature of the rate of reaction is deferred until Topic B22, and there is an introductory study of the reaction of magnesium with dilute hydrochloric acid.

</td></tr>
<tr><td>

Background knowledge

</td><td>

The reactions used as examples are, for the most part, familiar to pupils, with the possible exception of the precipitation of sulphur from sodium thiosulphate solution by dilute hydrochloric acid.

</td></tr>
</table>

Collected experiments offers additional experiments for this Topic in Chapter 14. *Handbook for teachers*, Chapter 21, provides background information on rates of reaction.
A detailed study of the magnesium–dilute hydrochloric acid system will be found in *Education in chemistry*, (1970), **7**, 20–24.

Film loop
2–13 'Catalysis in industry'

Overhead projection original
24 A chemical reaction considered as a collision process: two possible model processes *(figure A18.5)*

Film
'Catalysis' ICI Film Library

Handbook for pupils: Chapter 5 'Studying chemical reactions' and Chapter 10 'The world food problem'.

B13.1
Measuring the rate of a reaction: how does the concentration of a reactant affect the rate of a reaction?

The subject of reaction rates is introduced by reference to reactions which have been studied already and to everyday phenomena. The problem of how to measure the rate of reaction is discussed and the influence of the concentration of a reactant on the reaction rate for a specific chemical system is considered.

Objectives for pupils

1. Awareness that reactions proceed at differing rates
2. Recognition of the influence of the concentration of a reactant on the rate of a reaction
3. Ability to treat experimental results graphically

How do chemists establish whether a chemical change has taken place when substances are mixed together or are heated? From their earlier experiences, pupils should know that they must look for new compounds which have quite different properties from those of the reactants. They will also be aware that during a reaction reactants are used up and there is often an energy change. Make the point in discussion that reactions proceed at different rates. Many reactions, like the burning experiments, are rapid and take only a few seconds to complete; some are apparently instantaneous (as with the formation of a precipitate when two solutions react), and others (like the rusting of iron or the ripening of an apple) take several days. Temperature also effects the rates of many reactions. Indeed, some reactions don't seem to take place at all *until* the temperature is raised. Yet others require concentrated rather than dilute reagents. A number of these different effects can be demonstrated qualitatively and simply as occasion demands.

The effects of various factors (temperature, concentration of a reactant, and so on) are most easily studied by choosing situations in which just *one* factor is varied and all the rest are unchanged. In practice, however, this is not always easy to achieve since these factors interact. For example, if heat is liberated during the course of a reaction, the temperature of the reactants is raised and so the reaction may be speeded up. Unless some way is found of minimizing a temperature rise (such as the use of a large volume of dilute solution with a large heat capacity), it will be difficult to estimate the effect of temperature change on the rate. For our present purpose, most problems of this kind can be overcome by a careful choice of experiment.

Qualitatively, the effects studied in this section are very simple and easy to understand, and conditions can be adjusted so that a reaction goes faster or slower. Pupils could investigate whether a concentrated acid reacts more rapidly with certain metals than a dilute acid.

They could also find out whether a hot solution of sodium thiosulphate to which acid has been added decomposes more rapidly than a cold solution. For many pupils, it may be best to keep the initial discussion at this very simple level.

Before starting any investigations, some techniques for measuring the rate of reaction should be discussed. These should include the possibility of using changes of colour, changes in mass or in volume, and so on.

A reaction rate can be measured (1) by estimating the time it takes for a certain amount of reaction to happen, or (2) by finding the gradient of a line on a graph showing the change of some quantity (such as mass or volume) with respect to time. Both methods are useful, but it should be remembered that in method (1) the rate that is inferred is an average rate, whereas in (2) an instantaneous rate can be estimated, if the point at which this estimate is made is chosen with care. These matters are discussed in greater detail for individual experiments in this Topic. (Additional information is given in Chapter 21 of the *Handbook for teachers*.)

In the first experiment, pupils are asked to find out how long it takes for a known quantity of magnesium to react with an excess of dilute hydrochloric acid. By using different initial concentrations of acid for a series of experiments with a standard quantity of magnesium, some measure of the effect of acid concentration on the rate of this reaction can be obtained.

<table>
<tr><td>

Experiment B13.1a

Apparatus

Each pair of pupils will need:

Experiment sheet 103

Conical flask, 100 cm^3

Measuring cylinder, 50 cm^3

Stop-clock or access to a clock fitted with a seconds hand

(*Continued*)

</td><td>

What is the effect of the concentration of hydrochloric acid on the rate of reaction between magnesium ribbon and dilute hydrochloric acid?

Procedure
Details are given in *Experiment sheet* 103.

Experiment sheet 103
You will be provided with hydrochloric acid of six different concentrations obtained by diluting ordinary dilute hydrochloric acid (concentration represented by 1.0) to give concentrations which can be represented as 0.7, 0.5, 0.4, 0.3, and 0.25 (see table below).

</td></tr>
</table>

Supply of graph paper

6 lengths of clean magnesium ribbon (3 cm long in each case)

Hydrochloric acid of 6 different concentrations (see *Experiment sheet*), 200 cm³ of 2M acid

Put 50 cm³ of the most concentrated acid solution (1.0) into a 100 cm³ conical flask. Add a piece of clean magnesium ribbon of length exactly 3 cm and at the same time start a stop-clock (or note the time on a clock with a seconds hand). Find the time needed for the magnesium to dissolve. Repeat the procedure with the other five acid concentrations, using exactly 3 cm magnesium ribbon in each case. Record the results in the table.

Relative concentration of acid	Time taken for reaction/seconds
1.0 (50 cm³ dilute acid)	
0.7 (35 cm³ dilute acid + 15 cm³ water)	
0.5 (25 cm³ dilute acid + 25 cm³ water)	
0.4 (20 cm³ dilute acid + 30 cm³ water)	
0.3 (15 cm³ dilute acid + 35 cm³ water)	
0.25 (12.5 cm³ dilute acid + 37.5 cm³ water)	

Plot a graph of relative concentration of hydrochloric acid (vertical axis) against time (horizontal axis).

What do you learn from this graph?

Graphs of specimen results for this experiment are shown in figures B13.1 and B13.2. It is important that pupils appreciate that the shorter the time taken for the reaction to occur the greater its speed; that is, the greater its rate. It may help to compare this idea with the motion of a car. The less time the car takes to travel a certain distance the greater its speed must be. In the experiment, the 'distance travelled' is the 'amount of reaction' which has happened by the time the magnesium ribbon has all been 'used up'. Figure B13.1 shows how the time for complete reaction becomes shorter with greater concentration of the acid being used. Teachers will note that, for convenience, the concentration axis is labelled with two sets of units: the relative concentration term used for the experiment and the concentration expressed in moles per dm³. However, in Alternative IIB, this latter set of units should not be used since the pupils have yet to meet molarity or any use of the term mole.

Some teachers may also want their pupils to plot a graph of rate against concentration as shown in figure B13.2. In this case they will need to explain carefully why '$1/t$' is taken as a measure of the rate of reaction. It should be noted that it is the *average rate* of reaction which is estimated in this way (compare the situation with the car):

$$\text{average speed} = \frac{\text{total distance travelled}}{\text{total time taken}}$$

In this case, 3 cm length of magnesium corresponds to the distance travelled.

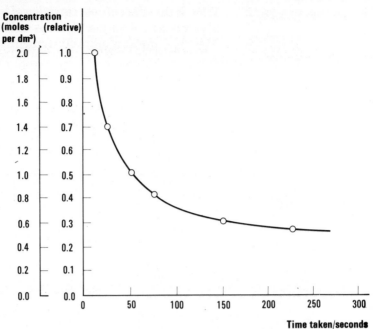

Figure B13.1
Sample results for
Experiment B13.1a.

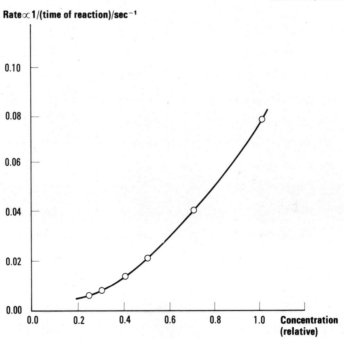

Figure B13.2
Sample results for
Experiment B13.1a.

Note. A detailed discussion of this reaction occurs in *Education in chemistry*, 1970, **7,** 20.

Next, the effect of the concentration of the acid on the reaction between calcium carbonate and hydrochloric acid is studied.

Note. This particular system has the advantage that the rate of reaction is almost unaffected by temperature and can be estimated by measuring the loss in mass of the reactants. From this study, a typical rate curve can be obtained and the effect of particle size and acid concentration can be determined.

What is the effect of concentration of hydrochloric acid on the rate of reaction between marble chips and hydrochloric acid?

This experiment is more conveniently carried out by the teacher.

Apparatus

The teacher will need:

3 conical flasks, 100 cm³

Stop-clock

Measuring cylinder, 100 cm³

Direct reading balance

Supplies of graph paper for the use of pupils

Cotton wool

Marble chips, medium size and free from dust*

2M hydrochloric acid

Procedure

The course of the reaction is followed by observing the change in mass which occurs when carbon dioxide is given off. The use of a direct reading balance is to be preferred and accordingly, the experiment will probably have to be demonstrated by the teacher. Pupils may help with weighing, timing, and in the recording of results.

Put 40 cm³ of 2.0M hydrochloric acid in a 100 cm³ conical flask and plug the end of the flask with cotton wool. Weigh 20 g of marble chips onto a watch glass, and place it and the flask on the balance pan. Note the total mass. Add the marble chips to the acid, replace the cotton wool, and start the stop-clock. Record the time taken for the system to lose successive masses of 0.10 g. Continue the experiment for at least 5 minutes.

The pupils should tabulate the results as follows:

Time in minutes	Total mass of apparatus and contents in grams	Loss in mass (grams) = mass of carbon dioxide given off
0.0		0.00
		0.10
		0.20
		0.30

Ask the pupils to plot a graph of loss in mass against time of reaction. (*Note.* The experiment can be extended and suggestions are provided in Alternative IIA, Experiments A18.1a and A18.3).

The effect of concentration of acid on the rate of reaction follows from the graph. The question arises as to how one can interpret the results of the experiment. It may be enough to say 'It is clear that the stronger acid has reacted more rapidly than the weaker acid' and then to infer that the greater the concentration, the more rapid the reaction.

Alternatively one might argue from the presentation of results used in figure B13.2. The results show that the rate of reaction is more rapid when the concentration of the acid is higher. Ask the class if they can explain the 'tailing off' of the reaction rate towards the end of the reaction. It looks as if the reaction rate decreases as the hydrochloric acid is used up. This is a further indication that the concentration of a reactant affects the rate of a reaction.

Note. Marble chips need to be washed with warm dilute hydrochloric acid, then with water, and then allowed to drain.

A cube of marble (calcium carbonate) of side 1 cm dissolves in dilute hydrochloric acid at an initial rate of 0.1 g per minute. At what initial rate would you expect a cube of marble of side 0.5 cm to dissolve in the same acid? Explain how you would obtain your answer.

B13.2
Investigating the influence of temperature on the rate of a reaction

After an introductory qualitative discussion about the influence of temperature on the rate of various reactions, the reaction between sodium thiosulphate and dilute hydrochloric acid may be introduced and studied in some detail.

A suggested approach

Objectives for pupils

1. Measurement of the effect of change in temperature on the rate of a reaction
2. Simple interpretations of the effect of temperature on reaction rate in terms of a simple collision model
3. The application of the simple collision model (designed to interpret one result) to explain the effect of the concentrations of reactants on reaction rates
4. Ability to treat results graphically.

Ask pupils what happens to the rate of a chemical change when we alter the temperature of the system. The effect of changing temperature on chemical reactions is a fact of everyday experience. We put a cake in the oven in order to speed up *desirable* chemical changes and we put milk in the refrigerator in order to slow down *undesirable* chemical changes. Simple comparative demonstrations of the effect of adding either a cold or a hot reactant to other reactants can be instructive. For example, the addition of either cold or warm 4M nitric acid to copper; the addition of either cold or hot oxalic acid solution to a solution of potassium permanganate (manganate(VII)) acidified with dilute sulphuric acid; the addition of either cold or hot hydrochloric acid to sodium thiosulphate solution; and so on. *No detailed chemistry is intended by the use of these systems. The effect of raising the temperature of a system on the rate of reaction is all that is required.*

Repeat the demonstration of the reaction of dilute hydrochloric acid and sodium thiosulphate to make the point that the appearance of sulphur occurs sharply. The time required for the appearance of sulphur may then be used as a measure of the rate of reaction, especially if some standard way of observing this change is adopted (see Experiment B13.2).

Experiment B13.2

Apparatus

Each pair of pupils will need:

Experiment sheet 74

Conical flask, 100 cm^3

Measuring cylinder, 10 cm^3

2 beakers, 100 cm^3

What is the effect of temperature on the rate of reaction?

Procedure
A simple method of estimating when a fixed amount of sulphur has been formed is to stand the reaction vessel on a piece of white paper on which a cross has been marked. On looking down onto the mixture, the cross will gradually become fainter as the precipitate forms and disappear when a certain amount of sulphur is present.

By using the same depth of liquid in each case, the same vessel, and the same piece of paper, the effect of temperature on the rate of this reaction can be determined. Details appear in *Experiment sheet* 74.

(Continued)

Stop-clock or sight of a
laboratory clock with a seconds
hand

Thermometer, -10 to
$+110 \times 1 \,^{\circ}C$

Bunsen burner and heat
resistant mat.

Tripod and gauze

Sheet of white paper

2M hydrochloric acid, about
$40 \, cm^3$

Sodium thiosulphate solution
containing $37 \, g/dm^3$
(approximately 0.15M), about
$60 \, cm^3$

Experiment sheet 74
The method for this experiment is the same as that for Experiment 73,*
except that in this case the concentrations of the reacting solutions are kept
constant and the temperature of the mixture is varied.

Mix $10 \, cm^3$ sodium thiosulphate solution with $40 \, cm^3$ water in a $100 \, cm^3$
beaker, and stand the beaker over the cross on the sheet of white paper. Add
$5 \, cm^3$ hydrochloric acid, with stirring, start the stop-clock, and read the
temperature of the mixture. Note the time for the cross to disappear.

Repeat the above procedure but this time heat the diluted thiosulphate
solution to just above 30°C before adding hydrochloric acid. Note
temperature and time as before. Then find the times required for
temperatures of about 40°C, 50°C, and 60°C.

Results
In each case, use a reaction mixture of $10 \, cm^3$ sodium thiosulphate solution,
$40 \, cm^3$ water, and $5 \, cm^3$ hydrochloric acid.

Temperature of mixture /° C	Time (t) /sec	1/t /sec^{-1}

Plot graphs of:
1. Temperature (vertical axis) against time (horizontal axis).
2. Temperature (vertical axis) against $1/t$ (horizontal axis).

What do you learn from these graphs?

Figure B13.3
See figure A18.5 or·OHP 24
which shows a chemical
reaction considered as a
collision process: two possible
model processes.

Discuss the results obtained and the meanings of the two graphs.
Then ask the pupils why a reaction might be expected to go faster at
a higher temperature than at a lower one. A simple interpretation
can be given in terms of the increase in the number of particle
collisions per second. Figure B13.3 provides two simple models
of chemical change and these can form convenient aids for a discussion. Further consideration of the collision theory model can be
raised in Topic B17.

This collision model can also be used to interpret the effect of the
concentration of a reactant on a reaction rate.

Suggestions for homework

1. Explain the use of the oven and the refrigerator in the kitchen to
control the rates of the changes involved in the cooking, storing,
and preparing of foodstuffs.
2. Potassium permanganate, which is purple, reacts with glycerol
(propan-1,2,3-triol), which is colourless, to give a mixture of colourless products. The reaction may be carried out in solution in water
at room temperature. The results in the table were obtained by
mixing equal volumes of glycerol and potassium permanganate
solution. The glycerol concentration was varied whereas the potassium permanganate concentration was kept constant. In each
experiment there was a considerable excess of glycerol. For each
mixture the time for the reaction to be completed was recorded.

*Experiment 73 is reproduced on page 229.

Concentration of glycerol % by volume	Time for reaction to be completed in minutes
5	5.0
10	2.5
15	1.7
20	1.2
25	1.0

a. Describe a practical procedure for obtaining these results. Explain how you would be able to tell when the reaction was complete.
b. Plot a graph of glycerol concentration against time for the reaction to be completed. What connection does this graph show between the rate of reaction and the concentration of glycerol?
c. Plot a graph of l/time against concentration of glycerol. What is the quantity l/time a measure of? Explain as far as you can the shape of the graph.
d. Describe the procedures you would adopt to find the effect of temperature on this system. What general result would you expect to obtain from such an investigation?

B13.3
What is a catalyst?

In this section, a series of qualitative experiments is used to introduce the phenomenon of catalysis. A pupil-directed investigation is then used to identify the characteristic properties of catalysts.

A suggested approach

Objectives for pupils

Knowledge of the phenomenon of catalysis and of the general properties of catalysts

The idea that a substance can affect the rate of a chemical reaction without itself being changed permanently may be new to most pupils. Catalysis is approached through a series of experiments which lead to a simple descriptive (operational) definition of this phenomenon. These demonstrations are followed by a detailed study of the effect of metal oxides on the decomposition of hydrogen peroxide.

The reaction of metals with dilute acids provides a simple introduction to catalysis. Thus, the reaction of zinc with dilute sulphuric acid can be shown to be relatively slow, but to be speeded up by the addition of a small quantity of copper(II) sulphate solution. (Compare with the experimental procedure used in Topic A7.4.) A separate study shows that when zinc reacts with copper sulphate solution, copper is formed. The addition of powdered copper to zinc and dilute sulphuric acid has some effect but less than when particulate copper is formed by the addition of copper sulphate solution to zinc and dilute sulphuric acid. The reaction of zinc with an excess of dilute sulphuric acid in the presence of copper from copper sulphate reveals that the copper remains 'unused' at the end of the reaction.

Experiment B13.3a

The reaction of zinc with dilute sulphuric acid
This experiment should be performed by the teacher.

The teacher will need:

6 test-tubes, 150×16 mm

Test-tube rack

Test-tube holder

3 beakers, 100 cm^3; or 3 conical flasks, 100 cm^3, fitted with bungs and delivery tubes plus trough filled with water

Bunsen burner and heat resistant mat

Tripod and gauze

Teat pipette

Filter funnel

Filter paper

Filter stand

Spatula

Splints

Battery, connecting wire, lamp, and holder

Zinc foil

2M sulphuric acid

0.5M copper(II) sulphate

Copper (powder)

Procedure

This demonstration falls into three parts:

1. The comparison between the reaction of zinc with dilute sulphuric acid alone and in the presence of some copper(II) sulphate solution, as used in Topic A7.4. This comparison needs to be carried out in such a way that pupils can 'see' differences in the reaction rates. Comparatively little copper sulphate solution will be needed.

If necessary, the time required by both systems to generate a given volume of hydrogen can be noted. Hence, the two forms of apparatus listed in the apparatus requirements: beakers may be used for studying the reaction initially; the flasks and associated gas collection apparatus can be used for more qualitative comparisons.

2. Investigation of the reaction of zinc with copper(II) sulphate solution. Warm some copper sulphate solution in a beaker and add several pieces of zinc foil. The production of a precipitate and the loss of colour in the copper(II) sulphate solution suggest that the precipitate is metallic copper. The separation of the precipitate by filtration enables the colour of the precipitate to be observed. Its ability to conduct electricity can also be tested, and possibly other properties can be demonstrated. The effect of adding powdered copper to a sample of zinc and dilute sulphuric acid can then be tried and the difference in its effectiveness when compared with copper(II) sulphate solution can be discussed. What is attempted in this part of the study must clearly depend on the pattern of discussion and demonstration.

3. Ascertaining whether copper remains unchanged at the end of the reaction of zinc with excess dilute sulphuric acid and copper(II) sulphate solution. Within the time available, only limited testing of this suggestion can take place. However, some residual copper can be collected and identified, as above.

Another striking demonstration is the effect of a platinum catalyst on the combination of hydrogen with oxygen.

Experiment B13.3b

Apparatus

The teacher will need:

Plastic safety screen

2 gas jars and cover slips

Tongs

Bunsen burner and heat resistant mat

Cylinder of hydrogen, connecting tubing, and a delivery tube

Platinized asbestos

Asbestos wool

The reaction between hydrogen and oxygen in the presence of some platinum
This experiment MUST be done by the teacher.

Procedure

Before the lesson, heat a tuft of platinized asbestos in a Bunsen flame for a few seconds to ensure that it is dry and then return it to the bottle.

Fill a gas jar with hydrogen and close the jar with a cover. Hold the edge of a tuft of platinized asbestos with some tongs; remove the gas jar cover, and hold the tuft over the mouth of the gas jar. As the hydrogen rises out of the jar, it mixes with the air and the tuft of platinized asbestos glows red. This ignites the residual hydrogen and produces the usual gentle explosion. Repeat the experiment without the tuft of platinized asbestos and in the presence of asbestos alone.

Discuss the meaning of the demonstration with the pupils and identify the general characteristics of catalysts – (a) their general effect on the rate of a reaction, and (b) the fact that they remain unchanged chemically at the end of a reaction although they *may* be altered physically.

Note. It is not intended that theories of the action of catalysts should be discussed.

Continue the discussion by adding a further example, the decomposition of hydrogen peroxide in the presence of a metal oxide. How could we find out if metal oxides used for this reaction are catalysts? What tests would we need to apply? Discuss the possibilities with the pupils and warn them of the hazards of using hydrogen peroxide (bleaching action) before allowing the pupils to try Experiment B13.3c.

Experiment B13.3c

Investigating the effect of metal oxides on the rate of decomposition of hydrogen peroxide

Apparatus

Each pair of pupils will need:

2 conical flasks, 100 cm³ – wide neck and with one such flask fitted with a bung and delivery tube

Measuring cylinder, 25 cm³

Stop-clock

Scrap paper

Spatula

Bunsen burner and heat resistant mat

Tripod and gauze

Teat pipette, with 1 cm³ graduation mark

Either syringe, 100 cm³ (glass) or 50 cm³ (plastic), syringe holder and stand; *or* burette, 50 cm³, stand, boss, and clamp, trough, delivery tube

Access to an oven, heated to 100 °C

Access to a top pan balance

Access to filtration apparatus

20 vol hydrogen peroxide, 10 cm³

Various oxides (for example, manganese(IV) oxide, copper(II) oxide, zinc oxide, sand

Acetone (propanone)

Procedure

The investigation is in three parts. Each will need to be discussed with the pupils.

1. Pupils collect oxygen formed by the decomposition of hydrogen peroxide in the presence of a metal oxide. The time taken to collect a known volume of gas is recorded. (The experiment is then carried out in the absence of a metal oxide, care being taken to use a clean reaction vessel.)

2. The reaction mixture is filtered and the residue washed with a little acetone (propanone) before being dried in an oven. The metal oxide which was used to prepare oxygen from hydrogen peroxide is compared with a fresh sample of that oxide. For some pupils, this study can be made into a quantitative experiment and it will be found that very little oxide is lost.

3. By repeating the investigation using another metal oxide (or by pooling the results obtained throughout the class), it will be found that not all metal oxides are equally effective in bringing about the decomposition of hydrogen peroxide.

After the pupils have completed their investigations and discussed their findings, the various terms and ideas used in the Topic can be reviewed. The film loop 2–13 'Catalysis in industry' can form an important contribution to this part of the lesson.

Suggestion for homework

Make a summary of the results of your experiments in this section. Describe clearly the effect of temperature, concentration of reactants, and catalysts on the rates of the reactions you have studied.

Summary

By the end of this Topic, pupils will have found that, in general, the rate of the reaction may be increased by increasing the concentration of one or more of the reactants and by increasing the temperature of the system. They will also have studied the effect of a catalyst on the rate of a reaction. If all the suggestions made in this Topic have been followed the case of a reaction involving a solid reactant will also have been studied, and it will be found that an increased surface area results in an increased reaction rate.

Topic B14

Chemistry and the world food problem

Timing

About five weeks should be enough to cover all aspects of this Topic. Some schools may find that more time is needed to discuss the social implications of a chemist's work.

Introduction to the Topic

The principal difference between the presentation of this Topic and that of Topic A22 lies in the emphasis given to the social implications of the widespread use of chemicals in food production and storage.

When discussing the world food problem, it is as well to know whether it has been discussed in general studies, geography, biology, or English lessons. Clearly it is useful to coordinate such studies in the school and make use of the different approaches to this theme. Constant repetition can lead to boredom and a lack of concern on the part of the pupil.

An outline of various issues and some relevant data appear in the *Further information* section of this introduction. Additional information appears in the *Handbook for pupils*, Chapter 11, 'The world food problem' and in *Chemists in the world*, Chapter 7, 'Fertilizers'.

Chemists are asked 'What can be added to the soil to make plants grow "better and bigger"?' Plants obtain nourishment from their environment and we know that plants grow better under certain conditions. To determine the constituents of soil which must be transferred to the plants, pupils need to be asked the question – 'What are the constituents of plants?' The thermal decomposition of grass, flour, and other similar materials are investigated. Pupils who have studied biology to this level will be aware that plant constituents include proteins. Hence, after thermal decomposition, one might expect to detect an amine, an animo acid, or even ammonia. Subsequent experiments enable pupils to identify the ultimate alkaline product of decomposition as ammonia. The analysis and synthesis of ammonia leads to the manufacture of ammonia and ammonium salts.

The development of insecticides and pesticides is another area in which the chemist can serve the community. The point should be made that education plays an important role in the prevention of the indiscriminate use of chemicals by society.

Alternative approach

There are two suggested routes through Topic B14: route 1 extends the general theme of acidity and basicity as well as showing the relevance of chemistry to society, whereas route 2 places Topic B14 as a sequel to a regrouped Topic B15 and Topic B17 (that is, B17, then B15.5–B15.8, B15.1–B15.4, B14, B16), see Part 1 Chapter 3.

Topic A22 offers a different emphasis, as indicated by the title 'Ammonia, fertilizers, and food production'.

Background knowledge

Pupils need to have studied the descriptive chemistry of acids, bases, and salts (Topics B11, B12, and B13) before attempting this Topic.

Teachers will need to know whether the 'world food problem' has been discussed in other sections of their pupils' curriculum, and, if so, what has been attempted.

Further information
for the teacher

'Who is undernourished?' Put simply, one-third of the world's population has enough to eat, one-third has enough food to meet energy requirements but only through a use of an unbalanced diet, and one-third has insufficient food for both work and growth. Thus, two-thirds of the world's population is undernourished and half of this number is at or near starvation level.

The following themes seem most likely to be raised in any discussion of the basic problem and the points made complement the material presented in Chapter 11 of the *Handbook for pupils*, 'The world food problem' and Chapter 7 of *Chemists in the world*, 'Fertilizers'.

1. *World population: current trends*
Table B14.1 and figure B14.1 illustrate two methods of presenting data on population trends. Another way of putting the data is to say that when compared to the population in 1967, world population is expected to double by AD 2000 and to double again by AD 2033. If this forecast is correct and if our present technical skills can solve the food problem by AD 2000, a similar but even greater problem requiring even greater achievements remains to be solved by AD 2033!

Table B14.1 World population 1 [OHP 32] (*Note.* Estimates before 1900 are tentative and based on a variety of historical evidence. Twentieth-century estimates and projected population figures are from Department of Economic and Social Affairs Statistical Office (1975) *Demographic Year Book 1974.* U.N.

World population in millions	Population increase over previous figure	Year	Time taken to build up to this figure/years
	? Half a million years BC – Man evolved		
10	—	7000 BC	?½ million years
200	190	Birth of Christ	7000
400	200	1600 AD	1600
700	300	1750	150
1000	300	1830	80
1550	550	1900	70
1907	357	1925	25
2501	594	1950	25
3890	1389	1974	24
6267	2377	2000	26

Figure B14.1
World population: forecasts for 1970 made in 1958 were lower than more recent estimates. Present world population growth rate is about 2.1 per cent per year, corresponding to a doubling time of 33 years.
(*After Bogue, D. J. (1969).* Principles of demography. *Wiley.*(OHP 33).)

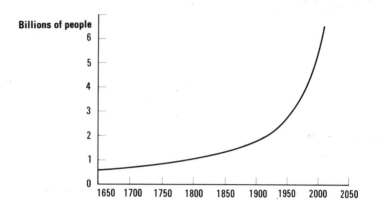

Any discussion of this point must necessarily raise questions about population control and a need for people to accept a responsible attitude towards the size of the average family. Mention should also be made of the increase in life span for people in the western world, largely through a balanced diet and advances in medical and chemical sciences.

The issues in this section are not limited to science but can appear as religious issues, political issues, or even as educational issues.

Economic considerations also influence courses of action, whether they relate to birth control or to medical aid in the fight against disease. The examples used in this section should indicate the key role of the chemist either in a research role or in monitoring products.

2. *Food production: the cultivation of more land*
Figure B14.2 illustrates a simple way of showing that only 10% of the land area is at present cultivated. If deserts were watered, poor soil fertilized, and all other possible areas used, the total area under cultivation could be increased to 50%. However, each one of these changes requires the development of knowledge and the use of resources, such as the desalination of sea water (a costly procedure); the use of plastic soil to retain moisture in sand – plastiponics – (a process dependent on good oil supplies); or the use of a petroleum

Figure B14.2
Utilization of the Earth's land
areas. (OHP 34A)

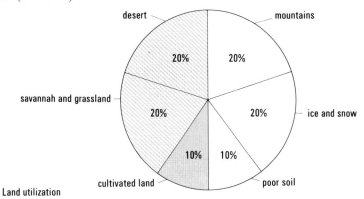

Land utilization

mulch to retain moisture and to prevent the sand shifting in desert places (again dependent on oil supplies).

3. *Food production: better cultivation of the land*
The better use of land requires not only a careful use of fertilizers but also the use of high yielding varieties of crops. Examples are given in the *Handbook for pupils*.

4. *Food production: the development of plants and animals that contain a greater proportion of edible material*
It is feasible for scientists to develop improved varieties of plants which can survive under adverse weather conditions – such as high winds and low rainfall – and still produce food. Also, hormone treatment may be used to regulate the growth and size of plants or even modify the growing period of a specific crop. In the case of animals, hormone treatment can be used to modify the proportion of fat, lean meat, bone, and hair produced.

5. *Food production: the manufacture of food from materials that could not otherwise be eaten*
Protein production is important to both people and animals. It can be made by bacteria grown on hydrocarbon residues, using nitrogen from the air. The problem of persuading people to eat such protein remains. However, it can be used without real difficulty to supplement existing animal foodstuffs.

Another example is the manufacture of margarine from a selection of otherwise inedible fats and oils. This process provides no problems of palatability.

6. *Food production: farming the sea*
Approximately two-thirds of the earth's surface is covered by water. As the pressures on land use grow, some investigators are examining the possibility of using the sea to provide more food for people. It is estimated that the amount of fish taken from the sea could be increased two and a half times without endangering future supplies. Experiments in farming fish have been undertaken, using fenced areas and the warm water from power stations. Algae farms to supply protein to fish or animals have also been suggested. It has

been shown that a given area of pond water can produce as much food as the same area of land.

Underwater experiments have been carried out in which artificial reefs (made from car bodies and the like) are constructed so that sea creatures can take refuge to eat and to multiply. Indeed current research is in the direction of breeding, cultivation, and cropping, and away from 'hunting' specific fish.

Once more there is a problem of education as many people are not accustomed to eating fish. Plankton also forms an excellent food, but it is tasteless and unattractive to humans.

7. *Food storage: removing pests and fungi*
It is estimated that some twenty per cent of all the food grown in the world is lost because of pests, diseases, and competition from weeds. Every year insects alone destroy enough food to feed one hundred and fifty million people. Chemicals have been developed which will destroy pests and weeds, and eliminate diseases. However, their use needs to be carefully controlled if we are to avoid polluting our environment. We need to mention the existence of a variety of food chains in nature and the dangers of interfering with such chains.

8. *Food transportation*
In North America, people suffer and some die from overeating. In Southern Asia, people starve. The teacher could present the transfer of food from one country, such as America, to another, such as India, as an 'obvious' solution to the food problem. It is not a practicable solution, since the cost of transporting food usually makes it uneconomic. Also, if the food is donated, it tends to undermine the self-reliance of the receiving country, and to depress selling prices of locally grown food so that local farmers are deprived of a living. A much better solution would seem to be to provide the means to raise local production. Thus, one could suggest the provision of expertise and capital to build fertilizer factories, or to pass on knowledge of modern farming methods.

Each of these aspects of the world food problem needs to be discussed at some time during this Topic. The teacher will need to ask the question 'How can the chemist help?' on more than one occasion.

Subsequent development

One development of this Topic is the chemistry of large molecules, such as the natural products sugar and starch. This forms part of Topic B15. Other large molecules considered in B15 are synthetic materials, such as plastics and fibres. A more detailed consideration of acid–base systems, including an elementary consideration of the Brønsted-Lowry theory of acids and bases, forms part of Topic B20.

Further references

for the teacher

Nuffield Advanced Physical Science (1974) *Sourcebook*, Part 11: Pollution.
Bogue, D. J. (1969) *Principals of demography*. Wiley.
Lowry, J. H. (1970) *World population and food supply* (especially Chapter 15). Edward Arnold.
Park, C. W. (1965) *Population explosion*. Heinemann Educational.
Nuffield Biology (1966) *Text III*, 59 *et seq*.
Revised Nuffield Biology (1975) *Text 2*, page 115.

Film loops or film strips
2–11 'Ammonia manufacture'
2–12 'Ammonia – uses'
2–13 'Catalysis in industry'
3–4 'Giant molecules – proteins'

Films
'Licence to grow' ICI Film Library
'Ammonia' ICI Film Library
'Cereals in the 70s' ICI Film Library
'Weeding by spraying' ICI Film Library
'Nothing to eat but food' Unilever Film Library

Overhead projection originals
32 World population 1 *(Table B14.1)*
33 World population 2 *(figure B14.1)*
34 Earth's land areas *(figure B14.2)* & Production and uses of ammonia
    (figure B14.9)
35 World fertilizer consumption *(figure B14.10)*
36 The nitrogen cycle *(figure B14.11)*

Reading material

for the pupil

B14.1
Identifying the problem and the ways in which the chemist can help

In this section, the world food problem is introduced and the ways in which chemists can help are discussed. As a first step towards solving the problem, pupils find out about the substances from which plant materials and foods are made.

A suggested approach

Objectives for pupils

1. Awareness of the world food problem and possible ways of solving it
2. Knowledge of the decomposition of foodstuffs by heat
3. Knowledge of the use of activated charcoal for absorbing gases

The lesson can be opened with a brief statement by the teacher about the world food problem and its possible solution along the lines suggested in the 'Further information' section of this Topic. The overall approach should not be forced in any one direction and care should be taken not to duplicate material which may feature in other courses. It may be appropriate to delay a detailed discussion until later and to merely direct attention to the use of fertilizers as *one* technical solution for obtaining better and bigger plants. The decomposition of foodstuffs and plants may offer a way to identify those factors which are needed by plants and which are added to the soil to promote their growth.

Through questioning, pupils can be encouraged to suggest the action of heat as a general method for breaking down complex chemicals into simpler ones. Such a series of tests can prove a useful starting point for further investigation.

It is suggested that a variety of materials be tested and that these should include grains of wheat, flour, meat, dried milk, and so on. Grass clippings can be used since grass is eaten by cows to produce milk and also meat. Some protein-containing foods should be included. The school kitchen is a good source of items to be tested and pupils may react readily to the suggestion that they should 'analyse a school dinner'!

Experiment B14.1a

Apparatus

Each pair of pupils will need:

2 hard-glass test-tubes, 100 × 16 mm

Test-tube holder

Test-tube rack

Bunsen burner and heat resistant mat

Spatula

Indicator papers (litmus, and Full-range Indicator)

Cobalt chloride (stored in desiccator)

Anhydrous copper(II) sulphate – freshly prepared and stored in a stoppered test-tube

Splints

Teat pipette

Soda lime

Limewater

Plants and foodstuffs suggested by pupils – for example: grass, dried egg, milk powder, flour, wheat, meat

Investigating foodstuffs (1)

Procedure

A plant material or a foodstuff is selected and heated in a hard-glass test-tube. Gases are evolved and tested with a lighted splint, litmus (or Full-range Indicator), limewater, and by smelling them cautiously. Carbon dioxide, water vapour, and an alkaline gas may be detected. Usually, the gases have an unpleasant smell, but in some cases the pungent odour of ammonia can be detected. Charring is also likely to occur.

As preparation for this, it will be necessary to discuss practical problems such as quantities of reagents to use, tests to be tried, and so on, and to indicate a way of presenting the results of an investigation.

A second series of tests may follow using a small quantity of test material with *at least* twice the bulk of soda lime.

Note. If the quantities of materials used are limited to one spatula measure, the odours evolved during the experiments are not too offensive!

When the pupils have tried a range of materials and tabulated their findings, call them together for a discussion of results. In those cases where an alkaline gas or gases are formed (in addition to water, carbon dioxide, and other products) ask whether the foodstuff contains something unusual. If need be, tell them it contains proteins, and proteins contain nitrogen. As suggested above, pupils might extend this experiment by heating a sample of the original material with some soda lime and comparing the result with a material known to contain nitrogen (from the bottle label), such as an ammonium salt (*or* a known protein food). Then ask whether the soda lime affected their results. Has the soda lime had the effect of intensifying the characteristic smell of the alkaline gas? Alternatively, it may prove more convenient to carry out this part of the investigation as a demonstration.

Variations in the odour of the alkaline gas can occur, and so we will need a filtering device to 'sort out' the gases formed during this thermal decomposition. Tell pupils about gas masks produced for use during the 1939–45 war. The masks contained a filter box filled with activated carbon (charcoal). This material is used in industrial gas masks today and has the property of absorbing dense poisonous fumes and of allowing light gases to filter through it. *Experiment sheet* 91 provides experimental details. Pupils might be asked to suggest ways whereby the experimental procedure could be modified to allow activated charcoal to separate the gases formed when foodstuffs are heated.

Investigating foodstuffs (2)

Apparatus

Each pair of pupils will need:

Experiment sheet 91

Hard-glass test-tube,
100×16 mm

Test-tube holder

'Calcium chloride' drying tube,
filled with activated carbon
(charcoal) and plugged with
cotton wool

Cork or bung to fit the test-tube
and the drying tube

Bunsen burner and heat
resistant mat

Indicator paper

Glass rod

Spatula

Soda lime granules, 6 spatula
measures

Gelatine, 3 spatula measures

Milk powder, 2 spatula
measures

Concentrated hydrochloric
acid

Procedure
Details are given in *Experiment sheet* 91.

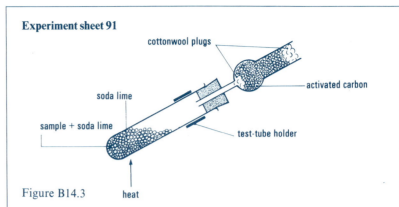

Experiment sheet 91

Figure B14.3   heat

In the test-tube put 2 measures of the sample mixed with 2 measures of soda
lime and cover this with a further 2 measures of soda lime. Heat the mixture
until a gas comes out of the activated carbon tube (smell cautiously). Test
the gas with (a) wet indicator paper; (b) hydrogen chloride gas, from a glass
rod dipped in concentrated hydrochloric acid.

Record your results and conclusions below.

After the pupils have tried this experiment and have detected at
least a trace of alkaline gas in the gases given off, tell them that this
gas is called ammonia. An experiment may then be demonstrated
in which the ammonia produced is unmistakable. The decision
whether or not to do this experiment *must* depend on the time available and on the success of the last two experiments. It need not
necessarily be part of the course.

Apparatus

The teacher will need:

Kjeldahl flask

'Calcium chloride' drying tube,
containing activated charcoal
and plug of cotton wool, fitted
with a cork (or bung) to fit flask

Beaker

Bunsen burner and heat
resistant mat

Tripod and gauze

Stand, boss, and clamp

Measuring cylinder, 25 cm³

Anti-bumping pellets

(Continued)

The detection of ammonia as a product of the decomposition of a protein
This experiment *must* be done by the teacher.

Procedure
The teacher should do this experiment in a fume cupboard.

Place 5 grams of egg albumen in the Kjeldahl flask and cautiously
add 15 cm³ of a solution containing a spatula measure of potassium
hydrogen sulphate in concentrated sulphuric acid. Boil the contents
of the flask in a fume cupboard for about ten minutes. Allow to cool
and then add the residue very carefully to a beaker containing 10 cm³
of water. Return the solution to the flask and add sufficient 8M
(or solid pellets) of sodium hydroxide to make the solution strongly
alkaline. Put 3 or 4 anti-bumping granules into the flask and fit the
drying tube on the top. Place a pad of damp cotton wool on the
open end of the drying tube. After 5 minutes boiling, the cotton
wool pad will be saturated with ammonia. Pass it round the class
for the pupils to smell.

Indicator papers

Spatula

Access to a fume cupboard

Egg albumen (white of egg)

Potassium hydrogen sulphate

Concentrated sulphuric acid

8M sodium hydroxide

At this point in the lesson sequence, it is suggested that ammonia be prepared and that its properties be fully investigated. Point out to pupils that ammonia is only one of the products of the thermal decomposition of proteins in the presence of soda lime. (They may appreciate the need to know something of its properties and its composition.)

Supplies of ammonia gas can be readily obtained by heating a few cm^3 of '0.880' ammonia solution in a large test-tube and collecting the dried gas in test-tubes or in gas jars.

The density of ammonia compared with that of air can be established qualitatively. The solubility of ammonia in water may be regarded as almost self-evident from its source of supply. Even so, pupils may find it of interest to invert a test-tube of ammonia in a beaker of water. As the gas dissolves in the water, the water rises up the tube. Some teachers may wish to show the high solubility of ammonia in water by means of a fountain experiment. (However, teachers should be aware that the technicalities of this experiment can produce a barrier to those pupils who fail to appreciate the point of such a demonstration.)

The fact that ammonia does not burn in air is very worth while establishing, as is the fact that ammonia will burn in oxygen. This last reaction is of value in a subsequent discussion of the constitution of the gas.

The reaction of ammonia with hydrogen chloride is also worth demonstrating and can be made spectacular!

Experiment B14.1d

Apparatus

Each pair of pupils will need:

Experiment sheet 104

Test-tube, 150×25 mm, fitted with a bung to carry a calcium chloride tube

'Calcium chloride' tube, filled with glass wool and a *few* potassium hydroxide pellets (say 5–8 pellets) *or* lumps of calcium oxide

Delivery tube fitted with bung (see figure B14.4 for details)

4 test-tubes, 150×25 mm, fitted with corks

Test-tube rack

Beaker, $250 \, cm^3$

Stand, boss, and clamp

Bunsen burner and heat resistant mat

(Continued)

Investigating the properties of ammonia

Procedure

Details are given in *Experiment sheet* 104. It may be necessary to demonstrate the technique of heating strong ammonia solution to obtain a steady supply of gas. It is important *not* to allow the ammonia solution to 'boil over' into the drying tube.

Experiment sheet 104

Use the apparatus shown below to obtain four test-tubes of ammonia gas.

Concentrated ammonia solution must be treated with caution; it gives off considerable amounts of ammonia gas which is unpleasant to breathe and has choking properties.

Keep your mouth and nose well away from concentrated ammonia solution at all times and take care not to inhale ammonia gas during this experiment.

Use about 10 cm^3 concentrated ammonia solution in the lower test-tube of the apparatus. Warm this very gently with a tiny Bunsen flame; ammonia gas is given off and pushes the air out of the apparatus. After about 30 seconds warming remove the upper test-tube and cork it while the open end is still downwards. Immediately replace it with a fresh tube and place the filled and corked tube in a rack. Allow 15 seconds for the new tube to fill, remove and cork it (open end downwards), and fit a new tube to the apparatus. Continue until you have four corked test-tubes containing ammonia gas. Stop the warming and place the apparatus in a fume

---

Splints

'0.880' ammonia, 10 cm³ (maximum) *or* ammonium chloride, 5 spatula measures, and calcium hydroxide, 5 spatula measures)

Litmus solution

The teacher will need:

Gas jar full of dry ammonia gas

Gas jar full of hydrogen chloride gas

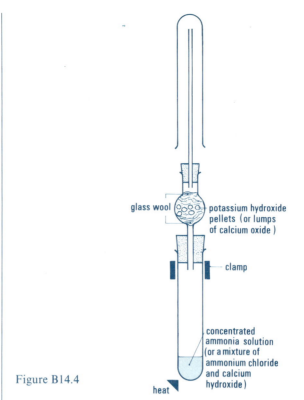

glass wool — potassium hydroxide pellets (or lumps of calcium oxide)

clamp

concentrated ammonia solution (or a mixture of ammonium chloride and calcium hydroxide)

heat

Figure B14.4

cupboard (if available), or out of harm's way, and allow it to cool before dismantling it.

Use the four tubes of gas for the following tests.

1. Place the corked end under water in a beaker and remove the cork. What happens and what does this show?

2. Hold a tube with the cork upwards, remove the cork for *at least* 2 minutes, and then replace it. Do the same with another tube held mouth downwards. Now test both tubes with litmus solution.

Which is denser, ammonia or air?

3. Use the remaining tube to discover whether ammonia will burn or allow a taper to burn in it. What are your conclusions?

After the experiment, it is suggested that the teacher demonstrates the reaction between ammonia and hydrogen chloride either by using gas jars of the dried gases or by other convenient means.

To find out if ammonia will burn in oxygen

Apparatus

The teacher will need:

Glass tube (preferably Pyrex), approximately 4 cm diameter and 15–20 cm long, fitted with a cork carrying 2 pieces of glass tubing (see figure B14.5)

(Continued)

Procedure
Arrange the apparatus as shown in figure B14.5. Turn on the oxygen cylinder to provide a gentle flow of gas. The glass wool distributes the gas around the jet. Use a glowing splint to detect the presence of oxygen in the tube. Generate a little ammonia gas by heating ammonia solution and show the presence of gas in the tall tube using damp litmus paper. Apply a lighted splint to the ammonia

Splints

Litmus paper

Bunsen burner and heat
resistant mat

Connecting tubing

Glass wool

2 stands, bosses, and clamps

Oxygen cylinder

Ammonia generator (as in the
previous experiment)

jet. A greenish flame appears and this can be extended to some
15 cm by increasing the gas flow.

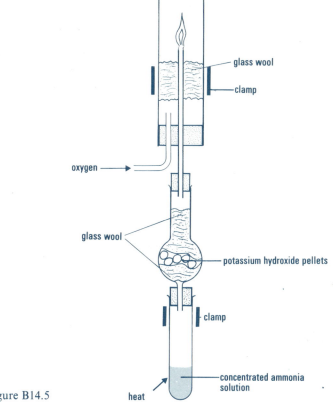

Figure B14.5

The characteristics of the alkaline gas formed when proteins are
decomposed by heating have now been established. If our original
considerations which led to the investigations are correct, then
ammonia could be used to promote the growth of plants. This can
be tested by lightly spraying 2M ammonia solution on a patch of
grass. (The grass will be noticeably greener and thicker after a
period of about two to three weeks depending on the time of year.)
While waiting for the results of this experiment ask the pupils to
consider the problem of making sufficient quantities of ammonia
for the large-scale production of fertilizers. (Heating dried milk or
other protein material is scarcely *the* answer!) To make ammonia,
we must know its composition. This is the concern of the next
section of this Topic.

Suggestions for homework

1. Suggest as many reasons as you can for the shortage of food in
the world.
2. Try to find out what other elements, apart from nitrogen, are
necessary for plant growth. How could these elements be provided
in fertilizers?

B14.2
Analysis of ammonia

Experiments are devised to find out what elements are present in ammonia. First, ammonia is found to contain hydrogen, and then it is shown to contain a relatively unreactive element, nitrogen.

Objectives for pupils

1. Knowledge of the composition of ammonia
2. Ability to design an experiment
3. Knowledge of the term analysis

The thermal decomposition of foodstuffs studied in section B14.1 showed that the products of reaction included water, carbon dioxide, and ammonia. Pupils may suggest that ammonia is an oxide since the other two products which they identified from the breakdown of foodstuffs were oxides! However, ammonia has been shown to combine with oxygen and it would seem very unlikely that an oxide formed by the combination of an element with oxygen will combine with even more oxygen. Gases which have previously been analysed, steam (in Topic A7) and hydrogen chloride (in Topic B11), were split up by passing them over an active metal, namely iron. This treatment removed one of the elements and so the other element was isolated (it was hydrogen in both cases). Discussion will reveal whether pupils will accept this line of argument and if so, one could suggest that the same approach be tried with ammonia.

Experiment B14.2a

Apparatus

The teacher will need:

Combustion tube, 10 cm × 1 cm diameter, fitted with bungs and delivery tubes

Delivery tube fitted with Bunsen valve (see figure B14.6)

2 Bunsen burners and heat resistant mats

2 stands, bosses, and clamps

Small trough

3 test-tubes, 100 × 16 mm, fitted with corks

Test-tube rack

Splints

Ammonia generator (as used in Experiment B14.1d)

Steel wool (freed from grease)

Can ammonia be broken down by passing it over heated iron?

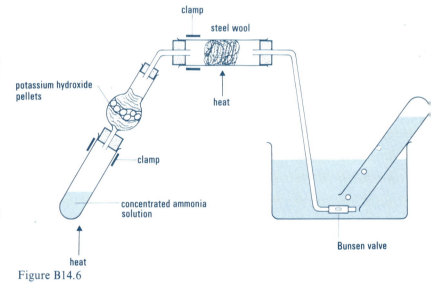

Figure B14.6

Procedure

Dry ammonia from the ammonia generator is passed over heated steel wool in a combustion tube as shown in figure B14.6. At least three test-tubes of the resulting gas should be collected. The lighted splint tests will show that hydrogen is present in this gas. The Bunsen valve is essential to avoid the danger of 'sucking-back'. The experiment shows that the ammonia contains hydrogen, as did steam and 'salt gas'. To discover what else is present in ammonia, hydrogen must be removed.

Topic B14 Chemistry and the world food problem

In the case of 'salt gas' this was achieved by supplying oxygen and removing hydrogen as water. Manganese(IV) oxide and potassium permanganate were used as sources of oxygen for this purpose. In this present problem it might well be that we need a less powerful oxidizing agent. Pupils may recall that we used copper(II) oxide in an attempt to remove hydrogen from 'salt gas' and that this was unsuccessful. Clearly, it will be wiser to try a less powerful oxidizing agent first before using a really powerful one such as manganese(IV) oxide or potassium permanganate. Pupils should be invited to make suggestions for solving the experimental problem before any attempt is made to remove hydrogen from ammonia.

Experiment B14.2b

To remove hydrogen from ammonia

Procedure
Details are given in *Experiment sheet* 92.

Apparatus

Each pair of pupils will need:

Experiment sheet 92

Hard-glass test-tube, 125 × 16 mm, *or* ammonia generator as in Experiment B14.1d

2 test-tubes, 100 × 16 mm

Hard-glass combustion tube

2 corks or bungs fitted with a straight delivery tube

Rubber connecting tubing

Delivery tube and Bunsen valve

Spatula

Trough

2 stands, bosses, and clamps

2 Bunsen burners and heat resistant mats

Mineral wool

Splints

Ammonium chloride, 2 g

Calcium hydroxide, 1.5 g, or 0.880 ammonia solution, few cm³

Calcium oxide (lumps), 5 g, or potassium hydroxide pellets, few pellets

Previously dried copper(II) oxide (wire-form is preferable; it is relatively expensive and should be reclaimed), 1 g

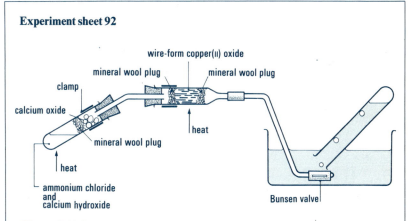

Experiment sheet 92

Figure B14.7

The apparatus is shown above. Ammonia, prepared by heating a mixture of ammonium chloride and calcium hydroxide and dried with calcium oxide, is passed over heated copper(II) oxide.

Mix 3 measures of ammonium chloride with an equal bulk of calcium hydroxide in the 125 × 16 mm test-tube. Add a plug of mineral wool (see diagram), and then add small lumps of calcium oxide. Put 3 measures of copper(II) oxide in the combustion tube.

Heat the bottom of the test-tube and the combustion tube. What happens to the copper oxide?

What does this tell you?

What do you notice on the cooler part of the combustion tube?

What does this tell you?

Try to identify the gas collected. Describe what you do and state your conclusions.

State briefly what you have learned from this experiment.

Since the gas extinguishes a lighted splint and has no odour, pupils will conclude that it is unreactive. So far, the only other non-reactive gas they have met is nitrogen. A shared lack of reactivity is *not* very convincing evidence that two samples of gas are identical, a point which needs to be made carefully.

Proof that the gas is indeed nitrogen could be obtained by gas density determinations, the production of spectra, or measurements using a mass spectrometer; these methods cannot be discussed with pupils since their background knowledge is limited. The burning of magnesium in nitrogen to produce magnesium nitride and the subsequent addition of water to produce magnesium oxide and ammonia could be confusing to pupils at this stage, and so no reference is made to this test. It follows that we must inform the pupils that other evidence *is* available to chemists to support the idea that the gas is nitrogen. However, if the pupils are familiar with the concept of density, it may be appropriate to mention that gas density determinations support the idea that the two samples of gas are identical.

We can therefore conclude *tentatively* that ammonia is a hydride of nitrogen. We do not yet know if ammonia contains other elements. The only way for us to be sure that ammonia contains only nitrogen and hydrogen is to follow the analysis of ammonia in Experiment B14.2b by a synthesis of ammonia – just as was done in the 'salt-gas' problem. The synthesis of ammonia is studied in the next section.

Suggestion for homework   Devise an experimental procedure for measuring the density of nitrogen.

B14.3
Synthesis of ammonia

The synthesis of ammonia from nitrogen and hydrogen is attempted and pupils are given some indication of the economic importance of this discovery before they consider the equivalent industrial process in some detail.

A suggested approach

Objectives for pupils

1. Knowledge of the synthesis of ammonia from its constituent elements
2. Understanding the social significance of this knowledge
3. Knowledge of the industrial process for the manufacture of ammonia
4. Understanding the application of catalysis to an to an industrial process

To support the suggestion that ammonia contains *only* nitrogen and hydrogen, we need to be able to synthesize the gas from these two elements. The difficulty at the end of the previous section was that there was no simple proof that the unreactive gas produced when ammonia was passed over heated copper(II) oxide consisted entirely of nitrogen. No *positive* evidence was produced to show that the gas was nitrogen. If appropriate, the teacher might now make a brief reference to the homework set at the end of B14.2 on gas density determination.

We need to synthesize ammonia from nitrogen and hydrogen. Tests show that these gases do not react together under ordinary conditions; clearly it would be useful if an experimental procedure could be devised whereby nitrogen and hydrogen could be passed

Topic B14 Chemistry and the world food problem

over a catalyst to form ammonia. Pupils might be asked why this would be advantageous and thereby revise the process of catalysis and their knowledge of the properties of catalysts (see Topic B13). The synthesis of ammonia can then be attempted by passing a mixture of nitrogen and hydrogen over freshly reduced heated iron wool.

Apparatus

The teacher will need:

2 syringes, 100 cm^3

2 syringe holders, bosses, and stands, *or* a syringe bench

Combustion tube, 15 cm long by 7 mm diameter

2 pieces of hard-glass rod about 2 cm long and of a diameter slightly smaller than the internal diameter of the combustion tube

Three-way stopcock with capillary tubing

Bunsen burner and heat resistant mat

Indicator paper (preferably litmus)

Thick-walled rubber tubing

Iron wool

Cylinders of compressed gases: nitrogen and hydrogen, with connecting tubing

Can ammonia be made from nitrogen and hydrogen?
This experiment should be performed by the teacher.

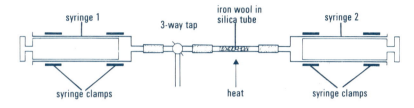

Figure B14.8 Apparatus for the synthesis of ammonia.

Procedure
Set up the apparatus as shown in figure B14.8.

When using compressed gases, it is very easy to break a syringe plunger by blowing it out of the syringe. The plunger should therefore be tied to the syringe barrel with string, and the following sequence of operations should be used:

1. Open the cylinder valve before connecting the cylinder to the stopcock. Adjust the fine control on the cylinder head so as to produce a slow, steady flow of gas. (This ensures that all air is swept out of the connecting tubing attached to the cylinder head.)
2. Hold the connecting tube from the cylinder lightly against the stopcock entry tube when passing gas into a syringe, thereby enabling the connecting tube to be removed quickly as soon as enough gas has been transferred to the syringe. Flush out the syringe and refill to ensure that no air (from the stopcock tubing) is present in the syringe.
3. When 'pure' gas has been transferred to syringe 1, turn the stopcock to connect syringe 1 to syringe 2 – via the rest of the apparatus.

The experimental procedure requires the absence of air in the apparatus. The apparatus is first flushed out with nitrogen and then with hydrogen. After emptying the apparatus, 100 cm^3 of hydrogen is transferred to syringe 1. The stopcock is closed to the surroundings and used to connect syringe 1 to syringe 2. The iron wool in the combustion tube is then heated strongly using a Bunsen burner and the gas is passed to and fro over the hot iron wool. This reduces any iron oxide which may be present. The process is continued until no further reduction in gas volume occurs. Remove the Bunsen burner and then discharge the residual gas from syringe 1.

Flush out the entire apparatus with nitrogen. Transfer 20 cm^3 nitrogen and 60 cm^3 hydrogen to syringe 1. Heat the combustion tube strongly before passing the gas mixture two or three times over the hot iron. Turn out the Bunsen burner, open the stopcock, and eject the gases onto a piece of damp red litmus paper. It will turn blue, showing the presence of ammonia.

The reaction may be summarized by the word equation:

nitrogen + hydrogen $\xrightarrow[\text{catalyst}]{\text{red hot iron}}$ ammonia

The chemical equation for the reaction *may* also be used, but without detailed explanation:

$$N_2(g) + 3H_2(g) \xrightarrow[\text{catalyst}]{\text{red hot Fe}} 2NH_3(g)$$

The teacher must indicate the features and possible development of this laboratory experiment:

1. Ammonia can be synthesized from nitrogen and hydrogen in the presence of an iron catalyst at a high temperature.

2. The quantity of ammonia produced is small. It will change the colour of an indicator *but* the smell of ammonia when the gas is discharged into the laboratory is not noticeable.

3. The equation for the reaction (quoted earlier) tells us that four particles or molecules of gas (one of nitrogen and three of hydrogen) react to form two particles or molecules of gas (two of ammonia). By operating the process at a higher pressure we force gas particles or molecules closer together and might 'encourage' this contraction in the number of particles taking part in the process to occur. Indeed, this is done in an industrial plant by operating it at a high pressure (150–200 atmospheres) and at a moderately high temperature (380–400 °C). (Topic B21 offers a further opportunity to consider this (and other systems) from another viewpoint, that of chemical equilibria.)

Next, indicate to the pupils how the demonstration can be scaled up to produce ammonia commercially. One might begin by recalling possible sources of supply for the reactants. Nitrogen can be obtained from air and pupils might be asked how this could be achieved. Hydrogen exists in compound form in both water and methane, both of which are relatively abundant.

Pupils know that hydrogen can be obtained from water (by passing steam over red-hot iron). They will appreciate that such a process could be improved. It would be even better if hydrogen could be obtained from steam using some catalytically controlled process.

The scale of operation required for an industrial process is difficult to convey in simple terms. Some indication can be given by listing alternative raw materials required to make 1 tonne of ammonia under typical conditions:

about 2100 m^3 of hydrogen;
or   1.5 tonnes of coke or coal;
or   800 m^3 of natural gas;
or   2500 m^3 of refinery gas;
or   0.7 tonne of fuel oil.

So to produce 600 tonnes of ammonia per day – the capacity of the ICI plant at Immingham – the space required for the plant is considerable. Most of the area is devoted to the production of synthesis gas (a mixture of nitrogen and hydrogen) and only a small part is occupied by the ammonia converter.

In recent years, natural gas (mostly methane) has grown rapidly in importance as a raw material. Synthesis gas is made from methane by reacting it with steam in the presence of a nickel catalyst at a high temperature to produce carbon monoxide and hydrogen:

$$CH_4 + H_2O \xrightarrow[\text{Ni catalyst}]{700-850\,°C} CO + 3H_2$$

Subsequently, carbon monoxide is converted to carbon dioxide by steam in the presence of a catalyst at a high temperature:

$$CO + H_2O \xrightarrow[\substack{\text{Fe}_3\text{O}_4 \text{ and then} \\ \text{Cu catalyst}}]{500\,°C} CO_2 + H_2$$

Carbon dioxide is removed by washing the products of this reaction with water. (Pupils can be reminded of this property of carbon dioxide by referring to 'fizzy' drinks and soda water; both use carbon dioxide dissolved in water under pressure.) The scale of this washing process can be conveyed by the fact that for every $2100\,m^3$ of hydrogen (needed to produce 1 tonne of ammonia) more than 100 tonnes of water are required to remove the carbon dioxide produced in the second stage of the synthesis gas process. The water used must be of high quality and since such quantities are not usually available 'on tap', the water has to be stripped of its carbon dioxide and recirculated.

Eventually, the dry hydrogen is compressed, mixed with compressed nitrogen from the air, and passed into the converter where the synthesis of ammonia occurs:

$$N_2 + 3H_2 \xrightarrow[\substack{\text{iron catalyst} \\ 150-200 \text{ atmospheres}}]{380-400\,°C} 2NH_3$$

The reaction does not go to completion: conversions of only 25–50% of the feed nitrogen and hydrogen are obtained. The unconverted gas must be separated from the ammonia and recirculated. All of these chemical engineering problems have been solved and diagrams showing a schematic layout of an ammonia plant can be used to add interest to the work. (Film loops 2–11, 2–12, 2–13 or other visual presentations may be used to convey essential details.)

Ask the pupils in what other ways the industrial process plant will differ from the simple laboratory demonstration. The following points should emerge:
1. Metal replaces glass as the construction material.
2. The pipes used have to be very strong to withstand high pressures. (Indeed, the cross-sections of the pipes look very much like cotton reels – a thick wall about a very small bore.) Also joints need to be sealed carefully to avoid leakages.
3. Pumps are needed to compress the gases and to move them from one part of the plant to another.
4. A continuous flow process is clearly better than a batch process (a one-off process like the laboratory demonstration).

Each of these points required separate investigation. Originally, Haber's laboratory work indicated the possibilities of the process. Bosch required an additional four years to translate the process to a large-scale plant. The industrial plants we have today are improvements on this first plant.

From this discussion pupils should gain some insight into the scale of operation used and learn something of the relationship between the chemist and the chemical engineer. The discussion may be summarized by using the film loop 2–11 'Ammonia manufacture' (if this has not been used earlier).

At the end of the lesson, the teacher should emphasize the basic reaction used in the production of ammonia and the conditions necessary for its successful use on a large scale.

Suggestion for homework Read Chapter 7 'Fertilizers' of *Chemists in the world.*

B14.4
Making fertilizers

Pupils use ammonia solution to prepare an ammonium salt and discuss the importance of ammonium compounds as fertilizers.

A suggested approach

1. Preparation of ammonium sulphate or ammonium nitrate
2. Revision of the terms acid, alkali, and salt
3. Knowledge of the use of ammonium salts and other nitrogen compounds as fertilizers

In the previous section, the manufacture of ammonia was considered in some detail. Most of the ammonia produced in this way is used for the production of ammonium salts, some is used in the manufacture of nitric acid. Figure B14.9 shows a typical breakdown.

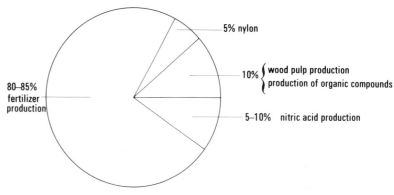

5% nylon

10% { wood pulp production
      production of organic compounds

5–10% nitric acid production

80–85% fertilizer production

Uses of ammonia

Production of synthetic anhydrous ammonia (U.S.A.)

1950	1421 000 tonnes
1960	4371 000 tonnes
1965	8046 000 tonnes
1970	11882 000 tonnes
1973	15093 000 tonnes

Figure B14.9
Production and uses of ammonia. (OHP 34B)

Pupils know that ammonia solution is alkaline and that when an alkali is used to neutralize an acid, a salt and water are the usual products of reaction. Ammonium chloride, NH_4Cl, ammonium nitrate, NH_4NO_3, and ammonium sulphate, $(NH_4)_2SO_4$, are common laboratory chemicals. (It will be necessary to mention both nomenclature and formulae of ammonium compounds.) Ammonium sulphate and ammonium nitrate are used as fertilizers. Discuss possible procedures for making these compounds on a large scale and indicate some of the difficulties which need to be overcome (for example, mixing reagents on a large scale so as to lessen the high cost of evaporation of solutions; large-scale filtration processes; danger of explosions with ammonium nitrate and the use of chalk to nullify its explosive nature). Now, ask the pupils to prepare samples of either ammonium sulphate or ammonium nitrate. Two procedures are suggested and the advantages and disadvantages of each procedure can then be discussed.

Experiment B14.4a

Apparatus

Each pair of pupils will need:

Experiment sheet 93

Evaporating dish or beaker, $100\,cm^3$

Glass rod

Bunsen burner and heat resistant mat

Tripod and gauze

Measuring cylinder, $25\,cm^3$

Microscope slide

Specimen tube

Filter flask, filter funnel, filter paper, and filter pump

M sulphuric acid, $20\,cm^3$

2M ammonia solution, $25\,cm^3$

Preparation of ammonium sulphate

Procedure
Details are given on *Experiment sheet* 93. Pupils may wish to retain a sample of the product in a specimen tube.

Experiment sheet 93
Put about $20\,cm^3$ M sulphuric acid into an evaporating basin. Add 2M ammonia solution, a little at a time, with stirring, until the mixture has a definite smell of ammonia. Evaporate the solution to about one-fifth of its original volume and set aside to cool. How will you tell when the solution will crystallize?

When crystallization is complete, filter off the crystals and dry them on blotting paper or filter paper. Keep them for the next experiment. Write the equation for the reaction.

Why is it not necessary to add exactly the correct amount of ammonia solution?

Experiment B14.4b

Apparatus

Each pair of pupils will need:

Experiment sheet 105

Beaker, $100\,cm^3$

Burette, $50\,cm^3$

Burette stand

Glass rod

(*Continued*)

Preparation of ammonium nitrate
The teacher will need to demonstrate the technique of using a burette before pupils start this experiment.

Procedure
Details are given in *Experiment sheet* 105.

Experiment sheet 105
You will be required to wear safety spectacles when doing this experiment.
Wash out a $50\,cm^3$ burette with water and then with ammonia solution. Fill the burette with ammonia solution, make sure the jet is filled with liquid, and adjust the liquid level to the zero mark.

Bunsen burner and heat resistant mat

Tripod and gauze

Filter paper

Measuring cylinder, 25 cm^3

Specimen tube

2M ammonia solution, 50 cm^3

2M nitric acid, 40 cm^3

Litmus solution

Use a measuring cylinder to put 20 cm^3 dilute nitric acid in a 100 cm^3 beaker and add a few drops of litmus solution. Place the beaker under the burette and run ammonia solution into the nitric acid, 1 cm^3 at a time, stirring between each addition until the litmus turns blue. Note the volume of ammonia solution used.

Empty and wash the beaker. Measure another 20 cm^3 portion of dilute nitric acid into it and then run in the same volume of ammonia solution as that used before. Do not add litmus this time.

Stir the contents of the beaker and heat it carefully on a wire gauze over a Bunsen burner until the remaining solution *only just* covers the bottom of the beaker. Allow the beaker to cool. Collect the crystals formed, dry on filter paper, and store in a small specimen tube.

After discussing the results of these experiments, the reactions may be summarized by chemical equations and the state notation 'aq' introduced:

$$2NH_3(aq) + H_2SO_4(aq) = (NH_4)_2SO_4(aq)$$

$$NH_3(aq) + HNO_3(aq) = NH_4NO_3(aq)$$

The question may now be put: 'Do these reactions require us to revise the meanings of the terms acid or salt?'

It is helpful to show pupils samples of ammonium nitrate that has been kept in stock for some time. Pupils may notice that their own samples of ammonium nitrate crystals become damp on keeping. This property of ammonium nitrate becomes serious when hundreds of tonnes of it are kept in store. To enable it to be used as a fertilizer, the compound must flow freely. Accordingly, the manufacturing process is adapted to produce a granulated product by coating the crystals with a very fine dust of china clay or chalk.

Ammonium nitrate (provided that it flows freely) is both a convenient material to handle and a highly concentrated nitrogenous fertilizer. Indeed, it finds greater use as a fertilizer than ammonium sulphate in the United Kingdom.

An indication of world fertilizer consumption appears in figure B14.10. It will be noted that total use is more than five times that during World War II. Pupils who produce about 1 gram of ammonium nitrate from Experiment B14.4b may be told that to match the potential output per day of the largest production plant for ammonium nitrate in the United Kingdom, this experiment would need to be repeated by every single person on Earth! And the output of, say, 3600 tonnes per day of ammonium nitrate from one industrial plant is only a *minute* fraction of the total world production of ammonium nitrate.

Pupils may be told that plants require three main ingredients for healthy growth – nitrogen, potassium, and phosphorus. Fertilizers are of two sorts:
1. 'Simple' fertilizers which contain nitrogen only;
2. So called 'compounded' fertilizers which contain all three constituents: nitrogen, potassium, and phosphorus.

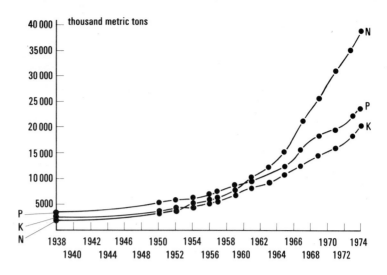

Figure B14.10
World fertilizer consumption.
N = nitrogenous fertilizer,
P = phosphatic fertilizer,
K = potash fertilizer.
(*Department of Economic and Social Affairs* (1957, 1967, 1975). Statistical Yearbook, United Nations.) (OHP 35)

Nitrogen compounds commonly used as fertilizers include ammonium nitrate, urea, ammonium sulphate, and ammonium phosphate.

Suggestion for homework

Read *Handbook for pupils*, Chapter 11 'The world food problem'.

B14.5
The balance of nature

The consequences of using fertilizers and other chemicals in the production and storage of food are considered briefly, and the role of science in society is discussed in simple terms.

A suggested approach

Objectives for pupils

1. Ability to organize information and to discuss matters of common concern
2. Awareness of the role of the scientist (and the chemist in particular) in society
3. Awareness of both the advantages and the limitations of synthetic aids (such as fertilizers, and so on)

The meeting of one basic human requirement, such as the production of adequate food supplies, can lead to unexpected difficulties in other apparently unrelated situations.

Consider the use of simple fertilizers, such as ammonium sulphate or ammonium nitrate, on the land. Indiscriminate application could make the soil acidic (compare the properties of solutions of the salts with Full-range Indicator and note the effect of adding small quantities of 'lime' to such mixtures). Consider also the effect of water draining off the land resulting in the transfer of fertilizer into a stream or river. The fertilizer promotes the growth of aquatic plants and the concentration of dissolved oxygen in the water falls. This means that less oxygen is available to fish and other creatures. Micro-organisms, such as fungi, bacteria, and protozoa, concerned with the decomposition of organic waste from dead animals and plants, are also affected. They all require dissolved oxygen to break down and remove such matter as phosphate and nitrate, thereby enabling some natural recycling of nitrogen and phosphorus to occur. It follows that waste matter builds up and the problem gets worse. In an extreme situation, there is a serious depletion of dissolved oxygen and a consequent rise in the concentration of ammonia, phosphate, and nitrate particles in solution. (A similar situation occurs when untreated sewage escapes into a river.) The result is a decrease in higher animal life in the water and an increase in algae. This state of affairs is called *eutrophication*. It is occurring on a small scale in many places and on a larger scale in the Great Lakes of America where so much effluent (from many different sources) has passed into the lakes that many fish and aquatic animals have died. Some lakes are no longer fit for bathing (for example, parts of Lake Michigan). In other situations, the solutions of nitrates render the water poisonous and useless for drinking.

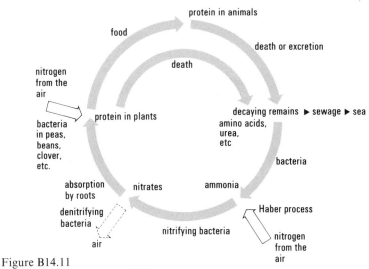

Figure B14.11
The nitrogen cycle. (OHP 36)

It may be appropriate for the teacher to provide some information about the nitrogen cycle (see figure B14.11). The point should be made that this particular cycle is but one of several such processes in nature whereby material is used again and again. Agriculture is a man-made activity imposed on a series of naturally occurring cyclic processes. Any man-made change can lead to one or more of these cycles being affected and in ways which are not always obvious or predictable. If such a network of natural processes is broken in enough places, the cycle will collapse. Eutrophication is the result of the collapse of cyclic processes, the nitrogen cycle included.

With an increasing world population, there are conflicting demands between those responsible for food production and those concerned with maintaining an unpolluted environment. The production of food for human consumption also provides a concentrated source of nutrients for a variety of pests – insects, birds, and mammals – which frequently devour a high proportion of crops.

Chemists have devised a series of remedies to lessen this competition and the term 'pesticide' is used to cover the wide range of chemicals used for this purpose.

In general, these substances are highly toxic and it is desirable that they have a long period of residual toxicity to eliminate the need for a repeated dosage. Some classes of compounds which are employed (for example, DDT, dieldrin, aldrin, and so on) pose special problems because they become more and more concentrated in food chains. Accordingly, man and predatory animals at the peak of the food pyramid are at special risk. The search continues for new insecticides of lower toxicity to man which still retain the capacity for killing pests. Arguments for and against the banning of pesticides might include the following points:

For the ban:
1. Balance of nature – cyclic processes upset.
2. Concentration of chemicals in food chains.
3. Species of pests can become resistant to a pesticide.
4. Unpredictable long-term effects.
5. Some insecticides kill insects, etc., without distinction.
6. Biological control of pests is more effective.

Against the ban:
1. Balance of nature – all agriculture interferes with the environment.
2. Elimination of disease (for example, malaria).
3. Control of pesticides *is* effective: tests are usually made over a six- to nine-year period.
4. Reduction of crop wastage: losses can be up to 50 per cent of the total crop.

Such a discussion can range over many issues and include an indication of the role of the chemist within our society.

Suggestions for homework

Write a brief account of the arguments for and against the use of chemicals (fertilizers and pesticides) in modern farming. (Refer to *Chemists in the world* for discussions on the nitrogen problem, nitrogen cycle, the Haber process, and fertilizers.)

Summary

By the end of this Topic, pupils should be aware of the factors which influence the world food problem, and of the role that the chemist continues to play in seeking its solution. Pupils should also be able to relate the chemistry of ammonia to the production of fertilizers and know something of the need to use fertilizers and pesticides with care. They should have considered the factors influencing the production of ammonia on a large scale.

Everyday materials: large molecules and metals

1. To emphasize the importance of chemistry in daily life.
2. To characterize the properties of foodstuffs.
3. To study the composition of crude oil.
4. To introduce the processes of cracking, polymerization, and hydrolysis.
5. To compare the properties of polymers with those of wood and of metals.
6. To provide a simple review of the plastics industry.
7. To review some of the mechanical properties of metals.

Contents

B15.1 Food
B15.2 Starch – an example of a carbohydrate
B15.3 Breaking down glucose
B15.4 Breaking down fats and oils
B15.5 What compounds are there in crude oil?
B15.6 Breaking down big molecules in crude oil
B15.7 Making and using plastics
B15.8 Metals and their importance to engineers

Timing

The key ideas and experiments in this Topic need about 7 or 8 weeks. Sections B15.1 and B15.5 each require about one week; B15.2, B15.3, and B15.4 require a little more than one week each. B15.6 and B15.7 together need about three weeks and the last section, B15.8, may require up to two weeks. Of course, these timings will require considerable modification if pupils have covered some of the work in their Biology course.

Introduction to the Topic

The Topic is intended to illustrate the contribution of chemistry to daily life in western society. Four themes are explored, and there is sufficient supplementary material to enable teachers to expand this Topic through project work, should this be required.

The previous Topic considered the world food problem. This Topic opens with a short review of the classification of foodstuffs and some tests to enable pupils to identify each group of substances. Molecular models may be used to indicate characteristic features of carbohydrates, proteins, fats, oils, etc. Pupils then investigate the composition and properties of starch, using knowledge and techniques established in earlier parts of the course. They investigate the breakdown of starch into glucose and may also consider the synthesis of starch from glucose. The breakdown of glucose is attempted next, and leads to the formation of ethanol by fermentation.

The second theme concerns oils and fats. It includes the preparation of a soap and a soapless detergent. The properties of soaps and soapless detergents may be compared and the use of biodegradable detergents is discussed.

The third theme is the study of some aspects of the oil industry. The fractionation of crude oil is revised and is then used to introduce the section on plastics. Samples of perspex and nylon are made, and their properties are compared. The properties of plastics are also compared with those of wood and of metals.

The final theme is the chemistry and physics of metals. A brief review is made of the isolation of typical metals (such as iron, copper, and lead) and is followed by an assessment of their more important mechanical properties on a simple comparative basis. The preparation of crystalline samples of several metals reveals the need to investigate the structure of metals in some detail. In addition, evidence is considered which supports the existence of an atomic rather than a molecular structure for these materials.

Subsequent development

The need for pupils to know something about atoms leads on to the next Topic, B16 'Atoms and the Periodic Table'. Matters relating primarily to the arrangement of atoms in elements and compounds receive further consideration in Topic B17, and again in somewhat greater detail in Topic B23. Topic B18 includes the mole concept and the quantitative determination of the formulae of some single compounds. Related Stage III Options include Option 3 'Drugs and medicines', Option 4 'Metals and alloys', and Option 5 'Plastics'.

Alternative approach

Three approaches are possible using Stage II materials. Thus, in Alternative IIA, the material presented in Topic B15 appears in Topics A14 and A21. In Alternative IIB, two routes are suggested for Topics B14, B15, B16, and B17 – as indicated in Part 1 Chapter 3 (pages 15–19). The essential difference between these last two suggestions rests with the part Topic B17 is seen to play for pupils of this age group. Thus route 1 follows the numerical sequence B14 to B17; route 2 requires Topic B17 to follow B13, then B15.5–B15.8; B15.1–B15.4; and finally B14 and B16.

Further references

for the teacher

The following organizations publish information on the production of materials from crude oil (as well as drilling, distillation, etc.). This information is only available to teachers, and they must apply on headed school notepaper.
Information Service Department, Institute of Petroleum, 61 New Cavendish Street, London W1M 8AR
Education Department, Shell UK Ltd, PO Box 148, Shellmex House, Strand, London WC2R ODX
BP Educational Service, PO Box 21, Redhill, Surrey
Educational Services, Public Affairs Department, ESSO Petroleum Company Ltd, ESSO House, Victoria Street, London SW1
Dodds, C. (June 1971) 'An introduction to the teaching of synthetic plastics and fibres in secondary schools.' *School Science Review*, **52**, No. 181, 812–33.
Gordon, J. E. (1968) *The new science of strong materials*. Penguin.

Supplementary materials

Slides and film loops
The following figures are available in a series of slides:
Figure B15.2 Models of molecular structures: carbohydrates and protein
Figure B15.6 Models of molecular structures: fats and oils
Figure A14.12 (see section B15.8) Photographs to illustrate the crystalline nature of metals

2–9 'Cracking hydrocarbons'
2–10 'Plastics'
3–3 'Metals: mechanical properties'

Overhead projection originals
37 Approximate percentages of food constituents of some foods *(figure B15.1)*
38 Effect of detergent *(figure B15.7)* & Fractionation of crude oil *(Table B15.1)*
39 Initial distillation of a crude oil and a possible demand pattern *(Table B15.2)*
40 Reactions of ethene *(figure B15.9)*
41 Using substituted ethenes in making polymers *(figure B15.10)*

Films
'Metallurgy: a special study' ESSO. Films for Science Teachers No. 28
'Polymers' ICI Film Library
'What is soap?' Unilever Film Library
'Chemistry of soapless detergents' Unilever Film Library
'Outline of detergency' Unilever Film Library
'Hard water' Unilever Film Library

Reading material

for the pupil

Chemists in the world, Chapter 6 'Polymers from petroleum'.
Handbook for pupils, Chapter 10 'The chemical industry' and Chapter 12 'Man, chemistry, and society'.

B15.1
Food

This section serves as a bridge between Topic B14 and the main theme for Topic B15, everyday materials. The principal constituents of food are classified as carbohydrates, fats, proteins, water, mineral salts, and vitamins, and the compositions of some common foods are given. Simple tests for the detection of carbohydrates, fats, and proteins in foodstuffs are introduced. The use of molecular models to represent chemical species is also discussed.

A suggested approach

Objectives for pupils

1. Awareness of the composition and classification of foodstuffs
2. Awareness of the use of molecular models to represent chemical species
3. Awareness of the role of models in science

Pupils may have completed an elementary study of foods and nutrition during an introductory science or Biology course. This section can then be used to revise and extend the work. If the second route (see page 19) is used and this Topic comes after Topic B17, then some changes in both tone and timing may be necessary.

In the previous Topic, we found that nitrogenous fertilizers were necessary to promote the growth of crops. Remind pupils that the production of crops depends on more than one factor: plants require a 'balanced' set of conditions – moisture, warmth, sunlight, nitrogenous and other fertilizers. Plants are an important source of our foodstuffs. In most instances, we use only part of a plant as food and we grow plants with specialized food storage parts. For example, plants which store food in their swollen roots include carrots and turnips; those which store food in stem tubers include the potato; those which store food in leaves include lettuce, cabbage, and so on.

Pupils should be able to supply a variety of examples of the use of plants as foods. They may be told that scientists have been able to indicate the basic requirements of a balanced diet.

Pupils should be aware of the classification of foodstuffs into carbohydrates, fats and oils, and proteins, and should know that an adequate supply of water, certain mineral salts, and vitamins is necessary too. Thus, red blood cells contain iron which is needed in very small quantities in a diet. Bones and teeth contain calcium, one reason for drinking 'hard' rather than 'soft' water; and so on. The various vitamins are needed in even smaller quantities and are usually found in a diet which includes fresh fruit. Vitamin A occurs in vegetables, oranges, and milk and is needed for healthy growth. Vitamin B1 occurs in cereals and in egg yolk, and its absence leads to a disease known as beri-beri. Vitamin B2 occurs in milk, kidney, and liver, and its absence can cause dermatitis. Vitamin C occurs in fresh fruit and in certain vegetables and its absence leads to a variety of troubles such as swollen gums and brittle blood vessels, both symptoms of scurvy, a disease which was much more common in earlier times particularly among sailors.

Figure B15.1 shows the approximate composition of some common foods expressed as percentages. Attention should be drawn to the scale used to present these analyses; reference to the presence or absence of vitamins in these foods is not practicable using this presentation.

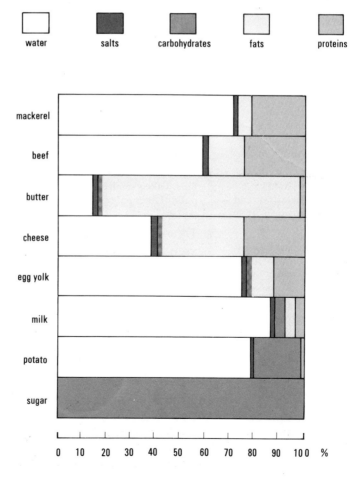

Figure B15.1
Approximate percentages of food constituents of some foods. (OHP 37)

Experiment B15.1 contains some simple tests for carbohydrates, fats, and proteins. In each case, it is suggested that the appropriate test be carried out on samples of starch, glucose, cane sugar, lard, and a suitable protein before it is tried out on the foodstuff under test.

Experiment B15.1

Apparatus

The teacher and/or each pair of pupils will need:

Experiment sheet 106

6 test-tubes, 150 × 16 mm

Test-tube rack

Test-tube brush

Test-tube holder

Glass rod

Bunsen burner and heat resistant mat

Teat pipette

Pestle and mortar

Filter paper

Iodine solution (12.7 g iodine and 20 g potassium iodide in 40 cm³ water, diluted to 1000 cm³), 0.5 cm³

Fehling's solution A (34.6 g crystalline copper(II) sulphate in water, diluted to 500 cm³), 1.0 cm³

Fehling's solution B (173 g sodium potassium tartrate and 50 g sodium hydroxide in water, diluted to 500 cm³), 1.0 cm³

Either Millon's reagent, 0.5 cm³, *or* concentrated nitric acid, 0.5 cm³, and 2M ammonia solution, 10 cm³

One or more of each of the following foodstuffs:

1. Starch, bread, potato, cereal, 1 spatula measure
2. Glucose, fresh fruit juice, 1 spatula measure or 1 cm³
3. Cane sugar, sugar beet extract
4. Lard, castor oil seeds, 1 g
5. Meat, beans, egg white, 1 cm³

Food tests

Procedure

Details are given on *Experiment sheet* 106. In each case, the tests should be tried first on a sample of the appropriate chemical before being tried on a foodstuff.

Experiment sheet 106

In this experiment you will test certain food materials for the presence of starch, certain simple sugars (called reducing sugars), proteins, and fats. Most of the tests described can be carried out in test-tubes.

Starch. To about 1 cm³ starch solution add 4–5 drops iodine solution. What happens?

Reducing sugars. Mix about 1 cm³ Fehling's solution A with an equal volume of Fehling's solution B. To the mixture add 10 drops glucose solution (dissolve 1 measure glucose in about 2 cm³ water). Shake the mixture and heat until it boils, *be careful*, it tends to spurt out of the test-tube. Describe all that happens.

Proteins. Your teacher will tell you which of the following tests to use.

a. To about ½ cm³ egg white (which is rich in protein) add about 5 drops Millon's reagent. Shake well and heat to boiling.

b. To about ½ cm³ egg white add 5 drops of concentrated nitric acid. *(Care: concentrated nitric acid is corrosive and causes painful skin burns.)*

Now add ammonia solution until the mixture is alkaline. *(Care: add the reagent a little at a time and mix well before adding the next portion.)*

Write the letter of the test you used, on the line below, and describe all that you saw.

Fats. Take a small piece of lard on the end of a glass rod and rub it across a piece of filter paper. Hold the paper up to a window or a light. What do you see?

Try to remove the smear by washing with water. Are you successful?

Try the tests on the samples of food materials with which you are provided. Record the results in a table with the headings below, using a tick to indicate a positive result.

Food material used	Contains			
	Starch	Reducing sugar	Protein	Fat

After testing various foods for the presence of starch, sugars, fats, and proteins, compare the results with figure B15.1.

Make the point that foodstuffs are made of complex mixtures of compounds having large molecules. Chemists find it convenient to represent the molecules of these substances by means of models. It

Figure B15.2
Models of molecular
structures: carbohydrates and
protein.

Note: these are available in a
series of slides and are
intended *only* to illustrate
arrangements of atoms in
molecular structures.
Photographs, Unilever.

may be helpful to show models of the molecules of typical carbohydrates (for example, starch made of a series of $C_6H_{10}O_5$ units;
glucose, $C_6H_{12}O_6$; cane sugar $C_{12}H_{22}O_{11}$; etc.); of fats (for
example, the main fat of olive oil which is glycerol trioleate); and of
part of a protein molecule (part of a polypeptide chain). Examples
of these models are shown in figure B15.2.

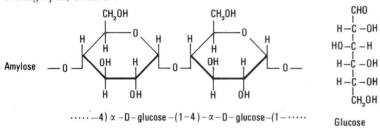

Amylose

$\cdots\cdots -4)\ \alpha-D-glucose-(1-4)-\alpha-D-glucose-(1-\cdots\cdots$

Glucose

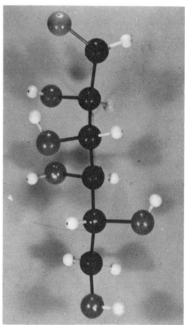

(*b*) Glucose, a model to
illustrate the simplest type of
representation.

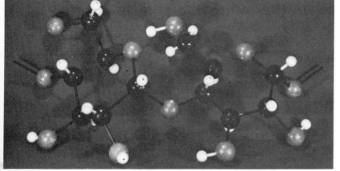

(*a*) Amylose, one of the two glucose polymers found in starch.

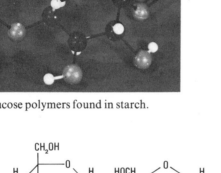

$\alpha-D-glucopyranosyl-(1-2)-\beta-D-fructofuranose$

(*c*) Sucrose, ordinary household sugar.

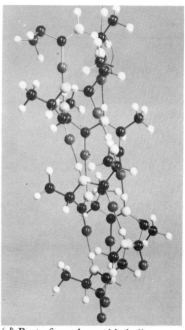

(*d*) Part of a polypeptide helix
to illustrate part of a typical
protein structure.

A discussion about models might be introduced by talking about the use of models in other areas – such as aircraft models or even architectural models, both of which enable us to visualize the appearance of the end product. Molecular models help chemists to visualize the structure of a molecule and to say something about its behaviour. They even enable the chemist to make predictions about the way in which different molecules behave in chemical reactions. Models always differ from the real thing in a number of respects: the model aircraft might be made of wood rather than metal. Similarly, molecular models are not precise: they represent atoms as hard spheres although, in the light of modern knowledge, we tend to think of atoms as being made up of electrical charges. Molecular models can vary quite a lot and serve different functions. For example:

a. Space-filling models – most accurately represent the shape of a molecule but are rather difficult to make;

b. Tangential models – represent the shape of a molecule fairly accurately and are relatively easy to make;

c. Ball-and-spoke models – show the location of centres of atoms and enable the chemist to 'see into' a molecule;

d. Dreiding models – enable the chemist to 'see into' a molecule best of all and are extremely useful in making molecular models of very large molecules.

A variety of different models of one or two simple molecular substances, such as water or methane, could be used to make different points. (See also figure B15.3.)

Figure B15.3
Examples of structure models. *Back row:* demountable ball-and-spoke (*Crystal structures*); space-filling, ionic structures (*Catalin*); miniature models (*Beevers*); Linell ball-and-spring (*Gallenkamp*). *Front row:* Stuart-type organic set (*Griffin and George*); Dreiding stereomodels (*Rinco, USA*); space-filling covalent set (*Catalin*).

The teacher can then return to the larger molecular models and identify various atoms and groupings. The identification of such features forms a basis for investigations in later sections of the course. (If this section is studied *after* Topic B17, in route 2 (page 19) some of these suggestions will have to be adapted.)

1. Make a collection of pamphlets, charts, and advertisements issued by various food manufacturers. Some of these give very valuable information on recent advances in the study of nutrition – but read them critically!
2. List the ingredients of ten package foods in order of decreasing content. Identify the food content and the additives separately.

B15.2
Starch – an example of a carbohydrate

Starch is used as an example of a substance made of large molecules. It is hydrolysed by saliva and by dilute hydrochloric acid to form glucose.

A suggested approach

Objectives for pupils

1. Knowledge that starch is a polymer
2. Ability to use molecular models to summarize the overall course of a reaction
3. Understanding of the term hydrolysis
4. Knowledge of the relationship of starch to maltose and of maltose to glucose

Molecular models of several large molecules were used in the previous section. Show pupils a model of starch. Ask them how they would show that starch contains carbon and hydrogen (remind them of their tests on foodstuffs in B14.1). A more effective way of obtaining carbon dioxide and water from starch is to mix the starch with an excess of copper(II) oxide.

Experiment B15.2a

Apparatus

The teacher will need:

2 test-tubes, 150×16 mm (previously dried)

Test-tube holder and rack

Teat pipette

Spatula

Bunsen burner and heat resistant mat

Starch

Dry copper(II) oxide stored in a desiccator

Limewater

Sample of anhydrous copper sulphate in a stoppered test-tube

Detecting the presence of carbon and hydrogen in starch

Procedure
Mix 1 spatula measure of starch with 3 measures of copper(II) oxide in a test-tube. Add a further two measures of dry copper(II) oxide and heat the tube. After a few minutes, remove a sample of the 'atmosphere' from the test-tube using a teat pipette and bubble this through a little limewater to detect the presence of carbon dioxide. Test any moisture formed on the side of the test-tube with anhydrous copper sulphate.

Direct evidence for the presence of oxygen in starch is less easy to provide.

Starch occurs in foodstuffs (for example, in bread and in potatoes) and pupils may be aware of its importance as a source of energy for the body. Since starch is a foodstuff which is not found in the human body, the process of digestion must convert it into other chemicals. Perhaps we can 'break up' the starch molecule by subjecting some starch to a digestive process? Begin by asking a few of the pupils to suck a piece of bread for several minutes and to report any change in taste.

Meanwhile a close examination of a molecular model of starch will show that it consists of a series of repeating units, each unit containing six carbon atoms, ten hydrogen atoms, and five oxygen atoms (starch is a polymer of $C_6H_{10}O_5$). The molecular model of glucose shows it to consist of six carbon atoms, twelve hydrogen atoms, and six oxygen atoms. The two models may then be contrasted and differences in the two unit formulae $C_6H_{10}O_5$ and $C_6H_{12}O_6$ recorded on the board or overhead projector – the difference being two atoms of hydrogen and one atom of oxygen.

Saliva contains water and other chemicals. Chewed bread tastes sweet. Glucose tastes sweet. We may *speculate* that *maybe* when starch reacts with saliva the $C_6H_{10}O_5$ units acquire two hydrogen atoms and an oxygen atom and become glucose. Saliva contains a catalyst for this reaction (see Topic B13).

Maybe starch does not break down completely into glucose in this way; the reaction might proceed by the production of a number of other compounds. This can be forecast using models. Compounds made of one, two, or even three basic units could be formed.

The breakdown of starch to form a mixture of products is possible and it will be difficult to separate such a mixture and to identify all the various components. Chromatography is a technique whereby we can do both.

The teacher should now explain the chromatographic technique to be used. Pupils may then attempt Experiment B15.2b. It requires careful planning.

Experiment B15.2b

Identification of the breakdown products of starch using a chromatographic method of analysis

Apparatus

Each pair of pupils will need:

Experiment sheet 85

Microscope slide

Test-tube, 125×16 mm, with side arm and fitted with bung and delivery tube as a 'cold-finger condenser'

Test-tube, 150×25 mm

2 lengths of rubber tubing

4 test-tubes, 100×16 mm

Test-tube rack

Test-tube holder

Glass rod

Stand, boss, and clamp

Measuring cylinder, 25 cm^3

(Continued)

Procedure
Details are given in *Experiment sheet* 85.

Experiment sheet 85
This is a long experiment, but an interesting one. You will have to work carefully if it is to be successful.

You will need to be able to test for the presence (or absence) of starch and certain kinds of sugar. For these you will use iodine solution and Fehling's solution (a rather complicated mixture containing copper compounds). Try these two reagents out on solutions of starch, glucose, and maltose, as follows.

1. Put separate drops of starch, glucose, and maltose solutions on a microscope slide (make it slightly greasy first by wiping it with your finger) and add one drop of iodine solution to each. What colour is:
the mixture containing starch?
the mixture containing glucose?
the mixture containing maltose?

2. Put five-drop portions of starch, glucose, and maltose solutions into separate test-tubes. Add five drops of each of the two Fehling's solutions to each and boil the mixtures over a small flame (use a test-tube holder). What happens:

Rectangle of Whatman No. 1 paper – to form a cylinder inside gas jar

Tripod and gauze

Bunsen burner and heat resistant mat

4 teat pipettes

2 paper clips

Thermometer, − 10 to + 110 × 1 °C

Gas jar and cover

Litmus or Full-range Indicator paper

In addition, access to 2 or more sets of apparatus for concentrating solutions under reduced pressure using filter pumps

1 (or more) plastic troughs (or shallow beakers) to contain solution of locating agent

2M hydrochloric acid, 2 cm³

2M solution hydroxide, 2 cm³

Fehling's solutions A and B (see Experiment B15.1), 2 cm³ of each

0.01M iodine in 0.1M potassium iodide solution, 2 cm³

10% glucose solution, 1 cm³

10% maltose solution, 1 cm³

1% starch solution freshly prepared, 20 cm³

Mixture of propan-2-ol, acetic (ethanoic) acid, and water in the ratio 3:1:1, 25 cm³

Locating agent: *either* aniline, acetone, diphenylamine, phosphoric acid; *or* p-anisidine hydrochloride, butan-1-ol; *or* m-phenylene-diamine, tin(II) chloride, ethanol; *or* saturated aqueous aniline oxalate solution

with starch?
with glucose?
with maltose?

You should now know how to distinguish between starch and the two sugars, glucose and maltose.

The action of dilute acid on starch
Have ready several separate drops of iodine solution on a microscope slide.

Put 10 cm³ starch solution into a 150 × 25 mm test-tube and add 10 drops 2M hydrochloric acid. Fit the test-tube with a 'cold-finger' condenser (see diagram) and heat the mixture to boiling over a small flame. Regulate the size of the flame so that the mixture is just kept boiling and no more. The condenser reduces water loss by evaporation. After 5 minutes boiling, remove the Bunsen burner, lift the condenser, and with a glass rod transfer one drop of the mixture from the boiling tube to a drop of iodine on the microscope slide.

Does the mixture still contain starch?

If so, continue the boiling for 2 minutes, stop the heating, and again test for starch. Continue doing this until no starch can be detected in the mixture. Then stop the heating and remove the condenser.

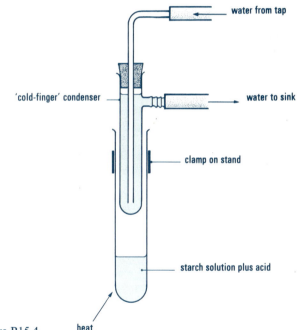

Figure B15.4

Using a teat pipette, transfer 5 drops of the final solution to a test-tube, add 5 drops 2M sodium hydroxide to make it alkaline, and then 5 drops of each of the two Fehling's solutions. Boil the mixture for one minute (test-tube holder). What do you see?

What has happened to the starch?

The unused solution has now to be neutralized with sodium hydroxide solution and evaporated to one-twentieth of its volume for further investigation. Your teacher will arrange for this to be done.

The action of saliva on starch
Have ready several separate drops of iodine solution on a microscope slide as

before. Half fill a 250 cm^3 beaker with water and warm it to 40 °C (no higher). Mix 10 cm^3 starch solution with about 1 cm^3 saliva in a 150 × 25 mm test-tube and place it in the warm water. After 5 minutes test for starch as before and continue this at 2-minute intervals until no more starch is present in the mixture. Boil 5 drops of the mixture with 5 drops of each of the two Fehling's solutions for one minute. What do you see? What has happened to the starch?

The unused remainder of this mixture must also be handed to your teacher to be concentrated (under reduced pressure) for further investigation.

You should now suspect that when starch is mixed with acid or saliva a sugar is produced. You will not know whether the sugar is glucose ($C_6H_{12}O_6$) or maltose ($C_{12}H_{22}O_{11}$). You can find this out by using paper chromatography. Your teacher will explain this in more detail. The following notes are for general guidance.

Use the rectangle of filter paper provided. Rule a pencil line 2 cm from the bottom of the paper and make four small vertical marks at 2 cm intervals along this line, two on either side of the middle point. Label the marks, *in pencil*, A, B, C, and D, underneath the line. At mark A put one small drop of glucose solution, from a narrow glass tube, and *allow it to dry*. Continue this until five small drops have been added, allowing each one to dry before the next is added.
At mark B, repeat this process with maltose solution.
At mark C, with the concentrated solution from the action of acid on starch.
At mark D, with the concentrated solution from the action of saliva on starch.

Roll the paper into a cylinder which will stand in a gas jar without touching the sides and fasten the edges with clips. Put a mixture of propan-2-ol, acetic acid (ethanoic acid), and water (3 : 1 : 1 by volume) in the gas jar to a depth of 1 cm and stand the cylinder of paper in it, with the pencil line and spots at the bottom. Put a cover on the gas jar and leave it for 6–12 hours.

Remove the paper cylinder, open it out, and hang it up to dry in a stream of warm air. While the cylinder was in the gas jar, the solvent mixture will have moved upwards through it, taking the substances in the spots with it. You have now to find the new positions of the spots. Your teacher will explain which method you will use for this. Describe it briefly on a separate sheet of paper. Also, draw a diagram to show the final appearance of the paper.

Complete the following statements.

starch $\xrightarrow{\text{acid}}$ ....

starch $\xrightarrow{\text{saliva}}$ ....

After the experiment, discuss and summarize the findings. The essential points are:

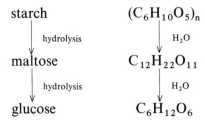

starch $(C_6H_{10}O_5)_n$
    $\downarrow$ hydrolysis $\downarrow$ H_2O
maltose $C_{12}H_{22}O_{11}$
    $\downarrow$ hydrolysis $\downarrow$ H_2O
glucose $C_6H_{12}O_6$

Some teachers may prefer to use molecular models as an aid to the summary.

Pupils should be aware of the process of photosynthesis in plants and it may be convenient to summarize the basic process at this point. Carbon dioxide and water are converted into sugars by plants with the help of sunlight and a catalyst. Grass stalks taste sweet when they are chewed. A sugary solution can be squeezed from a sugar cane plant. In addition, we know that these plants contain starch. It is possible that carbon dioxide and water are first built into glucose by the plant, and that this is subsequently synthesized into starch. Is it possible to use a plant to synthesize starch from glucose? We try to find the answer in the following experiment.

Experiment B15.2c

To attempt a synthesis of starch from glucose

Apparatus

Each pair of pupils will need:

Experiment sheet 107

Test-tube, 100×16 mm

2 Petri dishes or 100 cm³ beakers

Beaker, 250 cm³

Tripod and gauze

Crystallizing dish or Petri dish

Bunsen burner and heat resistant mat

Tongs or forceps

Glucose solution (5%), 20 cm³

Iodine solution, 20 cm³

Ethanol

The teacher will need:

A healthy geranium plant

A second geranium plant that has been kept in the dark for 4 days

Cork borer

Procedure
Details are given in *Experiment sheet* 107.

Experiment sheet 107
Your teacher will provide separate discs of geranium leaf from: *(i)* a plant which has grown normally, *(ii)* a plant which has been kept in the dark for 4 days. Remove a small notch from the leaf from plant *(i)* so that you can identify it. Treat both pieces as follows.

Dip each piece (use tongs) into boiling water in a 250 cm³ beaker for 10 seconds; this kills the leaf cells. Put the pieces into a test-tube. Turn out the burner under the beaker. Add about a 3 cm depth of ethanol to the test-tube containing the pieces of leaf and rest this in the beaker of hot water. The boiling ethanol removes the green colour from the leaves. When the pieces are white, remove the test-tube from the boiling water, pour off the ethanol, and wash the leaves in cold water to make them less brittle. Transfer the pieces of leaf to a shallow glass dish and just cover them with iodine solution. After 5 minutes pour off the iodine solution and wash the leaves with tap water. What do you observe and what does this show?

Take two more discs from the geranium plant which has been kept in the dark. Place one in 5 per cent glucose solution (in a shallow glass dish) and the other in water. Store both in the dark for 3 days. Test each for starch as described above. State the results and explain them.

Teachers should note that starch can be synthesized from glucose-1-phosphate using enzymes from potatoes. This is described in Nuffield Biology *Text III*, page 101, Revised *Text 2*, page 164.

After reviewing the experimental results the theme suggested as homework for this section could be introduced.

Suggestions for homework

The element carbon occurs naturally in many compounds, which are found in plants, animals, coal, and oil. It is also present in air and water as carbon dioxide. Draw a diagram to show how carbon can move from one naturally occurring compound to another. Name the different processes involved: such as burning, breathing, etc.

B15.3
Breaking down glucose

Glucose solution is fermented with yeast, and ethanol is separated as a product of the fermentation. The social uses and abuses of ethanol could be discussed.

A suggested approach

Objectives for pupils

1. Awareness of fermentation as a biochemical change
2. Knowledge of fractional distillation
3. Awareness of the social uses and abuses of ethanol

From section B15.2, pupils should know that starch, glucose, and maltose are examples of carbohydrates. They were shown that carbon and hydrogen in such compounds can be detected either by burning the compound in air or by heating it with copper(II) oxide to form carbon dioxide and water. Other less violent chemical processes enable us to convert carbohydrates into other chemicals. Fermentation is one such process.

In this section, pupils study the fermentation of glucose solution by yeast and isolate ethanol as one of the products of fermentation (Experiment B15.3a).

Experiment B15.3

Apparatus

Each pair of pupils will need:

Experiment sheet 86

Conical flask, 100 cm³, fitted with cork or bung and a delivery tube

2 test-tubes, 100 × 16 mm

Spatula

Stand, boss, and clamp

Glucose, 5 g

Limewater, 10 cm³

Bakers' yeast, fresh or granulated yeast or fresh brewers' yeast, 1 g

For the distillation, the teacher will need:

Round-bottom flask, 500 cm³

Fractionating column, 20 cm long, filled with pieces of glass rod or tubes

Liebig condenser

2 lengths rubber tubing

Thermometer, −10 to +110 × 1 °C

(Continued)

To break down the glucose molecule by fermentation

Procedure

Teachers should note that the fermentation requires between 3 and 7 days at about 21 °C. If the fermentation goes well, the mixture will froth and bubbles of carbon dioxide will escape through the limewater in the test-tube. The limewater will go milky and then become clear again. Details are given in *Experiment sheet* 86.

Experiment sheet 86

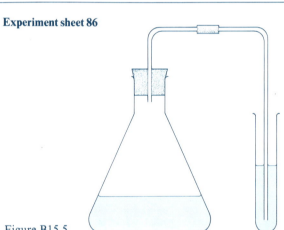

Figure B15.5

Put 5 glucose in the flask and dissolve it in 50 cm³ water. Mix two measures of baker's yeast with a little water to form a thin paste and add this to the glucose solution. Half fill the test-tube with limewater and stand the apparatus in a warm place (near a radiator in winter). Observe it at intervals for a few days. What do you see?

Name one product of the reaction.

To investigate the presence of any other product, the mixture remaining in the flask has to be distilled. This is best done on a larger scale by combining the mixtures from the whole class, so your teacher will take over here.

2 stands, bosses, and clamps

Anti-bumping granules

Bunsen burner and heat resistant mat

Tripod and gauze

Test-tube rack

Test-tubes, 100×16 mm

Corks to fit the flask and the fractionating column

Cork borers

Receiving flask or beaker

Draw a diagram of the apparatus used for the distillation on a separate sheet of paper.

Describe briefly what is done, giving the results of tests on the distillate.

Complete the statement:

glucose $\xrightarrow[\text{in yeast}]{\text{zymase}}$ .... + ...

Write an equation for the reaction.

Many pupils may be aware of fermentation as a process for producing alcohol from sugar. If not, then the widespread use of the process needs to be discussed in simple terms. During the fermentation experiment pupils should find that after three or four days the fermented product neither smells nor tastes like alcohol nor does it seem to be like glucose solution. Ask pupils how alcohol can be extracted from the mixture. How can we separate a mixture of liquids? It is essential for the teacher either to decant or filter the solution from the residual material before attempting to distil it.

Collect the fermented liquors prepared by the pupils and distil some 40 to 50 cm^3. The first few cm^3 of the distillate can be ignited and its properties compared with the laboratory supply of ethanol.

As the distillation proceeds, the working of the fractionating column can be discussed. It allows many successive distillations to be carried out using the same piece of apparatus. The hot vapours rising from the distillation flask redistil the liquid that has already condensed higher up the column. This condenses further up the column where it is re-distilled again by the hot rising vapours. (If pupils followed Alternative IA they will not have met a fractionating column before.) The variation of the temperature of the vapour at the top of the fractionating column with time shows that different substances are coming off and provides a useful way of identifying the distillate.

The discussion after this experiment may be enlivened by referring to the production of wine and spirits. The following suggestions may be appropriate for discussion, for projects, or for homework, as time permits:

1. The production of a map or chart showing the sources of starch and sugar that are fermented and the names of the fermented liquors and spirits produced from them in various parts of the world. (For example, in the United Kingdom we use barley to produce beer and whisky; in France they use grapes to produce wine and brandy, and so on.)
2. Information relating to differences between beers, wines, and spirits, and their methods of manufacture.
3. Illustrations showing the different devices used for distilling spirits.
4. The physiological and addictive effects of alcohol and the social problems that arise from them.
5. A diagram showing the operations carried out in a brewery,

followed by a class visit to a brewery.

6. The methods used to detect alcohol in the blood and breath (for example the breathalyser, etc.). The construction of a model breathalyser by drawing air through a bottle of aqueous ethanol and passing it over silica gel crystals that have been soaked in acidified potassium dichromate solution.

Suggestion for homework Write an account of the work done in this section.

B15.4
Breaking down fats and oils

This sequence of lessons includes the preparation of a 'soap' and a 'soapless detergent'. The properties of soaps and soapless detergents are compared and the significance of using biodegradable detergents is mentioned.

A suggested approach

Objectives for pupils

1. Knowledge of the preparations of soaps and soapless detergents
2. Knowledge of the way in which detergents work

Remind pupils that a fat is a foodstuff. Can fat molecules be broken down by hydrolysis during digestion just like starch molecules? Tell pupils that fats are not broken down by saliva but by enzymes inside the body and so it is not quite so convenient to carry out this hydrolysis biochemically! What other ways are open to us to try?

Starch was broken down chemically by acid. Pupils now try using acids on a fat (castor oil) in Experiment B15.4. They also try the effect of an alkali.

Remind pupils of the properties of oils and fats. Butter melts on a hot day and becomes oily and so does cooking fat when heated in a frying pan. On cooling, these oils readily solidify. Indeed the chemical similarity between fats and oils may be emphasized by using molecular models as indicated by figure B15.6 to represent the principal compounds present in specific fats and oils.

Figure B15.6
Models of molecular structures: fats and oils.
(*a*) Tripalmitin, tripalmitoyl glycerol: a 'typical' fully saturated fat (especially animal fat).
(*b*) Triolein, trioleoyl glycerol: a 'typical' unsaturated fat/oil (especially in olive oil).
(*c*) 2-oleodipalmitin, 2-oleoyl-1, 3-dipalmitoyl glycerol: a 'typical' fat from palm oil and other vegetable oils.
Note: these are available in a series of slides and are intended only to illustrate the arrangements of atoms in these molecular structures.
Photographs, Unilever Research Laboratories, Colworth House, Sharnbrook, Bedford.

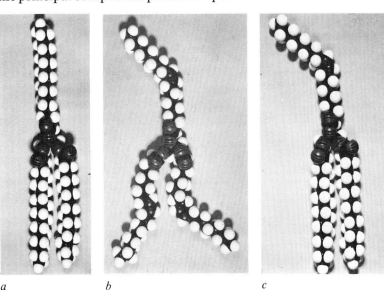

a   *b*   *c*

Emphasize the length of the carbon chain in each instance. Other similarities are apparent from the models. Attention may also be drawn to the fact that the oleate chain in olive oil is unsaturated but this need not be stressed.

In the experiment which follows, pupils will use castor oil whose main fatty acid constituent is ricinoleic acid:

$$CH_3(CH_2)_5CHOHCH_2CH:CH(CH_2)_7CO_2H.$$

The point can now be made that the similarity between most of these chemicals arises from their similarity in structure – that is, the way in which the atoms are linked in their formulae.

Thus, we may write the general formula of a fat as:

$$H_2CO.CO.R_1$$
$$|$$
$$HCO.CO.R_2$$
$$|$$
$$H_2CO.CO.R_3$$

where R_1, R_2, and R_3 represent the same or different groupings of atoms. A sodium soap can be written as $R.CO.ONa$. The reactions of castor oil (ricinoleic acid) with sodium hydroxide and with concentrated sulphuric acid are atypical: sodium hydroxide yields octan-2-ol and sodium sebacate (disodium decanedioate); and concentrated sulphuric acid yields a complex mixture of the hydrogen sulphate of ricinoleic acid in which the hydroxyl group is esterified and a compound in which the sulphuric acid has added to the double bond. Esterification and addition do not occur together in the same molecule of ricinoleic acid. Clearly this information will *not* be required by pupils and is only mentioned so that no over-simplification will be made.

Experiment 15.4

The reaction of castor oil with (1) sodium hydroxide; (2) sulphuric acid

Apparatus

Each pair of pupils will need:

Experiment sheet 89

Safety spectacles

4 test-tubes, 100×16 mm

Test-tube, 150×25 mm

Test-tube holder

Teat pipette

Measuring cylinder, $25\,cm^3$

2 beakers, $100\,cm^3$

Evaporating dish

Tripod and gauze

(Continued)

Procedure
Details are given in *Experiment sheet* 89.

Experiment sheet 89
You should wear safety spectacles for this experiment.
1. *Reaction with sodium hydroxide*
Put $2\,cm^3$ castor oil in an evaporating basin and add $10\,cm^3$ 5M sodium hydroxide solution (this solution is **very corrosive**, so be careful not to get any of it on your skin or on the bench). Place the basin on top of the beaker in which water is boiling and heat it in the steam for 10 minutes, stirring well.

Add to the mixture in the basin $40\,cm^3$ water in which 10 g sodium chloride has been dissolved, stir and heat for another 5 minutes. Allow the mixture to cool after removing the dish from the steam. A solid will separate on the top of the liquid; skim it off with a spatula and wash it with a little cold water in a small beaker. Pour off the water. Transfer a small quantity of the solid to a 100×16 mm test-tube, half fill the test-tube with water, and shake well. What happens?

What do you suppose the solid is?

Bunsen burner and heat resistant mat

Glass rod

Filter flask, funnel, filter papers, and filter pump

Spatula

Castor oil, 3 cm^3

5M sodium hydroxide, 10 cm^3

Common salt, 10 g

Concentrated sulphuric acid, 2 cm^3

2. *Reaction with sulphuric acid*

Put 1 cm^3 castor oil in a 100×16 mm test-tube and fasten this in a test-tube holder. Hold the test-tube over a beaker and add carefully 2 cm^3 concentrated sulphuric acid, keeping your face well away from the mouth of the tube. Stir the mixture with a glass rod. What happens?

When reaction appears to be finished pour the contents of the tube into a 3 cm depth of water in a 150×25 mm test-tube. Stir the mixture and carefully pour off the water into the sink. Wash the solid twice more with about the same amount of water, pouring off the liquid each time.

Shake a small quantity of the solid with half a test-tube of water. What happens?

What do you suppose the solid is?

Explain to the pupils that anything which produces a foam when it is shaken with water *and* has the various properties which make it useful in washing is known as a *detergent*. The only detergent known for many centuries was one form or other of *soap*. In recent years, a number of other compounds have become widely used and are known collectively as *soapless detergents* (or, in everyday speech, detergents.)

After the experiments, discuss the nature of the products obtained in each instance. Pupils have found that castor oil may be changed by the action of either alkali or acid. In the case of alkali, the solid produced is a soap and in the case of the acid the material produced is a soapless detergent, a sulphonated oil (Turkey red oil). Both products are atypical and less effective than industrially produced materials such as dodecyl benzene sulphuric acid (a detergent).

The Unilever Educational Booklet (Revised Ordinary Series No. 1) *Detergents* is a useful source of background reading on the history and chemistry of detergents.

Do soaps and soapless detergents serve the same purpose? What differences in properties exist for these two groups of materials? Everyday experience of using soaps and soapless detergents at home – particularly if the water is hard – can be reinforced by simple demonstrations. The reaction of calcium salts, for example, with soap to form insoluble calcium soaps can be summarized in a word equation. The result of such a demonstration should be retained and compared with that from a corresponding study using a soapless detergent in place of soap.

It is also convenient to show pupils the similarities and differences in the way that soaps and soapless detergents work. Run a layer of castor oil onto the surface of two beakers partly filled with water. Add detergent to one of the two beakers containing the oil *and* to a third beaker which contains water only. Shake up the contents of each of the three beakers. The detergent will be seen to break up the oil into very small droplets to form a cloudy solution. The contents of the other two beakers remain unchanged. Repeat the experiment using a soap solution in place of the soapless detergent solution.

Now tell pupils that chemists have found that the 'tails' of detergent molecules dissolve in the oil and the 'heads' dissolve in water. On shaking, the greasy or oily layer is broken up into droplets. Each droplet of oil or grease becomes surrounded by the tails of detergent molecules as shown in figure B15.7. Each droplet of oil then has a skin of detergent molecules around it which enables it to disperse in the water. In this way, water is able to remove grease and dirt out of clothes. Literally, the dirt 'falls' off!

The 'head' and 'tail' structure of a typical detergent molecule: the head is hydrophilic ('water-loving') and the tail is hydrophobic ('water-hating').

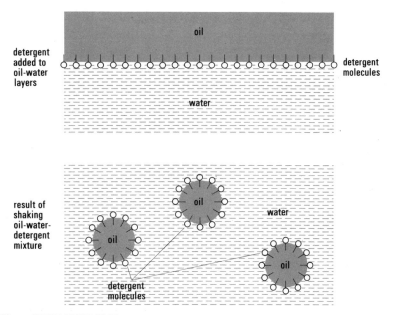

Figure B15.7 (OHP 38A)

All detergents (both soap and soapless detergents) also assist the cleaning process by allowing water to wet things better. Water does not wet things very well – as can be seen by allowing droplets of water to fall onto a hand, a tile, or even a cotton cloth. The addition of a drop of detergent to this water enables it to spread. It would seem that the 'skin' of the water droplets has been removed or broken by the detergent. The pressure of a skin on the surface of water can be demonstrated by floating a needle in a beaker of water. The addition of a drop of detergent (care needed!) to the surface of the water causes the needle to sink.

The properties of soap and soapless detergents could be tabulated, listing the advantages and disadvantages. Point out the effect of detergents on sewage. It is important that detergents should be broken down by bacteria if foams are not to form in streams and rivers. 'Hard' soapless detergents (those in which the carbon chains are branched) resist bacteriological action whereas 'soft' soapless detergents (those in which the carbon atoms are arranged in 'linear' or 'straight' chains) are readily broken down when they pass through the sewage plant. In the past, large masses of foam from hard soapless detergents were produced at sewage works and these tended to pass into rivers. Not only do these foams look horrible, they also reduce the efficiency of filter beds in the sewage works. Less is known about the biological effect of such foams although it is known that fish, such as trout, are killed if as little as one part detergent per million parts of water is present in a river.

Suggestion for homework

Write an account of the cleaning action of detergents. (See *Handbook for pupils*, Chapter 12 'Man, chemistry, and society'.)

B15.5
What compounds are there in crude oil?

The fractional distillation of crude oil is revised and extended.

A suggested approach

Objectives for pupils

1. Knowledge of the fractionation of crude oil
2. Knowledge of the properties of carbon compounds
3. Awareness that the physical properties of compounds are related to the size of their molecules

Pupils are reminded of the experiment in which they fractionally distilled crude oil in Stage I (A1.5 or B3.2), and if there is time the experiment could be demonstrated. Table B15.1 lists the results of the industrial fractionation of crude oil.

Name of fraction	Typical substance present in fraction	Boiling point °C	State
natural gas	methane, CH_4	−160	gas
bottled gas	butane, C_4H_{10}	0	gas
gasoline (petrol)	octane, C_8H_{18}	99	liquid
kerosine (paraffin)	decane, $C_{10}H_{22}$	174	liquid
diesel oil (fuel oil)	tetradecane, $C_{14}H_{30}$	250	liquid
lubricants (oil)	hexadecane, $C_{16}H_{34}$	300	liquid
		Melting point	
petroleum jelly (Vaseline)	octadecane, $C_{18}H_{38}$	28	solid
paraffin wax (wax)	pentacosane, $C_{25}H_{52}$	54	solid
bitumen (tar)	pentacontane, $C_{50}H_{102}$ and larger molecules + free carbon	92 (and higher)	solid

Table B15.1 Fractionation of crude oil [OHP 38B, in which the last column is left blank]

Show pupils the overhead projection original of Table B15.1. Ask them to deduce which compounds are solid, which are liquid, and which are gaseous. The fourth column can then be completed by the teacher. Ask about any trends in the table. They should report that the larger the molecule, the higher the boiling point of the fraction. In addition, they may appreciate the C_nH_{2n+2} relationship between the various formulae. They can also be asked why compounds with larger molecules have higher boiling points. Pupils can appreciate that 'more effort' will be required to separate such molecules than to separate smaller molecules. Similarly the point can be made that because larger molecules tend to 'tangle together more', these liquids are more viscous. The use of higher temperatures also leads to the breakdown of long-chain compounds and the formation of free carbon (charring). Bitumen contains a lot of free carbon. Also, hot lubricating oil darkens due to the formation of free carbon.

To introduce the idea of a homologous series, pupils could be asked to write structural formulae for, say, all the alkanes containing between one and ten carbon atoms. This exercise introduces the idea that successive members of the homologous series differ by a CH_2 group. Pupils could also make models of some of these alkanes. The point needs to be made that carbon atoms can join together to form chains consisting of very many carbon atoms and, indeed, can form many thousands of compounds. There are more compounds of carbon than compounds of all the other elements put together! Fortunately, the carbon compounds in a given homologous series possess similar (*but not identical*) properties.

The work of this section can be divided between class work and homework as appropriate. After a preliminary discussion, pupils could be given a table of petroleum fractions and then asked a series of questions about them. Alternatively, the entire section could be dealt with as a class lesson.

B15.6
Breaking down big molecules in crude oil

One of the products of oil fractionation, medicinal paraffin, is broken down into ethene (ethylene). This reaction is used to demonstrate an important industrial process.

A suggested approach

Objectives for pupils

1. Awareness of the effect of consumer demand on industrial production
2. Knowledge of the cracking of larger molecules to form smaller ones
3. Knowledge of the properties of the carbon–carbon double bond
4. Knowledge of the uses of ethene as a 'chemical building block'

The fractionation of crude oil yields a range of products which do not necessarily match consumer demand. The need to break large molecules into smaller ones may be introduced to solve the problem of over-production of the former and under-production of the latter. Pupils could be asked to study Table B15.2 and decide which products are over-produced and which are under-produced.

Product	Gases	Petrol	Naphtha	Kerosine	Gas oil	Fuel oil (includes wax)
No. of carbon atoms in molecule	1–5	5–10	8–12	11–16	16–24	25 up
Initial distillation of a crude oil expressed as a percentage	2	8	10	14	21	45 (from North Sea)
	2	5	9	12	17	55 (from Persian Gulf)
Product demand pattern expressed as a percentage	4	22	5	8	23	38

Table B15.2 Initial distillation of a crude oil and a possible demand pattern [OHP 39]
(*Note.* The percentages of the various constituents vary considerably in different crude oils; and market demand varies from place to place and season to season. As far as possible, therefore, a refinery will take in a crude oil or a mixture of crude oils to match the market demand, and the necessity for 'cracking' will be reduced.)

Pupils can be led to appreciate that more naphtha and wax are obtained during the initial fractionation than can be sold but that there is not enough gas, petrol, and gas oil to meet demands. However, if the large molecules in the naphtha or wax fractions could be broken into smaller molecules, then requirements could be met. How can this be achieved? Pupils may suggest the effect of heat. If we refer to molecular models, the effect of heat is tantamount to shaking a model about. If this treatment is sufficiently violent, the molecules should break up. In fact, chemists have solved the problem by a process called 'cracking' in which large molecules are heated and passed over a catalyst and are broken up into smaller ones. Related to this is the process of dehydrogenation, in which the saturated hydrocarbon molecules lose hydrogen and become 'unsaturated'. These points will require careful explanation, and it may help to use molecular models. Unsaturated hydrocarbons are reactive and extremely useful. Ethene (ethylene) is especially useful and is one of the most important 'building blocks' in the whole of the chemical industry. In the presence of a catalyst, it will react with water to form ethanol; and under other conditions it will polymerize to make polythene, a substance which should be familiar to everyone.

Pupils are now asked to 'crack' medicinal paraffin and to examine the properties of the reaction products.

Cracking petroleum products

Experiment B15.6

Apparatus

Each pair of pupils will need:

Experiment sheet 88

4 test-tubes, 150 × 25 mm

Hard-glass test-tube, 125 × 16 mm, fitted with a cork, delivery tube, and Bunsen valve (a piece of rubber tubing containing a slit and fitted with a piece of glass rod – see figure B15.8)

Bunsen burner and heat resistant mat

Mineral wool

Small pieces of porous pot, to fill a 125 × 16 mm test-tube

Stand, boss, and clamp

Trough

Teat pipette

Medicinal paraffin, 2 cm³

Dilute bromine water, 2 cm³

Procedure
A Bunsen valve is used to avoid the difficulties of sucking back and it will need to be explained. *Experiment sheet* 88 is reproduced below.

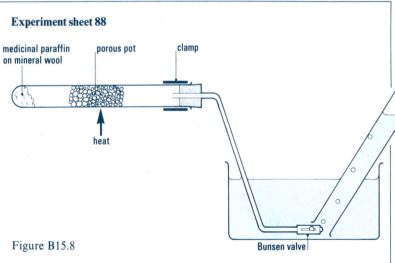

Experiment sheet 88

medicinal paraffin on mineral wool   porous pot   clamp

heat

Figure B15.8

Bunsen valve

Arrange the apparatus as shown in the diagram.
Pack the end of a 150 × 16 mm hard-glass test-tube with mineral wool to a depth of 2 cm, and saturate it with medicinal paraffin. Nearly fill the rest of the tube with pieces of porous pot. Clamp the test-tube in position and fit the delivery tube – as shown above.

Heat the porous pot strongly but do not heat the mineral wool until the tube is quite hot.

Allow for air to be expelled from the apparatus before collecting several test-tubes of gas.

1. Note the smell of the product.

2. Try to burn the gas.

3. Examine the reaction between the gas and bromine solution.

What do you think the gas is?

After the experiment, discuss the results. It is helpful to make a ball-and-spring model of the ethene molecule and to compare it with similar models of ethane and ethanol.

The ethene molecule is small and reactive and can be used as a building block to make other chemicals. Pupils could work out what is formed with bromine in the reaction they have carried out. Discuss their ideas using molecular models and introduce the term 'addition reaction'. They might then try to explain the reaction of ethene with water and deduce the formation of ethanol. It is helpful to build up a summary, either on an overhead projector or on a blackboard, of the various reactions of ethene (see figure B15.9).

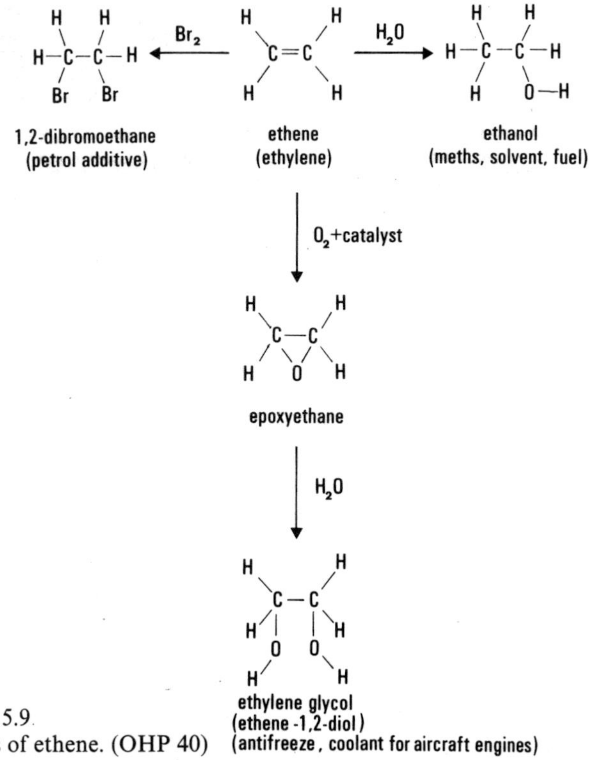

Figure B15.9
Reactions of ethene. (OHP 40)

This section can be concluded by showing the film loop 2–9 'Cracking hydrocarbons'. In 1970, the annual production of ethene reached one million tonnes in the U.K. (See BP Educational Services Booklet No. 3 *Ethylene – the building block*.) About three-quarters of the annual production was used for making plastics, which are studied in the next section.

Read *Chemists in the world*, Chapter 6 'Polymers from oil'.

B15.7
Making and using plastics

Pupils make some plastics and examine their properties. They become aware that the properties of the materials are related to their molecular structures.

A suggested approach

Objectives for pupils

1. Knowledge of the formation of polymers by (*a*) addition and (*b*) condensation processes
2. Knowledge that the properties of a polymer are related to its molecular structure
3. Knowledge of the preparation of some commonly used polymers
4. Knowledge of the distinguishing properties of thermoplastic and thermosetting polymers
5. Awareness of the need to relate the properties of a material to possible uses
6. Awareness of the use of plastic materials in everyday life

Begin the lesson by reminding the pupils of the structure of ethene (ethylene). Use a number of ball-and-spring models of ethene to make a model of polythene. Making polythene in this way is an example of *polymerization*. Ask the pupils how this might be achieved in practice. By arranging to push the ethene molecules together? By crowding the molecules together and then heating them? Polythene was first made in this way using a pressure two thousand times atmospheric and at a temperature of 170 °C. (This polymerization is catalysed by the presence of traces of oxygen but this need not be mentioned here.)

If the pupils have not yet deduced that polyethylene and polythene are the same substances, ask what they would expect the polymer to look like. They may expect it to be liquid, like medicinal paraffin, or to be a solid. We cannot produce the enormous pressures required to make the polymer in the school laboratory so show a sample to the class. It is a solid and pupils should deduce that probably it has bigger molecules than medicinal paraffin. It looks like a wax. In the next experiment, polythene is compared with a wax, paraffin wax.

Experiment B15.7a

Apparatus

Each pair of pupils will need:

Experiment sheet 108

Glass rod

Steel square, 3 inch side, 18 s.w.g. or thicker, *or* circular disc of comparable dimensions, fitted with 3 copper supporting wires

Tripod or stand, boss, and clamp

Bunsen burner and heat resistant mat

Asbestos paper or equivalent material

(*Continued*)

To compare the properties of polythene with those of wax

Procedure
The laboratory needs to be well-ventilated if combustion experiments are carried out by pupils. The experiment is in three parts: (1) melting behaviour; (2) burning behaviour; and (3) hardness and mechanical strength. Details are given in *Experiment sheet* 108.

Experiment sheet 108
In this experiment you will test two different kinds of polythene (the low and high density forms) and paraffin wax in order to compare their melting behaviour and ease of fibre formation, their burning behaviour, and their hardness and mechanical strength.

1. *Melting behaviour*. Place a metal plate about 5 cm above an unlit Bunsen burner. (Your teacher will give some help here.) On the plate put small samples of paraffin wax and the two types of polythene at roughly equal distances from the centre of the plate. Remove the burner, light it, make the flame just non-luminous, and adjust its height so that the tip will just touch the underside of the metal plate. Replace the burner so that the tip of the flame touches the centre of the plate. Note the time taken for each sample to melt.

Crucible tongs

Access to a hammer

Low density polythene (an offcut or granules)

High density polythene (an offcut or granules)

Paraffin wax (small lumps)

Paraffin wax melts in	................................seconds
Low density polythene melts in	................................seconds
High density polythene melts in	................................seconds

While the samples are melted use a glass rod to try to draw each out into a thread. Report on the ease with which they form fibres.

2. *Burning behaviour.* Heat small samples of the three substances in turn on folded asbestos paper, held in tongs, in a Bunsen burner flame. Report what happens in each case, including details of flame colour, smell, and charring.

Paraffin wax
Low density polythene
High density polythene

3. *Hardness and mechanical strength.* Compare hardness by scraping samples of the three materials with your finger nail, and mechanical strength by tapping them with a hammer (or a clamp). If you can obtain strips of the materials, compare the ease with which they can be broken by bending. Write a report of your results.

After the experiment, discuss the results. Point out that chemists can control the chain length of the polymers they build and so 'tailor' their properties. Another way to alter the properties of polymers is to 'dress up' the carbon chain in different ways. This can be done by using ethene molecules in which the hydrogen atoms are replaced by other atoms or groups of atoms. Figure B15.10 summarizes possibilities.

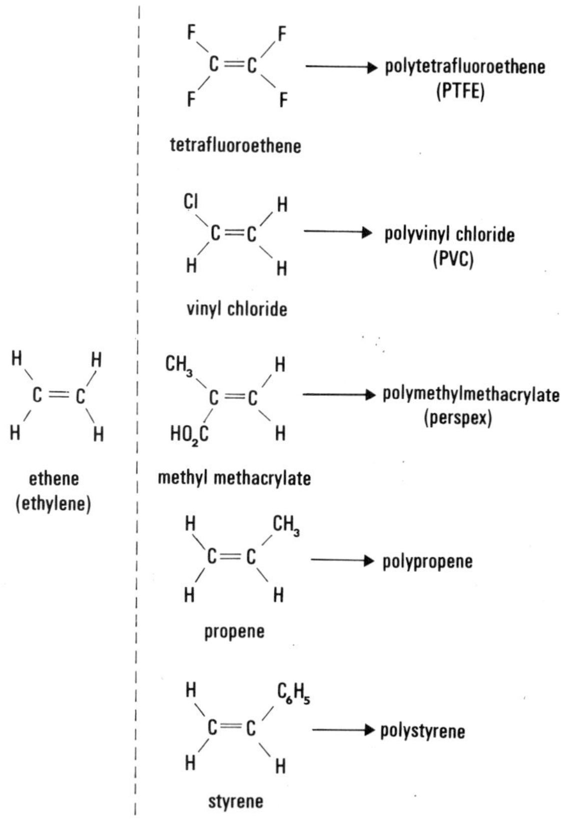

Figure B15.10
Using substituted ethenes in making polymers. (OHP 41)

Pupils can now try to make one or more of these polymers.

To make a sample of perspex

Apparatus

The teacher or each pair of pupils will need:

Test-tube, 100×16 mm, or a shaped metal mould

Methyl methacrylate monomer (with inhibitor removed), 5 cm^3

Lauroyl peroxide (enough to cover the end of a small spatula), 20 mg

Silicon grease or Vaseline

Note to teachers

To remove the inhibitor from methyl methacrylate monomer, extract it three times with one-fifth of its own volume of 40 per cent sodium hydroxide solution. Then wash it three times with half of its own volume of water. Test for the absence of free alkali in the product.

Procedure

5 cm^3 of purified methyl methacrylate monomer is mixed with 20 mg of lauroyl peroxide in a greased test-tube or shaped metal mould. The mixture is placed in an oven at 60 °C for about one hour. A solid piece of perspex is obtained and can be removed from the mould.

Alternatively, some small object such as a leaf or shell could be placed on the monomer while it is still setting. After solidification, a further 5 cm^3 of monomer is poured onto the mould taking care to avoid air bubble formation and to encase the object. The whole is then returned to the oven, as before. (Pupils may enjoy doing this since it enables them to use the product.)

Nylon is a quite different type of material and is not related to ethene. After a brief comment about its uses, ask pupils to prepare a sample.

To prepare some nylon

Apparatus

Each pair of pupils will need:

Experiment sheet 90

Crucible or a beaker, 5 cm^3

Forceps

Glass rod

Hexamethylenediamine (hexane-1,6-diamine) 5 per cent solution in water, 2 cm^3

Adipyl chloride (hexanedioyl dichloride), 5 per cent solution in carbon tetrachloride (tetrachloromethane), 2 cm^3

Procedure

Details are given in *Experiment sheet* 90.

Experiment sheet 90

You will be provided with a solution of adipyl chloride in carbon tetrachloride (tetrachloromethane) (solution A) and a solution of hexamethylenediamine (hexane-1,6-diamine) in water (solution B).

Put about 2 cm^3 solution A into a crucible or a very small beaker and carefully add 2 cm^3 of solution B *so that the two solutions do not mix*. A film will form where the two liquids meet; pull out a thread from this with a pair of tweezers and attach it to a glass rod. This is a thread of one kind of nylon. Rotate the glass rod above the liquid mixture so that a continuous thread is wound on to the rod. When you have collected about six feet of thread break it off with the forceps and wash it in a gentle stream of water. You can now handle it safely and test its strength. The process can be repeated until one of the solutions is used up.

After the experiment, compare the various samples of nylon that have been prepared. Ask the pupils to comment on the possible structure of nylon. Since it forms long fibres, it should be made up of long molecules. With an able group of pupils, it is possible to give them the structural formulae of the reactants and to ask how nylon might be formed. Demonstrate that hydrogen chloride is evolved when the reaction takes place, thereby assisting the discussion.

Note to teachers:
1. It should be appreciated that not all substances with long molecules are suitable for making fibres.
2. Polymers with strong inter-chain forces and symmetrical repeating units can be stretched or drawn to give fibres with a high modulus of elasticity.
3. There may not be enough time for pupils to prepare other types of large molecules or to explore the properties of polymers further at this stage. *Either* a 'circus' (rather than a series of class experiments) can follow with a full discussion of the work done, *or* the remainder of section B15.7 can be deferred until Option 5 'Plastics' in Stage III is attempted.

Optional

Experiment B15.7d

To make a sample of urea-formaldehyde resin

Apparatus

Each pair of pupils will need:

Experiment sheet 109

2 test-tubes, 100×16 mm

Glass rod

Filter paper

Formaldehyde solution, 5 cm^3

Urea, 3 g

Access to concentrated sulphuric acid (in a T.K. dropping bottle or in a bottle fitted with a teat pipette)

Procedure
Caution pupils about the use of concentrated sulphuric acid. This experiment can be done in a 'throwaway' container such as a small cream pot, covered by a small beaker to cut down the escape of fumes. Details are given in *Experiment sheet* 109.

> **Experiment sheet 109**
> Put 5 cm^3 formaldehyde solution into a test-tube. (*CAUTION: Formaldehyde gas is irritating to the nose and eyes, keep your face well away from the solution.*) Add powdered urea to the solution in small quantities, shaking between each addition, until no more will dissolve. Pour off the clear solution into another test-tube and add to it 2 drops of concentrated sulphuric acid (*CARE*). Stir the mixture until a solid polymer forms. Wash the solid with water, scrape it on to filter paper, and allow it to dry. Keep the resin you have made for use later.

If the pupils are given the formulae of the reactants, they may be able to deduce that water is the molecule eliminated during this condensation reaction.

urea   formaldehyde   water   'resin'

The mechanism of the formation of linear polymers and of the subsequent cross-linking is not fully understood. Teachers are therefore advised *not* to press for fine detail. Continued reaction results in cross-linking by free formaldehyde:

$$\left[\begin{array}{c} O\ H\ H \\ \parallel\ |\ | \\ \dots-N-C-N-C-\dots \\ | \qquad\quad | \\ H \qquad\quad H \end{array}\right]_n \qquad \left[\begin{array}{c} O\ H\ H \\ \parallel\ |\ | \\ \dots-N-C-N-C-\dots \\ | \qquad\quad\ | \\ \quad\qquad\quad H \end{array}\right]_n$$

$$O=CH_2 \qquad\qquad\qquad \rightarrow H_2O+ \qquad CH_2$$

$$\left[\begin{array}{c} H \qquad\quad H \\ | \qquad\quad | \\ \dots-N-C-N-C-\dots \\ \parallel\ |\ | \\ O\ H\ H \end{array}\right]_n \qquad \left[\begin{array}{c} \qquad\qquad H \\ | \qquad\quad\ | \\ \dots-N-C-N-C-\dots \\ \parallel\ |\ | \\ O\ H\ H \end{array}\right]_n$$

Using their experience of the properties of polymers, pupils should be able to anticipate that a substance with this type of structure will not form fibres easily. They might melt some of their product and test this prediction using a glass rod. The terms 'thermosetting' and 'thermoplastic' should be introduced (more detailed information can be given later, after Experiment B15.7e). Now investigate and compare the burning and melting behaviour of the samples of plastics that have been made, together with examples of other sorts of plastic material.

Optional

Experiment B15.7e

To compare the melting and burning behaviour of some common polymers

Apparatus

Each pair of pupils will need:

Experiment sheet 110

Circular steel disc, 10 cm diameter and fitted with three copper support wires

Stand, boss, and clamp

Bunsen burner and heat resistant mat

Asbestos paper or equivalent material

Crucible tongs

Samples of: polythene, PVC, polystyrene, nylon, urea-formaldehyde resin, perspex

Note to teachers: pupils should have their own samples of *at least* one of these materials.

Procedure
Caution the pupils about possible fire risks and dangerous fumes. Experimental details are given in *Experiment sheet* 110. A well-ventilated laboratory is essential if pupils are to attempt this investigation on a class basis. (*DO NOT use PTFE or acrylics* since toxic fumes are given off.)

Experiment sheet 110
This experiment should be carried out in a well-ventilated laboratory.
1. Using the method described in Experiment 108 (heating samples of plastics on a metal plate at roughly equal distance from the source of heat), compare the time taken for small amounts of the polymers supplied to melt. Also make any other observations that you think useful. Enter the results in the table below.

Polymer used	Time taken to melt /sec	Does it char?	Other observations

Topic B15 Everyday materials: large molecules and metals

Which of the polymers that you used are:
thermoplastic?
thermosetting?

2. Heat small amounts of each plastic in turn on folded asbestos paper, held in tongs, in a Bunsen flame. Look for colour of flame when the polymer burns, colour of smoke (if any), smell of vapours when burning *(take extra care here, some of the vapours are irritating or poisonous)*, and ease of burning ('good', 'average', or 'poor'). Enter the results in the table.

Polymer used	Flame colour	Smell	Colour of smoke	Ease of burning

The observations that you have made should enable you to attempt the identification of unknown samples of plastics. Your teacher will supply samples for this purpose. Remember to use *small* quantities only for each test, to be alert to fire risks, and to exercise caution when smelling the products of burning.

Tables B15.3 and B15.4 give sample results for these investigations. They may be used as a basis for the identification of plastics. It is suggested that once their results have been checked, pupils might use their results to identify the composition of some manufactured article, such as an old toy, the barrel of a pen, and so on.

Polymer used	Does it melt?	Does it char?	Other observations
polythene	yes	no	bubbles at the edges
PVC	yes	chars at the bottom	smokes – forms balls
polystyrene	yes	no	bubbles round the edge
nylon	yes	yes	
urea-formaldehyde resin	no	yes	fishy smell
perspex	yes	no	bubbles

Table B15.3 Effect of heat on polymers

Polymer used	Colour of flame	Smell	Colour of smoke	Ease of burning
polythene	blue	burning candle	none	average
PVC	yellow or green	pungent! CAUTION – hydrogen chloride	white	poor
polystyrene	orange/yellow	flowers (very faint)	black	good
nylon	blue (with yellow tip)	burning wool	none	poor
urea-formaldehyde resin	yellow with bluish green edge	fresh fish	none	poor
perspex	yellow	fruity	black	good

Table 15.4 Burning polymers

After discussing the experimental results, ask the pupils which materials are thermoplastic and which thermo-setting polymers.

Thermoplastic polymers are 'linear' polymers, soluble in many organic solvents: the materials soften on heating and become rigid on cooling. The process of heat softening, moulding, and cooling can be repeated and hardly affects the properties of the material. Typical thermoplastics include cellulose acetate, nitro-cellulose, polythene, and perspex.

Thermo-setting polymers are polymers whose molecules have bonds in each of three dimensions and are insoluble in any kind of solvent. They can be heat treated only once before they set – i.e. when they are formed. Further heating results in decomposition. It follows that these polymers cannot be reworked. Typical thermo-setting plastics include phenol-formaldehyde, urea-formaldehyde, and silicones.

Discuss the properties that plastics possess and the requirements of construction materials. Traditional materials include wood, metals, stone, glass, and ceramics. Plastics have been added to the list and like metals, ceramics, and glasses, they are the result of the efforts of the chemist. What sort of questions should we ask when choosing a material for a given purpose, say, bridge building, aircraft construction, road building, food storage, television tubes, etc.? *Experiment sheet* 111 lists a number of properties which could be investigated.

Ask pupils how they would answer their questions experimentally. If the teacher lists the questions on the blackboard or an overhead projector transparency it is sometimes helpful to restate the questions formally by referring to such headings as: corrosion resistance; impact strength; flexibility; tensile strength; hardness; density.

Experiment B15.7f

To compare the properties of some plastics with those of some metals and wood

Apparatus

Each pair of pupils will need:

Experiment sheet 111

Emery paper

Beaker, 100 cm³

Hammer

Retort stand

Metal-bending apparatus (see Experiment B15.8b)

Wire-stretching apparatus (see Experiment B15.8a)

Procedure

It is suggested that the experiment be divided among the members of the class. Details are given in *Experiment sheet* 111.

Experiment sheet 111

This is a class co-operative experiment in which different members of the class carry out different tests with some of the materials available, and the results obtained are exchanged between the working groups. Your teacher will tell you what you have to do.

For each material the questions to be answered are:

a. *Does it corrode?* Leave small samples of each material under water in separate test-tubes and observe over a period of two weeks.

b. *Is it attacked by chemicals?* Leave small samples of each material in contact with chemicals such as potassium permanganate solution; concentrated sulphuric acid *(CARE: highly corrosive)*; concentrated nitric acid *(CARE: highly corrosive)*; acetone (propanone) (cork the test-tube). Observe after at least one week.

(Continued)

Access to the following chemicals:

Concentrated sulphuric acid

Concentrated nitric acid

0.1M potassium permanganate solution

Propanone

Samples of the following materials: iron, copper, wood, polythene (low density), polythene (high density), urea-formaldehyde resin

c. *How hard can it be hit without breaking?* Place a small thin sheet of material against the base of a retort stand on the floor.

Figure B15.11

Hit it with a hammer (or a clamp), gradually increasing the force of the blow. Obviously the sheets used should be of about the same thickness. Use your judgement to compare the strength of the materials which you are given.

d. *How often can you bend it without breaking it?* There is a special piece of apparatus for this. Your teacher will explain how it is used.

e. *What mass can it support?* Again a special piece of apparatus will be supplied by your teacher.

f. *How hard can you pull it before it snaps?* Use the same apparatus as in (e) and measure the greatest increase in length for each sample before it breaks.

g. *How easily is it scratched?* Rub squares of different materials on different spots on a sheet of emery paper for 30 seconds. Compare the amounts of materials removed.

h. *How dense is it?* Compare the densities of samples of different materials, of the same volume, by 'weighing' in your hand or by seeing whether they float or sink in water. Comparisons only are required, for example lead is denser than wood.

When all the results have been collected from the class, record them in a table and attach it to this Experiment sheet. Below the table, write a few lines on the general conclusions you can draw about the different properties of metals, wood, and plastics.

After the experiment, discuss the findings and build up a table of results on the blackboard or on an overhead projection transparency as shown in Table B15.5.

Material	Corrosion	Chemical attack	Impact strength	Flexibility	Tensile strength	Hardness	Density
thermo-plastic polymer	no	little	high	high	very low	very soft	medium
thermo-setting polymer	no	little	low	low	low	hard	medium
metal	yes	rapid	high	high	very high	very hard	high
wood	no	rapid	low	low	high	soft	low

Table B15.5

In general, therefore, plastics are less strong (lower tensile strength), have a lower melting point, and are softer, than metals. However, plastics possess the advantages of comparatively low density, resistance to corrosion, and good flexibility. Plastics have a comparatively low melting point and this means that they are relatively easy moulded.

Quantitative data relating to the properties of these materials may now be considered (see Table B15.6).

Material	Density kg m^{-3}	Melting point °C	Tensile strength 10^6 N m^{-2}	Elongation % increase	Hardness (Mohs' scale)
polythene (low density)	0.92	137	13	600	2
phenolformaldehyde resin	1.30	chars	50	0.6	5
steel (piano wire)	7.7	1700	3000	0	7
wood (oak)	0.75	chars	—	0	3

Table B15.6

The chemist can, to some extent, 'tailor' a molecule to give the properties he wants. Some examples of this are listed below:

a. The length of a polymer chain can be controlled: longer chain molecules lead to higher melting points and to increased tensile strength.

b. Carbon chains can be altered considerably by the introduction of, for example, benzene rings: thus, polystyrene, which contains benzene rings, has inflexible chains and is a brittle material compared with polythene. The density of polystyrene is also greater than that of polythene.

c. Polymer properties may be modified by mixing them with other polymers. For example, PVC used to make gramophone records is often mixed with polyvinyl acetate. This improves the way in which the discs can be reproduced.

d. The degree of cross-linking between molecules controls the hardness of a material, as illustrated by the vulcanization of rubber.

In these ways, materials can be made with properties not only equal to, but for some jobs superior to, the traditional materials. For example, natural rubber loses its flexibility at very low temperatures whereas silicone rubbers, as used in aircraft engines, remain flexible not only at the very low temperatures of high altitude flying but also at the very high temperatures found inside the engines. Again, no natural material has the 'non-stick' properties of PTFE. Plastics can be made that can withstand a temperature of some 8000 °C for a limited period, such as two or three minutes. On the other hand, plastic wigs can be made to look similar to human hair, and have the added advantage of being moisture resistant so that curls will stay in even during a rain storm!

About two-thirds of the mass of all the organic chemicals produced are plastic materials. In 1970, twenty-three million cubic metres of plastics were produced compared with seventy-two million cubic metres of iron and steel. By 1980, it is estimated (1973) that the relative figures will be ninety-one and one hundred and fifteen, respectively.

The rapid growth in the production of plastics in recent years has arisen largely because of their cheapness. In addition, they are superior to traditional materials in several respects. Increased standards of living have made demands which traditional materials are unable to meet. Thus, there is insufficient wool and cotton for clothing, insufficient leather for shoes, not enough trees to make paper, and not enough rubber-laytex for car tyres. Eventually, there may not be enough steel produced to meet our everyday needs. Of course, if demands for plastics become excessive, there may be a shortage of materials to make plastics too!

Suggestion for homework

Draw up a list of objects commonly made of plastic material and state what they used to be made of before plastics were introduced.

Note: some suggestions as to how this might be done are shown in Table B15.7.

Object	Old material	New material	Advantages and disadvantages of the new material
bowls	enamelled steel	polythene	lighter, does not chip or dent, but scratches
picnic boxes	aluminium	polystyrene	lighter, cheaper, transparent but can be scratched
windows	glass	perspex	less brittle, can be moulded into shapes but can be scratched
radio cases	wood	PVC	lighter, and easier to make
drain pipes	cast iron	PVC	does not fracture or corrode; much easier to instal, does not require painting
electric cables	copper in a rubber sheathing	copper in PVC sheathing	sheathing does not perish
stoppers	cork	polythene	re-usable, non porous
stacking chairs	wood	polypropylene	lighter, moulded to body shape
wrapping material	paper	polythene	has a wet strength and is transparent but not easy to degrade
rope	jute	nylon	stronger, more elastic

Table B15.7

B15.8
Metals and their importance to engineers

In this section, the production of typical metals from their ores is revised and the mechanical properties of metals are investigated in some detail. The importance of metals in daily life is stressed and some preliminary studies on the structure of metals are suggested.

A suggested approach

In the previous section, the properties of plastics were compared with those of a typical metal and of wood. A more detailed study of the properties of metals accounts for their importance as a group of materials.

---

1. Knowledge of the extraction
of metals from their ores
2. Knowledge of the properties
of metals
3. Awareness of the structure of
metals

Pupils may recall something about the extraction of metals from their ores and the properties of metals from Stage I. (See Topics A1, A5, A6, A9, B7, and B10.) Film loops may be used to revise important ideas and matters of detail (for example, film loops 1–8 'Iron extraction', 1–9 'Copper refining'). Appropriate references to electroplating occur in Topics A8 and B10. It can be helpful to list the names of important metals, their ores, production figures, and principal uses, to provide a fresh frame of reference. Some mention of these matters occurs in the *Handbook for pupils,* Chapter 10 'The chemical industry'. One possible summary appears in Table B15.8.

Metal	Principal ores	Outline of extraction procedure	Production figures for 1973 tonnes	Principal uses of the metal	Approximate price of the metal £/tonne (1976)
Aluminium, Al (density 2700 kg m^{-3})	Bauxite, $Al_2O_3 2H_2O$ Corundum, Al_2O_3 Cryolite, Na_3AlF_6	Purified bauxite is dissolved in molten cryolite and then electrolysed between a carbon anode and an iron cathode.	12 020 000	Manufacture of alloys, e.g. magnalium which contains aluminium + 10–30% magnesium; duralumium which contains 95% aluminium, 4% copper, 0.5% magnesium, 0·5% manganese. Electrical conductors – in place of copper (e.g. overhead cables in the grid system). Domestic uses – saucepans, kettles, teapots, etc. Aluminium foil – for wrapping foods and making milk bottle tops.	1730
Copper, Cu (density 8940 kg m^{-3})	Copper pyrites, $CuFeS_2$	The ore is roasted to reduce the sulphur content, and this product is smelted with silica to remove iron as a slag. Finally it is fused and treated with air to produce crude metal. This 'blister' copper is refined by electrolysis using copper sulphate solution as the electrolyte and a pure copper cathode with the 'blister' copper as the anode.	Smelter production 7 670 000 Refined production 8 490 000	Wire for electrical circuits and appliances. Production of alloys – e.g. bronze, which is made of 90% copper and 10% tin; brass, which consists of copper with between 18 and 30% zinc. Copper is also used for making alloys for coinage.	850
Iron, Fe (density 7860 kg m^{-3})	Haematite, Fe_2O_3 Magnetite, Fe_3O_4	Reduction of oxide using coke in a blast furnace: the process requires the use of coke and limestone.	Pig iron and ferro alloys 512 000 000 Crude steel 683 800 000	Production of steel for construction purposes.	Pig iron 80

(Continued)

Metal	Principal ores	Outline of extraction procedure	Production figures for 1973 tonnes	Principal uses of the metal	Approximate price of the metal £/tonne (1976)
Magnesium, Mg (density $1740 \, kg \, m^{-3}$)	Dolomite, $CaCO_3$, $MgCO_3$ Magnesite, $MgCO_3$ Kieserite, $MgSO_4H_2O$	Electrolysis of fused magnesium chloride prepared from sea water, using a graphite anode and an iron cathode.	244 000	Alloys for aircraft and motor industries – e.g. magnalium which consists of aluminium with 10 to 30% magnesium; and magnesium zirconium alloy which consists of 96.3% magnesium, 3% zinc, and 0.7% zirconium.	1483
Sodium, Na (density $970 \, kg \, m^{-3}$)	Rock salt, NaCl	Electrolysis of molten salt, using a graphite anode and a steel cathode.		Manufacture of various sodium compounds and the production of titanium metal.	

Table B15.8

Pupils will appreciate that some metals and alloys are denser than others. The density of the metal is an important factor in the uses to which it may be put, such as in the manufacture of a vehicle. Thus, the metal used in the manufacture of a car or an aircraft needs to be strong but as light as possible.

It is suggested that the pupils investigate a variety of properties of metals which are likely to be of importance to engineers. Thus, the tensile strengths of metal wire can be compared by measuring the force needed to break such wires. During the loading of a wire before it breaks, pupils will notice the stretching of the wire in those metals which yield relatively easily.

The ductility of metals can also be compared. For example, pupils might determine the number of times a particular piece of metal in the form of a strip or a wire can be bent backwards and forwards before it fractures. Alternatively, a metal wire can be twisted until it fractures. In both experiments the actual number of bends and twists required for each metal to produce breaking is not important: it is the comparison between the samples of different metals which is significant. It is well worth while to compare hardened metals with some more ductile forms, as with hard-drawn copper and the softer, more ductile form. Alternatively, the comparison of a pure metal with one of its alloys can be made.

Another series of experiments might involve investigations of the hardness of metals. A variety of methods may be used. Thus, the fact that one material can scratch another which is softer enables us to arrange materials in order of increasing hardness. Alternatively, we might compare the size of indentations produced by allowing a hard material to fall from a standard height onto a series of metal samples. (Obviously, softer metals will be dented more readily than harder ones.)

How the lesson is arranged will depend on the resources available. Much can be achieved by sharing the various tasks among the pupils, and then asking them to report back at the end of a lesson. During a final discussion the effects of heat treatment on the properties of a metal might be raised and a variety of simple tests demonstrated. Details of the range of possibilities are given in the experiments below.

How strong are metals?

Apparatus

The teacher will need:

Standard lengths of 26 s.w.g. wire of some or all of the following metals: copper, brass, aluminium, steel, iron, nickel, nichrome, tin (fuse wire), etc.

Slotted weights

Wire stretching apparatus (see figure B15.12)

Pad of foam rubber

2 G-clamps

Procedure
Figure B15.12 shows the arrangement of the apparatus. A standard length of 26 s.w.g. wire (30 cm is a convenient length) is required for the experiment. Slotted weights are added to the wire and a pad of foam rubber is placed on the floor or the bench underneath the slotted weights to prevent damage. Tests should be carried out on at least two samples of the metal.

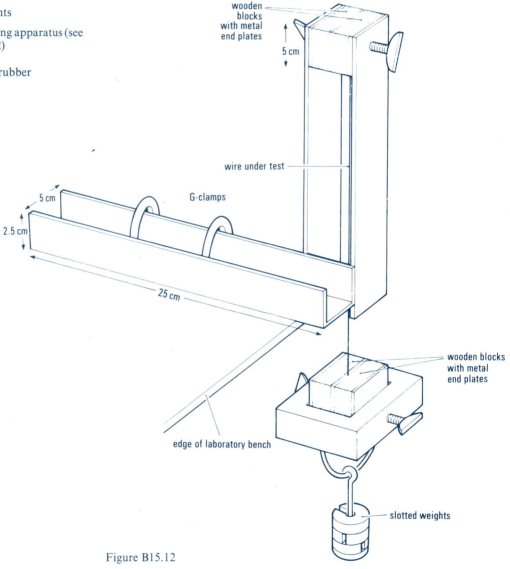

Figure B15.12

Comparing the ductility of metals

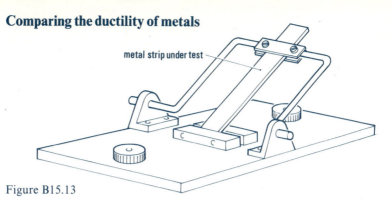

metal strip under test

The teacher will need:

Strips of different metals of the same thickness (e.g. 26 s.w.g.)

Metal-bending apparatus (see figure B15.13) or a metalwork vice

Figure B15.13

Procedure
Clamp the metal strip firmly in the metal-bending apparatus (figure B15.13). Move the handle backwards and forwards and count the number of bends needed for the metal to fracture. Repeat the experiment using a strip of a different metal of the same thickness and size as the first one. Tabulate the results.

An alternative procedure is to use the metalwork vice to hold each metal strip in turn and to bend the strip backwards and forwards about the point at which it is held. To ensure that bending does not occur over the length of the metal strip, a piece of rigid steel needs to be strapped to the metal under test at a point just above the clamp and secured along the entire length of the metal strip.

Investigations on the hardness of metals

The teacher will need:

Small pieces of various metals of approximately the same thickness

Apparatus for testing the hardness of metals (see figure B15.14)

Small wooden block

Stand, boss, and clamp

Hand lens or microscope

Procedure
1. *Using an apparatus for making indentations*
Arrange the apparatus as shown in figure B15.14. Allow the hardened centre punch to fall from a standard height onto the metal

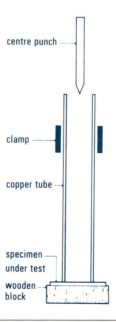

centre punch

clamp

copper tube

specimen under test

wooden block

Figure B15.14

---

sheet resting on the piece of wood. It is important to control the dropping of the punch by using the metal tube – as shown in the diagram. Examine the indentations produced in the samples of metals tested using a hand lens, and compare the hardness of the samples qualitatively.

2. *Scratch hardness*
The selection of metals available for testing may be arranged in an order of decreasing hardness by comparing the ease with which one metal can be used to scratch the surface of another. This test, although qualitative, can yield useful results.

Contrast the results obtained by the two methods.

When the pupils have completed their experiments and the results have been discussed, ask the question 'Why are the properties of metals so varied?' Some of the answers may indicate a need to know something about the structure of metals.

Pupils will be aware that very many of the metals used in their experiments are elements and so must be made of atoms. Differences in the ways in which the atoms are arranged *may* account for differences in the behaviour of metals. If we liken atoms to spheres, or even bubbles, then we may obtain some idea of possible arrangements of such particles.

Using polystyrene spheres to investigate how atoms may be arranged in a metal

Apparatus
Each group of pupils will need:

Bag containing 35 polystyrene spheres (preferred size 25 mm diameter)

Wooden triangle (size will depend on the size of the spheres selected; for construction details see page 160)

Thin glass rod

Tracing paper

Procedure
The main point of this exercise is for pupils to find out that there is more than one way in which the spheres can be packed together. They may then realize that there may be a number of ways in which atoms can be packed in a crystal of a metal. It is an advantage to use some preformed rafts of close-packed spheres rather than individual spheres. (This also helps pupils *not* to think of metal atoms being stacked in pyramids!)

When pupils have completed Experiment B15.8d, it is suggested that the polystyrene spheres and other pieces of apparatus are collected and then the experimental findings summarized. The important point to stress in the discussion is the choice of position of the spheres in the third layer so as to give 'ABA' and 'ABC' types of structure. The teacher may mention to the pupils that all metals have now been examined by special techniques, including X-ray diffraction methods, and that many of these elements have their atoms arranged in one or other of these kinds of structure. A more detailed statement may be included if this is felt to be appropriate (see Topic A14). Perhaps evidence for the orderly arrangement of atoms in metals can be obtained from a study of a metallic surface?

Making a bubble raft

There are several ways of performing this experiment. The original method used by Bragg and Nye (1947) is somewhat complex and is not recommended for routine use. What follows is a very simple method of carrying out this experiment with pupils. Alternatively, a tray holding many marbles may be used to produce the same effect as a bubble raft.

Apparatus

Each pair of pupils will need:

Beaker, $100 \, cm^3$

Petri dish

Glass tube drawn out to fine jet as shown in figure B15.15

Plastic syringe, $50 \, cm^3$

Rubber tubing to connect the syringe to the glass tube

Small quantity of solution of Teepol in water ($10 \, cm^3$ to $1 \, dm^3$ of solution)

Procedure
Fill the Petri dish with Teepol solution. Blow a large number of very small similarly sized bubbles, as shown in figure B15.15.

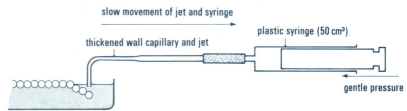

slow movement of jet and syringe

thickened wall capillary and jet

plastic syringe (50 cm³)

gentle pressure

Figure B15.15

Note. Experience will show that it is inadvisable to replenish the volume of air in the syringe without first disconnecting the syringe from the glass capillary tube. (It is all too easy to block the capillary with particles of dried-out detergent.) If no plastic syringes are available, connect the glass tube to the laboratory gas supply and adjust the flow to yield a convenient size of bubble when the tube is dipped into the detergent solution. *If gas is used, make sure that the laboratory is well-ventilated and that the gas supply is turned off as soon as the rafts have been made.*

It is suggested that the teacher demonstrate the technique before asking pupils to carry out the experiment. Figure B15.16 shows a typical bubble raft made by this procedure.

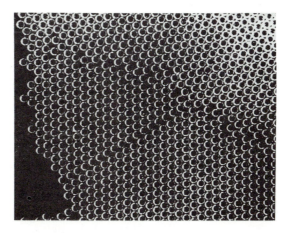

Figure B15.16
The presence of a dislocation can be detected using a ruler or by viewing the picture at a low angle and turning the page slightly.

While the pupils are carrying out the experiment, the teacher should go round the class discussing the various irregularities in the patterns of bubbles obtained. Two or three areas of relatively regular arrangement will probably be seen and it is often useful to stand back and view a dish from a distance of several feet and at a level approaching that of the bench so as to get a general impression of such areas. Such an inspection will show that boundaries between the areas result from bubbles getting out of their 'correct' positions. Occasionally, they will be derived from the occurrence of a few oversized bubbles.

When the pupils have observed these effects the application of the model to metal crystals may be discussed. Compare the packing of bubbles in the rafts to the close packing of spheres. What effect would discontinuities (grain boundaries) have on the properties of a metal? What happens in actual metal surfaces?

Experiment B15.8f

How do crystals form from molten tin?
This experiment may be done by the teacher or by pupils.

Apparatus

The teacher (or each pair of pupils) will need:

Microburner

Heat resistant mat

Tongs

Teat pipette

Access to sink

Piece of good quality tin plate – about 150 mm square

Concentrated hydrochloric acid

Procedure
Hold a piece of clean tin plate over the flame of a microburner for about 5 seconds. The area heated should be about that of a 1p piece. Remove the plate from the flame and allow to cool. Add concentrated hydrochloric acid to the surface of the plate and wash off the acid with water.

Repeat the experiment by heating a similar area in one of the corners of the piece of metal.

Record and discuss the results obtained. The rate of heat loss from the molten tin varies with direction and this effects the growth of crystals.

Experiment B15.8g

How do crystals form from molten lead?

This experiment may be demonstrated by the teacher *or* by one or two groups of pupils. Lead is easily melted and the shapes of the crystals formed on solidification are revealed by treatment with dilute nitric acid.

Apparatus

The teacher – or demonstrating group – will need:

Safety spectacles

Crucible (fused silica)

Flat smooth surface – such as a piece of copper sheet or stainless steel

Bunsen burner and heat resistant mat

Crystallizing dish, 10 cm diameter

Pipeclay triangle and tripod

Magnifying glass

(Continued)

Procedure
Caution is necessary – molten metals can be hazardous! Teachers and demonstrators are recommended to wear safety spectacles.

Melt the lead in the crucible but do not overheat. If the metal is dirty, a scum will collect on the surface but this may be removed using a nickel spatula, preferably after adding a little powdered charcoal. Pour the molten metal onto the flat surface so that it forms a pool 25 or 50 mm in diameter.

Tongs

Nickel spatula

30 to 40 g lead

2M nitric acid

Powdered charcoal

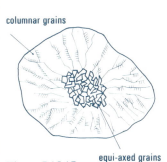

columnar grains

equi-axed grains

Figure B15.17
Crystals in a lead pancake (see also figure A14.11).

Immerse the cold lead 'pancake' in dilute nitric acid. Remove the specimen from the acid when the metal crystals reflect light like jewels. Wash the specimen with water and examine using a magnifying glass.

The use of acids to highlight the structure of metals is known as *etching*. The acid attacks the grain boundaries first. Deep etching results in some grain boundaries appearing darker than others.

Two types of crystal will be seen in the sample:
a. Elongated crystals which have grown from the outer edges of the specimen towards the centre (columnar crystals). Such crystals are produced by a steep temperature gradient.
b. Small crystals which are formed in the centre of the specimen where the rate of cooling is approximately the same over a comparatively large area (equi-axed crystals).

Experiment B15.8h

Apparatus

The teacher will need:

Emery cloth (various grades)

Sheet of plate glass

Metal polish

Crystallizing dish

Tongs

Metalwork vice

Magnifying lens
(multiplication × 10) or
metallurgical microscope

Several similar pieces of copper

Solution containing 5% iron(III)
chloride and 2% hydrochloric
acid in ethanol

The preparation of a metallic surface to show slip lines

Procedure
Polish a piece of freshly cut copper with an emery cloth resting on a plate glass surface. Use various grades of emery cloth and complete the task using metal polish. Etch the polished face for between 15 and 30 seconds in the iron(III) chloride–hydrochloric acid–ethanol solution. Repeat the process using other similar blocks of copper. Compress one block of copper using the metalwork vice until a slight crushing effect can be seen on the etched surface. Compare the polished and etched faces of two pieces of copper by viewing them either with a powerful hand lens or with a metallurgical microscope. Figure A14.20 shows typical results for this experiment.

The results from the last three experiments may be used to encourage the pupils to obtain additional evidence from a study of metallic crystals.

Experiment B15.8i

Apparatus

Each pair of pupils will need:

Experiment sheet 56

2 test-tubes, 100 × 16 mm, and a
loosely fitting cork

Access to a hand lens

(Continued)

Growing crystals of metals

1. *Lead crystals*
Procedure
Details are given in *Experiment sheet* 56 below.

0.1 M lead acetate solution, 10 cm³

Strip of zinc foil, 100 mm × 10 mm

15 cm length of copper wire (about 20 s.w.g.) or a strip of copper foil, 100 mm × 10 mm 0.1 M silver nitrate solution, 10 cm³

2. *Silver crystals*
Procedure
Details are given in *Experiment sheet* 56. Warn pupils about the need for caution when using silver nitrate solution.

Experiment sheet 56

You may not normally think of metals being crystalline because, as we usually see them, metals are highly polished so that the crystal boundaries are obscured. It is fairly easy, however, to grow crystals of some metals. In this experiment you will try to do this for lead and silver.

1. *Lead crystals*

Half fill a test-tube with lead acetate solution. Bend over the top 1 cm of a strip of zinc foil, so that it can be supported over the rim of the test-tube, and place it in the solution. Allow the test-tube to stand in a rack. Describe what happens, drawing a sketch if this will help.

2. *Silver crystals*

These can be made using silver nitrate solution and copper foil or copper wire. Silver nitrate solution should not be allowed to remain in contact with skin or clothing, as it produces black stains which are difficult to remove; if any is accidentally spilled, wash it off at once with plenty of water. **So be careful when using silver nitrate solution.**

If copper foil is used, the same procedure is followed as for lead crystals, see **(1)** above. If copper wire is used, shape the wire into a helix by winding it round a pencil. Open up the coil slightly so that each turn is separate from its neighbour. Place the wire in the silver nitrate solution and bend the top end over the edge of the test-tube. Allow the tube to stand in a rack.

Describe what happens.

After these experiments, the teacher should show pupils some crystals of metals that have been prepared previously or have been purchased as specimens. Photographs showing the crystalline nature of metals are available as slides and may be shown to the pupils. They are reproduced in figure A14.12.

Ask the pupils how they account for the regular shapes of metallic crystals. The point can be made that the use of X-ray diffraction techniques will be mentioned a little later in the course and such techniques have enabled chemists to confirm and extend the ideas put forward in this section. To appreciate such work, we will need to find out rather more about the atomic nature of matter – the subject of the next Topic.

Suggestion for homework

Compare the properties of materials studied in this Topic.

Summary

By the end of the Topic, pupils should be aware of the properties of a range of everyday materials. In particular, they should know something of the chemistry of starch, glucose, ethanol, crude oil, some typical plastics, and some metals. They should also appreciate the need to know something of the structure of these materials if a rational explanation of their physical properties is to be given.

Terms introduced in this Topic include cracking, polymerization, and hydrolysis. They were used in those sections dealing with the chemistry of crude oil, and of fats and oils.

Atoms and the Periodic Table

1. To extend the pupils' awareness of atoms, especially their very small size, and to introduce the idea of relative atomic mass.
2. To introduce pupils to the historical development of the Periodic Table.
3. To demonstrate the simpler patterns and trends in the Periodic Table.
4. To provide pupils with experience of the properties of the alkali metals, the inert gases, the halogens, carbon, and silicon.
5. To demonstrate the value of relating the properties of any element to its position in the Periodic Table.
6. To provide practice in the use of tables of data.

Contents

B16.1 Getting some idea of the size of atoms
B16.2 Getting some idea of the mass of atoms
B16.3 How can the elements be classified?
B16.4 The alkali metals
B16.5 The inert gases and the halogens
B16.6 Carbon and silicon
Appendix. Classifying elements (see section B16.3)

Timing

Between five and six weeks should be allowed for this Topic

Introduction to the Topic

So far in Alternative IIB, emphasis has been placed on a qualitative use of models: we have suggested that chemical knowledge can be understood through simple analogies – given the use of qualifying phrases such as '. . . it is as if . . .'. In this Topic, after establishing some idea of the size of atoms, ideas about atoms are extended quantitatively by introducing the idea of relative atomic mass.

The Topic starts by revising what pupils have learnt so far about 'atoms'. The idea of the existence of atoms leads on to various exercises designed to convey their extremely small size.

Pupils are then introduced to model atoms which are specially made so as to represent the *mass* of atoms to scale. A balancing exercise is used to develop the idea of relative atomic mass. Pupils will need to be aware of the older term 'atomic weight' if they are to use older books.

Relative atomic mass is then shown to be a useful idea by providing pupils with an opportunity to classify elements, using cards which show the symbol of a given element, its relative atomic mass, and, by means of colour tint, its chemical similarity to other elements. Use is also made of the Reference section in the *Handbook for pupils*.

In the remainder of this Topic, pupils examine the chemistry of vertical groups in the Periodic Table, in particular the alkali metals and the halogens. The inert (or noble) gases are mentioned, and carbon and silicon are studied in such a way as to complement other parts of the scheme (see Topic B15 on everyday materials, and Topic B24, in which structural and other matters related to the periodic classification are studied). A simple descriptive account of the abundance of the elements and the use of tables of data of the properties of typical elements complete this Topic.

Subsequent development

Periodicity as a theme reappears towards the end of Alternative IIB (see Topic B24) and is referred to at other points in the scheme, including Stage III Option 7 'Periodicity, atomic structure, and bonding'. Subsequent Topics continue the development of ideas about the basic concepts of chemistry, notably that of the atom and the chemical bond. As the pupil acquires his knowledge about atoms, molecules, and chemical bonds in Topic B18, he is gradually prepared for the discussion in Topic B19 on the relationship between atoms and ions and for the review of the properties of ions in Topic B20. Supplementary material on the structure of the atom, the arrangement of electrons in atoms, and on chemical bonding is provided in this volume (Stage II Supplement).

Alternative approach

The content of this Topic is similar to parts of Topics A11, A13, and A14. It will be recalled that in this Alternative, periodicity is considered not only in Topic B16 but also in Topic B24.

Topic B16 may be used in the numerical sequence B14 to B17 *or* in the sequence B17; B15.5–B15.8; B15.1–B15.4; B14 and B16. Topic B18 then follows, whichever route was followed earlier.

Background knowledge

This Topic follows from much of the work of Stage I and in particular from:

Alternative IA. Topic A5 'The elements' – the idea of an element is introduced and developed; elements are then divided into metals and non-metals. Topic A6 'Competition among the elements' – the idea of an order of reactivity or displacement series is introduced. Topic A10 'Chemicals from the sea' – the halogens are considered as a group of similar elements.

Alternative IB. Topic B7 'The elements' – the idea of an element is introduced and developed and metals and non-metals are distinguished from one another. Topic B8 'Further reactions between elements' – trends in the properties of oxides, sulphides, and chlorides are considered. Topic B10 'Competition among the elements' – the idea of an order of reactivity or displacement is introduced at this point and the halogens are then considered as a group of similar elements.

The earlier Topics of Alternative IIB (B11, B12, B14, and B15) also contain relevant background material. The treatment provided in this Topic makes allowance for pupils who join a chemistry course after following a broader introduction to science than that provided in Stage I.

Further references

for the teacher

Collected experiments, Chapter 6, contains further experiments on this Topic.
Handbook for teachers, Chapter 3, provides a detailed discussion on the presentation of this Topic.
Partington, J. R. (1957) *A short history of chemistry*, Macmillan, contains an authoritative account of the events leading to the periodic classification of the elements.
Spiers, A. and Stebbens, D. (1973) *Chemistry by concept: a comprehensive account of the principles of chemistry for G.C.E. O-level and C.S.E. courses*, Heinemann Educational, provides a number of examples on the use of bar charts to display periodicity in chemical data.
It may be helpful for teachers to refer to one or more of the following books before attempting section B16.6.
Hamilton, W. R., Woolley, A. R., and Bishop, A. C. (1974) *The Hamlyn guide to rocks, minerals, and fossils*, Hamlyn.
Geological Museum (1972) *The story of the earth*, H.M.S.O.
Turekan, K. K. (1972) *Chemistry of the earth*, Holt, Rinehart, and Winston.

Supplementary materials

Film loops
2–1 'Measuring the very small'
1–11 'Fluorine manufacture'
1–12 'Uses of fluorine compounds'
1–13 'Chlorine manufacture'
1–14 'Chlorine – uses'
1–15 'Bromine manufacture'
1–16 'Bromine – uses'
1–17 'Iodine manufacture'
1–18 'Iodine – uses'

Film
'Chemical families' CHEMStudy. Distributed by Guild Sound and Vision Ltd.

Overhead projection originals
A,B, & C Steps of ten
42 The structure of the Earth and temperature ranges of the Earth's layers *(figure B16.5 and Table B16.3)*
43 The distribution of the elements *(figure B16.6)*

Special teaching aids
'Classifying elements': cards for use with *Experiment sheet* 113 (Exercise B16.3)

Reading material

for the pupil

Handbook for pupils, Chapters 1, 'Periodicity', 2 'Finding out about matter', and 4 'Formulae and equations'.
Length and measurement (A SCISP Background book) provides a discussion on the measurement of the length of objects from those of a very large size to those requiring the use of an electron microscope.

B16.1
Getting some idea of the size of atoms

Atoms are used in this section as an accepted idea and some feeling for their extremely small size is given to the pupils.

A suggested approach

Objectives for pupils

Appreciation of the extremely small size of atoms

Pupils know that all matter is composed of substances which can be broken down into elements. So far in Alternative IIB, we have adopted the notion of an 'atomic' composition of materials since this idea can be equated with the pupils' everyday use of the term 'atom' in conversation.

In Alternative IIA (Topic A11) some evidence for the existence of particles is presented to pupils and a simplified oil film experiment is carried out to give an approximate idea of the size of particles. Furthermore, the atom is presented essentially as an idea, for which there is no proof, to explain many observations. In this alternative, we have accepted that the atom should be presented initially with very little of the discussion given in Topic A11 on the two theories of matter (that matter is continuous or discrete) and that emphasis should now be placed on developing a feeling for the size of very small particles.

Some pupils may, however, want to know what sort of evidence supports the idea of atoms. There is a good deal of evidence, but it is difficult to present it appropriately.

In Alternative IIA, the emphasis is mainly on diffusion; but the way this points to the atomic theory of matter is not always easy for pupils to appreciate. Other evidence which might be quoted and briefly discussed includes the direct observation of Brownian motion; a demonstration of radioactive decay (using a simple cloud chamber); a demonstration of the regular shape of crystals; the cleavage of crystals (e.g. calcite, mica); and so on. To enable pupils to acquire some feeling for the extremely small size of atoms, it is suggested that the teacher should show the pupils a series of specially prepared pictures of objects in a sequence of ten-fold reduction in size, as described in the following exercise.

Exercise B16.1

Apparatus

The teacher will need:

OHP transparencies 'A', 'B' and 'C' and a projector

Getting some idea of the size of atoms

It is suggested that *before* showing pupils pictures of very small objects, reference should be made to more familiar objects using specially prepared transparencies with an overhead projector. A convenient reference point is the height of a man, 1.8 m. A young boy or girl will be somewhat smaller, say 1 m. The height of a typical small house depends on its style and is likely to be between 7 and 10 m. The next step of ten brings us to 100 m and reference might be made to some well known building. For example, the spire of Salisbury Cathedral is 123 m high (1.23×10^2 m). This idea of 'steps of ten' can be conveyed relatively easily using *standard form* (scientific notation), i.e. by expressing any number as a number lying between 1 and 10 multiplied by some power of 10. When the scale reaches 10^4 m, we exceed the height of the highest mountain on Earth. Mount Everest is 8.85×10^3 m high. This use of standard form may be unfamiliar to pupils and require some simple discussion and the use of additional examples by way of illustration. The second stage of the discussion requires measurements going down in 'steps of ten'. Again the point of reference can be the height of a man or a child. The first four steps on this occasion keep distances to within the range of direct observation by the human eye. However, the use of negative indices may be novel to some pupils in so far as it may be necessary to relate, for example, 10^{-1} to $\frac{1}{10}$, 10^{-2} to $\frac{1}{100}$, and so on. Various types of microscope (first optical and then electron) can be used to examine small objects and the point can be made that microscopy does not enable us to see every small object. Chemists and other scientists use less direct means to 'see' relatively small groups of atoms and even individual atoms (cf. *Handbook for Pupils*, chapter 2). During this part of the lesson, reference may be made to the thickness of an oil film. Oleic acid molecules can form such a film and are about 2×10^{-9} m long (cf. page 43, Experiment All.2c). However, any account given should not distract attention away from the central theme of 'steps of ten'. The overhead transparencies provide a simple summary for the exercise. Atoms are very small. About ten thousand million (10 000 000 000) hydrogen atoms placed end-on measure one metre. Pupils may also be reminded that the smallest part of an element that retains the properties of that element is an atom.

Suggestion for homework

Try to find out the approximate sizes in metres of the following objects:
a. the highest mountain in Britain;
b. the diameter of the Earth;
c. the thickness of the thinnest sheet of paper you can find.

How many 'steps' are these objects from an object of size 1 m?

B16.2
Getting some idea of the mass of atoms

In this section, pupils are introduced to the relative atomic mass (atomic weight) of the elements by means of model atoms having correct relative masses.

A suggested approach

Objectives for pupils

1. Understanding of the meaning of the relative atomic mass of an element
2. Practice in the use of models which represent to scale the size and the mass of atoms
3. Knowledge of the symbols of the more common elements and their use to represent individual atoms of that element

Pupils should now realize that atoms are very small indeed and so are extremely 'light'; that is, that atoms possess a very small mass. Now is the time to introduce pupils to models of atoms* of common elements such as those of hydrogen, carbon, oxygen, magnesium, sulphur, copper, etc. If the models are made from spheres of different diameters, providing them with a correct relative mass, and if they are identified by means of a colour code, such models may be used to implant the idea that atoms of different elements have characteristic size and mass. The precise size of atoms is not of crucial importance for this Topic; *but* it is *desirable* for the spheres that are used to have the correct relative mass. The following table provides a suitable scale.

Number in set	Element	Mass of a sphere which represents 1 atom of the element /g	Colour
20	hydrogen	1	white
2	carbon	12	black
4	oxygen	16	red
1	magnesium	24	silver-grey
2	sulphur	32	yellow
1	copper	64	copper

Table B16.1

*See *Handbook for teachers*, Chapter 14 'Models and their uses' (page 202 *et seq.*), for sizes and methods of construction.

This can be most easily achieved by filling hollowed polystyrene spheres with an appropriate quantity of lead shot. It may be more convenient to make the spheres using a mass scale which is one-quarter that shown in the table but, of course, with a correct relative mass.

Getting an idea of the relative masses of some 'model' atoms
This exercise may be carried out by the teacher or by pupils.

Apparatus

The teacher or each group of pupils will need:

Experiment sheet 112

1 set of model atoms for the teacher or for each group of pupils

Simple beam balance with large pans to hold a number of model spheres

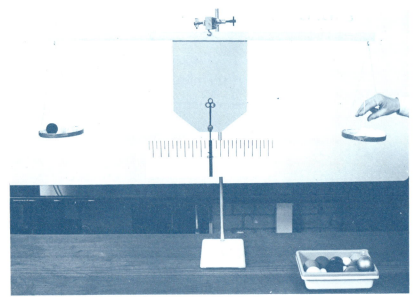

Figure B16.1

Figure B16.1 shows a typical arrangement for finding the relative mass of ballasted model atoms. *Experiment sheet* 112 is reproduced below.

Experiment sheet 112

You are provided with some models of the atoms of a number of common elements. These have been coloured for ease of identification using the following code:

model hydrogen atom	white
model carbon atom	black
model oxygen atom	red
model magnesium atom	silver-grey
model sulphur atom	yellow
model copper atom	copper

You will see that there is considerable variation in the size and mass of these models – just as there is in the atoms which they represent. The model atom for oxygen has been made sixteen times as heavy as each model hydrogen atom because it is known that oxygen atoms are sixteen times as heavy as those of hydrogen. The other model atoms have also been made to scale as far as their mass is concerned.

It is possible to express the mass of atoms in grams but it is so much more convenient to express the mass of an atom relative to the lightest atom known – that of hydrogen.

Topic B16 Atoms and the Periodic Table

Find out how heavy the models of atoms are, compared with each other, using the simple beam balance provided.

Complete the following table.

One model magnesium atom *balances* .... model carbon atoms

One model copper atom *balances* .... model sulphur atoms

One model copper atom *balances* .... model oxygen atoms

One model carbon atom *balances* .... model hydrogen atoms

You may have a little difficulty in deciding exactly how many hydrogen atoms balance a carbon atom *unless* your balance is very accurate and the model atoms have been made accurately.

After the pupils have spent a short time comparing the masses of model atoms, it is suggested that the results obtained could be summarized on the board or on an overhead projector transparency in the following way.

Element	Relative atomic mass	Colour code	Symbol
hydrogen	1	white	H
carbon	12	black	C
oxygen	16	red	O
magnesium	24	silver grey	Mg
sulphur	32	yellow	S
copper	64	copper	Cu

Table B16.2

This is a convenient place for the pupils to revise the symbols of some common elements. Tell them that for the time being the symbol H is a shorthand way of writing 1 atom of hydrogen. (Later, in Topic B18, they will be introduced to the idea that the symbol may also stand for 1 mole of hydrogen atoms.)

Summary

By the end of this section the pupils should realize that the masses of atoms of a given element are expressed by means of the scale of relative atomic masses. They should understand that atoms may be represented in two ways:
1. By means of models;
2. By means of symbols.

They should also realize that atoms of different elements have different sizes.

B16.3
How can the elements be classified?

In this section pupils build up a limited version of the Periodic Table by arranging elements in order of their relative atomic mass and by grouping the elements said to have similar chemical properties in the same vertical column. The positions of various types of elements are then noted.

It is desirable that any Periodic Table on the walls of the classroom or laboratory be removed or at least covered up during this section.

Objectives for pupils

1. Ability to classify the elements by 'trying out what Mendeleev did'
2. Awareness of the history of the development of the Periodic Table
3. Recognition of the Periodic Table as listing the elements in order of atomic mass, and of grouping together those elements with similar properties

Pupils should know from their work during Stage I (see Topics A10.4 and B10.5) that certain elements, particularly the halogens, show 'family resemblances'. This section follows directly from an acceptance of the term 'relative atomic mass'.

As in Alternative IIA, the Periodic Table may be introduced historically, without any reference to the electronic structure of elements. However, we suggest that this be done in a rather different way from Topic A13. Pupils are provided with a set of cards, each of which shows the symbol, name, and *approximate* relative atomic mass of an element. The cards are also coloured, and pupils are *told* that the elements with cards of the same colour have similar chemical properties. In order to reinforce this idea, one might start by asking pupils to pick out the cards representing the halogens. Then, tell them that the cards representing other elements, most of which are unfamiliar to them, are also coloured in such a way as to show family resemblances. Ask pupils to use the Reference section of the *Handbook for pupils* to verify this suggestion by examining at least some of the properties (for example, melting point, boiling point, density).

Exercise B16.3

How can elements be classified?

Apparatus

Each pair of pupils will need:

Experiment sheet 113

1 set of tinted Periodic Table cards for each pair of pupils (see Appendix, page 460, for details)

Handbook for pupils (towards end of lesson only)

Give out the cards in their envelopes. Pupils should be able to classify the elements without very much difficulty. Some of the arrangements they might suggest include those shown in figure B16.3. The Appendix lists the contents of the various envelopes. The Reference section of the *Handbook for pupils* lists melting points, boiling points, and densities of elements for use during this exercise.

Additional information appears in *Experiment sheet* 113 which is reproduced below.

Experimental sheet 113

In 1869 a Russian chemist, Dmitri Mendeleev, had an idea which caused a great deal of excitement at the time and which laid the foundations for much of modern chemistry. He developed a system which organized many facts of chemistry into a very simple pattern. This pattern still holds true today. Here is your chance to try out what Mendeleev actually did.

You have been given an envelope containing cards each of which is marked with a different name, symbol, and number. Each symbol represents a different element and the number is the atomic mass of that element based on the scale where the mass of one hydrogen atom is given as one unit.

For example:

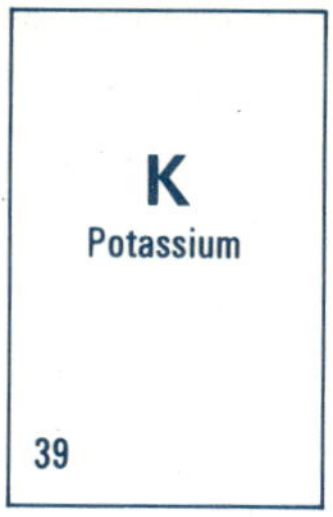

Figure B16.2

This card represents an atom of potassium (K) which is 39 times heavier than an atom of hydrogen(H).

Now, follow the instructions below carefully and use *only* the cards from envelope 1 'The elements'.

1. Arrange the cards which represent the elements in order of their increasing atomic mass.

You will have noticed that cards of elements with similar properties have been printed on cards of the same colour.

2. Now, without changing the order in which you have placed your cards, but by increasing or changing the number of rows, arrange the elements so that those with similar properties are placed below each other in the same vertical column. Remember that the elements must remain in the same order of increasing atomic mass as you read along the rows from left to right.

The arrangement that you have made is part of what is called the Periodic Table.

When you have done this, discuss your suggestions with your teacher. (There are several different arrangements which are equally good.)

How many different arrangements can you make using the same cards?

Questions on your Table
1. What is the name of the element with the lightest atoms? Write down its symbol and its atomic mass.

2. What is the name of the element with the heaviest atoms in your table? Write down its symbol and its atomic mass.

The vertical columns in your table are called *groups*.

3. How many groups are there in your table?

Your arrangement corresponds to the Periodic Table.

4. Fill in the missing word in the following sentence:

Elements that are in the same group of the Periodic Table have .... properties.

The groups in the Periodic Table are usually numbered from left to right so that the group that is headed in your table by the element hydrogen is called Group I.

5. What is the name for the group of metal elements found in Group I?

6. Write down some properties that the elements in Group I have in common.

7. Find another group of elements in your table which you have studied before.
a. What number is the group you have found?
b. What is the name of the group?
c. Name the elements in this group.

8. You will see that in the bottom row one element seems to be missing. The element that goes in that space was not known in Mendeleev's day but he predicted that such an atom should exist and described what its properties should be.

Fifteen years later, the element was discovered and Mendeleev was found to be correct in his predictions.

Mendeleev's predictions	**What the element is really like**
It will be a light grey metal	It is a dark grey metal
It will form a white oxide which has a high melting point	It forms a white oxide with a melting point greater than 1000 °C
The compound it forms with chlorine will have a boiling point of less than 100 °C	It forms a chloride with a boiling point of 86.5 °C

9. How do you think Mendeleev was able to predict the properties of the unknown element so accurately?

10. Can you predict what its atomic mass should be?

Now open envelope 2 marked 'missing element'
11. What is the symbol and atomic mass of the element that Mendeleev predicted should exist?

12. Draw the plan of the Periodic Table that you have now on a separate sheet of paper.

13. Shade in on your plan the area where all metallic elements are to be found.

The Periodic Table is not only useful for finding out about 'missing elements' but also when examining the relationship between the different elements. Elements in the same group have similar properties and also as we move from one element to another across the table, the sort of properties they have change gradually and in an orderly manner. Pupils should be asked to use the Reference section of the *Handbook for pupils* to verify this statement.

Note to teachers
It will be seen that the transition elements have been deliberately left out of the table. They are not discussed in this Topic. However cards for some of the first series of the transition elements are provided. Teachers who wish to use them should place them in a third envelope labelled 'other elements'.

Suggestion for homework

Read *Handbook for pupils*, Chapter 1 'Periodicity'.

Summary

By the end of this section, pupils should have some understanding of how Mendeleev classified the elements on the basis of their relative atomic masses. They should know that the metals are on the lefthand side and in the middle of the table; and that non-metals are on the righthand side of the table. Pupils should also be aware of the position of the halogens.

Topic B16 Atoms and the Periodic Table

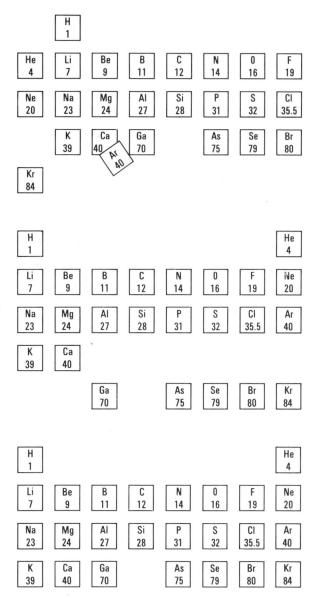

Figure B16.3
Three ways of arranging the cards in envelope 1.

B16.4
The alkali metals

The first group of the Periodic Table, the alkali metals, is studied in this section. The properties of sodium and potassium are investigated and pupils are then asked to predict whether lithium will be more or less reactive than sodium and potassium. They then test their predictions.

Objectives for pupils

1. Ability to identify a pattern of reactivity and to make predictions from the pattern
2. Knowledge of some reactions and of the comparative reactivity of the alkali metals
3. Ability to compare data for different elements

Experiment B16.4a

Apparatus

The teacher will need:

Safety spectacles

Chlorine generator made from a filter flask, 100 cm³, fitted with a bung carrying a tap funnel and delivery tube

Stand, boss, and clamp

2 gas jars and cover slips

2 combustion spoons for use in the gas jars

Bunsen burner and heat resistant mat

Filter paper

Asbestos paper or equivalent material

Beaker, 100 cm³

2 combustion tubes

Long-handled laboratory knife

Hexane or ether (ethoxyethane) (CAUTION: flammable liquids!)

Concentrated hydrochloric acid

Potassium permanganate (manganate(VII))

Vaseline

Sodium

Potassium

Full-range Indicator solution

Note. Chlorine should be generated in a fume cupboard.

In Stage I, pupils have come across sodium and potassium salts and may have seen sodium metal. The lesson could be started by asking the pupils about the two elements sodium and potassium. Then continue by demonstrating a series of experiments to show the typical properties of these metals.

To investigate the action of (1) air; (2) chlorine, on heated samples of sodium and potassium
This experiment should be done by the teacher.

Procedure
The teacher should wear safety spectacles during these experiments.

1. Remove a small piece of sodium (a cube of side 1 to 2 mm is suitable) from the bottle. Free it from the protective paraffin oil under which it is kept by washing it in hexane or other suitable solvent in a beaker and then pressing it gently between pieces of filter paper. Cut away any surface coating of oxide, and place the piece of sodium on asbestos paper in a combustion spoon. Heat the spoon carefully in a Bunsen burner flame and as soon as the sodium shows signs of burning, plunge the spoon into a jar of air and observe how vigorously it burns. When the burning has finished, heat the spoon again in the Bunsen flame to check that all the sodium has burned and then rinse the oxide into the gas jar with a little distilled water. Test the pH of the solution so formed with Full-range Indicator.

Repeat the experiment using a similar sized piece of potassium and compare the results obtained.

2. **Warning.** *Chlorine gas is extremely dangerous if inhaled. In addition, care must be taken to use only very small pieces of sodium or potassium, and to use only moderate heat to achieve a reaction.*

This part of the experiment *must* be carried out in a well-ventilated fume cupboard.

Chlorine of sufficiently pure quality for this experiment can be obtained by the dropwise addition of concentrated hydrochloric acid to potassium permanganate crystals in a filter flask, as shown in figure B16.4. Clean a piece of sodium as in part (1) of this experiment and place it on a piece of asbestos paper in the combustion tube. Assemble the apparatus (see figure B16.4). Add concentrated hydrochloric acid dropwise to the potassium permanganate in the flask to allow chlorine to pass through the apparatus. When the apparatus is full of chlorine, gently warm the sodium. Sodium will burn with a brilliant yellow flame in chlorine gas, and a white powder is left in the combustion tube. Leave this combustion tube to cool down, and prepare a second combustion tube using potassium in place of sodium.

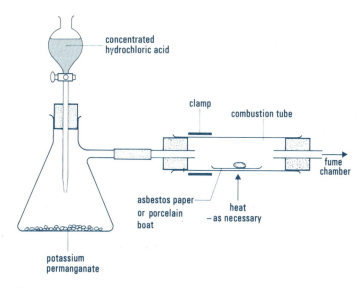

concentrated hydrochloric acid

clamp

combustion tube

fume chamber

asbestos paper or porcelain boat

heat – as necessary

potassium permanganate

Figure B16.4

Caution. *Under no circumstances attempt to use the previous tube which is hot and which contains free chlorine.*

When both tubes are cool and free from chlorine check that no metal remains in the tubes and then dissolve the products in water and test them with Full-range Indicator solution.

Experiment B16.4b

Apparatus

The teacher will need:

Safety spectacles

Safety screen

2 troughs

Beaker, 100 cm^3

Long-handled laboratory knife

Filter paper

Sodium

Potassium

Hexane

Full-range Indicator solution

To investigate the action of water on samples of sodium and potassium
This experiment should be done by the teacher.

Procedure
The teacher should wear safety spectacles and should carry out the experiment with a safety screen between the troughs of water and the class.

Place the two troughs side by side on the demonstration bench, each half filled with water. Remove a piece of sodium from the stock bottle, and free it from liquid paraffin by washing it in hexane or other suitable solvent in a beaker and then pressing it gently between pieces of filter paper. Cut away any surface coating of oxide. Place a cube of side 2–3 mm in the water in one trough.

Repeat the experiment with potassium using the other trough.

Draw attention to the fact that sodium melts and skates about the surface of the water. Potassium also melts, skates about the surface of the water, and a flame may appear.

Do *not* allow pupils to come too close to the demonstration bench! Finally, when the reaction has ceased, add some drops of Full-range Indicator to each trough of water and note the result.

Note. Small pieces of alkali metal which may be left over should not be returned to the stock bottle and must on no account be put into sinks or rubbish boxes. Such pieces of metal can be dealt with by putting them into an excess of ethanol (industrial methylated spirit) in a beaker. When the metals have finished reacting with the ethanol, the resulting solution can be poured down the sink and flushed away with much water.

After these experiments, the pupils should summarize (perhaps in the form of a table as below) their knowledge of the properties of sodium and potassium. The various reactions can also be summarized by words and chemical equations.

Reaction	Sodium	Potassium
Reaction with air Reaction with chlorine Reaction with water		

Point out that lithium is also placed in the same vertical group as sodium and potassium. The pupils should have their own Periodic Tables in front of them. What properties do they expect lithium to have? For example, how would they expect lithium to react with water? How would they expect it to react with chlorine? How would they expect it to react with air? What would the pH numbers of the solutions of the products be in each of these cases? Would they expect lithium to be a soft or a brittle metal? Would they expect it to have a high or a low melting point? After a discussion along these lines, carry out Experiment B16.4c. Teachers should note that the first part of the experiment is a demonstration. Pupils will need the *Handbook for pupils* for the last part of the experiment.

Experiment B16.4c

Apparatus

The teacher and each pair of pupils will need:

Safety spectacles

Experiment sheet 49

Beaker, 100 cm³

Asbestos paper or equivalent material

Pipeclay triangle

Tripod

Bunsen burner and heat resistant mat

Tongs

Handbook for pupils

(Continued)

What are the reactions of lithium?

Note to teachers. Lithium should be purchased as lithium shot. This preparation is of high purity and is supplied in liquid paraffin. The traces of liquid paraffin can be removed by using hexane or petroleum spirit and pieces of filter paper, but it is suggested that pupils should *not* be allowed to clean pieces of lithium themselves.

Caution. It is advisable not to exceed the quantity of lithium suggested for this experiment and to require participants to wear safety spectacles.

Procedure
See *Experiment sheet* 49 for details. It is important to note that the first part is a teacher demonstration.

Experiment sheet 49
You will be required to wear safety spectacles when doing this experiment.

1. *Your teacher will first demonstrate the effect of heating a small piece of lithium on dry asbestos paper.*

Topic B16 Atoms and the Periodic Table

Full-range Indicator solution

Small piece of lithium (1 mm^3)

Distilled water (wash bottle)

Does the lithium burn in air?
What is the colour of the flame?
Record the appearance of the residue.

Your teacher will then transfer the residue into a test-tube using distilled water and add drops of Full-range Indicator to the solution.

Record what happens.

2. Half fill a 100 cm^3 beaker with water and add a rice-grain size piece of lithium. What happens?

Add 10 drops of Full-range Indicator solution to the contents of the beaker.

What is the pH of the solution?

Write word equations for the two reactions you have studied.

lithium + oxygen $\longrightarrow$ ....

lithium + water $\longrightarrow$ ....

Find out as many similarities and differences as you can between the properties of lithium, sodium, and potassium, using the Reference section of the *Handbook for pupils*. Tabulate your results using the headings below.

Property	Lithium	Sodium	Potassium

After the pupils have completed their experiments, discuss their findings. Demonstrate the action of chlorine on heated lithium, using the same procedure as for sodium and potassium.

Lithium is seen to be very much like sodium and potassium but, in general, less reactive. In Group I of the Periodic Table, the elements tend to be more reactive as their relative atomic masses increase. Ask pupils to revise the formulae of the chlorides of these elements and to write equations for the reactions studied. Teachers should use their discretion about giving the formulae of the other compounds mentioned. The principal oxides, for example, do not all have similar formulae: they are Li_2O, Na_2O_2, and KO_2, respectively. The increasing oxygen content of this series might be used to illustrate the effect of the increasing affinity of these metals for oxygen.

During this discussion the Reference section of the *Handbook for pupils* can play a large part, particularly in obtaining the melting points and other physical properties of the alkali metals. Bar charts can give a real feeling for data, as mentioned in the mathematical note given at the end of section A13.2.

Summary

By the end of this section pupils should know that the alkali metals (Group I of the Periodic Table) have similar chemical properties but differ in degree of reactivity, which increases with relative atomic mass. These metals burn in air, combine with chlorine to give chlorides of the formula MCl, and react with water to give alkaline solutions. They should also know of some of the physical properties of these elements through the use of bar charts.

1. Use the Reference section of the *Handbook for pupils* to construct bar charts showing (*a*) melting points, (*b*) boiling points, (*c*) densities, of the alkali metals. What trends are shown by the bar charts?
2. From the densities of the alkali metals, work out the volume occupied by the relative atomic mass number expressed in grams for each of these elements. Plot the results as a bar chart. How do the results compare with one another? How do they compare with the values given for other metals (for example, iron or copper)?
3. In what ways would you expect rubidium and caesium to be different from or similar to the alkali metals?

B16.5
The inert gases and the halogens

In this section, the inert (noble) gases are mentioned and the halogens are examined more extensively than in Stage I. The pupils are asked to look for a 'reactivity trend' in the halogens and find that the elements of lower relative atomic mass are more reactive than those of higher relative atomic mass.

A suggested approach

Objectives for pupils

1. Ability to identify a pattern of reactivity and to make deductions from that pattern
2. Knowledge of some reactions and of the comparative reactivity of the halogens
3. Awareness of the properties and uses of the inert gases

The teacher should draw the attention of the class to the patterns within the Periodic Table: on the lefthand side of the table are the metallic elements: on the righthand side the non-metallic elements. On the extreme right of the table, another group of elements occurs known as the inert or noble gases. Some brief mention should be made of the properties and uses of these gases (see *Handbook for pupils*, Chapter 1).

However, the main purpose of this section is to examine the properties of the halogens. The pupils will know something about the halogens from Stage I, but this work did not go much beyond simple description. Methods for comparing the relative reactivity of the elements in the group should be discussed with the class, before the pupils do Experiment B16.5b. Experiment B16.5a is a demonstration.

Experiment B16.5a

Apparatus

The teacher will need:

Hard-glass test-tube, 100 × 16 mm

2 hard-glass test-tubes, 150 × 25 mm

Test-tube rack

Hard-glass test-tube, 125 × 16 mm with small hole near closed end and fitted with a cork carrying a straight delivery tube

(*Continued*)

Properties of the halogens
This experiment MUST be performed by the teacher and all the reactions must be carried out in a well-ventilated fume cupboard.

Procedure
Demonstrate the action of chlorine and bromine on water. Chlorine will be seen to dissolve in water if the gas is bubbled through a delivery tube into a 150 × 25 mm test-tube half full of water. Stand the test-tube in a test-tube rack in the fume cupboard with a piece of white card behind it. Ask the pupils what they can see.

Handle liquid bromine extremely cautiously when demonstrating the action of bromine on water. It is very dangerous to spill any bromine on the hands.* Use a teat pipette to add two or three drops

*Safety note. In the event of spilling bromine on the skin, drench with water immediately and then bathe with a dilute solution of ammonia or sodium thiosulphate in water. Seek medical attention.

Cork to fit a 150 × 25 mm test-tube

Stand, boss, and clamp

Teat pipette

Beaker, 250 cm³

Bunsen burner and heat resistant mat

Apparatus for producing chlorine (as shown in part of figure B16.4)

Connecting tubing

Bromine

Full-range Indicator paper

Iron wool

Sodium hydroxide solution

of bromine to a 150 × 25 mm test-tube half full of water. Cork the test-tube securely and shake it gently, avoiding the contamination of the cork by bromine. Be careful to ensure that no bromine escapes around the cork. The water will soon be coloured.

Test portions of both these solutions with Full-range Indicator paper. It will first indicate the presence of acid and then be bleached. Test further portions of these solutions with sodium hydroxide solution – which will remove the colours. Add sodium hydroxide solution to all solutions before disposal.

Pass chlorine over a plug of iron wool in a 125 × 16 mm test-tube fitted with a small hole near the closed end of the tube. Heat the iron wool to start the reaction and then remove the Bunsen burner flame. The iron wool will be seen to glow redhot and iron(III) chloride will be formed.

Demonstrate the action of bromine on iron wool by putting two or three drops of liquid bromine into a test-tube and pushing a tuft of iron wool halfway down the tube. Heat the iron wool until a reaction can be seen to take place. This experiment must be conducted in a fume cupboard.

When discussing the results of all of these experiments, make the point that bromine and chlorine react in a similar way but with different vigour. Summarize the reactions using word and chemical equations. Ask the pupils which they think the more reactive of these two elements – chlorine or bromine?

Ask the pupils to predict how they would expect iodine to react with water, sodium hydroxide, and iron wool. Let them test their predictions by trying Experiment B16.5b.

Experiment B16.5b

Investigating some of the properties of iodine

Apparatus

Each pair of pupils will need:

Experiment sheet 50

3 hard-glass test-tubes, 100 × 16 mm

Bunsen burner

Test-tube holder

Small crystals of iodine, 5 crystals

Iron wool, small tuft

2M sodium hydroxide solution, 5 cm³

Full-range Indicator, 5 cm³

Procedure
Details are provided in *Experiment sheet* 50 which is reproduced below.

Experiment sheet 50
You will have seen a number of experiments with the elements chlorine and bromine. The following experiments with iodine will enable you to find out whether it resembles chlorine and bromine in its properties and reactions.

1. Add a small crystal of iodine to a test-tube half filled with water; shake the mixture well.

Does the iodine appear to dissolve?

Test the liquid with Full-range Indicator paper. What happens?

2. Shake a small crystal of iodine with a quarter of a test-tube full of sodium hydroxide solution. **(Be careful, sodium hydroxide solution attacks the skin — do not put your thumb over the open end of the tube to shake it.)**

What happens?

3. Place a crystal of iodine at the bottom of a dry test-tube and wedge a small tuft of iron wool about half-way down the tube. Support the test-tube

---

horizontally in a clamp or holder and heat the iron wool strongly. Enough heat will probably reach the iodine to vaporize it so that purple iodine vapour will surround the iron wool. If this does not happen, warm the iodine for a second or two. Is there any evidence of a reaction between the iron and the iodine? If so, describe it.

Tell pupils that the reactions with iron produce compounds of formulae $FeCl_3$, $FeBr_3$, and FeI_2. They should appreciate that iodine is much less reactive than chlorine or bromine. The pattern for non-metals on the righthand side of the Periodic Table is thus the reverse of that of the metals on the lefthand side of the table. Finally, draw the pupils' attention to some physical properties of the halogens. As in the previous section, the use of bar charts to illustrate points may be helpful.

Suggestion for homework

1. Draw up a table of the chemical properties of chlorine, bromine, and iodine that you have observed. Point out any trends that you have noticed. Consult books in your school library to see if you can extend the range of properties covered in your survey.
2. Use the Reference section of the *Handbook for pupils* to construct bar charts showing (*a*) melting points and (*b*) boiling points of the halogens. What trends do you observe in the data?

Summary

By the end of this section pupils should know something of the properties and uses of the inert gases. They should also know that the halogens have similar chemical properties but differ in degree of reactivity, decreasing reactivity being found with an increase of relative atomic mass. Pupils should know that halogens are slightly soluble in water, bleach indicators, and react with sodium hydroxide solution or iron wool. Where appropriate, the reactions studied may be summarized by equations. Pupils should have some experience in comparing the physical data for the elements studied by the use of bar charts.

B16.6
Carbon and silicon

The left- and righthand sides of the Periodic Table have been examined in earlier sections. The pupils now examine the properties of two elements taken from the centre of the table: carbon and silicon.

A suggested approach

Objectives for pupils

1. Awareness of carbon as the key element in organic compounds

(Continued)

So far, our studies of the Periodic Table have been confined to the elements of Group I and Group VII. The attention of the class is now drawn to elements that occur in the middle of the table. There are a large number of elements which occur between the two extreme 'families' which have been studied so far. Carbon and silicon are examined because they occur in many everyday materials.

Use questions to remind pupils of the following characteristics of carbon: the occurrence of carbon in natural products (Topic B15);

2. Awareness of hydrogen as a common constituent in organic compounds
3. Revision of earlier work on the chemistry of carbon
4. Knowledge of the simple chemistry of silicon
5. Awareness of the similarities and differences in the properties of carbon dioxide and silicon dioxide
6. Awareness of the importance of silicon dioxide in the composition of the lithosphere

that carbon rods conduct electricity (see Stage I work on electrolysis); that carbon burns in oxygen to produce a gas, carbon dioxide, which dissolves in water to give an acidic solution (Topic B11). It may be necessary to demonstrate the fact that carbon rods (graphite) conduct electricity.

A large number of substances turn black when they are heated, suggesting the presence of carbon (*cf.* Experiment B14.1a). This is not a completely satisfactory test and a more convincing one needs to be found. This was attempted in Experiment B15.2a. Discussion will reveal whether or not pupils appreciate the use of copper(II) oxide in this context. Where appropriate, pupils may try heating organic compounds with copper oxide to revise this set of reactions. In the course of such an experiment, pupils should notice that not only is the copper oxide reduced and carbon dioxide produced, but also that droplets of liquid condense on the side of the test-tube. These drops can be proved to be water.

Point out that silicon, the element below carbon in the Periodic Table, is an element which might be expected to have properties similar to those of carbon. Introduce silicon as the 'element of the rocks'. How can we get a sample of pure silicon? One readily available source of silicon is ordinary sand, which is formed by the breakdown and weathering of rocks. Show the pupils some sand, tell them it is silicon dioxide, and ask them to suggest ways of getting the element silicon from it. Refer to the reactivity series. Pupils should also be able to suggest that a reactive metal or non-metal might remove the oxygen. They could be allowed to follow their own choice, for example by heating carbon and silicon dioxide together in a hard-glass test-tube. The obvious need for a more reactive element can lead to the use of magnesium in Experiment B16.6a.

Experiment B16.6a

To examine the reaction of magnesium with silicon dioxide (silica)
This experiment *MUST* be done by the teacher.

Caution. Since explosions have been known to occur during this experiment, it is *essential* to use *dry* sand in a *dry* tube. A safety screen and safety spectacles should be used.

Apparatus

The teacher will need:

Plastic safety screen

Safety spectacles

Hard-glass test-tube, 100×16 mm

Test-tube holder

Bunsen burner and heat resistant mat

Beaker, 100 or 250 cm^3

Watch-glass, large enough to rest on top of the beaker

Tripod and gauze

Funnel and filter paper

(Continued)

Procedure
About 2 g of an intimate mixture of magnesium powder (one part by volume) and dry purified sand (two parts by volume) are placed in a hard-glass test-tube and clamped horizontally. Heat the mixture at the end nearest the mouth of the tube. When the mixture glows, follow the glow with the flame to the bottom of the tube. A sample of impure silicon can be obtained as follows. When the tube is cool shake its contents onto a heat resistant mat for examination and then transfer them to a beaker containing about 20 cm^3 of 2M hydrochloric acid. This will react with the magnesium oxide and any magnesium silicide that may be present. Any gaseous hydrides of silicon formed will ignite on contact with air. (*The teacher should be prepared for these harmless explosions!*)

Funnel stand

Magnesium powder

Sand

2M hydrochloric acid

Sodium hydroxide solution

Limewater

Boil the contents of the beaker, filter the hot mixture, and wash the silicon with a little hot dilute hydrochloric acid followed by a small quantity of water. Dry the paper and silicon on a watch-glass over boiling water.

Point out the appearance of the product, test its solubility in water, and show there is no reaction with hydrochloric acid, dilute sodium hydroxide solution, and limewater.

Draw up a comparative list of the properties of silicon dioxide and carbon dioxide. Although carbon dioxide is a gas and silicon dioxide is a solid of very high melting point, similarities can be found. From analysis the formulae are shown to be similar – SiO_2 and CO_2. Pupils may remember that magnesium is able to remove oxygen from carbon dioxide in much the same way as magnesium was used to remove oxygen from silica. They may want to know why two oxides of elements next to each other in the Periodic Table which have similar formulae have such different physical properties. This can lead to a discussion about structure (Topic B17).

The teacher should also indicate that carbon dioxide and silicon dioxide are acidic oxides and will react with alkalis. Thus, the reaction between carbon dioxide and sodium hydroxide solution to form sodium carbonate solution can be demonstrated by standing a test-tube of carbon dioxide in a dish of sodium hydroxide solution:

$$CO_2(g) + 2NaOH(aq) \longrightarrow Na_2CO_3(aq) + H_2O(l)$$

However, the corresponding reaction of silicon dioxide with sodium hydroxide is less easy to carry out in a school laboratory. (To achieve a 'visual' indication of reaction taking place is far from simple.)

$$SiO_2(s) + 2NaOH(l) \longrightarrow Na_2SiO_3(l) + H_2O(g)$$

Pupils should appreciate that silicon occurs principally as silicon dioxide and as silicates. Earlier work (see Topic B12) made reference to the fact that silicon is a major constituent of the Earth's crust. Figure B16.5 may be used to illustrate the layer-like construction of the Earth. It will be found helpful to relate this presentation to the estimated temperatures for the various layers (and to their physical states) as in Table B16.3.

	Estimated temperature range /°C	Physical state
core	3000–5000	semi-solid
mantle	650–3000	liquid
crust	0–650	solid
atmosphere	−90–60	gas
hydrosphere	0–30	liquid plus polar ice caps

Table B16.3 [part of OHP 42]

The scale of this presentation can be conveyed by likening the Earth to a football. The Earth's crust (the rocky layer) can then be compared to the thickness of a postage stamp stuck onto the surface of the ball.

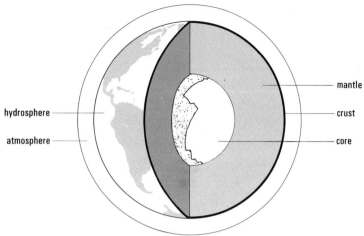

Figure B16.5
The structure of the Earth. (OHP 42)

The liquid material of the mantle (magma) is referred to as lava when it reaches the surface in a volcanic eruption. The extreme pressures exerted on the core by the other layers are considered to be responsible for it being solid rather than liquid.

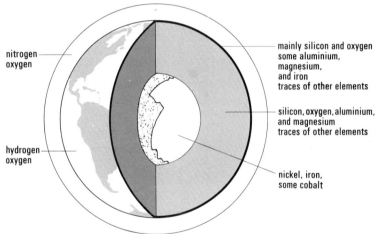

Figure B16.6
The distribution of the elements. (OHP 43)

Figure B16.6 shows the distribution of the most abundant elements in the various layers and striking differences in composition will be noted. Evidence for this summary comes from seismic surveys, analysis of meteorites, etc. When using figures B16.5 and B16.6, and Table B16.3, draw attention to:
the atmosphere, which contains abundant oxygen;
the hydrosphere, which contains abundant water;
the biosphere – defined as those parts of the Earth's crust, the hydrosphere and atmosphere, which contain living organisms – based on the element carbon;
the lithosphere, which contains abundant silicon in the form of rocks.

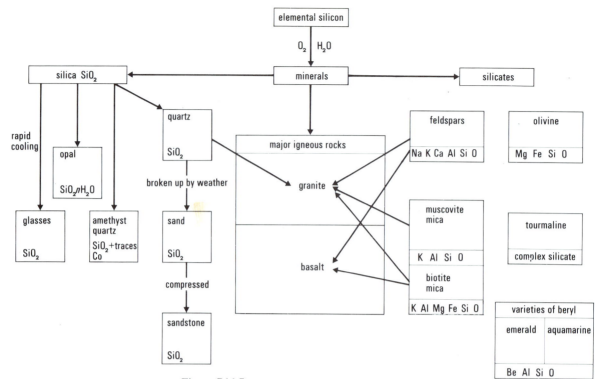

Figure B16.7
A layout for a display of silicon-containing minerals.

Figure B16.7 shows a possible layout for a display of silicon compounds. If good samples of minerals are not available, a series of photographs and slides could be used. If pupils are to handle specimens, hand lenses must be provided. The display should show that:
1. Practically all silicon compounds in the lithosphere are either forms of silica or are silicates.
2. Rocks are bulk structures composed of a mixture of minerals, that is, a mixture of chemical compounds.
3. Igneous rocks are considered to be the original rocks resulting from the solidification of magma from the mantle.

Arrange the display so that pupils can make sketches of the specimens. A discussion based on the pupils' observations can follow, and points which are likely to occur include:
1. Crystallinity and shape of mineral specimens, especially the hexagonal outline of quartz.
2. The two-dimensional sheet-like structure of mica.
3. The fact that quartz and mica can occur together in specimens of igneous rocks (and can be recognized by their crystalline shapes).
4. The non-crystalline structure of sandstone, due to fragmentation of quartz crystals and subsequent compression to form this sedimentary rock.
5. The poorly defined shape of opal (which may be considered to be a 'gel' of silica, $SiO_2 nH_2O$) and its characteristic display of colour.
6. 'Silica glass' has no definite shape since it was formed by being cooled too rapidly to permit crystal growth.

7. Silicate minerals are often coloured and such colours are due to the presence of specific metals as silicates, as shown in Table B16.4.

Mineral	Colour	Metal silicate present
olivine	green	iron, magnesium
biotite (brown mica)	yellow-brown	iron, magnesium, aluminium, potassium
muscovite (white mica)	colourless	no iron; aluminium and potassium
amethyst quartz	purple	manganese

Table B16.4

The formation of molten silicates in the Earth's mantle is speculative but must involve the formation of silicon dioxide and its subsequent reaction at high temperature with metal oxides. This could be discussed before going on to the 'crystal garden' demonstration. Experiment B16.6b is intended to show how the seeding of a solution of sodium silicate with other salts leads to a rapid growth of crystalline silicates of characteristic colours.

Making a 'crystal garden'

Apparatus

Each pair of pupils will need:

Glass jar (for example, a jam jar)

Watchglass (to cover the glass jar)

Glass rod

Sodium silicate solution

Hot distilled water

A few crystals each of cobalt nitrate, nickel nitrate, iron(III) nitrate, manganese sulphate, magnesium nitrate or magnesium sulphate

Procedure
Pour sodium silicate solution (a viscous liquid) into a glass jar to a depth of 30 mm. Add hot water to this solution, stirring well during the addition. The final depth of liquid required is about 120 mm. Stirring should continue until no separate silicate layer is visible.

Allow the solution to stand and when the liquid is quite still add the crystals. It is important *not* to allow the crystals to fall close to each other. Cover the jar and leave overnight – if possible.

When pupils have inspected their crystal gardens, remind them that the actual formation of silicates in the Earth's crust occurs in a more complex manner. The formation of silicates in the molten state from silica and metal oxides would be difficult to achieve in a school laboratory: the crystal garden enables pupils to 'see' the formation of coloured silicates and to enjoy the spectacle!

The section may be concluded by comparing the chemistry of carbon and silicon. Thus, carbon is present in compounds from plant and animal sources; such compounds on heating usually give a residue of carbon; and carbon dioxide and steam are usually formed when such compounds are heated with copper(II) oxide. Pupils should also be aware that silicon is an abundant element in many rocks, and that sand is silicon dioxide. Unlike carbon compounds, silicon compounds do not decompose on heating to form the element silicon. During the study pupils will have had some experience in comparing and contrasting the compounds carbon dioxide and silicon dioxide.

1. Write an account of the way in which rocks were first formed.
2. Take a piece of tracing paper and cut it to the size of your copy of the Periodic Table. Choose one physical property (such as melting point) and write the melting points of all the elements you have met so far on the piece of paper so that when you place it over

your copy of the Periodic Table, the melting points are just below the element symbol to which they refer. Comment on your findings.
3. Make a short summary of the work you have covered in this Topic.
4. Read *Chemists in the world*, Chapter 8 'Ceramics and glass'.

Summary

By the end of this Topic, pupils should be aware that the Periodic Table provides a convenient way of arranging a list of elements in order of their relative atomic masses, so that like elements fall into vertical groups. They should be aware that elements in vertical groups of the table show a simple gradation of properties. Thus, for the alkali metals of Group I, reactivity increases with increase in relative atomic mass. In the case of the halogens of Group VII, it decreases on going down the group. Pupils should be aware that metals occur on the lefthand side of the Table and that non-metals occur on the righthand side. From now onwards, they should relate the properties of any element they meet to its position in the Periodic Table. The work covered in this Topic should have included some reference to the alkali metals, the inert (noble) gases, the halogens, carbon and silicon.

Appendix

Classifying elements: cards for use with Experiment sheet 113
Each card shows the name, symbol, and relative atomic mass of one of 35 elements, as listed below. The cards are tinted to show similarities in chemical properties, as follows:

White	H	Li	Na	K				
Yellow	Be	Mg	Ca					
Red	B	Al	Ga					
Buff	C	Si	Ge					
Blue	N	P	As					
Pink	O	S	Se					
Grey	F	Cl	Br					
Green	He	Ne	Ar	Kr				
Orange	Sc	Ti	Cr	Mn	Fe	Co	Ni	Cu	Zn

Envelope 1: Some elements
H He Be B   C   N   O   F   Ne Na Mg Li
Al Si P   S   Cl K   Ar Ca Ga As Se   Br   Kr

Envelope 2: The missing element
Ge

Envelope 3: Some transition elements
Sc Ti Cr Mn   Fe Co Ni Cu   Zn

Note. Normally, envelope 3 will only be of value if the cards are used to illustrate Topic A13 or B24.

The arrangement of atoms in elements and compounds: an introduction to structure

Timing

About three or four weeks are required.

Introduction to the Topic

In this Topic, pupils examine some of the consequences of the particulate theory of matter and especially those relating to the structure of materials.

The Topic opens with a short section based on the differences between solids, liquids, and gases. The emphasis here is on providing an image of the role of atoms in chemical reactions. In the next section, pupils study the shapes of some crystals and are led to appreciate that such regularity provides strong evidence for the orderly way in which the constituent atoms are arranged. The term 'giant structure' is introduced and the structure of sodium chloride is discussed. The evidence for this interpretation of structure may be reviewed briefly and analogues of X-ray diffraction patterns may be used by the teacher in the presentation of this part of the Topic. No attempt is made to distinguish between atoms and ions at this point in the course.

In the next section the term 'molecule' is revised and here the emphasis is on developing an image of the structure of molecular substances. Evidence is presented to suggest that atoms are held together strongly in giant structures, whereas in molecular structures molecules are fairly readily separated. The subject of allotropy receives attention and allotropes of sulphur are prepared by the pupils.

Finally, pupils are introduced to the idea of a chemical equation as a means of summarizing the events which take place during a reaction.

The background knowledge provided by this Topic is used to pave the way for the mole concept in Topic B18 and for other work in the remainder of the course.

Alternative approach

This Topic may be used to complete the sequence to Part 2 of Alternative IIB, B14, B15, B16, and B17, *or* to introduce the same material in a modified order. Thus Topic B17 could follow from B13; then B15.5–B15.8; B15.1–B15.4; and finally Topics B14 and B16.

The material appears in quite a different order in Alternative IIA. Parts of Topics A14 and A15 correspond closely to B17, although it must be appreciated that energy considerations are deferred until Topic B23.

Further references
for the teacher

Handbook for teachers, Chapters 9, 10, and 11.
Collected experiments, Chapter 13 provides alternative experiments for the Topic.

Supplementary material

Film loops
2–3 'Sulphur crystals'
2–5 'Liquid–gas equilibrium'
2–6 'Solid/liquid equilibrium'
2–7 'Movement of molecules'
2–16 'The formula of hydrogen chloride'
2–17 'Solids, liquids, and gases'
Films
'Considering crystals' Unilever Film Library
'Crystal structure' ICI Film Library
'Exploring chemistry' (a film for teachers) Unilever Film Library
'The structure of matter' ESSO. Films for Science Teachers No. 18
'Molecular theory of matter' Encyclopaedia Britannica
'Properties of matter: Part 1. Solids, liquids, and gases' EFVA
'Properties of matter: Part 2. Atoms and molecules' EFVA

Overhead projection originals.
44 The structure of argon *(figure B17.1a)*
45 The structure of mercury *(figure B17.1b)*
46 The structure of copper *(figure B17.1c)*
47 The structure of sodium chloride and of lithium chloride *(figures B17.4 and B17.5)*
48 The structure of bromine *(figure B17.6)*
16 Changes in the arrangement of sulphur atoms on heating *(figure A14.27)*
17 Packing of S_8 rings *(figure A14.28)*
49 Reaction of hydrogen and oxygen *(figure B17.11)*
50 Reaction of magnesium and oxygen *(figure B17.13)*

Reading material
for the pupil

Handbook for pupils, Chapter 9 'The structures of elements and compounds'.
J. Ogborn (1973) Nuffield Physics Special *Molecules and motion.*

B17.1
The arrangement of atoms in solids, liquids, and gases

Differences between solids, liquids, and gases are related to differences in the arrangement of atoms. Examples studied in this section include copper, mercury, and argon.

Objectives for pupils

1. An awareness that differences in the properties of solids, liquids, and gases depend on differences in structure
2. The development of an image of how atoms are arranged in solids, liquids, and gases as exemplified by copper, mercury, and argon, respectively

Note for teachers
A very similar diagram to figure B17.1 is reproduced in Chapter 9 of the *Handbook for pupils.*

The presentation and development of this section will depend on the background knowledge of the pupils and on whether the teacher wishes to incorporate certain aspects of the kinetic theory of matter. The teacher may like to introduce the section with a simple qualitative discussion of, for example, the effect of heat on ice, water, and steam, and to enquire how pupils might account for differences in the properties of these forms of water. Pupils might be asked how they expect the particles comprising a solid, liquid, or gas to be arranged. They should be aware that solids and liquids are comparatively dense but that only liquids flow, and that gases possess lower densities than liquids or solids and are compressible.

The following examples are suggested. Figure B17.1 is in three parts and shows diagrams which are also available as overhead projection originals. Copper, mercury, and argon each consist of an arrangement of atoms rather than of other more complex particles. For the purposes of the discussion, we are not concerned to know how such information was obtained. We make use of the results of structure determinations to focus attention on differences in the structure of materials in the three states of matter. In a solid, such as metallic copper, the atoms are tightly packed and are held firmly together. In a liquid, such as mercury, the atoms are tightly packed but are free to move about within a definable and recognizable volume. Finally, in a gas, the atoms (or in other instances molecules) are widely separated and are free to move about to fill the space in which they are placed. This last point can be taken further by asking pupils to provide evidence to support this view. Reference to the release of, say, ammonia into a room full of air or of a coloured vapour, such as bromine, into a known volume of air or into a vacuum, can be a useful demonstration to aid discussion and lead to a better understanding of the kinetic theory.

Experiment B17.1a

Demonstration of molecular diffusion and motion

Apparatus

The teacher will need:

Gas jar

Coverslip

Crucible, the smallest size available

Teat pipette

Access to fume cupboard

Sheet of white card

Bromine

2M sodium hydroxide or 2M ammonia solution

Procedure
Transfer a few drops of bromine to the crucible. Stand the crucible on the coverslip and then place the inverted gas jar over the coverslip. Draw attention to the time required for bromine to diffuse into the air inside the gas jar. Place a sheet of white card behind the gas jar so that pupils can view the process.

After the demonstration transfer the apparatus to a well-ventilated fume cupboard and absorb the bromine in the apparatus by using either sodium hydroxide or ammonia solution and then water.

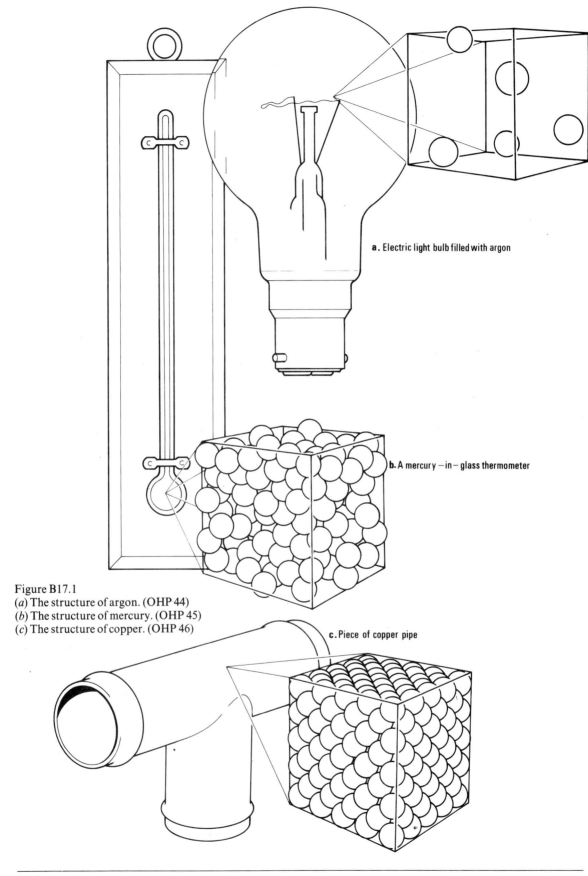

a. Electric light bulb filled with argon

b. A mercury − in − glass thermometer

Figure B17.1
(a) The structure of argon. (OHP 44)
(b) The structure of mercury. (OHP 45)
(c) The structure of copper. (OHP 46)

c. Piece of copper pipe

Demonstration of high speed molecular motion

Apparatus

The teacher will need:

2 bromine diffusion tubes (see Nuffield Physics; item 8)

1 pair of pliers

Access to a vacuum pump

Pressure tubing

Vaseline

Sheet of white card

2 clamps, bosses, and stands

1 pair of rubber gloves

2 ampoules of bromine (see Nuffield Physics; item 8)

Plastic bucket containing ammonia solution

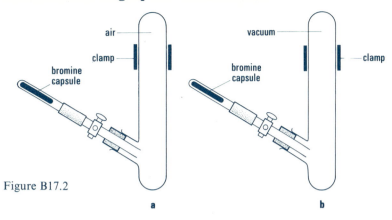

Figure B17.2

a   b

Procedure

Arrange two bromine diffusion tubes side by side on the bench – as shown in figure B17.2. To show the effect of air on the movement of bromine molecules in the tube, use a pair of pliers to break the ampoule of bromine in the delivery tube of the apparatus and open the tap to admit bromine into the tube.

Tube (b) is under vacuum. The delivery tube is attached as shown in figure B17.2. On breaking an ampoule of bromine in the delivery tube, and releasing the bromine into the tube, it will be seen that the tube is instantly filled with bromine vapour. It should be noted that this demonstration also shows the extremely high speed with which molecules move – as predicted by a much more advanced treatment of kinetic theory.

Note. At a convenient time, the apparatus may be cleaned by dismantling the various parts in a bucket of ammonia solution. After rinsing the parts with water and drying, the apparatus may be reassembled. Rubber gloves must be worn when cleaning the apparatus.

Discussion will be necessary to account fully for the behaviour of bromine in these experiments. By the end of this section, pupils should be aware of the basic differences between solids, liquids, and gases. (A detailed discussion of the kinetic theory appropriate to this level is provided in the Nuffield Physics Special *Molecules and motion*, by Jon Ogborn (1973), Longman/Penguin, and this may be appropriate for use by an able class.)

Summarize the important features of this section.

B17.2
How are atoms arranged in salts?

A revision of the appearance of crystals of a number of salts is used to lead to an understanding of the structure of a typical salt. The use of X-rays in obtaining details of the structure of crystals is mentioned and the term 'giant structure' is introduced.

First remind pupils of the experiments (section B12.3) in which they watched the formation of crystals from hot saturated solutions of salts on cooling to room temperature. For our present purpose, it is advisable to select only a few examples for discussion and to use compounds which yield crystals of regular shape fairly easily. In the experiment below pupils watch crystals growing. Alternatively, the teacher could demonstrate this, and follow it by passing round some crystal samples. The inspection of large and small crystals of, say, sodium chloride and potassium aluminium sulphate, enables the pupils to appreciate that regularity in shape is independent of the size of a crystal.

A suggested approach

Objectives for pupils

1. To study the shapes of some simple crystals
2. To infer from the regular shapes of these crystals that the atoms from which they are made are arranged in a regular manner
3. To understand the meaning of the term giant structure
4. Knowledge of the structure of sodium chloride

Experiment B17.2

Apparatus

Each pair of pupils will need:

Experiment sheet 55

3 watchglasses

1 teat pipette

Access to microscope or hand lens

Microscope slides

The teacher will need:

7 teat pipettes

7 beakers, 250 cm³

7 test-tubes, 16 × 150 mm

Test-tube rack

7 tripods and gauze

7 Bunsen burners and heat resistant mats

Saturated solutions of:
Ammonium chloride
Potassium chromate
Ammonium nitrate
Potassium nitrate
Potassium chlorate
Sodium hydrogen sulphate
(*Continued*)

Watching crystals grow

Procedure
Before the lesson, prepare warm saturated solutions from the salts listed above and check that the solutions will deposit the crystals in a reasonable time when poured onto a watchglass. These solutions may be kept at just below the boiling point while the teacher is introducing the lesson. The Bunsen burners should be extinguished and the solutions, in clearly labelled beakers, should be placed in accessible positions around the laboratory when practical work commences.

Remind the pupils to use warmed teat pipettes to transfer the hot saturated solutions to their watchglasses. (Warm the teat pipettes by using hot water from the beakers and allow the teat pipettes to drain thoroughly before transferring the hot saturated solutions of salts to the watchglasses.)

The teacher may demonstrate the crystallization of acetamide (ethanamide) from acetone (propanone) by placing a few drops on a microscope slide and allowing the acetone to evaporate. The use of microscope slides for crystallization facilitates the use of a microscope for observing the effects.

Experiment sheet 55
You will have grown some crystals earlier in the course. In this experiment you will concentrate on watching crystals grow rather than on the crystals that are formed.

Warm solution of acetamide
(ethanamide) in acetone
(propanone)

Supply of distilled water

Your teacher will provide hot, saturated solutions of a variety of substances. Using the teat pipette in one of the solutions transfer about 2 cm³ of it to a small watch glass. Stand the watch glass on the bench and observe carefully the formation of crystals as the solution cools, noting any points of interest below. Make a sketch to show the appearance of one or two of the crystals.

Repeat the observations and sketches for other solutions, using a clean watch glass in each case.

The following results are expected:
ammonium chloride – dendrites
potassium chromate – 'fluffy' crystals
ammonium nitrate – needles
potassium nitrate – needles
potassium chlorate – thin plates
sodium hydrogen sulphate – needles

Figure B17.3
See figure A14.3, which shows samples of (*a*) ammonium chloride, (*b*) potassium chromate, (*c*) ammonium nitrate, (*d*) potassium nitrate, (*e*) potassium chlorate, (*f*) sodium hydrogen sulphate. (Available in a series of slides)

Figure B17.3 illustrates the meaning of these specialist terms.

When practical work has been completed and results have been discussed, pass round some examples of large crystals for inspection. Pupils are usually fascinated by the shape and colour of crystals. They need very little encouragement to make sketches of them in their notebooks.

During the discussion, the relationship between regularity in crystal shape and internal structure can be explored. A four-sided pyramid might be constructed from cubic building bricks and this result compared with the effect of emptying onto the bench a large bag full of building cubes of identical shape and size to those used in making a pyramid. By repeating the process of emptying the bag onto the bench, the impossibility of reproducing the regular pyramid should make the teaching point using the model all the more obvious.

The *Handbook for pupils* contains several photographs of large crystals in Chapter 9.

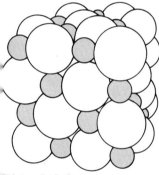

Figure B17.4
The sodium chloride structure (polystyrene spheres in contact). (part of OHP 47)

Without further explanation discuss with the pupils how the atoms are arranged in a giant structure such as that of sodium chloride* or some other crystalline salt. What are the possibilities? What are the simplest arrangements? If we assume that sodium chloride contains the same number of sodium and chlorine atoms, we can make several simplifications. With a little help, pupils may suggest either or both of the arrangements shown in figures B17.4 and B17.5.

Tell the pupils that the structure of sodium chloride has been found to correspond to figure B17.4. This structure was determined by using chemical analysis to find the composition of sodium chloride, and then an X-ray technique to form patterns (known as X-ray diffraction patterns) from which the structure could be worked out. The point to be made is that someone had to determine the structure of a crystal from information acquired during an experiment.

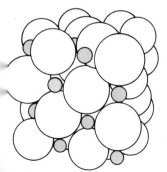

Figure B17.5
Lithium chloride structure. (part of OHP 47)

*It will be helpful if pupils become acquainted with the sodium chloride structure, as they will be meeting a similar arrangement of atoms in Topic B18.

The subject of X-ray diffraction patterns could form a natural sequel to this introduction to structure, and the approach put forward in Topic A14 can be adapted for use in this scheme. Alternatively, pupils could read an account in *Handbook for pupils*, Chapter 9 'The structures of elements and compounds'.

Suggestion for homework

Read the *Handbook for pupils*, Chapter 9 on crystal structure and the use of X-ray diffraction patterns.

Summary

By the end of this section, pupils should be aware that crystals have strikingly regular shapes and that this regularity provides strong evidence for a regular array of atoms in crystals. They should also know how atoms are arranged in sodium chloride and that this is an example of a giant structure.

B17.3
The idea of the molecule

The term molecule is presented formally in this section and differences in the properties of giant structures and molecular structures are explored experimentally.

A suggested approach

Objectives for pupils

1. Knowledge of the structures of molecular substances in gaseous, liquid, and solid states
2. Knowledge of the strong bonding between atoms in giant structures and of the weak bonding between molecules in molecular structures

So far, pupils following this alternative have not been introduced to the idea of the molecule. Historically it required a long period of time to establish currently accepted formulae for molecular materials like nitrogen, N_2; oxygen, O_2; water, H_2O; and so on. (*Note to teachers.* In Topic A14, pupils are provided with evidence about the atomicity of gases through applying Avogadro's hypothesis to information on the volumes occupied by a mole of molecules of each of the various gases.)

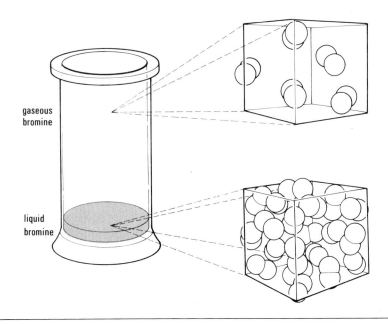

gaseous bromine

liquid bromine

Figure B17.6
Structure of bromine. (OHP 48)

The idea of the molecule in this section is seen at its simplest in gases and then through a consideration of the liquid state. Figure B17.6 (which is also available as an overhead projection original) distinguishes between the gaseous and liquid states, and can be used to introduce Experiment B17.3a.

Molecular models of liquid bromine and of solid iodine

Apparatus

The teacher will need:

Gas jar

Coverslip

Teat pipette

Enough models of the molecule of bromine or iodine to fill a transparent perspex cube (*Note*. Polystyrene spheres of 2.5 cm diameter are a convenient size to use for making these models.)

Large transparent sandwich box, cube, or large beaker

Watchglass

Spatula

2M sodium hydroxide

Bromine

Iodine

Procedure
First, tell the pupils that the models represent what is believed to be the smallest particle of bromine, called a 'molecule' of bromine.

1. Place a few of the molecular models in the cube and agitate the cube to simulate the structure of gaseous bromine. 'Pour' some of the model molecules into the cube so as to indicate the random arrangement of molecules in liquid bromine. Keep the molecules moving through agitation of the transparent cube. Finally, arrange some of the molecular models in a *regular* way (as in solid iodine) and explain that in the solid state, the molecules are less mobile than in the liquid state.

2. Compare the samples of bromine and iodine (contained in a gas jar and watchglass respectively). Note the odour (**caution!**) of solid iodine and compare this with that of bromine and with the relative ease of vaporization of bromine.

Having provided these simple images of molecular substances, contrast them with the giant structure of sodium chloride (used in the previous section). Through questioning, lead the pupils to the point that atoms in giant structures are held together very strongly whereas molecules in molecular structures are relatively easy to separate.

Apparatus

The teacher or each pair of pupils will need:

2 test-tubes

Test-tube holder

Test-tube rack

Bunsen burner and heat resistant mat

Spatula

Sodium chloride

Iodine

Comparing how easily the atoms may be pulled apart in sodium chloride (a giant structure) with how easily the molecules may be pulled apart in iodine (a molecular substance)

This experiment may be performed by the teacher or the pupils.

Procedure
Transfer a small quantity of sodium chloride into one test-tube and a crystal of iodine into the other test-tube. Heat each test-tube equally over a Bunsen burner flame and record differences in behaviour. (**Caution** needed with iodine experiment.)

The difference in behaviour of these two substances is very marked. The iodine is relatively easily vaporized whereas sodium chloride will probably not even melt under the circumstances. One may, at this stage, generalize and tell the pupils that one would expect molecular substances to have low melting points while one would expect substances having giant structures to have high melting points. A similar point can be made about the boiling point of a substance. Atoms in giant structures are held together strongly whereas molecules in molecular structures are relatively easy to separate although the atoms making up a molecule are strongly held together. With an able class, the terms for forces within a molecule, 'intra-molecular bonding', and forces between molecules, 'inter-molecular bonding', may be introduced.

Suggestion for homework

Make a list of twelve common substances from those listed in the Reference section of the *Handbook for pupils*. Record the properties of those substances and deduce which have molecular and which giant structures.

Summary

By the end of this section pupils should have an idea of the meaning of the term molecule as exemplified by gaseous and liquid bromine and solid iodine. They should also understand that atoms in giant structures, such as sodium chloride, are held strongly together. In addition, they should realize that in molecular substances the atoms making up the molecule are strongly held together whereas the forces between the molecules are relatively much weaker.

B17.4
Carbon and sulphur: some structural studies

Allotropic forms of carbon and of sulphur are studied and their properties are related to their structures.

A suggested approach

Objectives for pupils

1. Knowledge of the allotropes of carbon
2. Recognition of a giant structure
3. Knowledge of the allotropes of sulphur
4. Recognition of the idea that atoms of an element can join together to form molecules

Pupils should be aware that carbon is an element which occurs in all living things (Topic B16). They can now be told that the free element occurs in two very different forms – graphite and diamond. Slides showing good specimens of these allotropes could be shown (see figure B17.7) and their properties compared.

The main differences between graphite and diamond are listed in the following table.

Graphite	Diamond
Conducts electricity	Does not conduct electricity
Opaque to light	Transparent to light
Soft enough to mark paper and easily split into flakes	One of the hardest substances known: cleaved only with difficulty
12 g occupies 5.3 cm^3	12 g occupies 3.4 cm^3
Burns in oxygen to form carbon dioxide	Burns in oxygen to form carbon dioxide

Table B17.1 Properties of graphite and diamond compared

Figure B17.7
See figure A14.29 which shows (*a*) graphite, (*b*) diamond in rock, (*c*) diamonds in jewellery and a natural diamond. (Available in a series of slides)

This information suggests that differences between graphite and diamond must be due to differences in the arrangement of carbon atoms in the two structures. Reference may be made to the use of X-ray diffraction for the determination of crystal structures (see section A14.4 for detailed suggestions). X-ray diffraction analyses led chemists to believe that the carbon atoms in graphite are arranged in flat hexagons in layers; the atoms in diamond are arranged tetrahedrally so that each atom is joined to four others to give rise to an unending three-dimensional network. Models of these structures may then be shown to the class. (See Experiment A14.8 for details for making these models.) Pupils should be told that the existence of an element in two or more distinct forms in the same state is known as allotropy and that the various forms are called allotropes.

The next experiment investigates the effect of heat on graphite. The findings can be compared with those obtained in Experiment B17.3b.

Experiment B17.4a

Comparing how easily the atoms may be pulled apart in graphite (a giant structure) with how easily the molecules may be pulled apart in iodine (a molecular substance)

Apparatus

The teacher will need:

2 test-tubes

Test-tube holder

Test-tube rack

Bunsen burner and heat resistant mat

Spatula

Graphite

Iodine

Procedure

Transfer a small quantity of graphite into one test-tube and a crystal of iodine into the other test-tube. Heat each test-tube equally over a Bunsen burner flame, and record differences in behaviour. Compare the results obtained with those from the previous experiment, Experiment B17.3b.

Carbon is not unique in its ability to exist in the same state in more than one form, each having different physical properties. Sulphur is another element which exhibits allotropy.

Experiment B17.4b

Making various allotropes of sulphur

Note. The preparation of rhombic sulphur should be done by the teacher and not by pupils since the solvent used is toxic and flammable.

Apparatus

The teacher will need:

Test-tube, 100×16 mm

Watchglass

Spatula

Filter paper

Powdered roll sulphur *or a* pestle and mortar and a small lump of roll sulphur

Carbon disulphide (**Caution.** Keep away from naked flames and store in a fume cupboard when not in use. The vapour of this liquid is extremely toxic.)

Apparatus

Each pair of pupils will need:

Experiment sheet 57

Hard-glass test-tube, 100×16 mm

Beaker, 250 cm³

1. Preparation of rhombic sulphur

Procedure

Powder a piece of roll sulphur, about the size of a pea, *or* use a spatula measure of powdered roll sulphur, and add to it about 2 cm depth of carbon disulphide in a test-tube. Agitate the mixture (but do not warm it). When most of the sulphur has dissolved, decant the solution onto a watchglass in a fume cupboard. Allow the solution to evaporate slowly. (This may be done by covering the watchglass with a sheet of filter paper.) Crystals of rhombic sulphur will form after 10 to 20 minutes.

2. Preparation of monoclinic sulphur from liquid sulphur

This experiment and the one which follows, may be carried out by pupils at the discretion of the teacher. A well-ventilated laboratory is necessary since it is likely that some pupils will allow their samples of sulphur to catch fire, particularly during the second experiment.

(Continued)

Test-tube holder

Paper clip

Filter paper

Tongs

Bunsen burner and heat resistant mat

Sulphur

Procedure
Full details are given in *Experiment sheet* 57 but it is important for the teacher to emphasize the need for gentle and uniform heat during the experiment. Sulphur is a bad conductor of heat and some five or more minutes will be needed to achieve the desired effect by gentle and uniform heating with a Bunsen flame.

Experiment sheet 57
In this experiment sulphur crystals are obtained by allowing liquid sulphur to cool.

Three-quarters fill a test-tube with powdered roll sulphur. Fold a filter paper in the usual way and fasten it with a paper clip.

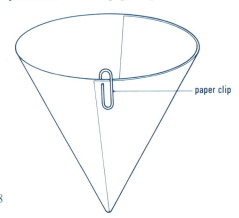

paper clip

Figure B17.8

Have ready a beaker of water.

Hold the test-tube in a holder and heat the sulphur very gently, keeping the test-tube moving all the time. It is important that the sulphur is only *just* melted, to a clear, amber-coloured liquid.

When all the sulphur has melted, hold the filter paper by the rim with a pair of tongs and pour the molten sulphur into it.

Allow the sulphur to cool until a crust has about half covered the surface and then pour out the remaining liquid sulphur into the water in the beaker. Immediately open the filter paper (be careful not to burn your fingers) and inspect the monoclinic sulphur crystals contained in it. Draw one or two of them.

Experiment B17.4c

Apparatus
Each pair of pupils will need:

Experiment sheet 58

Test-tube, 100×16 mm

Beaker, 250 cm^3

Test-tube holder

Bunsen burner and heat resistant mat

Postcard

Powdered roll sulphur

Observing what happens when sulphur is heated: making plastic sulphur

Procedure
Full details are given in *Experiment sheet* 58. The teacher should be prepared with a piece of card to hold over the mouth of any test-tube in which the sulphur catches fire. Again, pupils should be advised to use gentle and uniform heating until, on this occasion, the sulphur boils.

Experiment sheet 58
Gently heat a test-tube which is three-quarters full of powdered roll sulphur until all the sulphur melts, holding the test-tube in a holder. Observe carefully all that happens during the slow heating process and continue this

Topic B17 The arrangement of atoms in elements and compounds

until the liquid sulphur just boils. The vapour may catch fire, so be careful. (Your teacher will show you how to extinguish the flame if this does happen.)

Pour the liquid obtained into a beaker half-filled with cold water. Leave the sulphur in the water for a few minutes to cool and occupy the waiting time by describing below what you saw during the heating.

Now remove the sulphur from the water and examine its properties. It is called *plastic sulphur*. Knead a small piece between the finger and thumb for a few minutes and note any changes that take place. Record your observations below.

Figure B17.9
See figure A14.28 or OHP 17 which shows packing of S_8 rings.

After the experiments, show the pupils a model of an S_8 sulphur molecule. Tell them that X-ray crystallographers have found that in both rhombic and monoclinic sulphur, the sulphur atoms are arranged in groups of eight in rings just as in the model (see figure B17.9), but that the rings are packed together in a different way in the two allotropes. Point out the difference between these rings and giant structures. In the rings, the sulphur atoms are tightly bound together, so that even when the sulphur is molten rings of eight sulphur atoms still exist. The first liquid obtained on melting had a light colour and was comparatively non-viscous. This can be explained in terms of the ease with which the rings of sulphur atoms flow over each other. As the liquid was heated more strongly it became darker and more viscous and then finally less viscous just before boiling. This can be explained by supposing that the rings break open and then a number join end to end to form a large chain. The teacher could break open a model of an S_8 ring to form a chain and then join two or more S_8 chains. The effect of heat on sulphur is summarized in figure B17.10.

Figure B17.10
See figure A14.27 or OHP 16 which shows changes in the arrangement of sulphur atoms on heating.

These chains become intertwined and are unable to slide over each other very easily. Accordingly, the liquid becomes very viscous. The polymerization is greatest just below 200 °C when the largest chains are estimated to contain a million atoms. Between 200 °C and the boiling point of sulphur (444 °C), the chains tend to break into shorter lengths and so the viscosity of the liquid sulphur gradually falls. When almost boiling sulphur is poured into cold water, it is a mixture of chains and rings of sulphur atoms. This arrangement is 'frozen': there is not time for a general rearrangement of the atoms into S_8 rings as would occur with slower cooling.

Thus, plastic sulphur is a mixture of rings and chains of sulphur atoms. Further changes in structure in the solid state occur slowly. Details of the changes which occur when sulphur is heated appear in the *Handbook for pupils*, Chapter 9 'The structures of elements and compounds'.

Suggestion for homework

Write a short summary of the work done in this section. Include in your account a definition of the term allotropy.

B17.5
Chemical equations

The opportunity is taken here to suggest to pupils the ways in which chemical reactions may be summarized by using equations.

Objectives for pupils

1. Recognition of the way in which chemical reactions involve a rearrangement of atoms
2. The development of an image of chemical reactions using models of atoms, molecules, and giant structures
3. Recognition that a chemical equation provides a convenient method for summarizing a chemical reaction

From the previous sections of this Topic, pupils should have acquired some understanding of the distinction between giant structures and molecules. In earlier Topics, they were introduced to chemical equations. It is important that the writing of equations should be seen not as a purely algebraic exercise involving the juggling of symbols but rather as a convenient way of summarizing the way in which we think about the behaviour of atoms, molecules, and giant structures.

The writing of equations could be begun with the aid of models as follows:

1. *The reaction of hydrogen and oxygen*
Figure B17.11 (available as an overhead projection original) shows a way of summarizing this reaction.

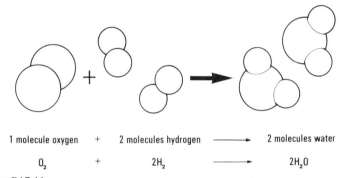

1 molecule oxygen	+	2 molecules hydrogen	$\longrightarrow$	2 molecules water
O_2	+	$2H_2$	$\longrightarrow$	$2H_2O$

Figure B17.11
The reaction of hydrogen and oxygen. (OHP 49)

Figure B17.12
See figure A14.42 which shows a magnetic board and hemispherical models illustrating chemical reactions.

It is possible to represent the rearrangement of atoms which takes place as the result of a chemical reaction using polystyrene spheres. However, it may be easier to do this using a magnetic board and hemispherical models such as those illustrated in figure B17.12 or described in *School Science Review* (for example, No. 170 page 126; No. 176 page 632; No. 178 page 120). When this has been done, explain that it is obviously easier to represent such a reaction by using symbols in place of models, for example:

$$H_2 \qquad\qquad\qquad H_2O$$
$$+ \quad O_2 \longrightarrow \qquad +$$
$$H_2 \qquad\qquad\qquad H_2O$$

or alternatively:

$$2H_2(g) + O_2(g) \longrightarrow 2H_2O(l)$$

using the state symbols as shown.

2. The reaction of magnesium and oxygen to give magnesium oxide

It is easy to let pupils acquire knowledge of such equations as:

$$2Mg(s) + O_2(g) \longrightarrow 2MgO(s)$$

without an adequate introduction. Similarly, it is easy to let pupils fall into the trap of thinking in terms of 'molecules' of magnesium oxide. Magnesium and magnesium oxide possess giant structures, as does sodium chloride. Accordingly before writing the equation to represent the combustion of magnesium and oxygen, it is highly desirable to present models of giant structures of magnesium and magnesium oxide. One way would be to represent the reaction as in figure B17.13.

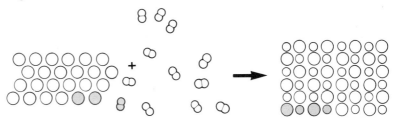

Figure B17.13
The reaction of magnesium and oxygen. (OHP 50)

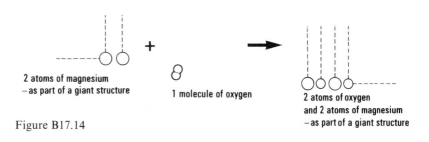

2 atoms of magnesium
– as part of a giant structure

1 molecule of oxygen

2 atoms of oxygen
and 2 atoms of magnesium
– as part of a giant structure

Figure B17.14

Point out that it would be tedious to draw such a large number of atoms on each occasion. We would be perfectly justified in reducing the number to those which have been shown in figure B17.13 as shaded 'atoms' – see figure B17.14.

If the pupils understand that when we write MgO, we are *not* implying that it is a molecule, we may now proceed to use the equation:

$$2Mg(s) + O_2(g) \longrightarrow 2MgO(s)$$

as a *convenient summary*. The significance of state symbols may be mentioned at this point. Equations for a few other reactions which pupils have already met either in Stage I or in earlier Topics in Stage II might now be discussed.

By the end of this section, pupils should be able to think about chemical reactions using chemical equations. A long time should *not* be spent on this matter at this stage. Pupils should *not* spend a long time on mere balancing of equations. Whatever is attempted should be limited to simple reactions.

Draw structures and write chemical equations for two or three simple reactions.

Summary

By the end of the Topic, pupils should be aware that the differences between solids, liquids, and gases are related to differences in the arrangement of atoms. They will know that chemists obtained further detailed knowledge about the structure of elements and compounds by using X-ray diffraction techniques. In addition, they will know that simple qualitative tests can give an indication of differences between giant and molecular structures. The Topic also includes detailed studies of the phenomena of allotropy in two instances, carbon and sulphur. A brief survey of chemical equations as reaction summaries concludes the Topic.

The mole and its use in the determination of the formulae of compounds

Timing

The whole Topic should not occupy more than five weeks.

Introduction to the Topic

The method of introducing the mole concept in this Topic is quite different to that used in the original Sample Scheme or in Alternative IIA. The main difference is that the pupils are working with a mental image not only of atoms but also of the ways in which atoms can combine. They have already seen a model of the sodium chloride structure as well as models of some simple molecules, such as those of water and iodine. Although this has not been stated explicitly, they have, to some extent, been led to expect simple whole number combining ratios.

In the first section of this Topic, the pupils look closely at the ballasted model of the magnesium oxide structure. They will see immediately that there is a 1:1 ratio of atoms. Then they carry out an exercise with this model which shows them that they could have found this ratio without direct visual counting by using the principle of counting by weighing. In this way, they are introduced to the essential logical processes involved in finding the formulae of compounds from combining masses by means of models which they can both see and handle – unlike the determination of combining ratios of actual chemical substances from experimentally determined masses, in which they are, of course, unable either to handle or to see the individual atoms.

When this work with ballasted models is completed, the pupils are introduced to the mole. They then determine the formulae of several compounds, and should be aware that their findings are subject to experimental error. Also, they will be aware of the fact that elements tend to combine in simple ratios by mass.

The Topic continues with a section in which pupils work out the volumes occupied by moles of various substances under laboratory conditions. This provides them with examples with which to exercise their knowledge of the mole concept. The Topic closes with two examples in which they determine the formula of a gas: hydrogen chloride, and ammonia.

The mole, which is introduced in section B18.3, is used throughout the remainder of Alternative IIB whenever masses of substances or concentrations of solutions need to be specified. Equations (see B17.4) also represent the molar quantities involved in chemical change, and this additional interpretation can be introduced at appropriate points in this and subsequent Topics. The pupils are introduced to molar solutions in Topic B20. Where possible, calculations should utilize the idea of the mole, and properties, such as heats of reaction, should be expressed per mole (mol^{-1}).

Note to teachers
The mole concept was introduced in the original Sample Scheme as part of the study of the particulate nature of matter. In this Alternative, the pupils are told the relative masses of the atoms and the masses of each element they will have to take to obtain the same number of atoms.

Many pupils find difficulty in understanding the mathematical relationships and chemical ideas in this part of chemistry. In Alternative IIB (and the alternative route through IIA) the introduction of the mole concept has been left much later than in the original Sample Scheme. This delay should alleviate some, but not all, of the difficulties experienced by pupils, since their ability to deal with relationships between abstract ideas is known to improve as they get older.

The careful use of models and graphs can help pupils to appreciate abstract concepts. There are some advantages to be gained from work with models prior to experimental work. First, exercises with models build on the work of previous Topics in which the external appearance of metals and simple compounds are related to their internal structure. Secondly, models provide pupils with a good visual image to help them to grasp the principles behind the calculation. Thirdly, the study of models provides a slower and more thorough introduction to the mole concept. Fourthly, conclusions from the experiments alone are open to question. Pupils are expected to infer two points from their experiments on the combining masses of elements: first that atoms are present in simple whole number ratios, and second, the values of these ratios. One major area of uncertainty stems from the inaccuracy of the pupils' own experimental results. It can be argued that it is unreasonable and undesirable to expect pupils to infer precise relationships from inaccurate results.

Of course, a discussion of the sources of experimental error can highlight the more obvious errors, but even so pupils may not be entirely convinced about the simple rules of chemical combination. The situation may be presented more convincingly by the combined study of models and of graphs in which the reacting masses are plotted (see figure B18.3). Arithmetical difficulties are reduced if a graph is used in which mass is plotted against moles of atoms (see figure B18.9). It is not easy, however, for some pupils to transfer the ideas which they have gained from looking at tangible model spheres to invisible atoms. It may help to ask how they would confirm the formula of magnesium oxide when the model structure is placed in a covered box.

Alternative approach

Topics A11 and A14 contain alternative presentations of this material. Two routes are provided, neither of which requires the use of ballasted spheres.

Background knowledge

In Stage I and again at the beginning of Alternative IIB, the meaning of the term element was explained. Pupils discovered that copper and magnesium react with oxygen to form their respective oxides and that copper oxide can be reduced to copper. So far in Alternative IIB, pupils have been shown that matter is made up of very small particles and have been given some idea of the mass of these particles. They have looked at several crystals, whose external appearance was related to their internal structure. The problem of trying to investigate particles that are too small to be seen has also been considered.

Much of the previous work, therefore, has been designed to give pupils a broad background to descriptive chemistry. Pupils will be familiar with some of the chemical changes in this Topic and will have seen models of the compounds used, such as magnesium oxide. They should now be ready to be introduced to the quantitative aspect of chemistry.

Further references

The *Handbook for Teachers* provides background information on this Topic. Chapter 4 deals with atoms, molecules, and ions, and Chapter 5 illustrates the mole concept and quantitative work. Additional help is provided in Chapter 9 which relates to the role of structure in the Sample Scheme.

Collected Experiments, Chapter 12, provides alternative experiments for use with this Topic.

Supplementary materials

Film loops
2–2 'The mole of atoms'
2–16 'The formula of hydrogen chloride'

Films
'Gases and how they combine' CHEMStudy. Distributed by Guild Sound and Vision Ltd.

Overhead projection originals
X Graph paper for use throughout the Topic
51 The reaction between zinc and hydrochloric acid *(figure B18.21)*

Reading material

for the pupil

Handbook for pupils, Chapters 3 'Atomic structure and bonding', Chapter 4 'Formulae and equations', and Chapter 9 'The structures of elements and compounds'.
Chemists in the world, Chapter 2 'Using ideas about the atom'.

B18.1
Examining a ballasted model of magnesium oxide

In Topic B16, the pupils acquired some understanding of the values of the relative atomic masses for certain elements, including magnesium and oxygen. This was done by using weighted or ballasted polystyrene spheres. In this section, they will study the structure of one compound, magnesium oxide, using a ballasted model. The section is divided into three parts: 'Counting spheres in a model'; 'The principle of counting by weighing'; and 'Using relative atomic masses'.

A   Counting spheres in a model

A suggested approach

Objectives for pupils

1. Calculating the simplest formula for magnesium oxide
2. Considering the meaning of the formulae MgO, Mg_2O, and MgO_2, using a model crystal of magnesium oxide

The ballasted magnesium and oxygen spheres, which were introduced during Topic B16, are used for construction of a special model of magnesium oxide (for details see Appendix, page 524).

Note to teachers. Experiment sheet 114 which follows is in three parts. Formula A relates to the 'complete' model and some pupils may appreciate that the ratio of magnesium to oxygen spheres is 1:1. They may find an exercise with a graph is helpful to the sequence. The two axes of the graph may be labelled (see figure B18.1) on the blackboard by the teacher and the pupils be asked to plot their findings. They will see that a straight line may be drawn through the various points.

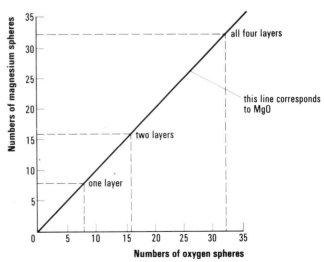

Figure B18.1
Establishing the relative numbers of magnesium and oxygen spheres.

If, therefore, the symbol Mg represents 1 sphere of magnesium and the symbol O represents 1 sphere of oxygen, then the simplest possible formula of magnesium oxide is MgO. Had there been twice as many magnesium spheres as oxygen spheres (formula B) then the simplest formula for magnesium oxide would be Mg_2O and the line would be in a different position. Formula C relates to the composition MgO_2.

Since there are thirty-two spheres of each kind in the first model supplied, some pupils may think that the formula is $Mg_{32}O_{32}$ – in which case it will be necessary for the teacher to explain that this is not the simplest formula for magnesium oxide! Later, in B18.3, the pupils will learn that the number of atoms in a mole of atoms is extremely large. Accordingly, in the first exercise it might be worth while to point out that there are equal numbers of magnesium and oxygen spheres however many atoms are present.

As a result of the exercise, three lines, each corresponding to a possible formula, can be obtained as shown in figure B18.2.

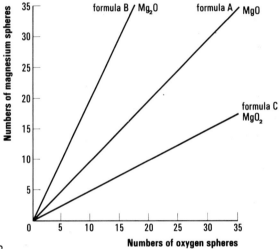

Figure B18.2
The lines for formulae B and C. (B is when half of the oxygen spheres are removed, C is when half of the magnesium spheres are removed.)

If a method can be devised by which spheres (that is to say, atoms) in a compound can be counted, then we have a direct method for establishing the formulae of compounds. Atoms are, however, very small, and it is not yet possible to count them directly. Nevertheless, the formula can still be obtained because it is only the relative or comparative numbers of atoms which are needed and not the absolute numbers.

Experiment B18.1a

Counting the spheres in a model of magnesium oxide

Procedure
Details are given in *Experiment sheet* 114.

Each pair of pupils will need:

Experiment sheet 114

1 complete model of magnesium oxide (see Appendix for construction details, page 524)

Experiment sheet 114

In the model of magnesium oxide with which you are provided, the gold spheres represent magnesium atoms while the red spheres represent oxygen atoms.

The framework of racks is simply a device to enable you to dismantle and reassemble the model quickly and easily. In magnesium oxide itself of course there is no such framework. You will learn about forces which hold the particles together a little later.

Before dismantling the model, study it carefully and note the positions of the gold and red coloured spheres.

Does each gold sphere make contact with another gold sphere?

After you have studied the arrangement of the spheres, dismantle the model. On a separate sheet of paper draw a diagram of each layer, as you dismantle the model, preferably colouring it in the same way as the model. There is no need to show the racks in your diagram – just draw the spheres. Your diagram will enable you to reassemble the model correctly.

Formula A
Complete the following table.

	Layer 1	Layers 1+2	Layers 1+2+3	Layers 1+2+3+4
No. of magnesium spheres				
No. of oxygen spheres				

Draw horizontal and vertical axes on a piece of graph paper. Label the horizontal axis 'Number of oxygen spheres' and the vertical axis 'Number of magnesium spheres'. Plot the numbers of magnesium spheres and oxygen spheres in the combined layers.

Join the points on the graph together. What do you get?

Formula A
If Mg represents a magnesium sphere and O an oxygen sphere, what is the simplest formula for magnesium oxide?

Write this formula at the top end of the line you have constructed.

Formula B
Remove half the oxygen spheres from each layer. Enter the numbers of magnesium and oxygen spheres present now in the table below.

	Layer 1	Layers 1+2	Layers 1+2+3	Layers 1+2+3+4
No. of magnesium spheres				
No. of oxygen spheres				

Plot these numbers on your graph and join the points together.

What would be the simplest formula for this compound?

Write this formula at the top end of the new line you have constructed.

Formula C
Re-assemble the model in its original form. Remove half the magnesium spheres from each layer. Enter the numbers of magnesium and oxygen spheres present now in the table below.

	Layer 1	Layers 1+2	Layers 1+2+3	Layers 1+2+3+4
No. of magnesium spheres				
No. of oxygen spheres				

Plot these numbers on your graph and join the points together.

What would be the simplest formula for this compound?

Write this formula at the top end of the third line.

Use your graph to complete the following table.

No. of oxygen spheres	No. of magnesium spheres		
	Formula B ( )	MgO	Formula C ( )
6			
10			
16			
20			
26			
50			
100			
1 000			
1 000 000			

By the end of this section, the formula of magnesium oxide has been established by direct counting. According to the model selected, it could be Mg_2O, MgO, or MgO_2. Some pupils will need additional exercises and suggestions are set out below either for use in class or as homework.

Additional exercises

1. Manganese and oxygen combine to form an oxide. Plot the numbers of manganese spheres against the number of oxygen spheres for the hypothetical formulae Mn_2O, MnO, and MnO_2 for 5, 10, 15, 20, 25, 30, and 35 oxygen spheres. When a model was inspected, it was found that the numbers of oxygen and manganese spheres were as follows:

Number of layers	Manganese spheres	Oxygen spheres
1	4	8
2	8	16
3	12	24
4	16	32

What are the *relative* numbers of manganese and oxygen spheres? What is the formula for this manganese oxide?

(*Note.* It should be made clear to pupils that manganese and magnesium are very different metallic elements. Manganese(IV) oxide was chosen as the first example because the model can be derived from that of magnesium oxide by using manganese atoms in alternate rows (see the Appendix at the end of this Topic).)

2. Iron and oxygen combine to form four main oxides. Plot graphs of iron atoms against oxygen atoms for the hypothetical formulae FeO, Fe_2O_3, FeO_2, Fe_3O_4.

The numbers of spheres of iron and oxygen in models for three of these materials are listed below. Which is the oxide that is NOT formed?

Numbers of layers	Compound I		Compound II		Compound III	
	Fe	O	Fe	O	Fe	O
1	8	8	—	—	6	8
2	16	16	—	—	12	16
3	24	24	16	24	18	24
4	32	32	—	—	24	32

3. Calcium and chlorine combine to form a chloride. The formula of the chloride is $CaCl_2$. Complete the following table showing the number of chlorine spheres (atoms) to be found in models containing the following numbers of calcium spheres (atoms).

Calcium spheres	Chlorine spheres
5	
10	
15	
20	
25	
50	
100	
1 000 000	
10 000 000	
1 000 000 000	

B The principle of counting by weighing

In the previous sub-section, the pupils established a formula for magnesium oxide by counting spheres. Now the principle of counting by weighing is introduced.

A suggested approach

Objectives for pupils

1. Understanding the principle of counting by weighing
2. Applying the principle of counting by weighing to a model

Since atoms are too small to be seen (see B16.1) it is necessary to devise an alternative method of counting. Some pupils may be able to suggest the first step in establishing the principle of counting by weighing. A few may have seen that coins are counted in a bank by weighing them but many more will have met the idea of counting by weighing in their primary or middle schools. If necessary, pupils might be given an exercise or two in which they count coins or slot weights using this principle. Having determined, for example, the mass of one, two, three and four similar coins, they might be asked to estimate the number of such coins in a light weight paper or plastic bag – before physically counting the coins.

When using polystyrene spheres, to make the arithmetic as simple as possible, it is helpful to keep the masses of each type of sphere to an integral number of grams. The number of spheres must depend on the type of balance used. The spheres used in the construction of the magnesium oxide model could be used for this preliminary exercise.

Note to teachers. Pupils determine the composition of magnesium oxide in section B18.2 and present their findings graphically, making use of the principles used so far.

The principle of counting by weighing

Apparatus

Each pair of pupils will need:

Experiment sheet 115

1 complete model of magnesium oxide

Access to balance

Procedure

Details are given in *Experiment sheet* 115.

Experiment sheet 115

In the last exercise, you established that there were equal numbers of magnesium and oxygen spheres in the model of magnesium oxide. Suppose that you had been unable to see the spheres. How then would you have found the formula of the compound represented in the model?

Perhaps you have seen people who work at a bank count coins by weighing them. By this means they are able to count coins of any type quite accurately even though the coins are wrapped in a bag and they are unable to see them. They use the following equation:

$$\frac{\text{total mass of coins}}{\text{mass of one coin}} = \text{number of coins}$$

You are going to apply the same method to counting model atoms (spheres) in the model of magnesium oxide and in other compounds.

Take the model apart carefully and weigh the magnesium and oxygen spheres in such a way that you can complete the following table.

	Mass of magnesium spheres/g	Mass of oxygen spheres/g
Layer 1		
Layers 1 + 2		
Layers 1 + 2 + 3		
Layers 1 + 2 + 3 + 4		
1 sphere only		

Your teacher will discuss with you how these masses can be used to confirm the formula of MgO.

What would the masses be if the formula were Mg_2O?

What would the masses be if the formula were MgO_2?

Plot a graph of the results shown in the table. Label the horizontal axis 'Mass of oxygen spheres' and the vertical axis 'Mass of magnesium spheres'.

Table B18.1 and figure B18.3 show the presentation of results from this exercise.

Mass of magnesium spheres/g	Mass of oxygen spheres/g	Number of layers
48	32	one
96	64	two
144	96	three
192	128	four

one magnesium sphere weighs 6 g
one oxygen sphere weighs 4 g

Table B18.1 Using the principle of counting by weighing to establish the formula of magnesium oxide

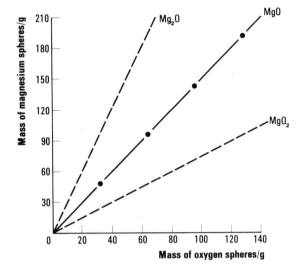

Figure B18.3

The theory underlying the graph requires discussion and application to the *real* but unknown structure of magnesium oxide. 'What would happen with a model made of more layers?' 'And even more layers?' 'Thousands of layers?' 'Would such a model be more like *real* magnesium oxide?'

By the end of this subsection the principle of counting by weighing will have been applied to the model of magnesium oxide. The pupils should realize that this principle could replace the direct method of counting the spheres. They will have been introduced to the type of graph (namely, mass of magnesium plotted against mass of oxygen) which they will use in the next section, B18.2.

Additional exercises
Models of three compounds were studied layer by layer and the masses of the metal and non-metal are recorded below. If the masses of the spheres are numerically equal to one-quarter (in questions 1 and 2) or one-third (in question 3) of the relative atomic masses as shown, what is the element in each case? Use the principle of counting by weighing and the graphical method to find the three formulae.

1. These masses are numerically equal to one-quarter of the relative atomic masses.

	Metal (A) /g	Non-metal (B) /g
1 layer	24	38
2 layers	48	76
3 layers	72	114
4 layers	96	152
1 sphere	6	4.75

The compound is MgF_2.

2. These masses are numerically equal to one-quarter of the relative atomic masses.

	Metal (C) /g	Non-metal (D) /g
1 layer	27.50	16
2 layers	55.00	32
3 layers	82.50	48
4 layers	110.00	64
1 sphere	13.75	4

The compound is MnO_2.

3. These masses are numerically equal to one-third of the relative atomic masses.

	Metal (E) /g	Non-metal (F) /g
1 pile	27	57
2 piles	54	104
3 piles	81	161
4 piles	108	208
1 sphere	9	6

The compound is AlF_3.

Note. The word pile has been introduced into this table as a suitable general term for when a model is not being used.

C   Using relative atomic masses

The principle of counting by weighing has been used already with the model structure. In this sub-section, pupils modify the principle and use the term *relative atomic mass*.

When spheres were counted by weighing, the equation:

$$\text{number of spheres} = \frac{\text{total mass of spheres}}{\text{mass of one sphere}}$$

was used. Can the same equation be used when considering the mass of atoms in real compounds? Since atoms are very minute, the mass of each individual atom is extremely difficult to determine. Atoms are in fact too small to be weighed directly. With some pupils, this point may be convincingly demonstrated by quoting the masses of individual magnesium and oxygen atoms. These may be given approximately as:

for magnesium $\dfrac{24}{6 \times 10^{23}}$ g

for oxygen $\quad\dfrac{16}{6 \times 10^{23}}$ g

Pupils may be impressed by this particularly if the long row of zeros is written out in full!

The calculation shows that the individual masses of atoms must be replaced by something else. When we establish the formula of a compound, we are primarily interested in comparing numbers of different types of spheres or atoms, that is, the relative numbers of spheres or atoms and *not* the exact number that are present. In the case of magnesium oxide, the formula (MgO) indicates that there are equal numbers of magnesium and oxygen atoms present in the compound. The formula CO_2 indicates that there are twice as many oxygen atoms as there are carbon atoms in carbon dioxide. The formula AlF_3 means that there are three times as many fluorine atoms as there are aluminium atoms in aluminium fluoride.

Using relative atomic masses

Apparatus

Each pair of pupils will need:

Experiment sheet 116

1 complete model of magnesium oxide

Procedure

Details are given in *Experiment sheet* 116.

Experiment sheet 116

Atoms are too small to be seen by the naked eye and, moreover, they have very small masses. This means that it will be inconvenient to use the individual atomic masses of atoms in expression such as

$$\text{number of atoms} = \frac{\text{total mass of atoms}}{\text{mass of one atom}}$$

However, in formulae such as MgO, MnO_2, and AlF_3, it is not always necessary to know the number of atoms. The formulae enable you to answer questions such as the ones which now follow:

1. In MgO, how many magnesium atoms are there for each oxygen atom?

2. In MnO_2, how many oxygen atoms are there for each manganese atom?

3. In AlF_3, how many fluorine atoms are there for each aluminium atom?

In the case of magnesium oxide, MgO, the relative atomic masses of magnesium and oxygen are 24 and 16 respectively. This means that twenty-four parts by weight of magnesium combine with sixteen parts by weight of oxygen to form magnesium oxide. Alternatively, one could write that for magnesium oxide to be formed:

$$\frac{\text{mass of magnesium used}}{24} = \frac{\text{mass of oxygen}}{16}$$

For example, in one experiment:

$$\frac{\text{total mass of magnesium}}{24} = \frac{192}{24} = 8$$

$$\frac{\text{total mass of oxygen}}{16} = \frac{128}{16} = 8$$

The number, '8' in both equations, indicates that for every eight magnesium spheres there are eight oxygen spheres in the model. Therefore, the simplest formula will be MgO (and *not* Mg_8O_8).

In the method of counting by weighing, the *relative* masses of the atoms are used and not the real masses expressed in grams.

Here are some more figures which relate to models of other compounds. Try to work out the formula of the compound in each instance.

1.

	Metal (M) with relative atomic mass 24 /g	Non-metal (Q) with relative atomic mass 19 /g
1 layer	24	38
2 layers	48	76
3 layers	72	114
4 layers	96	152

Simplest formula for compound:

2.

	Metal (M) with relative atomic mass 55 /g	Non-metal (R) with relative atomic mass 16 /g
1 layer	27.5	16
2 layers	55	32
3 layers	82.5	48
4 layers	110	64

Simplest formula for compound:

3.

	Metal (M) with relative atomic mass 27 /g	Non-metal (T) with relative atomic mass 19 /g
1 pile	27	57
2 piles	54	114
3 piles	81	171
4 piles	108	228

Simplest formula for compound:

Teachers should note that the problems in the *Experiment sheet* use the same figures as earlier and this will enable the pupils to compare the two sets of answers.

Pupils who benefit from graphical work may be shown the graph in which $\frac{\text{mass of magnesium spheres}}{24}$ is plotted against $\frac{\text{mass of oxygen spheres}}{16}$

The masses of the spheres in Experiment B18.1b were:

Mass of magnesium spheres/g	Mass of oxygen spheres/g	Number of layers
48	32	1
96	64	2
144	96	3
192	128	4

and the division of each pair of masses by the appropriate relative atomic mass clearly establishes the formula if the results are plotted as shown in figure B18.4.

The significance of figure B18.4 is that the pupils will see that there are lines which correspond to the formulae Mg_2O, MgO, and MgO_2. This presentation provides a good link with the 'mole of atoms' which is to be discussed shortly.

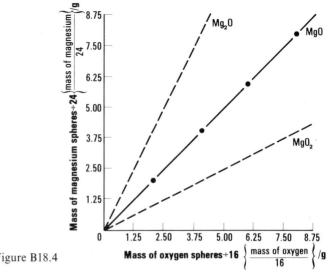

Figure B18.4

We can make the point that this is a good general method for establishing formulae; since if we know the relative atomic masses all we need to determine is the masses of the elements which combine together to form a compound.

By the end of this section pupils should be familiar with the use of relative atomic mass in simple calculations.

Suggestion for homework

In each case, piles of spheres were taken and weighed. Use the information provided to determine the formulae of the compounds concerned.

1.

Mass/g (relative atomic mass of element = 1) (*Use symbol H*)	**Mass/g** (relative atomic mass of element = 16) (*Use symbol O*)
2	16
4	32
6	48
8	64
10	80

The compound is H_2O.

2.

Mass/g (relative atomic mass of element = 64) (*Use symbol Cu*)	**Mass/g** (relative atomic mass of element $\doteq$ 16) (*Use symbol O*)
16	4
32	8
48	12
64	16
80	24

The compound is CuO.

3.

Mass/g (relative atomic mass of element = 200) (*Use symbol Hg*)	**Mass/g** (relative atomic mass of element = 35.5) (*Use symbol Cl*)
20	7.1
40	14.2
60	21.3
80	28.4

The compound is $HgCl_2$.

4.	Mass/g (relative atomic mass of element = 207) (*Use symbol Pb*)	Mass/g (relative atomic mass of element = 80) (*Use symbol Br*)
	20.7	16
	41.4	32
	62.1	48
	82.8	64

The compound is $PbBr_2$.

Note to teachers

The pupils could confirm the formulae of the above examples by plotting suitable graphs. In (1), if the relative atomic masses are 1 and 16 respectively these elements are hydrogen and oxygen. If the formula is HO, then 1 gram of hydrogen will combine with 16 grams of oxygen. If the formula is H_2O, then 2 grams of hydrogen will combine with 16 grams of oxygen, and so on. Since all of the points fall on the 'H_2O line', this is the formula of the compound concerned.

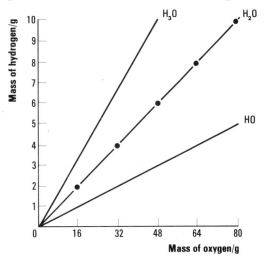

Figure B18.5

B18.2
Finding the formula of magnesium oxide by experiment

In this section, pupils try to find the formula of magnesium oxide by experiment, and apply relevant theory from B18.1.

A suggested approach

Objectives for pupils

1. Application of the ideas developed with the aid of a model to a real situation
2. Determining the masses of magnesium and oxygen which combine to form magnesium oxide
3. Presenting data graphically

(Continued)

The concept of a mole of atoms need not be introduced at this point. Pupils are asked 'What information is required from the experiment?' 'What experiment do you suggest?' From their work with the model of magnesium oxide in the previous section, pupils should know that they need to find the combining masses of magnesium and oxygen. Suitable reactions might be:

1. magnesium + oxygen $\longrightarrow$ magnesium oxide

or

2. magnesium oxide + carbon $\longrightarrow$ magnesium + carbon dioxide

Pupils may realize that reaction (1) is potentially simpler to use and

4. Appreciation of the fact that there is a fixed proportion of magnesium and oxygen present in magnesium oxide (that is, the law of constant composition)

more appropriate in this instance, a point which could be established in discussion.

Finding the formula of magnesium oxide by experiment

Apparatus

Each pair of pupils will need:

Experiment sheet 117

Crucible and lid (*Note*. Some teachers may prefer to prepare this apparatus before the lesson.)

Asbestos disc or equivalent material

Pipe clay triangle

Tongs

Tripod

Bunsen burner and heat resistant mat

Access to desiccator

Teat pipette

Emery paper

Access to a balance

2M nitric acid

Magnesium ribbon

Procedure
Details are given in *Experiment sheet* 117.

The crucibles to be used in this experiment should be lined with a disc of asbestos or equivalent material, and together with the crucible lids, heated to constant mass under the conditions to be used in the experiment and stored in a desiccator. There is a tendency for the magnesium to be incompletely converted to magnesium oxide during the experiment. Part of this is due to incomplete combustion and part due to the formation of magnesium nitride. The latter can be removed by adding a little dilute nitric acid (6 drops) to the cold ash. Magnesium oxide is re-formed in the crucible by heating cautiously until no further change in mass can be detected. Pupils need *not* be concerned with this detail.

Experiment sheet 117

You are going to measure the masses of magnesium and oxygen which combine together to form magnesium oxide.

Weigh the crucible, containing an asbestos disc, and a crucible lid. Place the apparatus on a pipeclay triangle on a tripod underneath which is an asbestos square to protect the bench from reflected heat. Heat the crucible gently by means of a Bunsen burner flame and then more strongly for several minutes. After allowing the crucible to cool, the crucible, lid, and asbestos inside the crucible are reweighed as a single unit of apparatus. This exercise is repeated until the three together are found to have a constant mass. Any one of these items may contain some water.

Why do you think we take this trouble to heat the apparatus to a constant mass?

The purpose of the asbestos disc is to protect the surface of the crucible from the action of burning magnesium.

Use a known length of magnesium ribbon for your experiment. Discuss the length you are to use for your experiment with your teacher. It will be between 10 cm and 30 cm long. Scrape the surface of the magnesium free from magnesium oxide by using some emery paper.

Why should you clean the surface of your piece of magnesium?

Coil the magnesium tightly and place it on the dried asbestos wool in your crucible. Replace the lid of the crucible and weigh the crucible and its contents as a single item of equipment.

Heat the crucible – just as you did first of all. When the magnesium ignites, heat it more strongly. Your Bunsen burner flame should never be too big otherwise the loss of magnesium oxide in the experiment is excessive. You may lift the lid CAREFULLY during the experiment to allow a little air (oxygen) into the crucible when the magnesium has ignited. To do this, the Bunsen burner should be moved well away from the crucible and the lid lifted slightly but kept level just over the rim of the crucible.

How would the loss of magnesium oxide alter your experimental result?

As soon as the magnesium ceases to flare up when the crucible lid is raised and appears to have finished burning, heat the crucible strongly but lift the crucible lid carefully a few times as in the previous part of the experiment. Allow the crucible and its contents to cool and reweigh them.

Add six drops of dilute nitric acid to the contents of the crucible. Replace the crucible lid and heat the crucible gently, carefully raising and lowering the lid of the crucible as in the previous part of the experiment. When the contents of the crucible are dry, heat the crucible more strongly. Allow the crucible and its contents to cool and then reweigh until a constant mass is obtained.

Why is it necessary to repeat the heating and reweighing to obtain a constant mass?

Results

Mass of crucible (asbestos and lid) + magnesium .... g

Mass of crucible (asbestos and lid) .... g

Mass of magnesium .... g

Mass of crucible (asbestos and lid) + magnesium oxide .... g

Mass of crucible (asbestos and lid) + magnesium .... g

Mass of oxygen .... g

Use the mass of magnesium and the mass of oxygen that you have determined by your experiment to obtain the formula of magnesium oxide. The relative atomic mass of magnesium is 24 and that of oxygen is 16.

Teachers may find that some pupils are confused with their calculations or need to be reminded constantly of the purpose of the experiment. Graphs can help considerably to limit not only the difficulties with arithmetic but also problems introduced by inaccurate experimental results. Pupils should be familiar with the idea that when a mass of magnesium spheres is plotted against the mass of oxygen spheres there will be a line which corresponds to a formula MgO. We need to ask the question 'Is this also the case for the real compound?' Labelled axes may be drawn up on the blackboard or on an overhead projector transparency on which pupils can plot their own results. The results can also be written down in a table so that all the class can plot a graph (a possibility for homework).

Figure B18.6 (and Table B18.2) show the results obtained by a class of sixteen pairs of pupils. Two of these failed to complete the experiment but most of the results can be seen to fall along the line corresponding to the formula MgO. The various 'formula-lines' can be drawn in either before or after the pupils have completed their experiments. Clearly, results A and B are questionable, and it was discovered that in case A there was a weighing error, and case B incomplete combustion.

When the position of the 'formula-line' has been confirmed experimentally, it might be convenient to mention that magnesium oxide always has the same formula. This can be used to introduce pupils to the law of constant composition, which they should certainly be aware of by the end of the Topic.

Group number	Mass of magnesium /g	Mass of oxygen /g
1	0.260	0.172
2	0.06	0.038
3	0.197	0.126
4	0.040	0.026
5	0.186	0.115
6	0.232	0.152
7	0.220	0.136
8	0.167	0.104
9	0.183	0.114
10	0.093	0.062
11	0.081	0.053
12	0.133	0.088
A 13	0.096	0.083
B 14	0.211	0.105
15	—	—
16	—	—

Table B18.2 Sample results for the determination of the formula of magnesium oxide

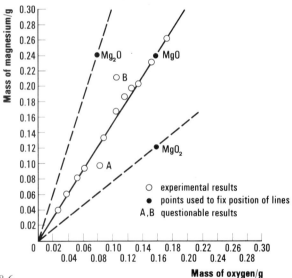

Figure B18.6
Sample results for the determination of the formula of magnesium oxide.

The pupils could now select a number of combining masses of magnesium and oxygen and confirm that:

$$\frac{\text{mass of magnesium}}{24} = \frac{\text{mass of oxygen}}{16}$$

Alternatively, the combining masses for magnesium found by the class (excluding obviously wrong results) could now be added together and the total mass divided by the number of determinations made. This can also be done for oxygen. These average figures may then be used to confirm the formula of magnesium oxide.

Plot the class results as a graph in which the mass of magnesium used is plotted against the mass of oxygen used.

(Alternatively, give the pupils the data shown in Table B18.2 and ask them to comment on these results.)

B18.3
The idea of a mole of atoms

A note to teachers

The word 'mole' is derived from the Latin word meaning a collection, mass, or pile. It is used whenever it is necessary to specify the quantity of a substance. It is defined in terms of the Avogadro constant, the value of which depends primarily on the choice of standard for atomic mass. Historically, the first scale was that relative to hydrogen (H = 1) but later oxygen (O = 16) became favoured since this element combined with so many of the others. Subsequently, it was found that atmospheric oxygen contains small variable amounts of oxygen-17 and oxygen-18 in addition to the predominant oxygen-16. Two series of atomic masses then came into general use. The so-called 'chemists' scale' was related to natural oxygen, whereas the 'physicists' scale' was related to the isotope oxygen-16. This unsatisfactory state of affairs was resolved by the IUPAC convention of 1961 when the standard mass was taken to be that of the isotope carbon-12. Accordingly the current definition of the Avogadro constant is 'The number of atoms of carbon in 0.012 kilograms of carbon-12'.

The accepted value for the Avogadro constant based on this definition is $6.022\,169\,(+ \text{ or } -0.000\,040) \times 10^{23}\,\text{mol}^{-1}$.

The *mole* is the amount of substance which contains as many elementary units as there are atoms in 0.012 kilograms of carbon-12.

The elementary units mentioned in this definition must be specified and may be an atom, a molecule, an ion, a radical, an electron, or any specified group of such entities. It is particularly important to state the entity present when there is likely to be ambiguity. For example:

1 mole of hydrogen atoms (H) has a mass of 1 g;
1 mole of hydrogen molecules (H_2) has a mass of 2 g;
1 mole of oxygen atoms (O) has a mass of 16 g;
1 mole of oxygen molecules (O_2) has a mass of 32 g;
1 mole of nitrogen atoms (N) has a mass of 14 g;
1 mole of nitrogen molecules (N_2) has a mass of 28 g;
etc.

When chemists refer to substances with ionic or covalent giant structures, the mass represented by the formula must be used. 1 mole of silicon(IV) oxide (60 g), which has a giant covalent structure, contains 1 mole of silicon atoms and 2 moles of oxygen atoms.

(After Topic B19, we will be able to go further: 1 mole of sodium chloride, which has an ionic structure, contains 1 mole of sodium (Na^+) ions and 1 mole of chloride (Cl^-) ions.) However, aluminium (III) chloride is present as the covalent dimer in both the solid and gaseous phases and so the need to avoid ambiguity is important when one wishes to refer to a specific entity in this case. Thus, 1 mole of the dimer, Al_2Cl_6, has twice the mass of 1 mole of the monomeric form, $AlCl_3$.

Terms such as 'gram-molecule', 'gram-equivalent', 'gram-ion' or 'gram-formula' are not used in this Alternative, although they are found in the original Sample Scheme II and in a number of text-books.

In its wider sense, the mole is used to establish formulae; to establish chemical equations; to calculate the percentage of an element in a given compound (for example, gravimetric analysis); to calculate the reacting masses of elements and compounds in chemical re-actions; to interpret the interacting volumes of gaseous substances, or the concentration of solutions, or the amount of substance de-posited in electrolysis. Each of these examples the pupils will meet in later Topics in Alternative IIB. Many properties such as heats of reaction or heats of combustion are given in energy units per mole (such as $kJ\,mol^{-1}$).

What makes the mole concept so useful arises from the fact that it tells a chemist the relative numbers of particles, such as atoms, that are present in a chemical system. Thus, he does not need to weigh out equal masses of particles but he does need to know the masses of an equal number of moles of different kinds of particles.

It is not intended that a formal definition of the mole should be given to the pupils at this point.

In this section, the mole concept is introduced and then used to solve simple problems. It is assumed that pupils have covered the work outlined in B18.1 and B18.2.

A suggested approach

Objectives for pupils

1. Consideration of the masses of different elements which contain the same number of atoms
2. Relating the masses of elements to the mole of atoms
3. Recognition that formulae can represent both the numbers of atoms and the numbers of moles of atoms which combine to form compounds

Pupils might be reminded of the results of the previous experiment as shown in figure B18.6. From this, 0.24 g of magnesium combines with 0.16 g of oxygen, and therefore 24 g of magnesium will com-bine with 16 g of oxygen. These quantities are numerically equal to the relative atomic masses expressed in grams of these two elements – but they have a much greater significance than this alone.

The formula, MgO, indicates that magnesium oxide contains an equal number of atoms of magnesium and oxygen. Therefore, 24 g of magnesium and 16 g of oxygen contain equal numbers of atoms. These masses are called 'moles of atoms'.

    1 mole of magnesium atoms weighs 24 grams;
    1 mole of oxygen atoms weighs 16 grams;
    and *both* contain the same number of atoms.

The idea of a mole of atoms can be applied to all elements. Ask pupils what they think the mass of 1 mole of each of the following elements is:

hydrogen, carbon, nitrogen, oxygen, magnesium, sulphur.

Statements of 'correct' replies might be tabulated, as in Table B18.3.

Stress the point that each of these quantities of different elements contains the *same* number of particles (atoms).

Alternatively, such information could be presented as a bar chart – such as that shown in figure B18.7.

Element	Mass /g
hydrogen	1
carbon	12
nitrogen	14
oxygen	16
magnesium	24
sulphur	32

Table B18.3 Moles of atoms

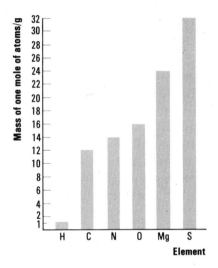

Figure B18.7
Moles of atoms.

Some pupils might benefit from a further exercise with ballasted spheres so that they can see that the masses of piles of spheres containing equal numbers of particles are in the same proportion as the relative atomic masses. Several spheres of different mass can be weighed and a table constructed using the results. In Table 18.4 the masses are equal to one quarter of the mass of 1 mole of atoms. Some teachers, however, prefer the mass of the spheres to be numerically equal to the mass of 1 mole of atoms (in which case 1 oxygen sphere would weigh 16 grams). A graph showing this relationship is given in figure B18.8.

	Hydrogen	Carbon	Nitrogen	Oxygen	Magnesium	Sulphur
1 sphere	0.25	3	3.5	4	6	8
$\frac{1 \text{ sphere}}{1 \text{ hydrogen sphere}}$	1	12	14	16	24	32
2 spheres	0.50	6	7	8	12	16
$\frac{2 \text{ spheres}}{2 \text{ hydrogen spheres}}$	1	12	14	16	24	32
3 spheres	0.75	9	10.5	12	18	24
$\frac{3 \text{ spheres}}{3 \text{ hydrogen spheres}}$	1	12	14	16	24	32
4 spheres	1	12	14	16	24	32
$\frac{4 \text{ spheres}}{4 \text{ hydrogen spheres}}$	1	12	14	16	24	32

Table B18.4 An exercise with spheres. (All masses are in grams)

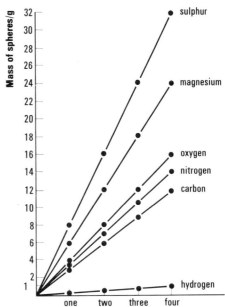

Figure B18.8
Masses of piles of spheres.

It might be worth while to ask 'What are the relative masses of piles of spheres containing 5, 10, 100, and 1000 spheres?' This may help to transfer the pupils' ideas from spheres to atoms.

Pupils might then weigh out – or at least be shown – examples of moles of atoms of a variety of elements. Each sample consists of 1 mole and therefore contains equal numbers of particles. Pupils should be reminded that the number of particles (atoms) in a mole of an element is very large because the particles themselves are very small.

If pupils are to use the idea of a mole of atoms to full advantage, they should be able to convert mass into moles of atoms. This can be done either by using a graph or by calculation.

A. *Graphical method for converting mass into moles*
Pupils could begin by considering something familiar such as magnesium. 1 mole of magnesium atoms weighs 24 g. One-hundredth of a mole of magnesium atoms weighs 0.24 g and one-thousandth of a mole of magnesium atoms weighs 0.024 g.

These last two points can be used to plot a graph of the mass of magnesium against number of moles of magnesium. The advantage of this approach is that mass can be quickly converted into moles of atoms without the pupils manipulating figures. (See figure B18.9.)

Similarly figure B18.10 can be used for converting mass of oxygen into moles of oxygen atoms.

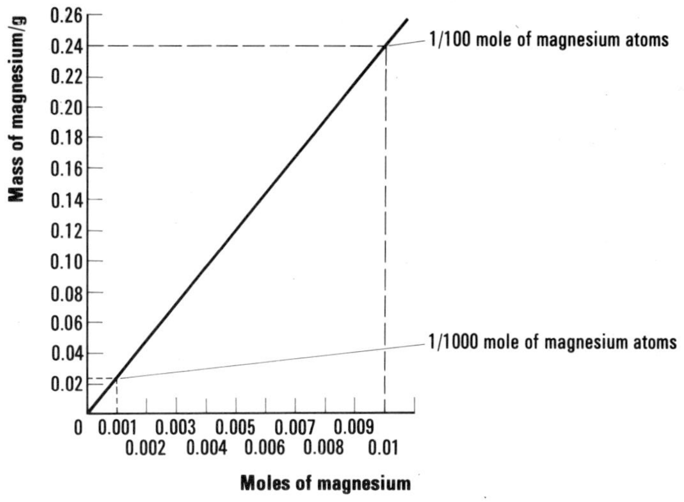

Figure B18.9
Converting mass of
magnesium into moles of
magnesium atoms.

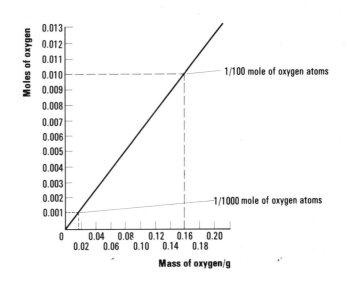

Figure B18.10
Converting mass of oxygen
into moles of oxygen atoms.

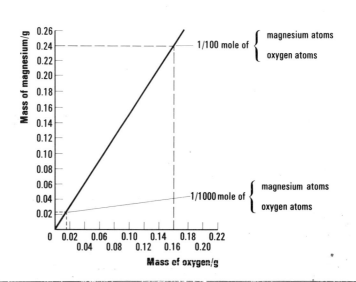

Figure B18.11

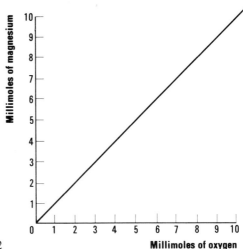

Figure B18.12
Moles of atoms in magnesium oxide.

By choosing an appropriate scale (up to a little more than one-hundredth of a mole of oxygen atoms), pupils convert their class results directly into moles (see figure B18.11 and figure B18.12). Pupils can see from the graphs, therefore, that one-hundredth of a mole of magnesium atoms combines with one-hundredth of a mole of oxygen atoms. This leads once more to the statement that one mole of magnesium atoms combines with one mole of oxygen atoms.

Note to teachers. The presentation used so far is compatible with the distinction made already between giant structures and molecules. The point to watch is that the formula MgO does *not* mean that one atom of magnesium and one atom of oxygen combine to form magnesium oxide, but that equal numbers of moles of atoms of magnesium and of oxygen combine to form magnesium oxide.

Similarly, the formula MnO_2 means that one mole of manganese atoms combine with two moles of oxygen atoms to form manganese(IV) oxide. Since manganese(IV) oxide has a giant structure rather than a molecular one, the formula MnO_2 *cannot* be used to represent the idea that one manganese atom combines with two oxygen atoms.

However in the case of carbon dioxide, a molecular compound, we can make the statement that 1 atom of carbon combines with 2 atoms of oxygen to form carbon dioxide, *and* that 1 mole of carbon atoms combines with 2 moles of oxygen atoms to from 1 mole of carbon dioxide molecules.

These points need to be made if the simplicity of the mole concept and its uses are to be appreciated.

B. *Calculation for converting mass into moles of atoms*
So far, pupils have seen that, for example:

$$\frac{\text{mass of magnesium}}{24} = \frac{\text{mass of oxygen}}{16}$$

The lefthand side of this equation indicates the number of moles of magnesium atoms present. Thus, 24 grams of magnesium corresponds to $\frac{24}{24} = 1$ mole of magnesium atoms. Similarly, 0.024 grams of magnesium would correspond to $\frac{0.024}{24} = \frac{1}{1000}$ mole of magnesium atoms. Similarly, the ratio $\frac{\text{mass of oxygen}}{16}$ on the righthand side of the equation gives the number of moles of oxygen atoms present. For any element, the following is true:

$$\text{numbers of moles of atoms} = \frac{\text{mass of element}}{\text{mass of one mole of atoms of the element}}$$

The mole of atoms

Procedure
Details are given in *Experiment sheet* 118.

Experiment sheet 118
You have been given six different types of sphere, each of which represents an atom. The white one represents hydrogen; the black one, carbon; the blue one, nitrogen; the red one, oxygen; the gold-coloured one, magnesium; the yellow one, sulphur. Weigh 1, 2, 3, and 4 spheres of each kind and record your results in tabular form.

	Hydrogen	Carbon	Nitrogen	Oxygen	Magnesium	Sulphur
mass of 1 sphere						
$\frac{\text{mass of 1 sphere}}{\text{mass of 1 hydrogen}}$						
mass of 2 spheres						
$\frac{\text{mass of 2 spheres}}{\text{mass of 2 hydrogens}}$						
mass of 3 spheres						
$\frac{\text{mass of 3 spheres}}{\text{mass of 3 hydrogens}}$						
mass of 4 spheres						
$\frac{\text{mass of 4 spheres}}{\text{mass of 4 hydrogens}}$						

By dividing each mass by the mass of an equal number of hydrogen spheres, we obtain the masses which are relative to hydrogen and which contain the same number of atoms. What do you notice about these relative masses?

How do you think they compare with relative atomic masses?

Plot a graph, of mass of spheres/g (vertical axis) against number of spheres (horizontal axis).

Join up the points for hydrogen, then join the points for carbon, oxygen, and so on until you have six lines, with one line for each of the six elements.

What are the masses of each element which contain four spheres?

How do these compare with the relative atomic masses of the elements?

Suggestions for homework

1. Plot graphs of mass against moles of atoms for the elements hydrogen, carbon, nitrogen, oxygen, magnesium, and sulphur. Use these graphs to write down the numbers of moles of atoms in 0.3, 0.25, 0.20, 0.15, 0.10, 0.05 grams of the elements.

2. Models of compounds were constructed and the total masses in grams of the different types of sphere are recorded below. The numbers in brackets are the masses of 1 mole of atoms of each of the different elements. Work out the formulae of each of the compounds represented. Use the symbols M and A for the first and second element respectively.

a. 55 g (55) and 32 g (16);
b. 64 g (64) and 16 g (16);
c. 32 g (32) and 32 g (16);
d. 32 g (32) and 48 g (16);
e. 14 g (14) and 3 g (1);
f. 2 g (1) and 16 g (16);
g. 2 g (1) and 32 g (16);
h. 55 g (55) and 32 g (32);
i. 64 g (64) and 32 g (32);
j. 28 g (14) and 16 g (16);
k. 14 g (14) and 32 g (16).

3. All chemicals have formulae although some are more complex than others. Look up the formulae of some simple compounds and write down the ratios of the moles of the constituent elements. In particular, study those compounds of the elements in the lithium group (the alkali metals), the fluorine group (the halogens), and the carbon group which were discussed in Topic B17. Present your data in a table with the headings shown below:

Name of compound	Formula	Ratio of moles of elements

4. What is the mass of one-tenth of a mole of atoms of each of the following elements?

(*a*) hydrogen; (*b*) helium; (*c*) lithium; (*d*) beryllium; (*e*) boron; (*f*) carbon; (*g*) nitrogen; (*h*) oxygen; (*i*) fluorine; (*j*) neon.

5. What is the mass of ten moles of atoms of each of the following elements?

(*a*) sodium; (*b*) magnesium; (*c*) aluminium; (*d*) silicon; (*e*) phosphorus; (*f*) sulphur; (*g*) chlorine; (*h*) argon.

6. For each of the eighteen elements listed in questions 4 and 5, give the fraction of moles of atoms in 1, 10, and 100 grams. Write your answers like this:

1 gram of nitrogen contains $\frac{1}{14}$ mole of nitrogen atoms.

7. Write down the number of moles in each of the following:

(a) 54 g of aluminium; (b) 8 g of chlorine; (c) 8 g of bromine atoms; (d) 20 g of hydrogen; (e) 24 g of carbon; (f) 0.016 g of oxygen; (g) 281 g of silicon; (h) 20.7 g phosphorus; (i) 32 g of sulphur; (j) 20 g calcium; (k) 5.2 g chromium; (l) 0.56 g iron; (m) 0.128 g copper; (n) 71 g chlorine.

8. Write down the mass of each element in the following:

(a) one-quarter of a mole of magnesium atoms
(b) half a mole of chromium atoms
(c) three-quarters of a mole of iron atoms
(d) one-seventh of a mole of nitrogen atoms
(e) one mole of barium atoms
(f) five moles of bromine atoms
(g) one-sixth of a mole of tin atoms
(h) one mole of carbon atoms
(i) 0.2 moles of chlorine molecules (Cl_2)
(j) one-sixteenth of a mole of sulphur atoms
(k) one-sixteenth of a mole of oxygen molecules (O_2)
(l) one-twelfth of a mole of carbon atoms

9. Write down the number of moles of atoms of each element which combine to form the following compounds:

CuO, MgO, H_2O, FeO, Fe_2O_3, Fe_3O_4, Al_2O_3,
Cu_2O, ZnO, CO_2, CO, SO_2, SO_3, N_2O_4,
HCl, CCl_4, NH_3, CH_4, C_2H_6, $NaCl$, $MgCl_2$,
Hg_2Cl_2, $HgCl_2$, $AlCl_3$, Cl_2, $CuCl_2$, $FeCl_2$, $FeCl_3$,
H_2S, CS_2, Na_2S, MgS, Al_2S_3, S_8, CuS,
FeS, FeS_2, $NaHSO_4$, Na_2SO_4, $CaCO_3$, $Ca(NO_3)_2$,
Na_3PO_4, H_3PO_4, NH_4Cl, NH_4NO_3.

Summary

In this section, pupils have been introduced to the idea of a mole of atoms in such a way that formal definitions of the mole and of the Avogadro constant have been avoided. The pupils should realize that the formula MgO, for magnesium oxide, means that one mole of magnesium atoms is combined with one mole of oxygen atoms. Pupils should also be aware that the mole of atoms will be used a great deal in subsequent Topics.

B18.4
Using the 'mole of atoms' to find the formulae of several compounds

In this section, various experiments are developed to find the formulae of simple compounds. (It is assumed that pupils will have completed sections B18.1, B18.2, and B18.3 before attempting this work.)

Present pupils with samples of copper oxide and encourage them to suggest possible formulae for this material. They can be given, or asked to work out, the positions of lines corresponding to the formulae Cu_2O, CuO, and CuO_2 on the graph (see figure B18.13) in which the mass of copper is plotted against that of oxygen. This graph could be placed on the blackboard and used for comparison with experimental results. The reacting masses can be plotted on this graph, and should correspond with the line for the formula CuO, after any obviously inaccurate results have been discussed and subsequently discarded.

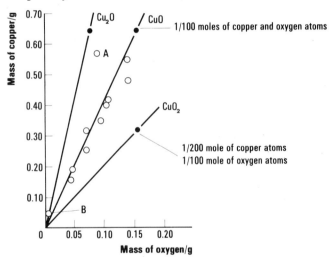

Figure B18.13
The composition of some oxides of copper.

The experimental result may be confirmed by the following calculation in which the masses of copper and oxygen may be taken from individual groups or from class average results.

$$\text{number of moles of copper atoms} = \frac{\text{mass of copper}}{64}$$

$$\text{number of moles of oxygen atoms} = \frac{\text{mass of oxygen}}{16}$$

This leads to:

$$\frac{\text{mass of copper}}{64} = \frac{\text{mass of oxygen}}{16}$$

If 0.55 g copper combines with 0.14 g oxygen to form black copper oxide, we can write:

number of moles of copper atoms $\quad = \frac{0.55}{64}\,(=0.086)$

and number of moles of oxygen atoms $= \frac{0.14}{16}\,(=0.088)$

Therefore, the formula of copper oxide must be CuO.

Pupils should also confirm the result obtained from the graph if they use a conversion scale (figure B18.14). They should know that one mole of copper atoms weighs 64 g, therefore they can work out the mass of one-hundredth of a mole and of one-thousandth of a mole.

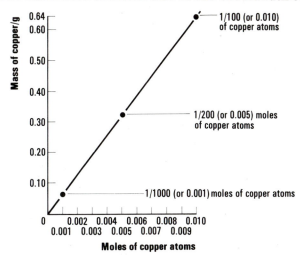

Figure B18.14
Converting mass of copper into moles of copper atoms.

The line drawn from the origin helps to confirm that there is a direct relationship between mass and the number of moles of copper atoms. A similar graph has been plotted earlier for oxygen atoms (see figure B18.10).

The formula for copper oxide can be confirmed by showing that equal numbers of moles of atoms of copper and oxygen combine.

Experiment B18.4a

To find the formula of black copper oxide

Apparatus

Each pair of pupils will need:

Experiment sheet 45

Hard-glass test-tube, 125 × 16 mm, with small hole near the closed end

Length of rubber tubing and glass delivery tube mounted in a cork to fit the test-tube

Bunsen burner and heat resistant mat

Note to teachers. It is important to use analytical grade copper(II) oxide for this experiment. 'Technical' copper oxide contains appreciable quantities of copper and gives poor results. For the best results, heat the copper oxide before the lesson in an open dish at 300–400 °C for several minutes to drive off any water. Allow the dish to cool in a desiccator containing either fresh anhydrous calcium chloride or silica gel.

Procedure
Town gas or natural gas is used for the reduction of copper oxide. Pupils should use between 0.5 g and 4.5 g copper oxide. Remind pupils that when the reduction is complete, it is essential to keep a

(Continued)

Access to 2 gas taps

Access to balance

Stand, boss, and clamp

Pure dry black copper oxide (use between about 0.5 and 4.5 g – analytical grade preferred)

Desiccator in which to store pure dry black copper oxide in a small dish

small stream of gas passing through the apparatus until it is quite cold: this prevents air from entering the apparatus and re-oxidizing the copper. Experimental details are given in *Experiment sheet 45*.

Experiment sheet 45
In this experiment, you will convert a known mass of black copper oxide to copper by heating it in a stream of natural gas (or town gas or hydrogen).

The apparatus to use is shown in the diagram.

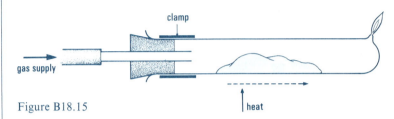

Figure B18.15

The test-tube must be dry. Find its mass (without the cork and glass tube) and enter it in the table overleaf. Add some pure, dry black copper oxide and find the mass of tube + oxide. Your teacher will tell you the exact mass of copper oxide to use. Fix the tube horizontally in a clamp (be careful not to spill any oxide) and insert the cork and glass tube. Connect the glass tube to the bench gas supply.

You need a gentle stream of gas through the special test-tube; if the gas flows too quickly it will blow the oxide out of the hole at the end of the tube. Place the palm of one hand just above the hole at the end of the tube and use the other hand to slowly turn the gas tap. You will feel a coolness on your palm when gas is flowing through the tube. When this happens leave the gas tap at this position, wait for ten seconds, and light the gas when it leaves the tube. Adjust the flame so that it is about 3 cm high.

Light a Bunsen burner, turn the gas full on, and adjust the air hole to give a roaring flame (this is for natural gas, your teacher will tell you how to adjust the burner if you use town gas). Heat the copper oxide at the point indicated by the arrow in the diagram, until a red glow appears in it. By slowly moving the flame to the right (dotted arrow) lead this glow through the copper oxide. Finally heat the whole of the solid residue in the tube by moving the flame to and fro along it for 3–5 minutes. The copper oxide should then have been changed to copper, a pink powder.

Turn out the Bunsen flame and allow the test-tube to cool with gas still passing through it. (Why?)

When the test-tube is cool enough for you to touch the under surface comfortably with your hand (CARE!), turn out the gas flame, remove the cork and delivery tube, and find the mass of the test-tube + copper.

Complete the table.

Mass of test-tube	.... g
Mass of test-tube + copper oxide	.... g
Mass of test-tube + copper	.... g
Mass of copper remaining	.... g
Mass of oxygen removed	.... g

Complete the calculation after discussing the method with your teacher.

Table B18.5 gives some class results for this experiment.

Mass of copper/g	Mass of oxygen/g	
0.155	0.043	
0.187	0.048	
0.253	0.070	
0.33	0.072	
0.35	0.095	
0.405	0.105	
0.415	0.108	
0.48	0.143	
0.55	0.14	
A0.57	0.09	weighing error
B0.046	0.01	too little copper

Table B18.5 Sample results for the determination of the formula of black copper oxide

Optional

Experiment B18.4b

The formula of lead bromide

This experiment is optional and follows the same general procedure as that set out in Experiment B18.4a. A preliminary discussion can be used to indicate possible formulae for lead bromide. A graph could again be plotted and pupils asked to deduce the lines corresponding to the compositions Pb_2Br, $PbBr$, and $PbBr_2$. When the pupils have completed their experiments and made appropriate calculations, the reacting masses can be plotted onto this graph and the formula of lead bromide established.

Procedure

Teachers may note that the experiment utilizes the fact that aluminium displaces lead from lead bromide. The equation for the reaction is as follows:

$$3PbBr_2 + 2Al \longrightarrow 2AlBr_3 + 3Pb$$

It follows that 1 gram of lead bromide contains approximately two-thirds of a gram of lead and so 1 gram of lead bromide needs about 0.1 gram of aluminium powder (about two spatula measures) which represents an excess of aluminium. In the experiment, aluminium powder is allowed to react with lead bromide in aqueous solution. Excess aluminium is removed using sodium hydroxide solution. The lead is washed and dried using acetone (propanone) (a flammable liquid).

This determination can be used to revise the pupils' ideas about reactivity series (and displacement reactions). Although many elements are above lead in the reactivity series (see Stage I), it is not easy to find alternatives to aluminium since any excess metal present

Apparatus

Each pair of pupils will need:

Experiment sheet 119

Wide-necked conical flask, $100\,cm^3$

Glass rod fitted with a rubber 'policeman'*

Bunsen burner and heat resistant mat

Tripod and gauze

Access to a balance

Spatula

Lead bromide, 1 g

Aluminium powder, 0.1 g

2M sodium hydroxide solution, $20\,cm^3$

Distilled water, $20\,cm^3$

Acetone (propanone), $20\,cm^3$

---

*A 'policeman' can be made from a short length of rubber tubing which is fitted over one end of the glass rod. (Specially prepared versions completely cover the end of a glass rod.)

in the system has to be removed with acid and this also reacts with lead. Lead does not react with sodium hydroxide and hence the choice of aluminium. (*Note*. The pupils will meet lead bromide again in Topic B19.)

Experiment sheet 119 is reproduced below.

Experiment sheet 119
In this experiment, lead is displaced from lead bromide by aluminium:

aluminium + lead bromide $\longrightarrow$ lead + aluminium bromide

Weigh a wide-necked conical flask (capacity $100 \, cm^3$). Reweigh the flask after adding about 1 g of lead bromide to it. Next add approximately $50 \, cm^3$ of distilled water to the flask and stir the mixture with a glass rod fitted with a rubber 'policeman'. Leave the glass rod in the flask and then stand the flask on a tripod and gauze. Bring the mixture to the boil using a medium-sized Bunsen flame.

Remove the flask from the tripod and gauze and gradually add a small excess of aluminium powder. This may be achieved by using two spatula measures of aluminium powder (total mass about 0.1 g). It is suggested that this addition is attempted *gradually*.

Boil the contents of the flask for about 10 minutes during which time the aluminium will displace the lead from the lead bromide.

The walls of the flask are then wiped with the glass rod tipped with the 'policeman'. Your teacher will demonstrate this technique to you.

Remove the flask from the tripod and gauze and stand it on the bench. Add about 20–$40 \, cm^3$ sodium hydroxide solution to remove any aluminium powder which remains.

Use your glass rod to press any precipitated lead together and, once the aluminium has been removed by the reaction with the sodium hydroxide, decant any liquid remaining in the flask.

Wash the residual lead at least twice with distilled water and then boil it with some more distilled water for a few minutes.

Decant off as much water as possible. Make sure your Bunsen burner is turned off. Wash the lead with a small amount of acetone (CAUTION— acetone is a flammable liquid) and pour off as much as possible. When the remaining acetone has evaporated, reweigh the flask and lead. Record all your results below.

Results

Mass of flask	.... g
Mass of flask + lead bromide	.... g
Mass of lead bromide	.... g
Mass of flask + lead	.... g
Mass of lead	.... g
Mass of bromine removed	.... g

Use these results to deduce the formula of lead bromide.

Figure B18.16 follows the format used in the previous section for the presentation of results. Table B18.6 shows some class results for this determination.

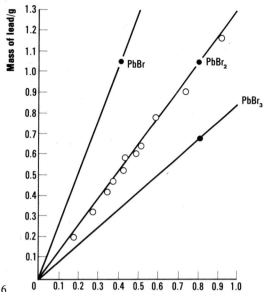

Figure B18.16
Some possible formulae for lead bromide.

Mass of lead/g	Mass of bromine /g
(1.04	0.80)
0.19	0.17
0.32	0.27
0.42	0.34
0.46	0.37
0.52	0.42
0.58	0.43
0.60	0.48
0.64	0.51
0.78	0.57
0.90	0.73
1.15	0.91

Table B18.6 Sample results for the determination of the formula of lead bromide

The formula of lead bromide found by graphical means should be confirmed by the following type of calculation in which the masses of lead and bromine used can be taken either from an average result for the class or from an individual result. Thus, for a determination which showed that 0.64 g of lead require 0.51 g of bromine, we can write:

$$\text{number of moles of lead atoms} = \frac{\text{mass of lead}}{207} = \frac{0.64}{207} = 0.0031$$

$$\text{number of moles of bromine atoms} = \frac{\text{mass of bromine}}{80} = \frac{0.51}{80} = 0.0064$$

Thus, for each mole of lead atoms there are two moles of bromine atoms and so the formula of lead bromide must be $PbBr_2$.

B18.4 Using the 'mole of atoms' to find the formulae of several compounds **509**

Conversion scales may be used by the pupils in the same way as in the previous section. They can look up the masses of one mole of atoms of lead and bromine and work out the mass of a hundredth or two-hundredth of a mole and so on. A graph can then be drawn of mass against number of moles. The pupils can look up some masses which are close to the lead bromide line and convert these to moles of lead and bromine atoms. Figure B18.17 shows a graph for converting the mass of lead to moles of lead atoms and figure B18.18 does the same for bromine.

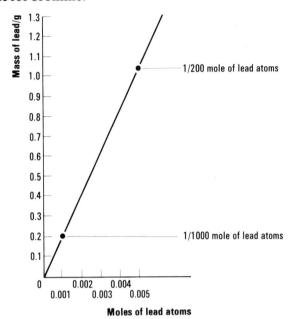

Figure B18.17
Converting mass of lead into moles of lead atoms.

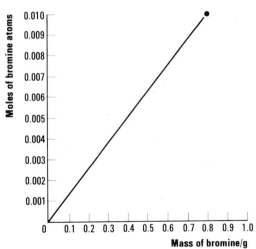

Figure B18.18
Converting mass of bromine into moles of bromine atoms.

Topic B18 The mole and its use in the determination of the formulae of compounds

Determination of the formula of mercury chloride

This experiment should be carried out by the teacher. The objectives and presentation are the same as those described in the previous experiments using copper oxide and lead bromide. Possible formulae should be discussed and the masses of mercury and chlorine which combine together plotted on a graph. The 'formula-lines' used to represent the composition $HgCl$, $HgCl_2$, $HgCl_3$, should be drawn, and the experimental determination then used to show that the composition of mercury chloride is represented by $HgCl_2$.

Apparatus

The teacher will need:

Beaker, $100 \, cm^3$

Water bath

Filter paper

Access to balance

Bunsen burner and heat resistant mat

Tripod and gauze

Measuring cylinder, $25 \, cm^3$

Glass rod

Acetone (propanone)

Mercury(II) chloride

Hypophosphorous acid, about 50 per cent strength

Procedure

Weigh out accurately $5 \, g$ of mercury(II) chloride into a previously weighed $100 \, cm^3$ beaker. Add $30 \, cm^3$ of distilled water and gradually add $10 \, cm^3$ hypophosphorous acid solution. Heat the mixture on a water bath. The mercury(II) chloride is reduced to mercury which, on further heating and stirring, collects into one or more globules at the bottom of the beaker. When this stage has been reached, the metal is washed by decantation with water and with acetone in succession. The last few drops of acetone (propanone) should be removed using a filter paper. Allow to stand for two to three minutes for the final traces of acetone to evaporate, then weigh the beaker and the mercury. Complete the calculation as in the previous examples.

Optional

Determination of the formula of water

The objectives and presentation are as for the earlier examples. Care and attention to detail are essential in this experiment.

Apparatus

The teacher will need:

Plastic safety screen

Safety spectacles

Combustion tube, about 15 cm

2 test-tubes, with side arm, $125 \times 16 \, mm$, fitted with bungs and delivery tube

Calcium chloride tube, bung, and jet

Glass wool

Access to balance

3 stands, bosses, and clamps

Bunsen burner and heat resistant mat

2 test-tubes

Splint

Cylinder of hydrogen

The formula of water can be discussed in much the same way as earlier examples. Most pupils will expect the formula H_2O to be confirmed. The masses of hydrogen and oxygen which combine together can be plotted on a graph. The lines for, say, HO, H_2O, and HO_2, are plotted and compared with the line obtained by experiment.

This experiment MUST be done by the teacher.

Note. It is *most* important that both the hydrogen and the copper oxide used should be dry. Before use in this experiment, the copper oxide should be heated to about 300–$400 \, °C$ for several minutes and then allowed to cool in a desiccator. It is also important to dry the apparatus in an oven beforehand, and to 'roast' the anhydrous calcium chloride just before use.

Procedure

Wire-form copper(II) oxide is reduced in a stream of hydrogen and the water formed is absorbed. The loss in mass of copper oxide gives the mass of oxygen in water; the gain in mass of the entire apparatus

(Continued)

B18.4 Using the 'mole of atoms' to find the formulae of several compounds   **511**

Concentrated sulphuric acid

Copper(II) oxide, wire form, 25 to 30 g

gives the mass of the hydrogen present in the water formed during the experiment.

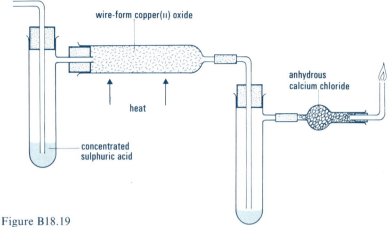

wire-form copper(II) oxide

anhydrous calcium chloride

heat

concentrated sulphuric acid

Figure B18.19

The apparatus is shown in figure B18.19. It consists of a piece of combustion tubing shaped as shown and connected to a test-tube fitted with a side arm. This tube holds concentrated sulphuric acid and in turn leads to a calcium chloride absorption tube filled with fresh anhydrous calcium chloride.

Place between 25 and 30 g of pure dry wire-form copper(II) oxide in the combustion tube. Secure the position of the copper oxide with tufts of glass wool, and weigh the tube (the weight of the copper oxide alone is not required). Then, connect the combustion tube to the rest of the apparatus and weigh the whole apparatus.

Pass hydrogen, from a cylinder fitted with an appropriate pressure reduction valve, through a side arm test-tube containing concentrated sulphuric acid to dry it, and then through the apparatus. (**Caution:** the careful regulation of hydrogen pressure is essential!) Collect samples of the gas and when you can show that all of the air has been displaced from the apparatus, ignite the excess hydrogen. Heat the combustion tube by means of a medium-sized Bunsen flame until the copper oxide is reduced to copper.

It is important to reduce *all* the oxide and to continue heating until no more moisture comes out of the combustion tube. Then leave the apparatus to cool and when quite cold stop the gas supply to the apparatus.

Reweigh the entire apparatus and then detach the combustion tube and reweigh it with its contents.

Optional

Experiment B18.4e

Determination of the formula of sodium oxide
Teachers should note that this is an optional experiment, which should be done by the teacher only.

Apparatus

The teacher will need:

Hard-glass test-tube, 150 × 25 mm

Stand, boss, and clamp

Bunsen burner and heat resistant mat

Glass wool

Beaker, 100 cm³

Filter paper

Oxygen cylinder

Sodium

Hexane

The reaction between sodium and oxygen to form sodium oxide, Na_2O is studied. To achieve this, sodium, which must be clean and dry, is gently warmed until it ignites in a stream of oxygen. The heating is then stopped to prevent sodium oxide being converted to sodium peroxide, Na_2O_2. The tube used should be clean and dry, and (to prevent oxidation of the sodium) should be weighed as soon as it is cool enough. The plug of glass wool is used to stop the loss of sodium oxide.

Figure B18.20 shows the now familiar formula-line linking the mass of sodium likely to react with a given mass of oxygen. Pupils can calculate the position of these lines. The formula for sodium oxide can be found by calculation. (A number of experimental determinations are shown in figure B18.20.)

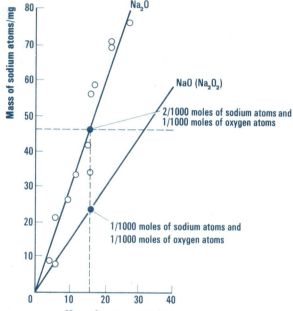

Figure B18.20
Some possible formulae for sodium oxide.

In a typical result, 0.26 g sodium combined with 0.016 g of oxygen:

number of moles of sodium atoms $= \frac{26}{23} \times \frac{1}{1000} = 0.0011(3)$

number of moles of oxygen atoms $= \frac{10}{16} \times \frac{1}{1000} = 0.0006(25)$

Therefore, for every mole of oxygen atoms our experiment shows that there are (approximately) two moles of sodium atoms. This also means that for each oxygen atom there are two sodium atoms and so the formula is Na_2O.

Procedure
Clamp a clean dry boiling tube lightly at an angle to the horizontal. Heat the tube with a Bunsen burner flame adjusted so that there is

yellow tint. Allow the tube to cool and flush the air out with oxygen. Push a plug of glass wool lightly into the tube and weigh tube and glass wool.

A freshly cut piece of sodium (less than 100 milligrams) is cleaned and dried using hexane and filter paper in the usual way. Place it in the boiling tube and reweigh. Then lightly clamp the tube and heat the sodium gently until it ignites. If the sodium burns with a 'pop', then it was not clean and the experiment should be repeated.

Once the sodium has been sufficiently ignited remove the Bunsen flame from the tube and reweigh the tube when it is sufficiently cool. The formula can be calculated by one of the methods outlined in earlier parts of this section.

Note. The experiment can be repeated using the same dry test-tube. Sometimes the tubes crack. This can be avoided if sodium is placed on a piece of asbestos paper. Fold a small piece of asbestos paper along its grain so that it easily slides into a test-tube. The binding starch is removed by heating the paper in a Bunsen burner flame. It is necessary to change the position of the tongs used to hold the paper to ensure that all the paper has been heated. The resulting paper is very brittle but, with care, may be inserted into a test-tube. The sodium is placed on the asbestos paper which may then be pushed to the bottom of the tube with a spatula.

It is suggested that the pupils record the results as below:

mass of tube + sodium	.... g
mass of tube	.... g
mass of sodium	.... g
mass of tube + sodium oxide	.... g
mass of tube + sodium	.... g
mass of oxygen	.... g

By the end of this section pupils should be familiar with some of the ways in which the numbers of moles of atoms in a solid or a liquid compound may be established. A chemical formula is taken to indicate the combining number of moles of atoms.

Suggestion for homework

Additional practice on problems of the type discussed in this section may be set as homework.

B18.5
The mole of molecules

In this section pupils are led to appreciate the significance of a mole of molecules before they consider the volumes occupied by moles of gases. Subsequently, they determine the formulae of gaseous compounds.

A suggested approach

Begin the lesson by directing attention towards the reaction of zinc with dilute hydrochloric acid to produce hydrogen gas and zinc

1. Understanding chemical
equations as reaction
summaries
2. Awareness of the application
of the mole concept to gases

chloride solution. This reaction could be demonstrated. Ask pupils how they would identify the products of reaction. An overhead projection transparency such as figure B18.21 could be used to summarize their responses and to present information.

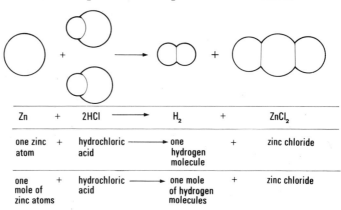

Zn	+	2HCl	$\longrightarrow$	H$_2$	+	ZnCl$_2$
one zinc atom	+	hydrochloric acid	$\longrightarrow$	one hydrogen molecule	+	zinc chloride
one mole of zinc atoms	+	hydrochloric acid	$\longrightarrow$	one mole of hydrogen molecules	+	zinc chloride

Figure B18.21
The reaction between zinc and hydrochloric acid. (This diagram is not to scale.)
(OHP 51)

Attention can then be directed to the fact that one zinc atom reacts with an excess of hydrochloric acid to produce no more than 1 molecule of hydrogen. Two zinc atoms would produce no more than 2 molecules of hydrogen under similar circumstances, and so on. In this way one can lead to the statement that 1 mole of zinc atoms produced 1 mole of hydrogen molecules and *not* 1 mole of hydrogen atoms. Moreover the equation shows that there are the same number of molecules in 1 mole of hydrogen molecules as there are atoms in 1 mole of zinc atoms. The problem now is: 'What is the mass of 1 mole of hydrogen molecules?' The pupils know that 1 mole of hydrogen atoms has a mass of 1 gram.

They should be able to deduce that 1 mole of hydrogen molecules (H$_2$) has a mass of 2 g.

Pupils could then work out the masses of the following:

1 mole of nitrogen atoms
1 mole of nitrogen molecules (N$_2$)
1 mole of oxygen atoms
1 mole of oxygen molecules (O$_2$)
1 mole of chlorine atoms
1 mole of chlorine molecules (Cl$_2$)

The reaction between hydrogen and chlorine to produce hydrogen chloride might be considered next. Figure B18.22 shows one type of overhead projection transparency which could be used. If necessary, the film loop 2–16 'The formula of hydrogen chloride' could be shown.

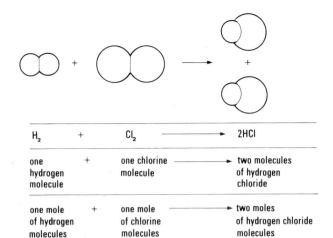

H_2	+	Cl_2	$\longrightarrow$	$2HCl$
one hydrogen molecule	+	one chlorine molecule	$\longrightarrow$	two molecules of hydrogen chloride
one mole of hydrogen molecules	+	one mole of chlorine molecules	$\longrightarrow$	two moles of hydrogen chloride molecules

Figure B18.22
The reaction between hydrogen and chlorine. (This diagram is not to scale.)
(OHP 52)

The discussion should reveal that 1 hydrogen molecule reacts with 1 chlorine molecule to produce 2 hydrogen chloride molecules. It follows then that if 2 molecules of each of the reactants react together then 4 molecules of hydrogen chloride will be formed, and so on. 1 mole of hydrogen molecules (2 g) must therefore react with 1 mole of chlorine molecules (71 g) and each mole contains precisely the same number of particles. Pupils might then list the masses of moles of molecules and of atoms. All of these contain the same number of molecules. The simplicity of this idea is its great virtue.

Table B18.7 lists some examples for use in such an exercise.

Moles of molecules

Name of gas	Formula	Mass of one mole /g
chlorine	Cl_2	71
fluorine	F_2	38
hydrogen	H_2	2
nitrogen	N_2	28
oxygen	O_2	32
ammonia	NH_3	17
carbon monoxide	CO	28
carbon dioxide	CO_2	44
ethane	C_2H_6	30
ethene	C_2H_4	28
methane	CH_4	16

Note. Each of these moles contains the same number of molecules: the number is also the same as that in a mole of atoms.

Moles of atoms

Name of gas	Symbol	Mass of one mole /g
argon	Ar	40
helium	He	4
krypton	Kr	84
neon	Ne	20
xenon	Xe	131

Note. Each of the above contains the same number of particles under the same conditions of temperature and pressure. It may be necessary to remind the pupils of the effect of temperature and pressure on the volume of a gas.

Table B18.7 Moles of molecules and of atoms

By the end of this section pupils should have learnt that 1 mole of atoms and 1 mole of molecules each contains the same number of particles.

Suggestion for homework In each of the following reaction summaries or equations, state how many moles of molecules of each gas are being used or produced.

a. $2Cu(s) + O_2(g) \longrightarrow 2CuO(s)$

b. $2Mg(s) + O_2(g) \longrightarrow 2MgO(s)$

c. $C(s) + O_2(g) \longrightarrow CO_2(g)$

d. $4Al(s) + 3O_2(g) \longrightarrow 2Al_2O_3(s)$

e. $HgCl_2(s) \longrightarrow Hg(l) + Cl_2(g)$

f. $2H_2(g) + O_2(g) \longrightarrow 2H_2O(l)$

Write down other reaction summaries which involve gases in the form of word equations and chemical equations and state the numbers of moles required for each gas which is a reactant or a product.

B18.6
What volumes do moles of gases occupy?

The pupils know that 1 mole of atoms contains the same number of particles as 1 mole of molecules. In this section they calculate the volume occupied by a mole of gas for a number of different gases and hence are led to Avogadro's law.

A suggested approach

In this short section, pupils are required to look up or work out the volume occupied by 1 mole of a given gas under standard conditions. The volume of 1 mole of gas is calculated by using the relationship:

Objectives for pupils

1. Awareness of the mole concept and its application to gases

$$\frac{\text{mass of 1 mole of gas}}{\text{density of gas}} = \text{volume}$$

2. Knowledge of the density and volume occupied by a known mass of gas

The relevant information is provided in the *Handbook for pupils* and also in Table B18.8 for the convenience of the teacher.

3. Knowledge of Avogadro's law

Name of gas	Formula	Mass of 1 mole of gas	Density of gas at 25°C and 1 atmosphere /g dm^{-3}	Volume occupied by 1 mole of gas /dm^3
argon	Ar	40	1.66	24.1
chlorine	Cl$_2$	71	2.99	23.8
fluorine	F$_2$	38	1.58	24.0
helium	He	4	0.17	23.5
hydrogen	H$_2$	2	0.08	25.0
krypton	Kr	84	3.46	24.3
neon	Ne	20	0.84	24.0
nitrogen	N$_2$	28	1.17	23.8
oxygen	O$_2$	32	1.33	24.1
xenon	Xe	131	5.5	23.8

(Continued)

ammonia	NH$_3$	17	0.71	23.9
carbon monoxide	CO	28	1.15	24.4
carbon dioxide	CO$_2$	44	1.81	24.3
ethane	C$_2$H$_6$	30	1.24	24.2
ethene	C$_2$H$_4$	28	1.14	24.6
methane	CH$_4$	16	0.72	24.2

Table B18.8 Volumes occupied by 1 mole of gas at 25°C and at 1 atmosphere pressure

The last column of Table B18.8 indicates that the volume occupied by 1 mole of atoms or 1 mole of molecules is approximately 24 dm^3 at 25 °C and atmospheric pressure. Moles of molecules of all gases present as molecules, and moles of atoms of all gases present as atoms, occupy about the same volume. It may be appropriate for the teacher to discuss the terms monatomic, diatomic, etc., used as adjectives to describe molecules of gases containing one, two, etc., atoms, at this point.

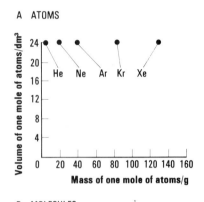

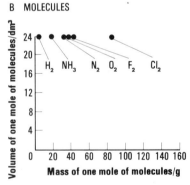

Figure B18.23
A. Volumes of moles of atoms (gases at 25 °C and 1 atmosphere).
B. Volumes of moles of molecules (gases at 25 °C and 1 atmosphere). (OHP 53)

Discussion of figure B18.23 shows the validity of our approximation. Then move to other questions. Thus, 'What do you remember about the number of particles in a mole of argon atoms? or a mole of chlorine molecules? or a mole of ammonia molecules?' Make the point that they all contain the same number of particles. Thus, 24 dm^3 of each gas must contain the same number of particles under the same conditions of temperature and pressure. It follows that we are now able to state Avogadro's Law: equal volumes of different gases under the same conditions of temperature and pressure contain the same number of particles.

Suggestion for homework How many moles are present in $2.4\,dm^3$ or $0.24\,dm^3$ or $24\,cm^3$ of gas at $25\,°C$ and at 1 atmosphere pressure?

B18.7
Using the 'mole of molecules' to determine the formulae of gases

At this point in the course it is convenient to provide at least one example of the application of the mole of molecules to the determination of the formulae of gases. Two examples are provided in this section: hydrogen chloride and ammonia.

A suggested approach

Objectives for pupils

1. Awareness of Gay-Lussac's law of combining volumes of gases
2. Application of the 'mole of molecules' to the determination of the formulae of gases

Remind pupils of the work of Topic B11 on the 'salt-gas problem'. Enquire how they might determine the composition of hydrogen chloride in quantitative terms. Some pupils may recall the information presented in figure B18.22, and, in any case, the properties of the reactants and products of the system will need to be recalled.

Experiment B18.7a may be demonstrated by the teacher. Hydrogen chloride gas is passed slowly over molten zinc and the volume of hydrogen formed is measured.

Alternatively, film loop 2–16 'The formula of hydrogen chloride' may be shown and discussed. The experiment shown in the loop is that used in the Sample Scheme (1966) and given in *Collected experiments*, Experiment E12.22.

Experiment B18.7a

Apparatus

The teacher will need:

2 syringes, $100\,cm^3$, in holders

Three-way stopcock with capillary tubing

Piece of hard-glass tubing, $12\,cm \times 7\,mm$ O.D.

Some glass beads, to fit inside the hard-glass tubing

Thick-walled plastic connecting tubing

Asbestos paper or equivalent material

Mineral wool

Test-tube

(Continued)

Finding the formula of hydrogen chloride
This experiment should be carried out by the teacher.

Procedure
Heat the piece of hard-glass tubing at its mid-point in a Bunsen flame and bend it through an angle of 5–10°. Allow the tube to cool. Cut small pieces of zinc foil, about 0.8 g, and roll them up so that they can be slipped into the tube. Add tufts of asbestos paper on either side of the zinc and pack them in using a glass rod. Next, pack the tube with glass beads such that the zinc is in about 2 cm of clear space in the middle of the tube. The glass beads reduce the volume of air which would otherwise occupy the tube. Additional pads of asbestos paper may be used to secure the beads in the tube before assembling the apparatus as shown in figure B18.24. The entire apparatus *must* be dry.

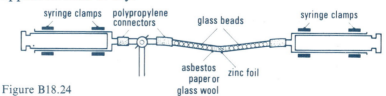

Figure B18.24

Plastic tubing

Small trough of water

Wooden splint

Bunsen burner and heat resistant mat

Zinc foil, in strips, 0.8 g

Apparatus for generating hydrogen chloride (as in B11.3)

Connect the apparatus to a hydrogen chloride generator. Admit hydrogen chloride to syringe 1, making sure that all air has been first displaced from the generator by flushing out the syringe and stopcock tubing once or twice with gas before filling syringe 1. Admit a little more than the required volume of gas, and disconnect the generator. Expel the additional gas until the syringe is exactly at the $50 \, cm^3$ mark. Turn the gas stopcock to connect the syringe to the hard-glass tube. Heat the tube under the zinc with a moderate, mobile, Bunsen flame and, as soon as the metal melts, *gently* pass the gas from syringe 1 through the tube to syringe 2. Continue passing the gas *gently* backward and forward between the syringes until there is no further reduction in volume. If the reaction is slow, raise the temperature of the metal, *but avoid overheating*, which will cause excessive volatilization of zinc chloride. (*Note.* The pads of asbestos paper are not intended to be heated.) The bend in the tube should be enough to keep the molten zinc in place, but there is some risk of it being blown onto one of the pads and perhaps penetrating through it if gas is not passed *gently* through the tube.

The final volume of gas left in the tube will be found to be close to $25 \, cm^3$. The glass syringe is likely to be sufficiently free running for the volume to be read off without using a manometer to check that the gas pressure remains at atmospheric pressure. The residual gas can be expelled from the apparatus and collected over water in the small test-tube. The gas is then ignited to convince the class that it is hydrogen.

Treatment of the result
The experiment shows that:

$50 \, cm^3$ of hydrogen chloride contains $25 \, cm^3$ of hydrogen.

It follows that $1000 \, cm^3$ of hydrogen chloride contains $500 \, cm^3$ of hydrogen. From the table of densities in the *Handbook for pupils*:

mass of $1000 \, cm^3$ of hydrogen chloride at about $25 \, °C$ and 1 atmosphere pressure $= 1.50 \, g$

mass of $500 \, cm^3$ of hydrogen at about $25 \, °C$ and 1 atmosphere pressure $= 0.04 \, g$

Hence, by difference, mass of chlorine present $= 1.46 \, g$

Thus, 0.04 g of hydrogen combines with 1.46 g of chlorine.

Since hydrogen has a relative atomic mass of 1.0, it is convenient to calculate how much chlorine will combine with this mass of hydrogen. Hence, 1.0 g of hydrogen combines with $\frac{1.46}{0.04}$ g of chlorine $= 36.5 \, g$ of chlorine.

This mass of chlorine approximates very closely to the relative atomic mass of chlorine expressed in grams. It follows – given our crude experiment – that it will correspond to 1 'relative-atomic-mass-worth' of chlorine (35.5 g). The formula of hydrogen chloride may thus be written H_1Cl_1, or, simply, as HCl.

After the experiment, point out the simple relationship between the initial and final volumes of gases used. Mention that (during the early part of the nineteenth century) Gay-Lussac noticed that gases reacted in volumes which were in a simple ratio and he proposed a 'law' to that effect. If time allows ask the pupils to comment on their understanding of the word *law* in chemistry (or in science): it is a record of nature rather than a command to nature!

In Topic B14, 'Chemistry and the World food problem', some mention was made of the analysis of ammonia. In B14.2 ammonia was analysed and shown to contain hydrogen and nitrogen. In the next experiment the volume composition of ammonia is demonstrated by the teacher.

Experiment B18.7b

The volume composition of ammonia
This experiment should be done by the teacher.

Apparatus

The teacher will need:

Either syringe bench arranged as shown in figure B18.26

Or 3 glass syringes, 100 cm³
3 syringe holders with stand, or special syringe bench

2 three-way stopcocks with capillary tubing

2 combustion tubes, 15 cm long × 7 mm diameter (ordinary hard-glass tubing is not satisfactory: transparent silica tubing is essential)

4 pieces of hard-glass rod about 2 cm long and of diameter slightly smaller than the internal diameter of the combustion tubes

Thick-walled rubber connecting tube

Apparatus for producing dry ammonia (see figure B18.25 and Experiment A22.3)

Bunsen burner and heat resistant mat

Indicator paper

Cylinders of compressed gases: hydrogen and nitrogen

Iron wool (clean and free from grease)

Copper(II) oxide, wire form

0.880 ammonia solution

Procedure
Set up the apparatus as shown in figure B18.26.

Make sure that the iron wool is freshly reduced by heating it in dry hydrogen (from the cylinder). Flush out the apparatus with nitrogen from a cylinder in order to remove all air. Fill syringe 1 with 40 cm³ of dry ammonia through stopcock (A) by gently warming about 10 cm³ of 0.880 ammonia solution in the test-tube with the drying tube attached. Heat the combustion tube containing the iron wool very strongly with a roaring Bunsen flame. Before passing the gas over the iron wool, the iron wool should be almost white hot. Have stopcock (B) set so that syringes 1 and 2 are connected. Pass the ammonia backwards and forwards over the heated iron wool from syringe 1 to syringe 2. Do this several times until no further increase in volume takes place. This operation may require several passes. Cool the combustion tube with a damp cloth, pass all gas into syringe 1 and read the increase in volume. The volume of gases (now a mixture of hydrogen and nitrogen) should have nearly doubled.

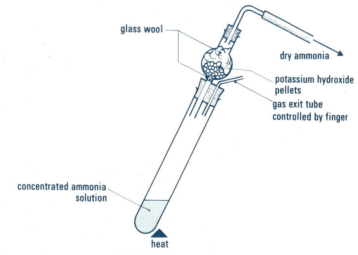

glass wool

dry ammonia

potassium hydroxide pellets

gas exit tube controlled by finger

concentrated ammonia solution

heat

Figure B18.25
Apparatus for producing dry ammonia.

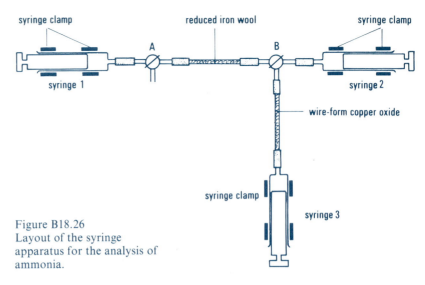

syringe clamp   reduced iron wool   syringe clamp

A   B

syringe 1   syringe 2

wire-form copper oxide

syringe clamp

syringe 3

Figure B18.26
Layout of the syringe
apparatus for the analysis of
ammonia.

Pass the gases into syringe 2 and turn the stopcock (B) so that syringes 2 and 3 are connected. Heat the combustion tube containing copper(II) oxide. Pass the gas from syringe 2 over the copper oxide into syringe 3 and back to syringe 2 until there is no further decrease in volume. The hydrogen will reduce the copper oxide to form copper and water, and the residual gas (nitrogen) should be now found to be just over a quarter of the volume of mixture of gases used. Hence, the result may be expressed as:

2 volumes of ammonia ⟶

1 volume of nitrogen + 3 volumes of hydrogen

Sources of experimental error
1. It is *most important* that the ammonia should be *dry*. Otherwise, some ammonia will dissolve in the water and the volume will more than double when the gas is passed over heated iron wool.
2. If there is (*a*) residual air in the apparatus or (*b*) iron oxide on the wool, some of the hydrogen will react with them and the overall results will be low.
3. In the second part of the experiment, water is produced. This means that the nitrogen remaining when the decomposition is complete is saturated with water vapour whereas the other two gaseous volumes were measured dry. The result will therefore be rather higher than the theoretical value.

Specimen results
40   cm^3 ammonia
79.5 cm^3 nitrogen and hydrogen
22   cm^3 nitrogen

The experiment shows that one volume of nitrogen and three volumes of hydrogen are produced from two volumes of ammonia. Accordingly, using Gay-Lussac's law and Avogadro's law, assuming the formula for nitrogen to be N_2 and hydrogen to be H_2, we can now write:

$$2N_xH_y(g) \longrightarrow N_2(g) + 3H_2(g)$$

The formula of ammonia is therefore N_1H_3 or NH_3.

Alternatively, one might argue:
2 volumes of ammonia yields 1 volume of nitrogen + 3 volumes of hydrogen, that is $40.0\,cm^3$ of ammonia yields $22.0\,cm^3$ of nitrogen and $57.5\,cm^3$ of hydrogen. Divide all numbers by 20 throughout, so that $2.0\,cm^3$ of ammonia yields $1.1\,cm^3$ of nitrogen and $2.9\,cm^3$ of hydrogen.

The syringes used are only accurate to within $1\,cm^3$ and so we may write within the limits of experimental error:

$2\,cm^3$ of ammonia yields $1\,cm^3$ of nitrogen and $3\,cm^3$ of hydrogen
or 2 molecules of ammonia yield 1 molecule of nitrogen and 3 molecules of hydrogen
or 2 moles of ammonia molecules yields 1 mole of nitrogen molecules and 3 moles of hydrogen molecules
or $2(ammonia) \longrightarrow N_2 + 3H_2$

from which it follows that the formula of ammonia must be NH_3.

Suggestions for homework

1. Summarize the ideas and procedures used for establishing the formulae of compounds.

2. What volume is occupied by (a) $\frac{1}{10}$, (b) $\frac{1}{100}$, (c) $\frac{1}{1000}$ of a mole of each of the following gases?
argon (Ar); hydrogen (H_2); hydrogen chloride; ammonia.

3. Write down the number of moles in each of the following:

a. 5.4 g aluminium atoms
b. 80 g bromine molecules
c. 8 g bromine atoms
d. 2 g hydrogen molecules
e. 24 g hydrogen molecules
f. 0.016 g oxygen molecules
g. 28.1 g silicon atoms
h. 207 g phosphorus atoms
i. 32 g sulphur (S_8) molecules
j. 71 g chlorine molecules

4. What are the volumes and masses of the following:

1 mole of chlorine molecules;
10 moles of chlorine molecules;
5 moles of oxygen atoms;
One-tenth of a mole of nitrogen molecules;
1 mole of nitrogen atoms.

Summary

By the end of the Topic, pupils will have used the mole of atoms and the mole of molecules to establish the formulae of a variety of compounds.

Appendix

Construction of the model of magnesium oxide
Cut four sheets of perspex into squares of side 20 cm. Alternate holes of diameter 2.9 cm and 3.5 cm are made in each sheet so there are eight holes of each kind in each sheet of perspex. The templates are shown in figure B18.27 and B18.28. It is possible to have fewer spheres in each layer, but the model illustrated does clarify the fact that magnesium oxide has a cubic structure.

Polystyrene spheres will need to be weighted or ballasted with screws and panel pins so as to represent the relative masses of magnesium and oxygen atoms. It is suggested that spheres of diameter 4.4 cm be ballasted with screws and pins to 4 or 16 grams and be painted red for oxygen; and spheres of 3.3 cm diameter can be ballasted to 6 or 25 grams and painted gold for magnesium. Figure B18.29 shows a partly completed model.

A model may be adapted to show the composition of manganese(IV) oxide. The templates required are shown in Figure B18.30.

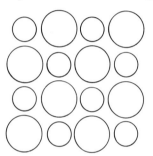

Figure B18.27
First and third layers.

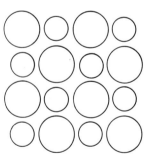

Figure B18.28
Second and fourth layers.

Figure B18.29
First three layers of the complete model of magnesium oxide.

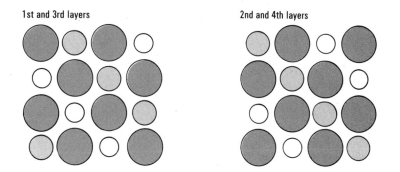

Figure B18.30
Model of manganese(IV) oxide. The same racks may be used for MnO_2 and MgO_2, but in the case of MnO_2 four manganese spheres are missing from each layer.

Topic B19

Electrolysis

Introduction to the Topic

Primarily, this Topic seeks to present the phenomenon of electrolysis so that pupils can appreciate the underlying principles. The Topic opens with a circus of investigations which are intended to revise and extend the work on electrochemistry already carried out during Stage I, or other introductory courses.

Pupils should be able to classify common materials as conductors, non-conductors, and electrolytes. They do various experiments to show that electrolysis can occur in solutions or in melts; that metals and hydrogen are liberated at the cathode and non-metallic elements are liberated at the anode during electrolysis. The terms used in this section include electrolyte, conductor, electrolytic cell, electrode, anode, and cathode.

An elementary study of electrostatics is included for those pupils who have not studied the subject in Physics. It serves as a basis for the explanation of electrolysis. The main theme of the Topic is a study of the migration of ions and this is followed by an explanation of electrolysis in simple terms, together with an elementary con-sideration of atomic structure and brief statements relating to molecular, giant, and ionic structures. The properties of ionic and molecular compounds are then reviewed through the use of examples.

The work of Faraday and Davy on this subject is introduced in the next section and the discussion on the properties of ions is used to illustrate the contributions of these two men. Faraday's laws are used to find the amount of charge on specific ions, and an experiment on the variability of charge on the copper ion completes the pupils' quantitative work. The Topic concludes with a review of the essential phenomena.

Subsequent development

The properties of ions form the theme of Topic B20 and are studied further in Topic B21 which deals with chemical equilibrium. The ideas and concepts developed in this Topic are needed in the Options, especially Options 6, 7, 8, 9, and 10. Option 10 gives pupils an opportunity to follow through the development of Davy's and Faraday's investigations of electrochemistry. Electron transfer processes at electrodes during electrolysis receive further consideration in Option 6.

Alternative approach

The content of this Topic appears in a modified sequence in Topic A16, Alternative IIA.

Background knowledge

Although it is assumed that pupils have followed Stage I of Nuffield Chemistry or another course using similar material, there is a limited opportunity to revise introductory work on electrolysis in this Topic. If no introductory work on electrolysis has been attempted, more time should be given to section B19.1. The length of time spent on section B19.2 on elementary electricity also depends on the pupils' previous experience.

Further references

for the teacher

Collected experiments, Chapter 5, provides additional experiments on electrolysis.
Handbook for teachers, Chapters 6 and 7, provides background information.
Nuffield Advanced Physical Science (1972) *Teachers' guide I*, Sections 4, 5, and 7 carry the subject further.

Other sources of information are:
Hockey, S. W. (1972) *Fundamental electrostatics*. Methuen Educational.
Chemical Education Material Study (1963) *Chemistry – an experimental science*. W. H. Freeman. (See Chapter 5.)

Supplementary materials

Film loops
2–8 'Electrolysis of lead bromide'
2–18 'The electrolysis of potassium iodide solution'

Films
'Electrochemistry' ESSO. Films for Science Teachers No. 24
'Electrostatics' ICI Film Library
'Static electricity in the Chemical Industry' ICI Film Library

Reading material

for the pupil

Chemists in the world, Chapter 1 'Finding out about the atom' and Chapter 4 'Davy and Faraday'.
Handbook for pupils, Chapter 7 'Electricity and matter' and Chapter 9 'The structures of elements and compounds'.

B19.1
What happens when an electric current is passed through various solutions and molten substances?

This section provides an opportunity for pupils to revise and extend the work on electrolysis which was carried out during Stage I (see sections A8 or B6).

A suggested approach

Objectives for pupils

1. Knowledge of the properties of a simple electrical circuit
2. Knowledge of the phenomenon of electrolysis
3. Knowledge of the terminology used in the study of electrolysis
4. Awareness of the differences between an ion and its corresponding neutral particle

It is suggested that pupils should start by revising simple electrical circuits. This provides the teacher with an opportunity to coordinate work done during Stage I or other courses. In view of the range of possibilities no specific guidance is given, but pupils should know, at least, how to use ammeters and lamps as current indicators, and that some materials can be classified as conductors and others not.

Experiment B19.1a

Apparatus

The teacher will need:

Power supply or battery

Lamp and socket

Ammeter – demonstration type preferred

2 probes

Connecting wires

Samples of various metals in rod and sheet form, non-metals (such as carbon rod and roll sulphur), plastics, and other everyday materials

An electrolytic cell

Electrodes made of copper, carbon, etc.

Some properties of a simple electrical circuit

Procedure

A circuit consisting of a power supply, lamp, ammeter, and two probes should be used to revise the properties of a simple series electrical circuit. The addition of other circuit elements – such as sizeable pieces of metal or plastic or an electrolytic cell – should be allowed for in the circuit design.

As the result of this demonstration, pupils should have a clear idea of the meaning of the terms conductor and non-conductor, and know how to use and set up a simple series electrical circuit.

The class may then be divided into groups to operate a 'circus' of experiments to illustrate the phenomena of electrolysis and to encourage discussion. This section is a necessary preliminary to subsequent parts of the Topic. Pupils should have an opportunity to attempt at least some of the suggested experiments and to report their findings at the end of the lesson to the class. It is *not* intended that each group of pupils attempt every possible variant suggested.

Teachers should note that the experiments are intended for pupils and will require some introductory discussion:
1. To distinguish between conduction and electrolysis.
2. To examine the effect of varying the distance between the electrodes in an electrolytic cell.
3. To note the effect of the relative sizes of the electrodes on the reaction in an electrolytic cell.
4. To observe how electrodes need to be placed in an electrolytic cell to obtain an even deposit of plating on the cathode.

5. To observe the need for controlled current density for quantitative work.

6. To become aware of the significance of the material used to construct electrodes.

7. To know that the concentration of an electrolyte can effect the products of electrolysis.

Some investigations of the properties of an electrolytic cell

Apparatus

Each pair of pupils will need:

Power supply

Ammeter, 0–1 A

Lamp (2.5 V 0.5 A) and lamp holder

Stand, 2 bosses, and 2 clamps

Open-ended glass tube, 40 cm long by 4.5–5.0 cm wide, fitted with a bung carrying a copper electrode

Copper electrode mounted on a long wire support: several types required – for example, circular disc electrode (4 cm diameter), angled electrode (to encourage uneven deposit), simple wire electrode, etc.

U-tube with side arms and related electrodes (e.g. copper, graphite, platinum)

Connecting wire

Test-tube, 150 × 16 mm

Teat pipette

Plastic syringe, 50 cm³

Beaker, 400 cm³

Measuring cylinder, 100 cm³

Splint

Ruler

Access to a hand lens

Stop-clock

Litmus paper

Limewater

Distilled water

1.0M copper sulphate solution

1.0M sodium chloride solution

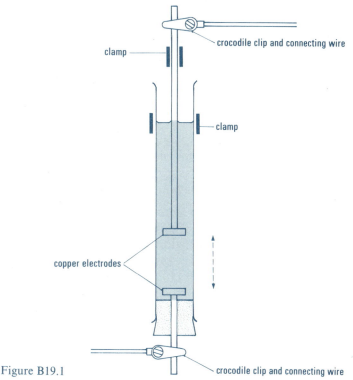

Figure B19.1

Procedure

A number of similar sets of apparatus will be needed, one for each group. Pupils may be asked to investigate one or more of the possible experiments listed. The special electrolysis cell is shown in figure B19.1. Pupils set up a series circuit consisting of the cell, an ammeter, and the power supply. The movable electrode should be connected to the negative pole of the power supply (that is, should be used as the cathode). They fill the cell with an electrolyte and use the circuit to carry out one or more of the following.

a. Investigating the effect of distance between electrodes on the current passed through the cell. Pupils use copper sulphate solution between copper electrodes, each of which is 4 cm diameter. Arrange to have the electrodes 4 cm apart and adjust the power supply to give a current of 1 A. The pupils now gradually increase the distance between the electrodes and note the current passed at a number of specific inter-electrode distances. They may present their findings in the form of a graph.

---

b. Investigating the effect of the concentration of the electrolyte on the current flow through the cell. Pupils note the effect on current flow of using 1.0M, 0.5M, 0.25M, and 0.125M copper(II) sulphate solutions. (Each of these diluted solutions is obtained by diluting 1.0M copper(II) solution.) The copper electrodes are set at a fixed distance (e.g. 5 cm or 10 cm) for this series of experiments. It will be found convenient to adjust the power supply to give a current of 1 A when 1.0M copper(II) sulphate is electrolysed. Again, pupils may find it convenient to present their results in the forms of a graph.

c. Finding the limitations of an electric lamp as a current indicator. Pupils use both an electric lamp (2.5 V 0.5 A) *and* an ammeter (range 0–1 A) as current indicators in this experiment. For a given set of conditions (e.g. electrodes at a fixed distance – say 5 cm; electrolyte of known concentrations – say 0.5M copper(II) sulphate solution; copper electrodes), the variation of current with distance between the electrodes can be studied. Start by adjusting the power supply to give a current corresponding to that needed for 'safe' maximum brightness with the lamp (i.e. 0.5 A). Record the effect of increasing the inter-electrode distance on the current flow. Note when the lamp 'goes out' and when the current flow actually ceases.

d. Investigating the effect of electrode shape. The effect of using a variety of different electrodes is noted when all other variables are held constant. Electrodes to be tested are a standard copper disc electrode (4 cm diameter) with either (*i*) a copper disc electrode 4 cm diameter; (*ii*) an angled copper electrode, at 80° to the horizontal (and so much longer than 4 cm); or (*iii*) a single stiff copper wire electrode. It is suggested that the suspended electrode is used as the cathode to enable pupils to see whether an uneven deposit of metal can be detected. Copper(II) sulphate solution is used as electrolyte and it will be convenient to use a fixed 'nearest' inter-electrode distance of 5 cm in each instance. Pupils can time the electrolysis and note when plating can be 'seen' to occur in each instance and whether it is 'even'. (Inspection by using a hand lens after 2–5 minutes electrolysis will be found to be convenient.) The significance of these observations for the particulate nature of matter and of electricity can be raised in discussion.

e. Exploration of the effect of electrode material on the products of electrolysis. In this case, the electrolytic cell needs to be a U-tube – not the special cell used earlier. The electrodes to be tested include copper, graphite, and platinum. Either copper(II) sulphate or sodium chloride solution can be used as the electrolyte. The conditions needed to achieve electrolysis must be recorded and observations made on the effects seen or detected – for example, changes in the electrode surfaces; gases evolved; effect of products on damp litmus paper, etc.

f. Exploration of the effect of concentration of electrolyte on the products of electrolysis. As in (*e*) above, the electrolytic cell needs to be a U-tube. It is suggested that carbon rod electrodes are used with a variety of concentrations of sodium chloride solution (e.g. 1.0M, 0.5M, 0.25M, 0.125M, 0.63M, etc.). Only qualitative studies are intended, with some provision for pupils to test the products of electrolysis (e.g. effect of products on damp litmus paper; collection – teat pipette or small plastic syringe attached to the side arm of the U-tube – and testing of gases evolved.) Discussion with pupils will encourage them to look for a variety of gases, e.g. chlorine, oxygen, and (when carbon electrodes are used with dilute solutions) carbon dioxide.

At the conclusion of the experimental work, the results obtained from each investigation need to be reported and discussed. In summing up, draw attention to the general findings: an increase in inter-electrode distance results in a decrease in current flow; a decrease in the concentration of an electrolyte also results in a decrease in current flow for electrolysis under otherwise comparable conditions; an electric lamp does not necessarily indicate whether a current is flowing in a circuit – an ammeter is a better current indicator; to achieve effective and even electroplating, the electrodes in an electrolytic cell need to be parallel to one another; the nature of the electrodes used during an electrolysis needs to be known; the concentration of the electrolyte solution can influence the products of electrolysis. Evidence to support the particulate nature of matter and of electricity from the investigations can also be discussed.

At the end of this discussion of experimental work, terms such as conductor, non-conductor, electrolyte, electrode, anode, cathode, and electrolytic cell, can be defined and reference made to the glossary of terms in the *Handbook for pupils*.

So far our investigations have used solutions of electrolytes. In Experiment B19.1c a molten electrolyte is used.

Experiment B19.1c

Investigating electrolysis of a molten salt

This experiment **must** either be carefully supervised by the teacher or be a demonstration.

Teachers are reminded that lead salts and the vapours produced during this experiment are toxic. Molten lead bromide can cause severe burns.

Note. Lead bromide must be of a good quality otherwise some bromine may be given off on merely melting this chemical. As part of the pre-lesson preparation, it is advisable to check that there are no nitrate ions in the sample.

Procedure

Figure B19.2 shows the general layout of the apparatus. The experiment is started by warming the bottom of the U-tube with a small flame and adding lead bromide till it reaches the level shown in the diagram. Then warm the electrodes and lower them into the melt. With a current flow of a little over 1 A, bromine vapour can be seen almost at once in the anode compartment of the U-tube. The elec-

Apparatus

The teacher will need:

Safety spectacles

Power supply, 10–12 V d.c.

Rheostat, 12 Ω, 5 A capacity

Demonstration ammeter with a scale giving a maximum reading of 3 A

Connecting wire

2 barrel terminals

Carbon rod (to serve as anode)

Steel rod (to serve as cathode)

U-tube, about 15 cm long and made from glass tubing of about 1 cm diameter

(*Continued*)

Stand, boss, and clamp

Bunsen burner

Heat resistant mat

Spatula

Lead bromide

trolysis can be run for a few minutes in a well ventilated laboratory but thereafter a fume cupboard should be used. Only just enough heat should be used to keep the lead bromide molten. After four or five minutes, switch off the current and remove the cathode and anode. Allow the pupils to examine the cathode to see some small beads of lead clinging to the end. This can be shown to be lead by the ease with which it can be cut and the fact that it marks paper.

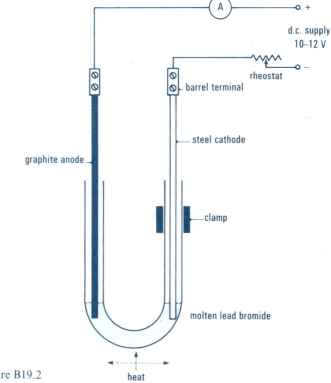

Figure B19.2

The following points should emerge from a discussion of the results:
1. When lead bromide is solid, no current flows and no decomposition takes place.
2. When molten lead bromide is used, an electric current will flow through it and simultaneously bromine is seen to be liberated at the anode.
3. At the end of the experiment, the steel cathode is found to mark paper – showing that lead was liberated at the cathode during electrolysis at the same time as bromine was liberated at the anode.
4. Electrolysis can be shown using either solutions of salts or molten salts.

Ions can be introduced at this point as a way of 'explaining' electrolysis. (If this is done, then the function of section B19.2 can be more readily understood by pupils and seen as a necessary prelude to section B19.3). 'If ions exist, then their charges must distinguish them from atoms and other bodies which may be present during electrolysis.' 'What are the unique properties which we can ascribe to charges?' In such a way, teachers can make a case for the study of electrostatics in B19.2.

Summarize the results of the experiments in this section. How do you explain the various phenomena shown?

B19.2
Elementary electricity

A knowledge of elementary electrostatics and the properties of simple electrical circuits is essential for an understanding of electrolysis. Much will depend on the background knowledge of the pupils. To prevent repetition, it is desirable to coordinate work done in this section with that in Stage I, in the previous section, and in the school physics department.

Objectives for pupils

1. Awareness of the electrical nature of matter
2. Awareness of the electrical basis of electrostatic phenomena
3. Knowledge of the laws of attraction and repulsion for electrically charged bodies
4. Knowledge that equal quantities of unlike charges cancel one another
5. Knowledge that a charged body placed in an electric field experiences a force
6. Knowledge of the characteristics of electrical conductors and non-conductors

By this stage of the course, pupils should be aware of a variety of electrical phenomena. Remind pupils that when an electric current is passed through a conductor (such as the element in an electric fire or the hot plate of a cooker) it becomes warm. When dry hair is rubbed with a plastic comb, the hair 'stands on end'. When we remove clothing made from man-made fibres on a cold dry night, sparks fly. These and other instances are so familiar to everyone that we tend to ignore them. Simple demonstrations of rubbed plastic rods attracting pieces of paper and of attracting and repelling other rods of rubbed plastic material can lead on to more detailed investigations such as those in Experiment B19.2a, b, and c. Electricity and matter can be seen to be inter-related.

The use of a source of high voltage for simple electrostatic phenomena is intended to shorten the sequence of demonstrations and to present what may be familiar effects in a new guise. In effect, mains electricity is used in a teacher demonstration to reproduce some of the effects and laws obtained by the more familiar route adopted by many physics courses and in Experiment B19.2a. Such an approach provides a simple link to circuit theory and to electrolysis.

Note to teachers. Practical points when using apparatus to demonstrate electrostatic phenomena
If modern plastics are used (for example, polythene, perspex, etc.), electrostatic experiments can be made to work under very adverse conditions. Ebonite rods may require cleaning to remove the oxidized surface. This can be done by rubbing such rods with very fine grade sand paper. Glass rods also need special treatment: they need to be dried in an oven before use and, ideally, used while they are still warm.

If possible arrange to have an electric reflector-type fire to throw a direct beam of radiant heat along the bench on which you propose to do your demonstrations.

Modern plastics acquire and retain electrostatic charges very easily. On occasions this can be a handicap, such as when the insulating support of a conductor becomes charged and produces misleading effects. Plastic materials can be discharged quite easily by passing them through the air immediately above a low Bunsen flame. However, it is advisable to keep the Bunsen flame away from a particular demonstration since the flame itself is a source of charged particles. It is equally important to see that draughts do not blow ionized air towards the apparatus!

Apparatus

The teacher will need:

Beaker, containing very small pieces of paper

Polythene rod

Perspex rod

Rods of other materials (for example, glass, ebonite, etc.)

Piece of fur

Piece of silk

Duster

Wire jockey

Nylon thread

Burette stand and clamp

Bunsen burner

Simple demonstrations of electrostatic phenomena

Procedure

Rub the polythene rod with a piece of fur or with the duster and show that the rubbed rod attracts small pieces of paper. Repeat the experiment using a perspex rod rubbed with silk or with foam plastic. One can ask the class 'Suppose you were the first to discover this strange effect, what would you try next?' Probably pupils will suggest trying another substance. If the class has already met some simple ideas from mechanics and has understood something of the nature of force, one might get a better suggestion. Thus, 'If one electrified rod exerts a force on a piece of paper, does one electrified rod exert a force on another electrified rod?' To test such a suggestion, we suspend a charged polythene rod in a wire stirrup and bring up a second charged polythene rod. The amount of repulsion seen will be small – perhaps too small to be noticed by pupils. We need to use something made of polythene with a relatively large surface area – and low moment of inertia – to achieve a measurable affect. A thin but rigid strip of polythene can be used with great effect. Alternatively, perspex sheet cut into strips can be used.

It is a good idea to mark one end of the selected suspended plastic strip and to charge that end only. The next step is usually easy: to test the rubbed polythene against rubbed perspex and so demonstrate attraction. It is important to show that a finger placed near a charged suspended rod also gives rise to attraction, as does an iron stand when placed on a bench near the suspended rod. We may therefore conclude 'The only sure test for electrification is repulsion'.

Simple experiments of this kind only go so far as to suggest the existence of two kinds of charge. The real problem which remains is to find out if there are more than two kinds of charge. We repeat earlier experiments using a range of different materials and so lead pupils to the statement 'Any substance that can be charged either repels charged polythene and attracts charged perspex or vice versa.' Therefore, it is necessary only to assume the existence of two kinds of charge. At this point the teacher could introduce the terms positive and negative:

positive charge, the term used for the charge that perspex acquires from silk;
negative charge, the term used for the charge that polythene acquires from fur.

Experiments with an electrometer

Apparatus

The teacher will need:

Perspex box, fitted with a copper-plated base (see figure B19.3)

2 conducting spheres suspended from terminals inside the box

Connecting leads

(Continued below)

Procedure

Set up the apparatus as shown in figure B19.3. Arrange for the shadows of the two conducting spheres to be visible to the class. Apply a potential difference to the spheres and note that unlike charges attract one another. Disconnect the power unit and note the effect of joining the spheres together: no charge remains, the spheres no longer attract one another. (See figure B19.4.) Rearrange the wiring as shown in figure B19.5a and demonstrate that like charges repel. Disconnect the power supply, and discharge the spheres before putting the apparatus away.

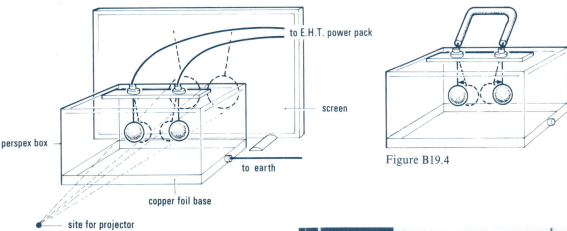

to E.H.T. power pack

screen

perspex box

to earth

copper foil base

site for projector

Figure B19.3

Figure B19.4

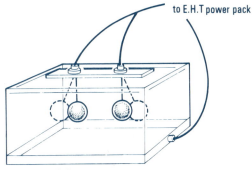

to E.H.T power pack

Figure B19.5a

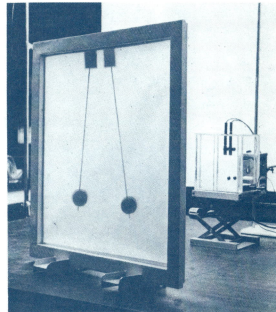

Figure B19.5b

EHT power supply (p.d. about 5000 V d.c.)*

Screen

Source of illumination (for example, a point source of light)

*Teachers are reminded of the restrictions relating to power units contained in D.E.S. regulations (see *Safety in Science Laboratories*, D.E.S. Safety Series No. 2, HMSO, 1978). Detailed specifications for such apparatus exist and include those put forward in Nuffield Physics Year 3.

The effect of placing a charged sphere in an electric field

Apparatus

The teacher will need:

EHT power supply
(maximum p.d. 5000 V d.c.)*

2 metal plates fitted with
insulating handles

Table tennis ball coated with
Aquadag

Reel of nylon thread

3 retort stands and bosses

1 clamp

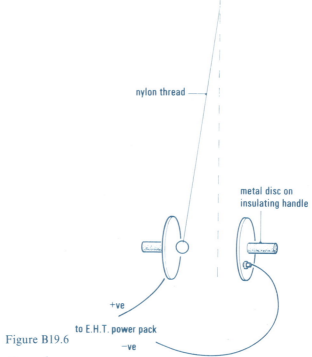

nylon thread

metal disc on
insulating handle

+ve

to E.H.T. power pack

−ve

Figure B19.6

Procedure

Arrange the apparatus as shown in figure B19.6. The two metal plates are connected to retort stands using the bosses and are set parallel to each other, about 10 cm apart. The plates are connected to the positive and negative terminals of the EHT power supply using crocodile clips attached to special lugs on the back of the plates. The table tennis ball coated with Aquadag is suspended by a nylon thread from the third retort stand so that it hangs freely between the plates. Allow the ball to touch the plate connected to the negative terminal of the EHT power pack. The power supply is now switched on (CAUTION!) and a potential difference of between three and four thousand volts applied to the plates. Note the effect on the ball. Repeat the experiment using a positive charge on the ball and again note the effect.

The demonstration may serve as a model of an ion moving in an electric field. By adjusting the potential difference across the plates and the distance separating the plates (CAUTION! High voltages), the ball may be made to move to and fro between them. The ball then represents a positive ion when moving in one direction and a negative ion when moving in the other direction. (This point requires emphasis.) In discussion the teacher can suggest that electrolysis depends on the presence of ions in an electrolyte. The discharge of ions at the cathode and anode yields the products of electrolysis. This assumes that ions move in an electric field during electrolysis. In the next section we test this assumption.

*See footnote on page 535.

Summarize the results of the demonstrations in this section. Indicate their significance for explaining the various effects obtained during electrolysis.

B19.3
Observing the migration of ions

In the previous section, it was suggested that during electrolysis ions moved between the electrodes inserted in an electrolyte. They must therefore carry the electric current. In this section, the movement of ions is demonstrated visually.

A suggested approach

Objectives for pupils

1. Increasing the familiarity of pupils with the properties of ions and charged particles
2. Awareness of the properties of electrodes during the process of electrolysis

We begin this section by reminding pupils that when a current is passed through an electrolyte we assume that it is carried by the ions present. The demonstrations in the previous section suggest that ions which carry a positive charge move towards the cathode and that ions which carry a negative charge move towards the anode. We call positively charged ions *cations* and negatively charged ions *anions*. The phenomenon of the movement of ions may now be demonstrated. Two experiments are described here: one is a teacher demonstration and the other a pupils' experiment. Additional experiments appear in *Collected experiments*, Chapter 5.

Experiment B19.3a

Apparatus

The teacher will need:

Buchner funnel

Buchner flask

Filter pump

Filter paper

Beaker 250 cm³

Wide bore U-tube

Wash bottle

Stand, boss, and clamp

Power supply capable of yielding 20 V d.c.

Pipette

Suction device for use with the pipette

Connecting wire

2 carbon electrodes

2M hydrochloric acid

1M copper(II) sulphate solution

1M potassium chromate solution

Urea

Looking at the migration of ions
This experiment should be done by the teacher.

Procedure
The teacher may demonstrate the movement of ions during electrolysis using the following method. Part 1 of the procedure is part of the pre-lesson preparation.

Part 1. Precipitate copper(II) chromate by mixing equal volumes (100 cm³) of 1M copper(II) sulphate solution and 1M potassium chromate solution. Filter the copper(II) chromate, using a Buchner funnel and flask and a filter pump, wash with distilled water, and then dry by suction.

Dissolve the product in a *minimum* quantity of 2M hydrochloric acid, and then dissolve as much urea as possible, in order to increase its density.

Partly fill the U-tube with 2M hydrochloric acid, and run in the copper(II) chromate solution very slowly and carefully from a pipette, delivering it at the bottom of the U-tube so that it forms a separate layer with a clean layer of hydrochloric acid above it on both sides of the U-tube. Withdraw the pipette carefully to avoid mixing.

Part 2. Insert a carbon electrode into each arm of the U-tube, so as to dip into the dilute hydrochloric acid layers, and connect a source of about 20 V d.c. Assemble the rest of the apparatus carefully and switch on the current. After about 5 to 10 minutes, the development of a green colour near the cathode and an orange colour near the anode is observed. After about thirty minutes, the blue copper(II) ion and orange dichromate ion boundaries are clearly seen.

The migration of ions

Apparatus

Each pair of pupils will need:

Experiment sheet 62

2 crocodile clips

Microscope slide

2 lengths of connecting wire

2 test-tubes, 150×16 mm

2 teat pipettes

Test-tube rack

Power supply (20 V d.c.)

Filter paper – cut into strips to fit the microscope slide

Small crystal of potassium permanganate

0.1M silver nitrate

0.2M potassium chromate solution

Supply of distilled water

Procedure
See *Experiment sheet* 62. Two experiments are suggested: the migration of permanganate ions, and the migration of silver ions and chromate ions, under a potential gradient.

Experiment sheet 62
Using the apparatus shown in the diagram you can find out whether the movement of coloured ions can be seen when a potential difference is applied to a solution containing them.

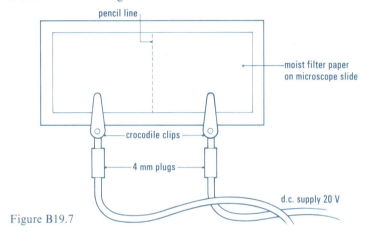

Figure B19.7

Draw a faint pencil line across the centre of the filter paper. Moisten the filter paper with tap water and fasten it to the microscope slide with the crocodile clips. Place a small crystal of potassium permanganate in the centre of the filter paper. Connect the crocodile clips to the terminals of a 20 volt d.c. supply, noting which clip is connected to the positive terminal and which to the negative terminal. Observe the filter paper for at least ten minutes. Describe what you see.

What can you deduce about the charge on the permanganate ion? Give your reasons.

Remove the filter paper and microscope slide from the clips, discard the filter paper, and clean the microscope slide. Use a fresh strip of filter paper on the slide, arranged in the following way.

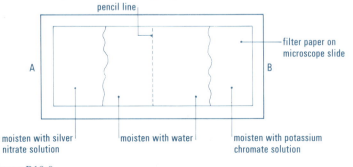

Figure B19.8

Using crocodile clips connect end A to the *positive* side of the 20 volt d.c. supply and end B to the *negative* side. Observe what happens during several minutes. Describe and explain all that you see.

In order to help your explanation, try adding 10 drops of silver nitrate solution to 10 drops of potassium chromate solution in a test-tube.

What is the coloured precipitate?

Suggestion for homework   Write a short account of the migration of ions during electrolysis.

B19.4
The structure of the atom; some implications

At this point in the course we consider the implications of the ionic theory for the structure of the atom.

A suggested approach

Objectives for pupils

1. To understand that ionic theory throws light on the structure of the atom
2. To provide a basis for a more detailed consideration of atomic structure at a later stage

It is clear from the work in this Topic so far, that atoms may gain or lose charge to form ions and that the converse process may also take place. Therefore atoms must possess charge. However, experience tells us that atoms of elements, such as lead or copper or hydrogen, are all electrically neutral. Therefore, atoms must also contain a zone of positive charge to neutralize the effect of this negative charge. At this point, the teacher may provide pupils with a simple model of the atom. Tell pupils that atoms are made of a positively charged centre or nucleus which is surrounded by a zone of negative charge which contains electrons. Clearly one should try to avoid creating an image of the electron as a planet rotating around a nucleus since this idea may prove difficult to eradicate later. It is not essential to develop the structure of the atom in greater detail at this stage although some teachers may wish to do so. (Suggestions for ways of doing this appear later in this book, see Supplement S1, page 634.) At this point, the important point to stress is the very big difference in properties between atoms and their ions. For example, one might remind pupils how vigorously metallic sodium reacted with water and how non-reactive sodium ions are when sodium chloride is placed in water.

This simple picture is adequate for most of the scheme, except during the discussion on radioactivity when some knowledge of the structure of the nucleus will be required.

Teachers will have noticed that the term 'valency' has not been mentioned during the course. It is, however, convenient to arrive at the formulae of ionic compounds and to see that a pattern is emerging. This can be done by treating electrolysis quantitatively and using Faraday's laws to determine the charges on ions. Given such information from experiment we may extend the argument. We know that salts are electrically neutral and so we can determine the charge on an ion from the formula for a given salt. Thus, analysis shows that silver sulphate has the formula Ag_2SO_4. Make the point that one mole of electrons (one Faraday) is required to deposit one mole of silver atoms. The silver ion has one positive charge and is represented by the symbol Ag^+. Hence the sulphate radical must have two negative charges if silver sulphate is electrically neutral. The sulphate ion is therefore written as SO_4^{2-}. In this way pupils should acquire a knowledge of the charges of common cations and anions, and appreciate that they have been found experimentally.

This simple treatment of atomic structure can be used to introduce a brief resumé on the structure of chemical substances. Atoms were discussed in Topic B11 and again in B16; molecules and giant structures in Topic B15; and ions have been introduced earlier in this Topic. Atoms and ions are the building blocks of matter with which chemists are concerned. These building blocks are used in different ways. Atoms are either arranged in molecules of a specific size or in giant structures of an indefinite size. Ions exist not only in giant structures in the solid state but also as ions in molten salts or solutions of salts. Pupils should be aware that very few substances exist as single atoms at room temperature and that atoms of most elements are usually extremely reactive – so much so that they rarely exist for more than a small fraction of a second on their own. When we consider the properties of an element, we usually consider the properties of its giant structure or of its molecules: we do *not* review the properties of its single atoms. Only in the case of inert (noble) gases are atoms unreactive. Ions, on the other hand, can exist indefinitely, and solutions of ions possess the ability of being able to transfer charge from one electrode to another during the process of electrolysis.

This outline provides a basis for a discussion of the composition and structure of a number of common substances. Ask the pupils if they remember the structures of such substances as polythene, argon, paraffin, common salt, copper. 'Are they made up from ions or from atoms?' 'How are the ions or atoms arranged – as molecules or as giant structures?'

It may be helpful to use the *Handbook for pupils*, Chapter 9, as a basis for any detailed discussion which may arise from their questions. The point can be made that an atom may gain or lose electrons to form ions. Towards the end of the lesson it is convenient to summarize the discussion as follows:

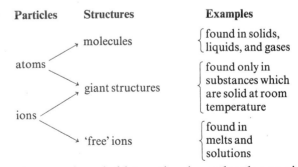

'How are atoms or ions held together in molecules or giant structures?' This is fairly easy for pupils to answer in the case of ions. They will know enough about electrostatics to understand that unlike charges attract. 'But what of the substances which are composed of atoms rather than ions?' Here we shall have to go cautiously since it is not intended that the electronic theory of bonding should be explained or that electron configuration should be discussed in any great detail. Pupils should have a mental picture of the atom as a positively charged nucleus surrounded by a negative charge in the form of electrons. We can therefore explain that two positively charged nuclei can be held together by a pair of electrons (one from each atom so that there is no redistribution of charge). Bonding between two atoms is thus an essentially electrostatic phenomenon.

The theme of atomic structure and bonding is taken up in *Handbook for pupils* in Chapter 3. The treatment required is elementary and includes a simple comparison of the properties of 'model' compounds which differ according to the particular type of predominant chemical bond.

Pupils will know that chlorine is a very reactive element which combines with most other elements to form compounds called chlorides. The formulae of the compounds formed between chlorine and the elements in the period sodium to argon could be listed. There is a definite pattern in the formulae of these compounds. The properties of the chlorides also show a similar pattern. So, certain physical properties of a compound such as:
1. electrical conductivity when molten,
2. melting point,
3. boiling point,
can provide information about the structure and bonding of the compound.

The Reference section of the *Handbook for pupils* provides data for such a discussion.

The chlorides of the elements in the period sodium to argon can now be divided into two types, the division being decided by the physical properties of the compounds: ionic chlorides and non-ionic chlorides.

Ionic chlorides	Non-ionic chlorides
crystalline solids at room temperature – never liquid or gas	frequently liquid or gas at room temperature
conduct electricity when molten	do not conduct electricity when molten
high melting point	low melting point
high boiling point	low boiling point

This could be discussed. For example, ionic chlorides are crystalline due to the regular arrangements of ions in the structure. Ionic chlorides conduct electricity when molten because the giant structure of ions is partly broken down on melting, so that each ion can escape sufficiently from the attraction of other ions of opposite charge and move towards the electrodes during electrolysis; and so on. Similarly, the properties of non-ionic chlorides can be related to their molecular structure.

Clearly the quantitative aspect of the structure of ionic materials depends to some extent on an appreciation of the early work by Faraday and Davy. This forms a suitable topic for homework and also an introduction to the next section.

Suggestion for homework

Read *Chemists in the world*, Chapter 4 'Davy and Faraday' or *Handbook for pupils*, Chapter 7 'Electricity and matter'.

B19.5
Some studies in electrolysis

The work of Faraday and Davy is introduced and interpreted in terms of the properties of ions. The section includes experiments which demonstrate the quantitative aspect of Faraday's contribution.

A suggested approach

This section could be opened with a brief review of the work of Faraday and Davy, using as a basis Chapter 4 in *Chemists in the world*.

Objectives for pupils

1. Awareness of the contributions of Faraday and Davy to the study of electrolysis
2. Understanding the quantitative aspect of electrolysis and especially that the number of coulombs needed to deposit 1 mole of an element is 96 500 coulombs or a simple multiple thereof
3. Understanding how these values lead to the values of charges on ions and how they support the particulate nature of electricity

It is suggested that pupils determine either the quantity of electricity needed to deposit 1 mole of copper from copper(II) sulphate solution or the quantity of electricity required to deposit 1 mole of silver from silver nitrate solution.

Experiment B19.5a

How much electricity is needed to deposit 1 mole of copper atoms and 1 mole of silver atoms from solution?

Apparatus

Each pair of pupils will need:

Experiment sheet 63

6 volt battery or alternative d.c. supply

Rheostat, 10 Ω

Ammeter, 0–1 A

Stop-clock or watch or view of laboratory wall clock if this is fitted with a seconds hand

2 beakers, 100 cm³

Procedure
The layout is shown in figure B19.9. Details are given in *Experiment sheet* 63.

(Continued)

2 copper foil electrodes (size 5 × 5 cm)

Wooden support to hold electrodes

Stainless steel basin (7 cm diameter)

Cork ring to support steel basin

Silver anode (10 cm of silver wire – 2 mm diameter)*

Filter paper and rubber band

Connecting wire

Stand, boss, and clamp

Cork

Access to steel wool, paper tissues, and to a balance reading to 1 mg

0.5M copper(II) sulphate, 100 cm^3

0.1M silver nitrate, 100 cm^3

Supply of distilled water

Access to acetone (propanone)

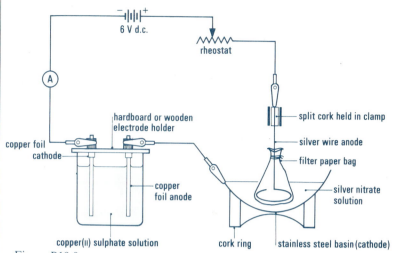

Figure B19.9

The completed apparatus for this experiment is shown in the diagram above. Assemble it in the following order:

1. Clean carefully the copper foil cathode and anode and the stainless steel basin by rubbing them with steel wool and then with paper tissues. Mark the upper end of the copper foil cathode, in pencil, with a C and the upper end of the copper foil anode with an A.

2. Find the masses of copper cathode, copper anode, and stainless steel basin, and enter them in the table below.

3. Three-quarters fill the beaker with copper sulphate solution and insert the copper electrodes, using the holder provided.

4. Place the stainless steel basin in the cork ring support and add silver nitrate solution to within 1 cm of the rim.

5. Enclose the silver wire anode in a bag by wrapping a filter paper round it and holding the edges together with a rubber band round the vertical portion of the wire. Fix the vertical portion of the wire outside the bag in a split cork held in a clamp attached to a stand. Lower the anode, in its bag, into the silver nitrate solution in the basin, so that the silver ring is just below the centre of the surface.

6. Complete the rest of the circuit, using connecting wires and crocodile clips, but do not attach the crocodile clip on the wire from the ammeter to the copper cathode. Check the rest of the connections, making sure that the copper cathode and anode are in their correct positions.

7. Attach the crocodile clip to the copper cathode, start the stop-clock (or note the time on a clock with a seconds hand), and adjust the rheostat to give a current of 0.20 amp.

Allow the current to pass through the solutions for 20 minutes, keeping it at 0.20 amp by adjusting the rheostat when necessary. At the end of 20 minutes disconnect the two copper electrodes and the stainless steel basin. Pour the silver nitrate from the basin into the stock bottle and rinse the basin twice with distilled water and twice with acetone (propanone). *Acetone is highly flammable*; keep it well away from flames. Be careful not to lose any silver during these operations. Allow the remaining acetone to evaporate and

*Some teachers may prefer to use a stainless steel strip in view of the relatively high cost of silver wire.

warm the basin high above a Bunsen flame to complete the drying. Find the mass of basin + silver and record the mass in the table below. Rinse the copper electrodes with water. The anode (A) will need a strong jet of water from the tap, followed by a firm wipe with a paper tissue to remove the film which collects on the surface. Rinse each electrode with acetone, wave in the air for a minute or two to evaporate most of the acetone, and dry by warming high above a Bunsen flame. Find the mass of each electrode separately and record the results in the table.

	Copper cathode /g	Copper anode /g	Steel basin cathode /g
Initial mass			
Final mass			
Change in mass			

Current used .... amp

Time current passed .... sec

Calculation
Quantity of electricity (coulombs) = current (amp) × time (sec)

Quantity of electricity used in experiment

= .... × .... = ... coulombs

.... g silver deposited by .... coulombs

108 g silver deposited by

.... × 108   coulombs

....

Quantity of electricity needed to deposit 1 mole of atoms (gram-atom) silver

is .... coulombs.

.... g copper deposited by .... coulombs

64 g copper deposited by

.... × 64   coulombs

....

Quantity of electricity needed to deposit 1 mole of atoms (gram-atom)

copper is .... coulombs.

Comment on the results obtained.

Figure B19.10
See figure A16.10 for class results for the electrolysis of copper sulphate.

Treatment of results (copper)
The results of the copper part of the experiment obtained by various groups can be presented graphically as shown in figure B19.10. It is then an easy matter to read from the graph that $\frac{1}{1000}$ of a mole (0.064 g) of copper will be deposited by 193 coulombs. Therefore, one mole (64 g) of copper will be deposited by 193 000 coulombs.

Also from this graph, it is apparent that:

(mass liberated during electrolysis) $\propto$ (quantity of electricity passed)

Topic B19 Electrolysis

This, of course, is the essence of Faraday's first law, which might be stated at this point – although there is no particular need to emphasize it.

Treatment of results (silver)
The results of the silver experiment can be treated in a similar way. If only one determination is made, then the result will need to be calculated as shown below.

Suppose it were found that 0.2 A deposited 0.34 grams of silver in twenty-five minutes, then:

quantity of electricity passed $= 0.2 \times 25 \times 60$ coulombs
$$= 300 \text{ coulombs}$$

0.34 g silver is deposited by 300 coulombs.

Therefore, 108 g silver will be deposited by $\dfrac{300 \times 108}{0.34}$ coulombs

$$= 95\,300 \text{ coulombs}$$

Note. This result is quite close to the generally accepted value of 96 500 coulombs.

At this point, the teacher might ask pupils about their criticisms of the experimental procedure.

Table B19.1 shows a useful summary of results for a number of elements.

Element	Number of coulombs required to liberate 1 mole of atoms of element
silver	96 500
copper	193 000
hydrogen	96 500
zinc	193 000
aluminium	289 500

Table B19.1

The pupils should be able to see that the number of coulombs required to liberate 1 mole of atoms of an element is either 96 500 coulombs or a multiple thereof, within the limits of experimental error. At this point, one may introduce the unit known as the Faraday as a quantity of electricity. Alternatively, one can eliminate the need for this unit by referring to a mole of electrons (see page 539). (Clearly the latter possibility possesses advantages when subsequent topics are presented.)

Table B19.2 sums up our findings so far.

Element	Number of moles of electrons = number of Faradays required to liberate 1 mole of atoms of element
silver	1
hydrogen	1
copper	2
zinc	2
aluminium	3

Table B19.2

The significance of these results can now be discussed. They provide indirect evidence for the values of charges carried by ions during electrolysis and suggest quite strongly that electricity itself is particulate. Indeed, these results will be difficult to explain without assuming that both matter and electricity are particulate!

Pupils should now be able to appreciate that the same quantity of electricity is needed to deposit either 1 mole of silver atoms or 1 mole of hydrogen atoms. We know that both of these quantities of elements contain the same number of atoms (6×10^{23}). It follows that the same quantity of electricity must be associated with 1 atom of either silver or hydrogen. In the case of copper, lead, and zinc, etc., the pupils should appreciate that each of these elements requires the same amount of electricity to deposit a mole of atoms, and that this quantity of electricity is just twice that required to liberate 1 mole of silver or hydrogen atoms. Since the mole of copper or zinc atoms contains the same number of atoms (6×10^{23}) as a mole of silver and hydrogen atoms, twice the quantity of electricity must be associated with each atom of copper and zinc. In the case of aluminium, three times the quantity of electricity is needed for each atom compared with one atom of hydrogen or silver. The conclusion that definite amounts of electricity are associated with ions becomes inescapable.

A quantitative study of the electrolysis of molten lead bromide
The experiment should be done by the teacher.

The teacher will need:

Safety spectacles

2 carbon rods mounted in a suitable holder

2 crystallizing dishes

d.c. supply, 12 to 14 V, accumulator preferred

Switch

Connecting wire

Ammeter (demonstration type 0–5 A)

Rheostat, 12 Ω, 10 A

Stop-clock

Balance

Spatula

Tripod and gauze

Stand, boss, and clamp

Bunsen burner and heat resistant mat

Tongs

Access to a fume cupboard

Lead bromide

Procedure
The experiment should be carried out in a fume cupboard. The teacher is advised to wear safety spectacles.

A somewhat greater current is required for this experiment than for that used by the pupils in Experiment B19.1a. As in the earlier experiment, allow the lead bromide to melt before inserting previously warmed electrodes into the electrolyte. Adjust the current to a steady value and maintain the current at this value for the period of electrolysis. After ten or fifteen minutes of electrolysis, switch off, withdraw the electrodes, and pour the molten electrolyte carefully into the second dry crystallizing dish.

Take care to retain the bead of lead in the first dish. Allow the dish to cool. The bead of lead may now be pressed cleanly away from the glass using a spatula. Any adhering lead bromide can be broken off from the bead of lead. Weigh the clean lead bead.

Specimen results
Current passed 1 A
Time for which current passed = 15 minutes = 900 seconds
Mass of lead deposited = 1.002 g

The pupils may need to be reminded that the problem is to find out how much electricity is required to deposit 207 g of lead.

The quantity of electricity is measured in coulombs. The coulomb is that quantity of electricity which flows when 1 ampère of electricity is passed for one second.

Thus, if one ampère is passed for 360 seconds, the quantity of electricity is 360 coulombs.
1.002 g lead is deposited by 900 coulombs
1 g lead is deposited by $\frac{900}{1.002}$ coulombs
So, 207 g lead will be deposited by $207 \times \frac{900}{1.002}$ coulombs
= 186 000 coulombs (approximately)

An alternative method of presenting this calculation is shown in section A16.4.

It is essential to discuss the validity of the procedure which has been adopted for the presentation. 'How could we improve our experimental determination?'

In the discussion that follows, list the suggestions made by the pupils before revealing to them the exact value for the quantity of electricity required to deposit 207 g of lead. If the demonstration is carried out in a separate lesson to that in which the silver and copper experiment is done, it may be necessary to go over the approach used.

The teacher will need

Film loop 2–8 'Electrolysis of lead bromide'

Film-loop projector

The lesson may be concluded by showing the film loop 2–8 'Electrolysis of lead bromide', giving emphasis to the significance of electrolysis when interpreting quantitative aspects of the structure of lead bromide: lead ions are represented by Pb^{2+} and bromide ions by Br^-.

Experiment B19.5c

Is the same quantity of electricity always required to liberate 1 mole of copper atoms?

Apparatus

Each pair of pupils will need:

Experiment sheet 64

6 V battery or alternative d.c. supply

Rheostat (to control current density of between 10 and 20 mA per square centimetre of electrode surface)

Ammeter, 0–1 A

2 beakers, 100 cm³

Tripod and gauze

Bunsen burner and heat resistant mat

Thermometer, −10 to +100 × 1 °C

4 copper foil electrodes and two supports (approximate size of electrodes 50 mm × 50 mm)

Connecting wire

(Continued)

Procedure
Experiment sheet 64 is reproduced below. It is important to control current density within the limit suggested. To avoid difficulties, the procedure suggests following the decrease in the mass of the anode rather than the increase in mass of the cathode for each of the cells. The temperature control for the electrolysis of an alkaline solution of sodium chloride between copper electrodes is not critical. Reasonable results are obtained when the temperature is maintained near 80 °C.

Experiment sheet 64
In this experiment you are going to pass a current through two different solutions, using copper electrodes.

The solutions to be used are:

1. Copper sulphate solution at room temperature.

2. Sodium chloride solution made alkaline with sodium hydroxide, at about 80 °C.

The object of the experiment is to find the change in mass of the two copper *anodes* when the same current is passed for the same length of time. This can be ensured by using two electrolysis cells connected in series, as in Experiment 63.

Paper tissues

Fine steel wool

Access to balance – to read to 1 mg

Stand, boss, and clamp

0.5M copper(II) sulphate, 100 cm^3

Solution containing 100 g of sodium chloride and 1 g of sodium hydroxide per 1000 cm^3, 100 cm^3

The actual value of the current need not be known but it must be kept near the value recommended below. A good deal of care is needed if the experiment is to be successful and accuracy in finding the masses of the anodes before and after the electrolysis is essential.

The circuit used is the same as that used in Experiment 63, except that the stainless steel basin, and its silver electrode, are replaced by a second beaker with two copper foil electrodes dipping into the alkaline sodium chloride solution. This is heated to about 80 °C, by a Bunsen burner, when supported on a tripod and gauze.

First clean the pieces of copper foil that are to be used as anodes, with steel wool, followed by paper tissues. Mark the upper ends '1' and '2' in pencil so that you can identify them later, and find the mass of each.

Connect the circuit as for Experiment 63 (supporting the leads to the alkaline solution by a clamp to keep them away from the flames). Leave the cathode in the first beaker (copper sulphate solution) disconnected. Heat the alkaline sodium chloride to 80 °C and adjust the height of the Bunsen flame to keep it at about this temperature.

Complete the circuit, adjust the rheostat to give a current of 0.1 amp, and allow electrolysis to proceed for 15–20 minutes. The anode in the second beaker (containing the hot sodium chloride solution) must be moved about from time to time, to prevent the orange coloured precipitate which appears from sticking to it.

Disconnect the circuit, remove the two anodes, wash them with a stream of tap water, and rub each firmly with paper tissues to remove any loose film. Rinse each twice with acetone (propanone), partially dry by waving in air for a minute or two, and complete the drying high above a Bunsen flame. Find the mass of each.

	Copper anode in copper sulphate solution /g	Copper anode in alkaline sodium chloride solution /g
Initial mass		
Final mass		
Change in mass		

Take a separate sheet of paper and comment on the results obtained.

From Experiment B19.5a, we know that in the electrolysis of copper sulphate solution, two moles of electrons (Faradays) were required to transfer one mole of copper from the copper anode to a copper cathode. The present experiment shows that when the determination is carried out simultaneously using two sets of different conditions, the same quantity of electricity is able to transfer twice as much copper in one cell as in the other. This statement is equivalent to saying that half as much electricity (namely, 1 mole of electrons) would transfer the same mass of copper in the cell containing the alkaline solution as that needed for the more usual electrolysis of copper sulphate.

Thus, in a beaker containing the alkaline solution, the following change might take place at the anode:

$$\text{Cu} \longrightarrow \text{Cu}^+ + \text{e}^-$$

these ions pass into solution   these charges return to the battery

In the beaker containing the copper sulphate solution, the following change occurs at the anode:

$$\text{Cu} \longrightarrow \text{Cu}^{2+} + 2\text{e}^-$$

Therefore the copper atoms must be capable of losing either one or two electrons in order to form copper(I) ions, Cu^+, or copper(II) ions, Cu^{2+}, respectively.

Pupils will now need to be reminded that copper is only one example of a special group of elements – the transition elements. At this point it is helpful to refer to the Periodic Table and to a display of compounds of the transition elements. This might include examples of compounds of the elements chromium, manganese, iron, cobalt, nickel, and copper. It should show that one of the characteristics of transition elements is that they produce more than one series of salts.

Experiment B19.5d

Investigating the electrolysis of solutions of electrolytes

Apparatus

Each pair of pupils will need:

Electrolysis cell (see figure B19.11 for details of construction)

2 test-tubes, 75 × 10 mm

2 connecting wires, each fitted with crocodile clip

Elastic band – and a hardboard divider

6 V d.c. supply

Stand, boss, and clamp

Indicator paper

Splints

Access to an ammeter and another length of connecting wire

One or more of the solutions:
0.5M hydrochloric acid
0.5M potassium iodide solution
0.5M copper(II) chloride solution

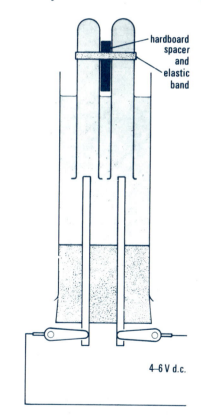

hardboard
spacer
and
elastic
band

4–6 V d.c.

Figure B19.11

Procedure

Pupils investigate the electrolysis of one or more of the solutions provided. They test the products of electrolysis.

Figure B19.11 shows the apparatus used. Test-tubes can be issued to enable gases to be collected. Advise the pupils to suspend the test-tubes above the electrodes – as shown – and not around the electrodes since this considerably reduces the rate of electrolysis. (This effect can be shown if an ammeter is included in the circuit.)

In the case of the electrolysis of dilute hydrochloric acid, hydrogen is liberated at the cathode and chlorine at the anode. The electrolysis of potassium iodide solution results in the formation of iodine at the anode (identified by colour) and hydrogen can be collected at the cathode. The electrolysis of copper(II) chloride results in the formation of a coating of copper on the cathode and the liberation of chlorine at the anode.

Under appropriate conditions, it is possible to obtain oxygen with the halogen in these instances. Using graphite electrodes, oxygen is difficult to identify. In addition, some hydrogen may be formed at the cathode during the electrolysis of copper(II) chloride.

When the pupils have completed their experiment, discuss their results. Tabulate the results on the blackboard or on an overhead projector transparency. Can the pupils offer a simple explanation for their findings? It looks as though current is carried by ions moving in aqueous solutions as in the case of fused electrolytes. The electrode at which the ion is discharged tells us whether the ion is positively or negatively charged. It is clear that hydrogen and copper ions are positively charged and that chlorine and iodine ions are negatively charged.

The teacher will need

Film loop 2–18 'The electrolysis of potassium iodide solution'

Film-loop projector

The question may be asked 'Why doesn't potassium appear at the cathode when potassium iodide is electrolysed?' Remind pupils of the fact that potassium reacts vigorously with water. However they should *not* be left with the impression that potassium is first formed and then reacts with water to form hydrogen. A demonstration of this electrolysis can be useful for this part of the lesson. The film loop 2–18 'The electrolysis of potassium iodide solution' can be used to review the ideas used.

By the end of the lesson, pupils should be aware that water can play a part in the electrolysis of solutions of electrolytes. The convention using the symbol 'aq' to indicate that ions are in aqueous solution – such as $H^+(aq)$ and $I^-(aq)$ – can be introduced.

Suggestion for homework

1. Summarize the important conclusions arrived at as the result of the studies you have carried out in this Topic.
2. Read *Handbook for pupils*, Chapter 1 'Periodicity'.

Summary

By the end of this Topic, pupils should be able to interpret the phenomenon of electrolysis, and have some idea of how to relate the charges carried by ions to the particulate theory of electricity. They should understand Faraday's quantitative studies on electrolysis and be able to interpret the structure of an atom simply, so as to provide information on bonding and the properties of compounds. The experimental work covered in this Topic includes both qualitative and quantitative studies of electrolysis.

Ions in solution and related chemical systems

Purposes of the Topic

1. To demonstrate the interplay between speculation and verification in chemistry.
2. To determine reaction stoichiometry in a few simple cases.
3. To provide opportunities for calculations associated with reacting quantities of materials and the writing of chemical equations.
4. To provide experience of reversible chemical reactions.
5. To provide an experimental basis for the Brønsted-Lowry theory of acids and bases.

Contents

B20.1 The mole as a unit
B20.2 Formation of precipitates
B20.3 Reactions involving gases
B20.4 Properties of ions in solution
B20.5 Reversible reactions
B20.6 Looking at the role that water plays

Timing

Section B20.1 needs a single period but other sections may need more than a week each. The entire Topic requires about six weeks.

Introduction to the Topic

The prime purpose of this Topic is to enable pupils to investigate the properties of ions in solution. The Topic opens with a brief section on the mole.

In Alternative IIB, we have delayed the introduction of the mole concept until Topic B18. Topic B19 referred to this concept only briefly. In the opening section of this Topic, we relate the Faraday to the mole of electrons and then introduce the use of the mole when referring to the concentration of a solute in a solution. The preparation of a solution containing a mole of solute in a definite volume provides a new means of 'counting particles' and introduces the symbol 'M' as a means of indicating concentration. (A solution having a concentration of $1.0 \, \mathrm{mol \, dm^{-3}}$ is described as 1.0M, etc.)

Section B20.2 concerns the formation of precipitates. A solution containing potassium and iodide ions is mixed with one containing lead and nitrate ions. Tests indicate that the precipitate formed as a result of this reaction is lead iodide, and that the residual, or spectator, ions are potassium and nitrate ions. An investigation follows to verify the prediction that lead iodide has the formula PbI_2. Other precipitation reactions are studied and the results are summarized by writing chemical equations. Reactions involving gases are studied quantitatively in the following section of the Topic.

The general theme of the properties of ions in solution is taken up again in section B20.4 by means of a quantitative study of the reaction between iron and copper(II) sulphate. This is followed in

Figure B20.1
A volumetric flask containing 1.0M sodium chloride solution.

BS1792
100 ml
C 20°C

section B20.5 by a series of qualitative investigations of reversible reactions which involve ions, providing the basis for a more advanced treatment of the properties of acids and bases in the section which follows.

So far, all of the reactions of ions which have been studied have depended on the presence of water as a solvent. Pupils may well question the role which water plays. Earlier work on acids, bases, and salts may be referred to at this point (see section B12.1). The properties of solutions of hydrogen chloride in water and in toluene are compared and used to establish the principles of the Brønsted-Lowry theory of acids and bases. The Brønsted-Lowry theory is then used to make predictions about the reaction between barium hydroxide and sulphuric acid, and these are tested by experiment.

Subsequent development

By the end of this Topic pupils have met the three basic units of chemical structure – atoms, ions, molecules – and have studied several reversible reactions. In the next Topic (B21) they use this knowledge to build up a theory of dynamic equilibrium.

Alternative approach

This Topic has no single equivalent Topic in Alternative IIA. The content is a mixture of material drawn from Topics A17, A19, and A20.

Further references
for the teacher

Collected experiments, Chapters 8, 12, and 15, provides further suitable experiments.

Supplementary materials

Overhead projection originals
25 Traffic-flow model *(figure A19.1)*
26 Formation of ions *(figure A20.2)*

Reading material
for the pupil

Handbook for pupils Chapter 5, 'Studying chemical reactions'.

B20.1
The mole as a unit

In this section, the use of the mole is extended to solutions of salts and other ionic compounds.

A suggested approach

Objectives for pupils

1. Knowledge of the mole – in another context – as a term referring to an amount of substance
2. Knowledge of the fact that the concentration of solutions can be expressed in $mol\,dm^{-3}$ and of the 'M' notation

Pupils have already met the mole in Topics B18 and B19. It may be useful to emphasize that a mole of magnesium atoms contains the *same number* of particles as a mole of oxygen atoms, or a mole of electrons, and so on. They will need to be introduced to the new idea of a mole of a substance with a giant structure in order to equate the term mole with the mass of the formula of a substance expressed in grams. Thus, 1 mole of magnesium atoms has a mass of 24 g; 1 mole of the compound sodium chloride has a mass of 58.5 g; 1 mole of the compound potassium iodide has a mass of 166 g; 1 mole of the compound magnesium oxide has a mass of 40 g; and so on.

In this Topic we are going to investigate reactions between substances in solution. We find it useful to use solutions containing a measured concentration of solute expressed in mol dm^{-3}.

A molar (1M) solution contains 1 mole of solute dissolved in 1 dm^3 solution.

Preparation of 1.0M solution of sodium chloride

The teacher will need:

2 volumetric flasks (each of capacity 1 dm^3)

1 beaker, 100 cm^3

Glass rod

Wash bottle

Filter funnel

Measuring cylinder, 100 cm^3 maximum capacity

Access to balance

Sodium chloride, 58.5 g in bottle labelled '1 mole sodium chloride = 58.5 g sodium chloride'

Distilled water

Procedure

Sodium chloride has been selected for this experiment since it is familiar and cheap. Pupils should be aware that the mass of 1 mole of sodium chloride (NaCl) is 58.5 g.

In front of the class transfer 58.5 g of sodium chloride to the volumetric flask using a stream of water from a wash bottle. After mixing the solution, make up the solution to the mark in the flask. The resulting solution is then labelled as 1.0M solution of sodium chloride. (See figure B20.1.)

The point should be made that 1 dm^3 of 1.0M solution of sodium chloride contains 1 mole of NaCl.

Transfer 100 cm^3 of solution to a second 1 dm^3 volumetric flask and ask how many moles of sodium chloride are in each of the two flasks. Next make up the solution in the second flask to the 1000 cm^3 mark with water and repeat the question. The exercise can be repeated until the pupils grasp the point that the terms 'quantity of solute' and 'concentration of solute' have different meanings. The convenience of working with molar solutions (or convenient multiples thereof, such as 2.0M or 0.1M) is that the reacting volumes are greatly simplified, as pupils will appreciate in the next section of this Topic.

B20.2
Formation of precipitates

In this section some typical precipitation reactions are studied. Pupils speculate about the equations for the reactions on the basis of their knowledge of the chemical formulae of the reactants and of the charges on their ions. These equations are verified by experiment.

Objectives for pupils

1. Familiarity with precipitation reactions
2. Ability to predict possible products for such reactions given the properties and formulae of reactants

(Continued)

So far, pupils have been encouraged to write equations as summaries for reactions. In Topic B18, evidence for the formulae of such binary compounds as magnesium oxide, copper(II) oxide, and mercury(II) chloride was obtained. The concept of a molecule was developed in Topics B15 and B17, and in Topic B19 evidence for the existence of ions was presented. Pupils know that when we write Na$^+$, Pb^{2+}, etc., the charges on these ions have been determined by experiment.

Remind pupils of precipitation reactions which they met in earlier parts of the course. Ask them to name reactants and products. Then

3. Ability to test such predictions qualitatively
4. Ability to formulate a possible equation for such a reaction from a knowledge of symbols, formulae, and ionic charges of the reactants and products
5. Ability to verify quantitatively that an equation is correct by measuring the quantities of reactants

ask pupils to *predict* the reaction between lead nitrate solution and potassium iodide solution. What products might we expect? Which of these products is likely to be soluble and which insoluble? What properties of potassium nitrate and of lead iodide do we know which support these suggestions? After discussion, show what happens when lead nitrate and potassium iodide solutions are mixed. Compare a solution of potassium nitrate made from stock supplies of potassium nitrate with the residual liquor from the experiment. Crystals prepared from both solutions are needle-shaped (see figure A14.3). The precipitate should also be investigated and shown to contain both lead and iodine. Pupils may suggest that electrolysis of a molten sample of this substance would provide evidence of its composition (compare with electrolysis of lead bromide, Experiment B19.1c).

Whatever is attempted, the following points need to be borne in mind:
1. The formulae of lead nitrate, $Pb(NO_3)_2$, and potassium iodide, KI, have been determined by using techniques like those used to determine such formulae as that of copper(II) oxide in section B18.4. The formulae of lead nitrate and potassium iodide have to be known in order to make up solutions of known molarity.

2. The charges on the lead and iodide ions are also known. They have been determined by finding the number of moles of electrons (Faradays) required to discharge 1 mole of lead ions and 1 mole of iodide ions (see section B19.5).

Thus, after some class discussion, pupils should be able to predict the following equation:

$$Pb^{2+}(aq) + 2I^-(aq) \longrightarrow Pb^{2+}(I^-)_2(s)$$

They should be able to explain what happened to the spectator ions – the nitrate ion, $NO_3^-(aq)$, and the potassium ion, $K^+(aq)$.

The purpose of experiment B20.2a is to see whether this prediction is verified experimentally.

Experiment B20.2a

Each pair of pupils will need:

Experiment sheet 65

2 burettes and stands*

4 test-tubes, 100 × 16 mm

Test-tube rack

Glass rod

Teat pipette

Watchglass

Beaker, 250 cm³

(*Continued*)

To verify the composition of the precipitate formed during the reaction between lead nitrate solution and potassium iodide solution

Procedure
Experiment sheet 65 provides full details. It is suggested that the teacher demonstrate the technique of using a centrifuge beforehand.

*Some teachers prefer to use proprietary constant volume pipettes in place of burettes.

Bunsen burner and heat resistant mat

Tripod and gauze

Access to centrifuge

Access to ethanol (I.M.S. grade), few drops

1.0M lead(II) nitrate solution, 7 cm³

1.0M potassium iodide solution, 10 cm³

Experiment sheet 65

Using a burette, measure 5.0 cm³ 1.0M potassium iodide solution into a clean, dry 100 × 16 mm test-tube. Add to this 0.5 cm³ 1.0M lead nitrate solution from another burette. Mix the two solutions with a thin glass rod. Centrifuge the mixture for 30 seconds (do not forget to balance the tube with another containing an equal volume of water). Measure the height of the precipitate, from the bottom of the test-tube, as accurately as you can. Record this in the table described below.

Add another 0.5 cm³ lead nitrate solution, stir, and again centrifuge for 30 seconds (remember to add another 0.5 cm³ water to the balancing tube). Measure the new height of the precipitate.

Repeat this process, using successive quantities of 0.5 cm³ lead nitrate solution, until the height of the precipitate does not change with further additions. Record your results in a table using the headings below.

Volume 1.0M KI(aq) /cm³	Volume 1.0M Pb(NO₃)₂(aq) /cm³	Height of precipitate /cm

What volume of 1.0M lead nitrate solution ($Pb(NO_3)_2(aq)$) just reacts with 5.0 cm³ 1.0M potassium iodide ($KI(aq)$)?

Does this result verify your predicted equation for the reaction? Give your reasons.

Make up a mixture of 5 cm³ 1.0M potassium iodide solution and exactly the right quantity of 1.0M lead nitrate solution for complete reaction. Add 2–3 drops of alcohol and centrifuge to compact the precipitate. What is the precipitate?

What does the solution above it contain?

Devise experiments to verify your answers. On a separate sheet of paper, describe how you carry them out and state what the results are. (Class discussion will help you with this.)

When the pupils have completed their experiments, their results should be discussed. The experiment shows that, within the limits of experimental error:

5 cm³ of 1.0M potassium iodide reacts with 2.5 cm³ of 1.0M lead nitrate

i.e. 5 cm³ 1.0M KI reacts with 2.5 cm³ 1.0M $Pb(NO_3)_2$

but, 1 dm³ of 1.0M solution contains 1 mole of solute

and so, 2 moles of KI react with 1 mole of $Pb(NO_3)_2$

or, 2 moles of I^- react with 1 mole of Pb^{2+}.

This statement verifies the equation which was predicted for the reaction:

$$Pb^{2+}(aq) + 2I^-(aq) \longrightarrow Pb^{2+}(I^-)_2(s)$$

This investigation may be carried out as a class experiment. Alternatively, Experiments B20.2a, B20.2b, and B20.2c could be attempted by different groups of pupils at the same time, and results pooled afterwards.

To verify the composition of the precipitate formed during the reaction between barium chloride and sodium carbonate solutions

Apparatus

Each pair of pupils will need:

Experiment sheet 66

2 burette stands and burettes

4 test-tubes, 100×16 mm

Test-tube rack

Glass rod

Teat pipette

Watchglass

Beaker, $250 \, cm^3$

Bunsen burner and heat resistant mat

Tripod and gauze

Access to thermometer, $0-110\,°C$

26 s.w.g. copper wire

1.0M barium chloride solution, $10\,cm^3$

1.0M sodium carbonate solution, $12\,cm^3$

Procedure

Show the pupils how to use the copper wire to make a frame to hold a test-tube in the beaker. Further details appear in *Experiment sheet* 66.

Experiment sheet 66

Your teacher will show you how to make a wire frame to hold several test-tubes in a beaker.

Place a 100×16 mm test-tube in a wire frame so that it is immersed to a depth of about 8 cm in a beaker of water kept at a temperature of 80 to $90\,°C$ (small Bunsen flame under gauze on tripod stand). Remove the test-tube, and add to it $5\,cm^3$ 1.0M barium chloride solution and $1\,cm^3$ 1.0M sodium carbonate solution using burettes. Stir the mixture with a glass rod and replace it in the wire frame. Heat the test-tube in hot water for 5 minutes. Centrifuge the mixture in the tube observing the same conditions as in the previous experiment. Measure the height of the precipitate.

Repeat the process with additional portions of $1\,cm^3$ sodium carbonate solution, measuring the height of the precipitate after each addition. Continue until the height of the precipitate does not change with further addition. On a separate sheet of paper record the results in a table using the headings below.

Volume 1.0M $BaCl_2(aq)$ /cm^3	Volume 1.0M $Na_2CO_3(aq)$ /cm^3	Height of precipitate /cm

What volume of 1.0M sodium carbonate solution ($Na_2CO_3(aq)$) just reacts with $5.0\,cm^3$ 1.0M barium chloride solution ($BaCl_2(aq)$)?

Does this result verify your predicted equation for the reaction? Give your reasons.

Make up a mixture of $5\,cm^3$ of 1.0M barium chloride solution and exactly the right quantity of 1.0M sodium carbonate solution for complete reaction. Stir and allow the precipitate to settle while the test-tube is immersed in hot water. What is the precipitate?

What does the solution above it contain?

Devise experiments to verify your answers. Describe how you carry them out and state what the results are. (Class discussion will help you with this.)

To verify the composition of the precipitate formed during the reaction between barium chloride and potassium chromate solutions

Procedure

See *Experiment sheet* 67.

Experiment sheet 67

The procedure in this experiment is similar to that described in Experiment 66, except that potassium chromate solution is used instead of sodium carbonate solution. The precipitate is slow to settle but you will find that there is a colour change in the reaction mixture when precipitation is approximately complete. Make up your own table for this experiment and write notes similar to those for the previous two experiments.

Apparatus

Each pair of pupils will need:

Experiment sheet 67

2 burette stands and burettes

6 test-tubes, 100×16 mm

Test-tube rack

Glass rod

(*Continued*)

Teat pipette

Watchglass

Beaker, 250 cm^3

Bunsen burner and heat resistant mat

Tripod and gauze

26 s.w.g. copper wire

0.5M barium chloride

0.5M potassium chromate

Note to teachers
Small variations in the particle size of precipitates affect the volumes occupied. A number of other factors can also influence the volume occupied by a precipitate. By using a sample of one reactant and adding known quantities of the second reactant, mixing, and centrifuging the precipitate, it is possible to see when a further addition fails to achieve precipitation. As will be appreciated, a centrifuge greatly assists the speed with which such investigations can be made.

The careful study of precipitation reactions and verification of the composition of precipitates is intended to emphasize the need to check expectations by experiment. The composition of precipitates can be 'unexpected' and can vary with conditions used, for example, the reaction between copper(II) sulphate solution and sodium hydroxide solution. Because of the introductory nature of the course, it is important not to introduce unnecessary variations too soon and so the choice of systems for study has been restricted to a few examples.

Suggestion for homework

Questions may be set based on experiments B20.2a, B20.2b and B20.2c.
For example:

5 cm^3 of 1.0M lead nitrate solution were found to react completely with 10 cm^3 of 1.0M potassium bromide solution. Use this information to formulate an equation for this reaction. What additional tests would you need to carry out in order to complete the investigation?

B20.3
Reactions involving gases

Three systems involving gases are investigated: (A) the reaction between sodium carbonate and dilute hydrochloric acid; (B) the reaction between magnesium and dilute hydrochloric acid; (C) the reaction between gaseous ammonia and gaseous hydrogen chloride. The volumes of the gaseous components are measured and then used to find out the reacting quantities and hence the equations for the reactions.

A suggested approach

Objectives for pupils

1. Familiarity with reactions in which gases are either reactants or products
2. Ability to predict probable equations to summarize such reactions
3. Ability to verify such predictions experimentally
4. Knowledge of the use of Avogadro's law

A. *The reaction of sodium carbonate with dilute hydrochloric acid*
The section opens with a study of the reaction between solutions of sodium carbonate and hydrochloric acid. Pupils have already met reactions between acids and carbonates. A demonstration shows that a gas is evolved and a solution remains in the reaction vessel. Pupils should recognize that the gas is carbon dioxide from its reaction with limewater.

They should be able to suggest that:

sodium carbonate + hydrochloric acid $\longrightarrow$ carbon dioxide
$+$ solution of ?

or go further – for example:

sodium carbonate + hydrochloric acid $\longrightarrow$ carbon dioxide

+ solution of sodium chloride

The opportunity can be taken to revise and extend the work done in Topic B12. Pupils can then write:

$$Na_2CO_3(aq) + HCl(aq) \longrightarrow CO_2(g) + \quad ?$$

and the state symbols can also be revised if need be.

In Experiment B20.3a, pupils are asked to determine the volume of 1.0M hydrochloric acid which reacts completely with a known volume of 1.0M sodium carbonate solution. Their results are used to determine whether the lefthand side of the above reaction summary is correct.

Experiment B20.3a

Apparatus

The teacher or each pair of pupils will need:

Experiment sheet 68

2 burettes and stands

Conical flask, 100 cm^3

1.0M sodium carbonate solution, about 25 cm^3

1.0M hydrochloric acid, about 25 cm^3

To determine the volume of 1.0M hydrochloric acid which reacts completely with 10 cm^3 1.0M sodium carbonate

Procedure
Full details are given in *Experiment sheet* 68 reproduced below.

> **Experiment sheet 68**
> Use a burette to measure 10 cm^3 1.0M sodium carbonate into a 100 cm^3 conical flask. From another burette add 1 cm^3 1.0M hydrochloric acid and shake the mixture gently. A gas is given off. What is it?
>
> Continue adding 1 cm^3 portions of acid, shaking the flask between each addition, until there is no gas evolved. (Remember that a few air bubbles are trapped when any liquid is shaken. It is the evolution of gas when shaking is finished that you have to look for. If any gas is given off, wait until the evolution is complete before adding the next portion of acid.)
>
> What volume of 1.0M hydrochloric acid just reacts with 10 cm^3 1.0M sodium carbonate solution?
>
> How many moles of hydrochloric acid react with one mole of sodium carbonate?
>
> What is present in the final solution?
>
> Test your answer by experiment and report on this below.
>
> Can you now write an equation for the reaction between hydrochloric acid and sodium carbonate?

The procedure does not give results with a precision greater than 1 cm^3. They find that *about* 20 cm^3 1.0M hydrochloric acid reacts completely with 10 cm^3 1.0M sodium carbonate. Variations in this result will require discussion and verification. Discussion during the experiment will have led the pupils to the fact that the residual solution contains sodium chloride. Some pupils may require help in completing the equation correctly. Thus, the summary:

$$Na_2CO_3(aq) + 2HCl(aq) \longrightarrow CO_2(g) + 2NaCl(aq) + \quad \ldots$$

leaves 2 atoms of hydrogen and 1 atom of oxygen unaccounted for. Some pupils will probably assume that water is formed during this reaction and it is necessary to point out the difficulty of proving this assumption. The equation:

$$Na_2CO_3(aq) + 2HCl(aq) \longrightarrow CO_2(g) + 2NaCl(aq) + H_2O(l)$$

would be of greater value if we could predict and subsequently verify the quantity of carbon dioxide formed. The equation suggests that 1 mole of sodium carbonate, $Na_2CO_3(aq)$, produces 1 mole of carbon dioxide, $CO_2(g)$ – i.e. 24 dm^3 at room temperature. Thus, if 0.002 mole of sodium carbonate were used in an experimental determination, we would expect 0.002 mole ($24\,000 \times 0.002 = 48\ cm^3$) of carbon dioxide to be liberated. Experiment B20.3b provides a procedure for checking this prediction.

Experiment B20.3b

To measure the volume of carbon dioxide formed when a known amount of sodium carbonate reacts with dilute hydrochloric acid

Apparatus

The teacher or each pair of pupils will need:

Experiment sheet 69

100 cm^3 glass syringe or 50 cm^3 plastic syringe

Test-tube, 150×25 mm, with bung and right-angled delivery tube

Test-tube, 75×10 mm, or specimen tube

Spatula

Stand, boss, and clamp, or syringe holder

Rubber connecting tubing

Measuring cylinder, 25 cm^3

Access to balance – weighing to 1 mg

Anhydrous sodium carbonate*

5M hydrochloric acid, 5 cm^3

Procedure

Details are given in *Experiment sheet* 69 which is reproduced below.

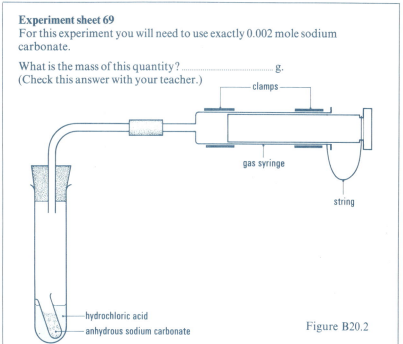

Experiment sheet 69

For this experiment you will need to use exactly 0.002 mole sodium carbonate.

What is the mass of this quantity? .... g.
(Check this answer with your teacher.)

Figure B20.2

Assemble the apparatus shown in the diagram. In the larger test-tube put 5 cm^3 5.0M hydrochloric acid, using a measuring cylinder. Add a measure of sodium carbonate to the acid and wait until all of this has reacted; the purpose of this addition is to saturate the acid with carbon dioxide. Why is this necessary?

Weigh out 0.002 mole sodium carbonate into the smaller test-tube. Slide this tube into the larger tube so that its contents do not come into contact with the acid (see diagram). Connect the larger test-tube to the gas syringe by inserting the rubber bung. Record the reading of the syringe piston. Tilt the larger tube so that the acid makes contact with the sodium carbonate. Carbon dioxide is evolved and pushes out the piston of the syringe. Make sure that all the solid reacts. Record the final reading of the syringe.

*It will be found convenient to prepare a number of samples of 0.002 mole (0.212 g) of anhydrous sodium carbonate before the lesson in corked test-tubes (75×10 mm).

Results

Amount of sodium carbonate used .... 0.002 mole

Initial reading of syringe .... cm^3

Final reading of syringe .... cm^3

Volume of carbon dioxide evolved .... cm^3

At room temperature 24 000 cm^3 of carbon dioxide contains 1 mole.

Therefore .... cm^3 of carbon dioxide contains

.... mole.

Therefore 1 mole of sodium carbonate gives .... mole carbon dioxide.

You now have all the information required to write the complete equation for the reaction between hydrochloric acid and sodium carbonate.

The final part of this sub-section is concerned with re-writing the reaction summary (equation) in terms of ions. Are solutions of hydrochloric acid, sodium carbonate, and sodium chloride ionized? If so, they will conduct electricity.

Experiment B20.3c

To find whether hydrochloric acid, sodium carbonate, and sodium chloride solutions are electrolytes

Apparatus

The teacher or each pair of pupils will need:

Experiment sheet 70

1 pair of carbon electrodes mounted in a holder

4 test-tubes, 150 × 25 mm

Ammeter, 0–1 A

6 V d.c. supply

Connecting wires

Test-tube rack

1M hydrochloric acid, 25 cm^3

1M sodium carbonate, 25 cm^3

1M sodium chloride, 25 cm^3

Distilled water

Procedure

Full details are given in *Experiment sheet* 70.

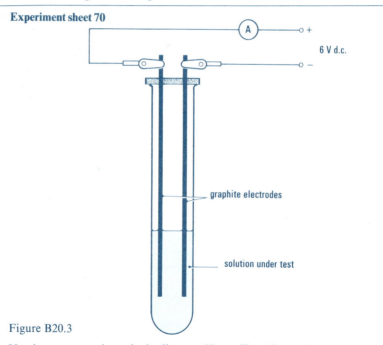

Experiment sheet 70

6 V d.c.

graphite electrodes

solution under test

Figure B20.3

Use the apparatus shown in the diagram. How will you know whether the solution tested is an electrolyte or not?

Try distilled water first, then the three solutions to be tested. Remember to wash the electrodes thoroughly with distilled water between each test. What do you find?

Topic B20 Ions in solution and related chemical systems

The overall results of this investigation may now be written as an ionic equation:

$$2H^+(aq) + 2Cl^-(aq) + 2Na^+(aq) + CO_3^{2-}(aq)$$
$$\longrightarrow 2Na^+(aq) + 2Cl^-(aq) + CO_2(g) + H_2O(l)$$

This can be simplified by removing 'spectator' ions – i.e., those ions which do not take an active part in the reaction – from both sides of the equation: thus,

$$2H^+(aq) + CO_3^{2-}(aq) \longrightarrow CO_2(g) + H_2O(l)$$

B. *The reaction of magnesium with dilute hydrochloric acid*

Open the section by demonstrating the reaction of magnesium with dilute hydrochloric acid to form a gas, which may then be identified in the usual way as hydrogen. Then review the information for writing the equation by questioning the class. From the earlier example, and from previous experience (compare pages 353 and 363), the argument may be summarized in the following way:

a. magnesium + hydrochloric acid $\longrightarrow$ product in solution(?)

$$+ \text{hydrogen}$$

b. $Mg(s) + H^+(aq) + Cl^-(aq) \longrightarrow \quad ? \quad + H_2(g)$

c. $Mg(s) + 2H^+(aq) + 2Cl^-(aq) \longrightarrow Mg^{2+}(aq) + 2Cl^-(aq) + H_2(g)$

d. $Mg(s) + 2H^+(aq) \qquad \longrightarrow Mg^{2+}(aq) + H_2(g)$

To verify equation (*d*), we need experimental evidence. If it is a correct prediction, then:

1 mole Mg would liberate 1 mole H_2
i.e. 24 g magnesium would liberate $24\,000\ cm^3$ hydrogen at room temperature
i.e. 0.024 g magnesium would liberate $24\ cm^3$ hydrogen at room temperature.

In Experiment B20.3d the volume of hydrogen produced is measured. Experiment B20.3c can be adapted for the detection of ions in magnesium chloride solution.

| Experiment B20.3d | **To measure the volume of hydrogen produced when 0.024 g magnesium reacts with dilute hydrochloric acid** |

Apparatus

The teacher or each pair of pupils will need:

Experiment sheet 71

Syringe, $100\ cm^3$ glass or $50\ cm^3$ plastic

Stand, boss, and clamp or syringe holder

(Continued)

Procedure

Pupils will need some help in measuring out exactly 0.024 g magnesium and this may be discussed before starting the experiment. In addition, pupils will need to ensure that their apparatus is gas-tight. Other details are given in *Experiment sheet* 71 which is reproduced below.

Experiment sheet 71
The apparatus and procedure are those described in Experiment 69, using magnesium ribbon instead of sodium carbonate. You will not need to saturate the acid with gas beforehand, the solubility of hydrogen in dilute

Test-tube, 150 × 25 mm, with bung and right-angled delivery tube

Rubber connecting tubing

Measuring cylinder 10 cm³

Emery paper

5M hydrochloric acid, 5 cm³

Magnesium ribbon

hydrochloric acid is small enough to be neglected. Your only problem is to obtain exactly 0.024 g magnesium ribbon. How will you do this?

Results

Mass of magnesium used is 0.024 g. This is .... mole.

Initial reading of syringe .... cm³

Final reading of syringe .... cm³

Volume of hydrogen evolved .... cm³

This is .... mole.

What is the equation for the reaction between magnesium and hydrochloric acid?

Use the method described in Experiment 70 to find whether the magnesium chloride solution is an electrolyte. (You know already about hydrochloric acid in this respect.) What do you find?

C. *The reaction between gaseous ammonia and gaseous hydrogen chloride*

This sub-section is merely intended to show a reaction between two gaseous reactants. The production of ammonium chloride as a white smoke from separate gas jars of dry hydrogen chloride and dry ammonia – both colourless gases – is quite dramatic (*c.f.* B14.1). The argument will then follow the general pattern established in earlier sub-sections:

1. The statement of the word equation for the reaction:

ammonia + hydrogen chloride $\longrightarrow$ 'white smoke'

2. The identification of the white smoke as a chloride.
3. An experiment to demonstrate that equal volumes of gaseous ammonia and hydrogen chloride react together to yield only one product: a white smoke, which is ammonium chloride.
4. The formulation of the equation as:

$NH_3(g) + HCl(g) \longrightarrow (NH_3)HCl(s)$

5. A demonstration of the fact that ammonium chloride, $(NH_3)HCl$, dissolves in water and – like sodium chloride – is also an electrolyte.
6. Finally, the formula for ammonium chloride is amended so that it reads $NH_4^+Cl^-$. Show pupils samples of ammonium chloride and other ammonium salts in order to familiarize them with the ammonium ion NH_4^+.

The use of $(NH_3)HCl$ initially as the formula of ammonium chloride is seen as part of a sequence of ideas. Ammonium chloride solution may be shown to conduct electricity and to contain chloride ions in just the same way as any other solution of a chloride. Accordingly $(NH_3)H$ is amended to NH_4 and thence to the ionic form $NH_4^+(aq)$, thereby stressing the distinction between the names 'ammonia' and 'ammonium'.

One way of finding out the relative number of reacting moles of gases for this system is described in Experiment B20.3e. It requires the use of Avogadro's law which was introduced somewhat earlier in the course (Topic B18).

To verify the equation for the reaction between gaseous ammonia and gaseous hydrogen chloride

Apparatus

The teacher will need :

2 syringes, 100 cm^3, glass

Three-way stopcock – with capillary tubing

Plastic connecting tubing

2 stands, and syringe holders or syringe bench

Indicator paper

Apparatus for producing dry hydrogen chloride (see page 343)

Apparatus for producing dry ammonia (see page 521)

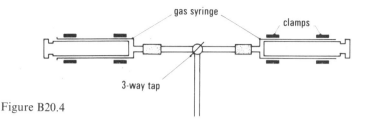

Figure B20.4

Procedure

Set up the apparatus as shown in figure B20.4. Fill one syringe with 40 cm^3 of dry ammonia through the three-way stopcock, flushing the syringe out with the gas two or three times first. Similarly, fill the other syringe with 50 cm^3 of dry hydrogen chloride. Turn the three-way stopcock so as to connect the two syringes but to isolate them from the atmosphere. Push the hydrogen chloride through the stopcock into the ammonia. The gases will react, forming a white powder, ammonium chloride. There remains 10 cm^3 of gas that is not reacted. It is relatively easy to show that this gas is hydrogen chloride by ejecting it over a piece of damp indicator paper.

The result of this experiment shows that 40 cm^3 of ammonia reacts with 40 cm^3 of hydrogen chloride. From Avogadro's law, this means that 1 mole of ammonia molecules reacts with 1 mole of hydrogen chloride molecules. We can therefore write the lefthand side of the equation as :

$$NH_3(g) + HCl(g) = \ldots .$$

As far as the righthand side of this equation is concerned, we know that a white solid is formed. Since ammonium chloride is known to be a white solid of formula NH_4Cl, it is reasonable to assume that this is what is formed – although this argument is *not* a proof for the equation. The equation may then be written as :

$$NH_3(g) + HCl(g) \longrightarrow NH_4Cl(s)$$

or, in the ionic form :

$$NH_3(g) + HCl(g) \longrightarrow NH_4^+ Cl^-(s)$$

if evidence is offered to support this view (see above).

Questions may be set based on reactions involving gases. For example :

20 cm^3 of methane, CH_4, were mixed with 80 cm^3 of oxygen in a suitable apparatus and the mixture exploded. After cooling, the volume of the residual gas was 60 cm^3. This gas was then washed with dilute potassium hydroxide solution. The volume of unreacted gas amounted to 40 cm^3 and consisted of oxygen. Use this information to verify the equation for the reaction between methane and oxygen.

B20.4
Properties of ions in solution

In this section investigations of displacement reactions are used to extend the pupils' experience of the properties of ions in solution.

By now, pupils should understand how chemists are able, from a knowledge of such data as the value of charge on an ion, to predict what an equation is likely to be for a given reaction. They will have also learnt that it is essential to verify an equation by means of experiment.

There are, however, cases where it is not possible to predict in advance what the equation for a reaction will be. One example is the replacement reaction using iron and a solution of a copper salt. If a known mass of iron is added to a solution of copper(II) sulphate (in excess), the iron displaces the copper. One might ask the pupils to predict the equation for this reaction. They might predict – or the teacher may suggest – *either*:

$$Fe(s) + Cu^{2+}(aq) \longrightarrow Fe^{2+}(aq) + Cu(s)$$

or:

$$2Fe(s) + 3Cu^{2+}(aq) \longrightarrow 2Fe^{3+}(aq) + 3Cu(s)$$

It should be noted that there is no way in which pupils can decide which of these equations represents the reaction, *on the basis of their knowledge at this point.*

Suppose that 0.01 mole of iron (0.56 g) is used. If the first equation is correct, then 0.01 mole of copper (0.64 g) will be produced. If, on the other hand, the second equation is correct, then 0.015 mole of copper (0.96 g) will be produced.

The experiment may now be performed to see which of these two equations is the correct one.

Objectives for pupils

1. Ability to determine an equation for a reaction which the pupils are unable to predict before doing experimental work
2. Familiarity with replacement reactions requiring the use of a solid and a solution
3. Knowledge of gravimetric procedures

Experiment B20.4

An examination of the reaction between iron and copper(II) sulphate solution

Apparatus

Each pair of pupils will need:

Experiment sheet 72

Either 2 test-tubes, 100 × 16 mm, or 1 test-tube and beaker, 100 cm³

Test-tube rack

Test-tube holder

Glass rod

Teat pipette

Beaker, 100 cm³

(Continued)

Procedure
See *Experiment sheet* 72.

> **Experiment sheet 72**
> Weigh exactly 0.56 g iron powder into a 100 cm³ beaker and add to it at least 15 cm³ 1.0M copper(II) sulphate solution. Heat the mixture just to boiling, stirring well all the time, and allow it to boil for one minute. Now allow the contents of the beaker to cool and the precipitate of copper to settle. Pour off as much of the liquid as you can, being careful not to lose any copper.
>
> Add distilled water (about one-third fill the beaker), and stir the mixture. Allow the copper to settle and pour off as much liquid as possible. Repeat the washing with distilled water.

Spatula

Bunsen burner and heat resistant mat

Tripod and gauze

Access to balance

Access to centrifuge

Iron, fine powder (not filings)* 1 g

1.0M copper(II) sulphate, 15 cm³

Acetone (propanone)

Distilled water

Add about 20 cm³ acetone (propanone) and stir the mixture. Allow the precipitate to settle and pour off the acetone. *(Be careful, acetone is highly flammable!)* Allow the beaker to stand for 5 minutes so that most of the acetone clinging to the precipitate can evaporate. Drive off the remainder by heating the beaker in an oven at 100 °C for 5 minutes. Allow it to cool and find the mass of beaker + copper.

Results

Mass of beaker	.... g
Mass of beaker + iron powder	.... g
Mass of beaker + copper	.... g
Mass of iron powder used	.... g
This is .... mole.	
Mass of copper obtained	.... g
This is .... mole.	

Which of your two predicted equations is correct?

It will be found that, if the copper has been properly dried, it weighs approximately 0.60 g. This may be taken to confirm the first equation.

Suggestion for homework

1. Write a brief account of the assumptions which you think have been made when investigating the reaction between iron and copper sulphate solution. Indicate how you might investigate such assumptions.
2. Summarize the work you have done so far on chemical equations. List some of the important uses for equations. Have there been any underlying assumptions in the investigations made so far?

B20.5
Reversible reactions

In this section some examples of reversible reactions which the pupils have already met are discussed and other examples introduced.

A suggested approach

Objectives for pupils

1. Further knowledge of the nature of chemical change
2. Familiarity with the idea that certain reactions can be reversed
3. Ability to explain observations
4. Knowledge that the direction in which a reversible reaction proceeds depends on the relative concentrations of the reactants and products

Of the reactions studied so far, those which have received detailed consideration in this and earlier Topics have been of a relatively simple kind: each has appeared to proceed to completion from reactants to products. Pupils may guess that this need not always be the case. They may recall that when copper(II) sulphate crystals are heated water and a white powder (anhydrous copper(II) sulphate) are produced. The reverse reaction can be achieved by adding water to the anhydrous copper sulphate. Similarly, they may recall the decomposition and synthesis of mercury(II) oxide or the formation and decomposition of calcium hydroxide, and so on.

*Iron metal powder reduced by hydrogen (BDH) is suitable. Coarse powder or filings are not suitable, as they will not be completely converted to copper.

In each instance, pupils should be able to write chemical equations for the changes described. Thus, in the case of the reversibility of the reaction of anhydrous copper sulphate and water, we could summarize the two reactions:

$$CuSO_4 5H_2O(s) \underset{\text{add water}}{\overset{\text{heat}}{\rightleftarrows}} CuSO_4(s) + 5H_2O(g)$$

The equilibrium sign, $\rightleftharpoons$, is better avoided at this stage since it is used to describe a reaction in equilibrium. The concept of chemical equilibria is developed in the next Topic, B21.

One might continue this discussion of 'reactions which go both ways' by demonstrating the reaction between chlorine and iodine monochloride.

Experiment B20.5a

A study of the reaction between chlorine and iodine
This experiment should be done by the teacher.

Apparatus

The teacher will need:

Chlorine generator (see page 348)

U-tube

Corks or bungs to fit U-tube, each carrying a right-angled delivery tube

Connecting tube

Stand, boss, and clamp

Piece of white card

Access to fume cupboard

Spatula

Iodine crystals

Procedure
Place a few crystals of iodine in the bottom of a U-tube and then pass chlorine through it. The experiment should be performed in a fume cupboard. Tell the pupils to watch the U-tube carefully. They will see the iodine turn into a brown liquid with a brown vapour above it. The bottom of the U-tube gets hot. On passing more chlorine through the tube, yellow crystals appear on the walls of the U-tube.

Now, detach the chlorine supply and remove the U-tube from the fume cupboard. Tip the U-tube slowly until it is almost upside down and then turn it upright again. The yellow crystals disappear and the brown vapour will be seen again. Pass more chlorine through the U-tube and the yellow crystals will reappear. This experiment may be repeated several times. It is helpful to hold a large white card behind the U-tube so that pupils can see clearly what is happening.

The experiment should now be discussed with the class. Since no reaction took place until chlorine was passed over the iodine, it follows that the reaction must involve chlorine and iodine and their compounds only. The brown vapour and liquid must be a compound of chlorine and iodine. That such a compound is possible may be unexpected since pupils are not used to compounds being formed between elements within the same group. However, they should realize that if atoms of chlorine can combine with themselves to form Cl_2, then one chlorine atom might also combine with an atom of iodine to form ICl, iodine monochloride. This is the compound formed.

When more chlorine was passed through the tube, yellow crystals were seen. Then, the chlorine supply was detached and the U-tube removed from the fume cupboard. It was then tipped slowly until it was almost upside down and then it was turned upright again. The yellow crystals disappeared and the brown vapour reappeared. On passing more chlorine through the U-tube, the yellow crystals reappeared. Some pupils may have seen the comparatively dense chlorine gas being tipped out of the tube. It is worth repeating the experiment to ensure that the effect of tipping the U-tube is understood. Then we may deduce that chlorine is one of the components of this reversible reaction: thus,

chlorine + brown vapour $\longrightarrow$ yellow solid

What can we say about the yellow solid? Since it is a solid, we would expect it to have a higher molecular mass than the volatile liquid which gave the brown vapour. Pupils can see that the yellow solid is a compound of chlorine and iodine monochloride and must contain more chlorine atoms than iodine atoms. Tell them the yellow solid is iodine trichloride, and sum up the overall reaction thus:

$$Cl_2 + ICl \underset{\text{remove chlorine}}{\overset{\text{pass chlorine}}{\rightleftharpoons}} ICl_3$$

(Avoid using the equilibrium sign, $\rightleftharpoons$, since an equilibrium condition may not have been obtained in this demonstration and the use of the sign is deferred until B21.)

This reaction is a very good one for demonstrating the reversibility of chemical reactions. At this point we stress the reversibility of the reaction and leave for the moment the concept of chemical equilibrium.

Pupils could explore the idea of reversibility by attempting a variety of simple experiments for themselves. For example, they may try the effect of acid and alkali on bromine water and on potassium chromate solution. In each case, the reaction will be seen to be reversible. When pupils try their experiments, ask them if they can find a 'point-of-balance' in either of the systems.

| **Experiment B20.5b** | **Two examples of reversible reactions** |

Apparatus

Each pair of pupils will need:

Experiment sheet 76

Beaker, 100 cm³

White tile or sheet of white paper

2 teat pipettes

Bromine water, 10 cm³

2M sulphuric acid, 100 cm³

2M sodium hydroxide, 100 cm³

0.1M potassium chromate, 10 cm³

Distilled water

Procedure
See *Experiment sheet* 76.

Experiment sheet 76

In this experiment you are going to investigate two reactions that can be made to go 'either way'.

1. Put 10 cm³ bromine solution into a 100 cm³ beaker and stand this on a white surface. Add sodium hydroxide solution, five drops at a time, from a teat pipette stirring between each addition, until no further change takes place. What do you observe?

Now add dilute sulphuric acid similarly; what happens now?

By adding alkali or acid, in smaller portions, try to establish a position between the two extremes of excess acid and excess alkali. What is the colour of the mixture now?

2. Repeat the above procedure, using potassium chromate solution instead of bromine solution, but *adding the acid first*.

Effect of acid:

Effect of alkali:

Colour of intermediate stage:

Write equations for the two equilibria which you have investigated. (Your teacher will help you with these.)

During these experiments it is suggested that the teacher should move around the class suggesting ways in which pupils can find a 'point-of-balance' for a given system. Ask how they could identify the components other than by colour. The teacher may suggest that there may be traces of all components present in a given reaction system even when the reaction is complete and over to one side.

Figure B20.5
See figure A19.1 or OHP 25 which shows a traffic flow model.

After the experiments, discuss the findings for the two systems. Since both reactants and products are particulate, the chemical situation may be likened to that which develops when traffic flows between two towns joined by roads of different load-carrying capacities. The traffic flows more readily and at different speeds from Town A to Town B on a motorway than via a secondary road (see figure B20.5). If the motorway only carried traffic flow from A to B and the secondary road only carried traffic from B to A, then the model would not be too far from the situation when a reversible reaction occurs with a greater tendency to go from A to B than from B to A.

Pupils could be given equations for the two systems. Thus, when bromine and water are mixed:

$$\underbrace{Br_2(aq) + H_2O(l)}_{\text{reactants}} \xrightarrow[\text{gain of products}]{\text{loss of reactants}} \underbrace{H^+(aq) + Br^-(aq) + HOBr(aq)}_{\text{products}}$$

As this reaction proceeds, there is a loss of the reactants from the system and a simultaneous formation of products. This process is called the *forward reaction*. However, as soon as the products have formed they begin to react:

$$\underbrace{Br_2(aq) + H_2O(l)}_{\text{reactants}} \xleftarrow[\text{loss of products}]{\text{gain of reactants}} \underbrace{H^+(aq) + Br^-(aq) + HOBr(aq)}_{\text{products}}$$

This process is called the *back reaction*.

The effect of adding alkali or acid on the reaction can be considered and summarized as follows:

$$Br_2(aq) + H_2O(l) \underset{\text{addition of acid: addition of } H^+(aq)}{\overset{\text{addition of alkali: removal of } H^+(aq)}{\rightleftarrows}} H^+(aq) + Br^-(aq) + HOBr(aq)$$

Similarly, equations can be given for the second reaction:

$$2CrO_4^{2-}(aq) + 2H^+(aq) \underset{\text{addition of alkali: removal of } H^+(aq)}{\overset{\text{addition of acid: addition of } H^+(aq)}{\rightleftarrows}} Cr_2O_7^{2-}(aq) + H_2O(l)$$

It should be noted that aspects of these and other systems relating to the theory of dynamic equilibrium are deferred until Topic B21.

Suggestion for homework

1. What experiments could you advise to confirm the formulae of the brown liquid (iodine monochloride) and yellow solid (iodine

trichloride) in the reaction between iodine and chlorine? Outline some possibilities. If possible, give equations for the reactions you describe.

2. Make a list of all the reactions you have come across which go both ways. Describe in each case what conditions are necessary to make the reaction go in each direction.

B20.6
Looking at the role that water plays

In this section, the effect of the solvent on the properties of a solute is studied with the view to establishing an explanation for acidity in terms of ions. This is then extended to provide a simple introduction to the Brønsted-Lowry theory of acids and bases.

Objectives for pupils

1. Awareness of the role of the solvent in a study of reactions in solution
2. Knowledge that acidity and alkalinity can be interpreted in terms of the properties of ions
3. Knowledge of the Brønsted-Lowry theory of acids and bases
4. Familiarity with acid–base reactions
5. Ability to make and test a hypothesis

The teacher will need:

Thermometer, −10 to +110 × 1 °C

2 gas jars of dry hydrogen chloride gas

Toluene

Water

In Topic B12 pupils found that water was needed if the property of acidity was to be shown by substances called acids. An experiment with hydrogen chloride allows this interplay between water and acidic substances to be investigated more precisely. Dip a thermometer into some toluene and then into a gas jar of dry hydrogen chloride. The thermometer shows no rise in temperature. Take the thermometer out, dry it, and then dip it into water before replacing it in the jar of hydrogen chloride. On this occasion, there will be a temperature rise. We may postulate therefore that this change is due to the fact that a reaction has taken place between the hydrogen chloride and the water. We now investigate this hypothesis more carefully by dissolving hydrogen chloride in water and in toluene and examining the products. The details are described in Experiment B20.6a.

A comparison of the properties of hydrogen chloride in water and in toluene solution

Apparatus

The teacher will need:

Gas generator (for supplying dry hydrogen chloride, see page 347)

Corked flask for drying toluene

Procedure
Part 1: before the lesson
Prepare a saturated solution of dry hydrogen chloride in dry toluene. (To ensure that the toluene is absolutely dry, it should stand overnight at least in a corked flask and be in contact with anhydrous calcium chloride.)

(Continued)

Access to a fume cupboard

6 dry test-tubes, 100×16 mm

2 test-tubes, 150×25 mm, fitted with corks

2 beakers, $250\,cm^3$

Beaker, $100\,cm^3$

Test-tube rack

2 teat pipettes

6 V battery or supply of d.c.

2 steel electrodes mounted in a suitable holder

Hofmann voltameter fitted with carbon electrodes

Connecting wire

Paper tissues

Indicator paper stored in a desiccator

Limewater

Marble chips

Magnesium ribbon

Distilled water

Concentrated hydrochloric acid

Concentrated sulphuric acid

Sodium chloride

Anhydrous calcium chloride

Toluene

2M hydrochloric acid in bottle labelled 'solution of hydrogen chloride in water'

The hydrogen chloride from the gas generator should be dried by bubbling it through concentrated sulphuric acid before passing it through the dry toluene. The solution of hydrogen chloride in dry toluene should be kept corked until immediately before use. It is suggested that 2M hydrochloric acid be used as the 'solution of hydrogen chloride in water'.

Part 2: the demonstration
Parts A, B, C, and D should be carried out with both solutions of hydrogen chloride. It is helpful to build up a summary on the blackboard of the evidence obtained as the demonstration proceeds. Parts E and F follow as required.

A. The effect of the solutions of hydrogen chloride on dry indicator paper.

B. Reaction of the solutions of hydrogen chloride with marble chips and the detection of any gas given off.

C. The reaction of the solutions of hydrogen chloride with magnesium ribbon and the detection of any gas given off.

D. The effect of immersing two steel electrodes into samples of each solution when the electrodes are connected to a d.c. supply and an electric bulb. (*Note*. The electrodes should be placed in the toluene solution first to avoid transporting any water to the toluene.)

E. Then transfer about $10\,cm^3$ of the toluene solution to a corked 150×25 mm test-tube and shake it with an equal volume of distilled water. After removing the upper layer of toluene, each of the tests A, B, C, and D should be applied to the aqueous layer.

From these tests, it should be clear that some at least of the hydrogen chloride has passed into the aqueous layer and in doing so it has developed different properties.

While the experiment is being carried out some discussion of the various parts should take place. The evidence shows that:
1. A reaction took place between water and hydrogen chloride (but not between toluene and hydrogen chloride).
2. Ions are produced when hydrogen chloride dissolves in water.
3. Acidic properties are associated with the solution containing ions.

F. We need to know something of the nature of the ions formed when hydrogen chloride reacts with water. The electrolysis of a strong solution of hydrogen chloride (concentrated hydrochloric acid) can show that hydrogen is evolved at the cathode and chlorine at the anode. (*CAUTION*. Do not attempt this electrolysis in the presence of direct sunlight or artificial light.) Thus, one might predict the existence of hydrogen cations, H^+, and chloride anions, Cl^-.

Such a demonstration can be very effective and is best done by electrolysing concentrated hydrochloric acid in a Hofmann voltameter fitted with carbon electrodes. Equal volumes of gases are collected only when the acid has become saturated with chlorine.

Now let us look more closely at the formation of hydrochloric acid from hydrogen chloride and water. If a reaction occurs between these compounds, one would expect to find a temperature change during the dissolution of the gas. If this were due to the simple breaking of the hydrogen–chlorine bond to form hydrogen ions and chloride ions, then energy would be required to achieve this and heat would be absorbed. We would expect the temperature of the system to drop. The original demonstration in this section showed that heat was evolved when hydrogen chloride dissolved in water. We may therefore conclude that reaction a is unlikely but that reaction b is likely:

a. $HCl \longrightarrow H^+ + Cl^-$

b. $HCl(g) + H_2O(l) \longrightarrow H_3O^+(aq) + Cl^-(aq)$

If need be, the teacher might argue 'If the change is merely:

$$HCl(g) \longrightarrow H^+(g) + Cl^-(g)$$

why doesn't this occur with toluene or by itself? It doesn't. It follows then that the reaction must require the presence of water. Hence reaction b is likely.'

(A more detailed explanation extending the initial energy argument given above can be deferred until B23.)

Although it is likely that the hydrogen ion is surrounded by a number of water molecules rather than by one, the choice of the representation $H_3O^+(aq)$ enables pupils to appreciate the simple interpretation that a hydrogen chloride molecule has given a hydrogen ion to a water molecule.

No real proof can be offered at this stage for the existence of $H_3O^+(aq)$. However, the following argument may seem to be reasonable at this level. The proton, H^+, is extremely small and is much smaller than the hydrogen atom. Further, its positive charge is likely to attract electrons associated with the oxygen atom in a water molecule. (This point can be made very simply by using scale models or scale drawings on an overhead projection transparency – see figure B20.6.)

Figure B20.6
See figure A20.2 or OHP 26 to illustrate the formation of ions.

The fact that reaction b is reversible can be shown readily by boiling a little concentrated hydrochloric acid and identifying the hydrogen chloride gas evolved:

$$HCl(g) + H_2O(l) \rightleftharpoons H_3O^+(aq) + Cl^-(aq)$$

At this point, it might be appropriate to state that although acidic solutions are said to contain ions, H_3O^+, it is quite usual to refer to them as hydrogen ions (or merely as protons). Chemists accept the fact that such ions are normally hydrated in solution and that the symbol $H^+(aq)$ also implies the interaction of H^+ with water.

Concerning the definitions of the terms we use

We are now in a position to define the terms acid and alkali by explaining the way in which chemicals react: that is, to employ *conceptual* definitions. An acid is a substance containing hydrogen which in solution yields hydrogen ions while an alkali is a substance containing hydroxide ions or hydroxyl groups which on solution in water yields hydroxide ions.

Pupils know that heat is generated when hydrogen chloride and water react together:

$$HCl(g) + H_2O(l) \longrightarrow H_3O^+(aq) + Cl^-(aq): \quad \text{heat evolved}$$

Add some concentrated nitric acid to water and show that this system behaves in a similar manner:

$$HNO_3(l) + H_2O(l) \longrightarrow H_3O^+(aq) + NO_3^-(aq): \quad \text{heat evolved}$$

Similarly, if $50\,cm^3$ of concentrated sulphuric acid is added to $1000\,cm^3$ of cold water carefully and with stirring, a considerable temperature rise is obtained:

$$H_2SO_4(l) + H_2O(l) \longrightarrow H_3O^+(aq) + HSO_4^-(aq): \quad \text{heat evolved}$$

$$HSO_4^-(aq) + H_2O(l) \longrightarrow H_3O^+(aq) + SO_4^{2-}(aq): \quad \text{heat evolved}$$

Pupils will know that, in each case, the acids produce hydrogen ions and that these ions can be written as either $H_3O^+(aq)$ or $H^+(aq)$.

For the system:

$$HCl(g) + H_2O(l) \rightleftharpoons H_3O^+(aq) + Cl^-(aq)$$

we may think of the hydrogen ions being pulled in different directions.

Thus:

$$Cl^- \longleftarrow -- H^+ \longrightarrow OH_2$$

The molecules of water win a major share of the hydrogen ions and the solution is strongly acidic. When this idea is discussed, the teacher may point out that there is an even more general definition of the terms acid and base. This is that an *acid* is a proton donor and a *base* is a proton acceptor.

Such definitions lead to the somewhat surprising conclusion:

$$HCl(g) + H_2O(l) \rightleftharpoons H_3O^+(aq) + Cl^-(aq)$$
acid   base   acid   base

On the basis of this definition, water may be regarded as a base since it is acting as a proton acceptor.

It is equally important at this point to indicate how alkalis give rise to the presence of hydroxide ions: thus:

$$Na^+OH^-(s) \xrightarrow{\text{solution in water}} Na^+(aq) + OH^-(aq)$$

$$Ba^{2+}(OH^-)_2(s) \xrightarrow{\text{solution in water}} Ba^{2+}(aq) + 2OH^-(aq)$$

$$NH_3(g) + H_2O(l) \rightleftharpoons NH_4^+(aq) + OH^-(aq)$$

Point out that hydroxides such as sodium hydroxide or barium hydroxide are largely ionized in the solid state. Ammonia gas is not ionized; when ammonia dissolves in water, ammonium ions and hydroxide ions are formed.

It is helpful to use the example of ammonia solution to extend this definition of acid and base. Ask pupils to identify which substance acts as a proton donor in this case. Water may be considered to act as an acid *in this instance*:

$$NH_3(g) + H_2O(l) \rightleftharpoons NH_4^+(aq) + OH^-(aq)$$
base   acid   acid   base

Acceptance of the (conceptual) definition of the term acid leads to the idea that substances such as water which are not acids or bases in the commonly accepted sense (that is, according to operational definitions) can nevertheless exhibit both acidic and basic properties.

All of these interpretations have their uses. In particular, the *idea*, expressed by the relationship:

$$acid \rightleftharpoons base + proton$$

helps to clarify our thinking about acids and bases. However, it would be far too confusing for chemists to interpret this idea rigidly and to say, for example, that water is an acid. Even chemists *cannot* altogether dispense with operational definitions! (Later, we will need to modify our interpretation after accepting the implications of dynamic equilibrium.)

Pupils should now be aware that we consider the hydrogen ion or the hydrated proton to play a vital role in acid–base reactions. 'What happens to an acid when it loses its acidic properties through the addition of a base or an alkali?' Clearly it will help to study just one system in great detail and then to apply the findings to other systems to see whether our ideas are of general application.

We have found that when an acid is dissolved in water, ions are formed, and this also happens for alkalis. We know that acids react with metal oxides and metal hydroxides to form salts (see Topic B12). We also know that there are ions in aqueous solutions of salts (see Topic B19). Therefore, if we use a simple conductivity meter (electrode, a.c. source, current indicator) to follow the course of a reaction between an acid and a base, it will be profitable to choose a reaction in which the product of reaction is insoluble. Such a product of reaction would not interfere with the conductivity of the reactants. One such reaction is that between dilute sulphuric acid and barium hydroxide solution.

Two possibilities now arise: the teacher may choose to follow this preliminary discussion with either:
1. the class experiment (Alternative 1) or

2. the demonstration experiment (Alternative 2)
using a source of a.c., a current indicator, electrodes, and an acid–base indicator.

Either experiment can show pupils that at the end-point the solution is non-conducting, thereby providing evidence to support the equation for the reaction. Whichever approach is used, the apparatus, the concentration of solutions, and the specification of electrical equipment, need to be adhered to closely if satisfactory results are to be obtained.

Experiment B20.6b

To find the ratio in which barium and sulphate ions combine

Apparatus

Each pair of pupils will need:

Experiment sheet 80

Test-tube, 100×16 mm

Teat pipette

Beaker, 100 cm^3

2 carbon electrodes in a holder

3 lengths of connecting wire

Bulb holder and 6.5 V 0.06 A bulb or a.c. ammeter (0–100 mA)

Crocodile clips

Source of 12 V a.c.

Glass rod

Burette, $50 \times 0.1 \text{ cm}^3$

Burette stand

Measuring cylinder, 25 cm^3

Phenolphthalein solution (10 drops)

0.05M barium hydroxide, 60 cm^3: standardized (*CAUTION:* solutions of barium compounds are poisonous when taken by mouth.)*

0.5M sulphuric acid, 20 cm^3: standardized

Distilled water

Procedure
See *Experiment sheet* 80. Advise the pupils to exercise care with the solutions.

Note to teachers
1. The acid is 10 times the concentration of the alkali so that the volume of liquid in the beaker does not change appreciably. Accordingly the electrode contact does not increase during the titration and changes in concentration due to this effect do not arise.
2. If greater precision is required in relating end-point and current flow, then the bulb must be replaced by an a.c. ammeter (0–100 mA) as a current indicator.

> **Experiment sheet 80**
> You are provided with 0.05M barium hydroxide solution and 0.5M sulphuric acid. Using an electric bulb and an alternating current you are going to investigate the changes in conductivity when the sulphuric acid solution is added to the barium hydroxide solution.
>
> As a preliminary experiment, put about a 1 cm depth of the barium hydroxide in a test-tube and add a few drops of the dilute sulphuric acid. What do you see?
>
> What do you think it is?
>
> Use your beaker and test each solution for conductivity. Wash out the beaker *and* the electrodes with distilled water before testing the other reactant. Record your results.
>
> Dilute sulphuric acid:
>
> Barium hydroxide solution:
>
> For the main part of the investigation use the apparatus shown in the diagram below.
>
> Use a measuring cylinder to put 50 cm^3 of the barium hydroxide solution into the 100 cm^3 beaker and add 5 drops of phenolphthalein solution. Put the sulphuric acid solution into the burette, see that the jet is filled with solution, and fix the burette in a stand so that it is above the beaker. Put the two electrodes in the solution vertically and connect them in series with the a.c. supply and bulb.

*First aid: *if swallowed*, give two tablespoonfuls of magnesium sulphate in water and then induce vomiting *either* by giving salt water *or* by tickling the back of the throat with two fingers or a spoon. Obtain medical attention. Keep the patient warm.

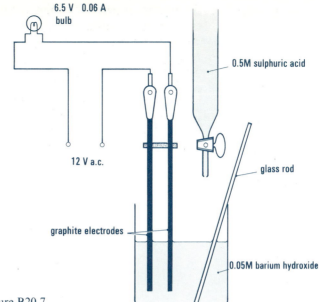

Figure B20.7

Run the sulphuric acid solution into the beaker 1 cm^3 at a time, stirring between each addition. Note the appearance of the bulb and the mixture in the beaker after each addition. When the bulb starts to become dim add the acid one drop at a time. When there is enough sulphuric acid to react with all the barium hydroxide, the bulb will go out and the phenolphthalein will lose its pink colour.

How much sulphuric acid has been added when this happens?

.... cm^3

Continue to add acid; what happens to the lamp now?

Remembering that, according to theory, the current that lights the lamp is carried by ions, what conclusion can you draw about the ions in solution at the 'end-point' of the neutralization, that is, when the lamp goes out?

The volume of 0.05M barium hydroxide solution used is 50 cm^3 and this contains $\dfrac{0.05 \times 50}{1000} = 0.0025$ mole barium ions.

The volume of 0.5M sulphuric acid used is .... cm^3.

and this contains $\dfrac{0.5 \times \text{........................}}{1000} =$ .... mole sulphate ions.

How many moles of sulphate ions react with one mole barium ions?

.... mole.

Write the equation for the reaction between barium ions and sulphate ions.

What has happened to the hydrogen and hydroxide ions?

Alternative method of finding the ratio in which barium and sulphate ions combine

Apparatus

The teacher will need:

Test-tube, 100×16 mm

Beaker, 100 cm^3

2 carbon electrodes in a holder

4 lengths of connecting wire

Measuring cylinder, 100 cm^3

Burette stand and burette

a.c. milliammeter reading 0–10 mA (demonstration model preferred)

Crocodile clips

Source of 4 V a.c.

Rheostat, $20 \, \Omega$

Magnetic stirrer

Phenolphthalein solution

0.01M barium hydroxide solution, made from solid and standardized

0.1M sulphuric acid standardized to match the barium hydroxide solution

Procedure

Transfer 50 cm^3 of barium hydroxide solution into a 100 cm^3 beaker using a measuring cylinder.* Place the two electrodes in the solution, and connect them in series with a rheostat, a.c. milliammeter, and the 4 V a.c. supply (see figure B20.8). Adjust the rheostat so that the milliammeter reads 10 mA to start with. A few drops of phenolphthalein solution can now be added to the beaker.

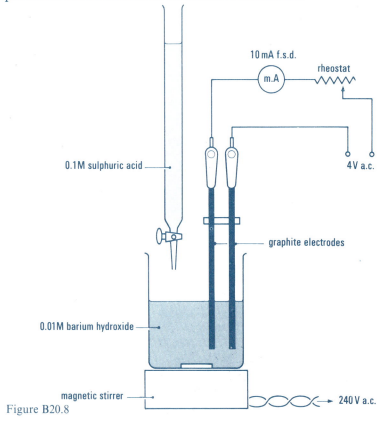

0.1M sulphuric acid

10 mA f.s.d.

m.A

rheostat

4V a.c.

graphite electrodes

0.01M barium hydroxide

magnetic stirrer

240 V a.c.

Figure B20.8

Now set up the burette containing the sulphuric acid (10 times the concentration of the barium hydroxide solution so as to prevent a large volume change in the beaker). Add the sulphuric acid 1 cm^3 at a time to begin with and record the volume added and the corresponding milliammeter readings. Stir the solution before recording the milliammeter reading.

When the milliammeter reading is less than 2 mA, add the acid in 0.5 cm^3 portions and as the meter reading approaches even closer to zero, use even smaller portions of acid. The indicator turns colourless as the meter reading reaches a minimum value, due to the lack of ions in solution. The meter reading increases again when an excess of acid has been used. Specimen results are shown in figure B20.9.

Figure B20.9
See figure A20.5 which shows some specimen results for the experiment.

*Note: barium hydroxide solution is very poisonous; it should NOT be transferred from one vessel to another using a pipette.

The experiment confirms our expectation and so we can express the lefthand side of the equation as:

barium hydroxide + sulphuric acid =
$$Ba^{2+}(aq) + 2OH^-(aq) + 2H^+(aq) + SO_4^{2-}(aq) =$$

Since the reaction results in the removal of ions, we can write:

$$H^+(aq) + OH^-(aq) \longrightarrow H_2O(l)$$
and
$$Ba^{2+}(aq) + SO_4^{2-}(aq) \longrightarrow BaSO_4(s)$$

The teacher may go on to define a neutral solution as one which contains an equal number of hydrogen and hydroxide ions. We say it has pH 7 (see Stage I, A1.4 and B2.1). The reaction between an acid and a base is sometimes referred to as neutralization. This term suggests that it is a process which gives rise to a neutral solution and, of course, this is not necessarily so. For this reason the term neutralization has been avoided in discussion.

By the end of this section, pupils should appreciate that when solutes are dissolved in a solvent, the resulting solution may contain a variety of ions, and ions from an ionic solid may become solvated. A solution of a molecular substance such as hydrogen chloride may contain solvated ions as well – as when hydrogen chloride is dissolved in water. It is not intended that the teacher should go more deeply than this into the role of the solvent in ionic solutions.

Suggestion for homework

Write ionic equations (including state symbols) for a variety of reactions including those between acids and bases.

(Some teachers may find that it is necessary to spend a little time in class introducing this work before setting the homework.)

Summary

By the end of this Topic, pupils should be able to distinguish between electrolytes and non-electrolytes and interpret the properties of ions in solution. They should be capable of summarizing the reactions of ions by writing ionic equations and of interpreting the properties of acids and bases according to the Brønsted-Lowry theory.

Topic B21

Chemical equilibrium

Purposes of the Topic

1. To provide an indication of the dynamic nature of equilibria.
2. To demonstrate the effect of concentration change on the position of equilibrium.

Contents

B21.1 Characteristics of chemical reaction
B21.2 A study of systems in equilibrium

Timing

Section B21.1 requires a single period, B21.2 about two double periods. Thus almost two weeks are required for this Topic.

Introduction to the Topic

The first section of the Topic may be used to review principles which have been used throughout the course so far. From the previous Topic, pupils will be aware that certain chemical reactions can be reversed and they may be able to formulate some of the conditions needed to do this. The situation may then be developed by asking the pupils 'Do all chemical reactions necessarily go to completion?' By comparing chemical systems with model situations pupils can be led towards an appreciation of the idea of the dynamic nature of chemical systems in equilibrium. The use of radioactive tracers is mentioned in the distinction that is made between 'simple' reversibility and dynamic equilibrium.

Subsequent development

To appreciate the usefulness of radioactive tracers pupils need to know something of the techniques of radiochemistry. Some teachers may wish to use this opportunity to introduce radioactive tracer techniques whereas others may prefer to come back to this at an appropriate point in Topic B22. Much will depend on individual taste and on the interest shown by pupils in the possibility of using tracer techniques. The idea of dynamic equilibrium plays a significant role in almost all aspects of chemistry and so is assumed in the remaining Topics of Alternative IIB and in the Options.

Alternative approach

Topic A19 offers an alternative presentation of B20 and B21.

Further references
for the teacher

Handbook for teachers provides background information on this Topic – see Chapters 20 and 22.
Collected experiments, Chapter 15, offers alternative experiments.
Nuffield Advanced Chemistry (Topics 12 and 15 in *Teachers' guides I* and *II*), and Nuffield Physical Science (Sections 3, 4, and 5 in *Teachers' guide I*) offer advanced treatments of the Topic.

Supplementary materials

Film loops
2–5 'Liquid–gas equilibrium'
2–6 'Solid/liquid equilibrium'
2–19 'Dynamic equilibrium'

Films
'Equilibrium' CHEMStudy. Distributed by Guild Sound and Vision Ltd.
'Chemical equilibrium' ESSO. Films for Science Teachers No. 23

Overhead projection original
54 Fish-bowl analogy *(figure B21.1)*

Reading material

Handbook for pupils, Chapters 5 'Studying chemical reactions' and 10 'The chemical industry'.

B21.1
Characteristics of chemical reaction

This section starts by reviewing the characteristics of chemical change which have been developed so far in the course. Then the idea of the dynamic nature of chemical equilibrium is introduced.

A suggested approach

Objectives for pupils

1. Ability to recognize chemical reactions
2. Ability to classify chemical reactions
3. Ability to identify reversible reactions
4. Awareness of the nature of scientific enquiry: the formulation of ideas about an equilibrium process

At this stage of the course, pupils should be able to identify and subsequently classify a wide range of chemical reactions. It is rewarding to ask pupils to express their views simply and to justify any assertions which they may make. The examples selected for study may be drawn from any of the Topics studied so far, for example, acid–base reactions; oxidation–reduction; reversible reactions; ion combination reactions, etc. The previous Topic introduced reversible reactions and included some studies of the conditions needed to reverse specific reactions, and these could be revised. Alter the direction of the discussion by stating that in all of the systems so far, we have assumed that reactions tend to go to completion. Ought we to question this assumption? Ought we to speculate about other possible interpretations? What happens when 'reactants' react to form products in a reversible reaction? Do the 'products' formed react again to re-form the 'reactants'?

In any discussion about reversible reactions, there will be a need to discuss and then define the meaning of the terms used. Thus, in everyday usage, the word 'equilibrium' means a 'state of balance'. So, we might refer to the balance achieved when a weighing operation is carried out using a beam balance, and show that a state of balance is only reached when two static forces become equal. Alternatively, the precarious balance achieved by a tight-rope walker could be discussed. There is a common idea of some form of constancy obtained by the opposition of forces or processes irrespective of their nature.

When we are discussing the possibility of setting up a system in equilibrium, we will need to consider matters at two levels: the macroscopic and the microscopic. In the former we are concerned whether there will or will not be a net change in composition. In the latter case, we will be concerned at the molecular or ionic level with whether there is a continual change in the location of particles. The analogy of 'two towns joined by two roads' used in the previous Topic can be helpful in such a discussion. One road carries cars from A to B and the other carries cars from B to A. The flow of cars can be counted in each instance. Equilibrium will be achieved for this system when no net change in the total number of cars in Town A or in Town B can be measured. Actually, in terms of individual cars, movement can be continuous: one car can replace another. At this point, it may be practicable to discuss one or more of the experiments given in the previous Topic (e.g., those in section B20.5) to illustrate the points made by the analogy. In addition, the characteristics of a chemical system in equilibrium can be established. These may be stated as:

1. Starting either with the substances on the lefthand side or with those on the righthand side of the equation used to represent a reaction, it should be possible to detect changes in composition.

2. After a period of time, which varies with the reaction studied, all changes appear to cease.

3. Under specific physical conditions, the same composition is achieved no matter whether the starting materials are those on the lefthand side or the righthand side of the equation which summarizes the reaction.

B21.2
A study of systems in equilibrium

This section provides pupils with an opportunity of studying chemical systems in such a way as to lead to a theory of dynamic equilibrium.

A suggested approach

Objectives for pupils

1. To show that certain reactions do not proceed to completion, but reach a 'point of balance'
2. To show that the point of balance to which reactions proceed depends on the relative concentrations of the reactants and products
3. To provide a theory of chemical equilibrium and to show that it gives a satisfying explanation of the systems investigated

The distribution of iodine between two immiscible solvents, potassium iodide solution and chloroform, forms the subject of the first investigation. It is suggested that the system be represented by the equation:

$$I_2(s) + I^-(aq) \longrightarrow I_3^-(aq)$$

The effect of mixing such a system with another solvent enables iodine to be removed from the system and the reaction to be reversed:

$$I_2(s) + I^-(aq) \longleftarrow I_3^-(aq)$$

Experiment B21.2a

The distribution of iodine between two immiscible solvents

Procedure
Details are given in *Experiment sheet* 75.

Apparatus

The teacher or each pair of pupils will need:

Experiment sheet 75

2 test-tubes, 100×16 mm

Teat pipette

Spatula

2 corks

Chloroform (trichloromethane), $20 \, \text{cm}^3$

1M potassium iodide solution, $20 \, \text{cm}^3$

Iodine crystals

Experiment sheet 75

1. Put potassium iodide solution into a test-tube to a depth of about 2 cm, add a *small* crystal of iodine, and shake the tube until the iodine has dissolved. What is the colour of the solution?

Now add about an equal volume of chloroform (trichloromethane) and shake the test-tube gently. What do you see?

2. Put chloroform (trichloromethane) into a test-tube to a depth of about 2 cm. Add a small crystal of iodine as nearly the same size as that used in (**1**) as you can manage. Shake the tube until the iodine has dissolved. What is the colour of the solution?

Add about an equal volume of potassium iodide solution and shake the test-tube gently. What happens?

3. Cork the test-tubes used in (**1**) and (**2**) and shake both vigorously for one minute. Allow them to stand until the two layers of liquid in each have separated. Compare their appearances.

4. What do you think will happen if you remove the potassium iodide layer from one test-tube, replace it by a fresh solution, and shake the tube vigorously again?

Now try it, using a teat pipette to remove the potassium iodide layer. Were you right?

Account as fully as you can for what you have seen during this experiment. (Class discussion should help here.)

The significance of this experiment and the explanation of the results in terms of dynamic equilibrium need to be discussed. The phenomena can only be explained in terms of a dynamic interpretation of equilibrium, and the symbol $\rightleftharpoons$ should be introduced:

$$I_2(s) + I^-(aq) \rightleftharpoons I_3^-(aq)$$

The discussion of dynamic equilibrium must necessarily raise a number of key points. These include:
a. the fact that changes in composition will occur when the substances on the lefthand side or on the righthand side of an equation react;
b. the fact that catalysts (see Topic B13) influence the rate of change of either reaction under given conditions;
c. the fact that, after a certain interval of time dependent on the composition of the system and on the temperature, all changes appear to cease;
d. the fact that the same composition of reaction mixture is reached no matter whether we start from the substances on the lefthand or righthand side of the equation, or whether a catalyst is used or not.

Thus, it would seem that the position of equilibrium is not controlled by the actual rates of the reactions that take place. Also, the concentration of the catalyst used has no influence on the final composition of the equilibrium mixture. However, the condition of equilibrium requires an equality of the overall forward and backward rates of reaction.

Two other examples of systems in chemical equilibrium

Apparatus

Each pair of pupils will need:

Experiment sheet 77

5 test-tubes, 100×16 mm

4 test-tubes, 150×25 mm

3 corks to fit 100×16 mm test-tubes

Test-tube rack

Glass rod

Spatula

3 teat pipettes

Access to centrifuge

Bismuth(III) chloride

Concentrated hydrochloric acid, $2 \, cm^3$

0.1M silver nitrate, $10 \, cm^3$

0.1M iron(II) sulphate, $10 \, cm^3$ – freshly prepared

1.0M iron(III) nitrate, $10 \, cm^3$

Freshly prepared potassium hexacyanoferrate(III) solution

0.1M potassium thiocyanate solution

Procedure

See *Experiment sheet* 77.

Experiment sheet 77
Caution. Care is needed when handling the chemicals mentioned below.

1. *Hydrolysis of bismuth chloride*
Put a spatula measure of bismuth chloride in a test-tube and, using a teat pipette, add concentrated hydrochloric acid until the solid has *just* dissolved. You will need about $2 \, cm^3$ of the concentrated acid.

a. Two-thirds fill a 150×25 mm test-tube with water and add a few drops of the bismuth chloride solution. What happens?

b. To study this equilibrium more carefully, two-thirds fill three other 150×25 mm test-tubes with water. To the first add 5 drops concentrated hydrochloric acid, to the second 10 drops, and to the third 15 drops. Mix the contents of each tube thoroughly with a glass rod. Add 5 drops of bismuth chloride solution to each tube. Note any differences between the three precipitates and between the speed with which they are formed.

Tube 1 (5 drops conc. acid).

Tube 2 (10 drops conc. acid).

Tube 3 (15 drops conc. acid).

Write an equation for this equilibrium.

2. *Reaction of silver nitrate and iron(II) sulphate solution*
To study the second reaction (between iron(II) ions, $Fe^{2+}(aq)$, and silver ions, $Ag^+(aq)$), you will need to know how to test for iron(II) ions and iron(III) ions, $Fe^{3+}(aq)$.

Iron(II) ions, $Fe^{2+}(aq)$, give a dark blue colour or precipitate with potassium hexacyanoferrate(III) solution.

Iron(III) ions, $Fe^{3+}(aq)$, give a deep red colour with potassium thiocyanate solution.

Put iron(II) sulphate solution into a 100×16 mm test-tube to a depth of 2 cm. Add an equal volume of silver nitrate solution. Cork the tube and shake well. What happens?

Transfer a few drops of the solution with a teat pipette to another test-tube and test for the presence of $Fe^{3+}(aq)$ ions. What is the result?

Transfer a few drops to another test-tube and test for $Fe^{2+}(aq)$ ions. What is the result?

Do you think a state of equilibrium has been reached? If so, write an equation to represent it.

Now try to reverse the reaction. Centrifuge the mixture, remove the upper liquid layer, wash the precipitate (add water, centrifuge, remove liquid), and add iron(III) nitrate solution to a depth of about 4 cm. Cork the tube and shake well for 2 minutes. Allow any precipitate to settle, remove a few drops of liquid, and test for $Fe^{2+}(aq)$ ions. What is the result?

After the experiment, discuss the results and check that pupils have been able to formulate equations for these systems. Thus, the hydrolysis of bismuth chloride may be represented by:

$$BiCl_3(aq) + H_2O(l) \rightleftharpoons BiOCl(s) + 2H^+(aq) + 2Cl^-(aq)$$

  bismuth   bismuth oxide
  chloride   chloride

The reaction of silver and iron(II) ions may be represented by:

$$Ag^+(aq) + Fe^{2+}(aq) \rightleftharpoons Ag(s) + Fe^{3+}(aq)$$

Note. Iron(III) nitrate solution was used instead of iron(III) sulphate solution to avoid the formation of a precipitate of silver sulphate.

The overall effect of the experimental work is to emphasize the need to use some form of labelling so that we can see the dynamic nature of equilibria. Figure B21.1 provides a convenient illustration of this point. It shows two large vessels filled with water and joined together by a wide glass tube. One can imagine placing two kinds of fish in the apparatus. We might put four of one kind of fish in one bowl and four fish of another kind in the second bowl. If these actions were carried out simultaneously, we could say that the system obtaining at the beginning of the 'reaction' was that shown in (I). After a few minutes, the system might look like that shown in the (II). Clearly, we need some form of labelling of this kind to let us see what happens to chemicals under similar circumstances.* In addition, the film loop 2–19 'Dynamic equilibrium' may be used to recapitulate the main points of a discussion on the silver ion and iron(II) ion reaction system.

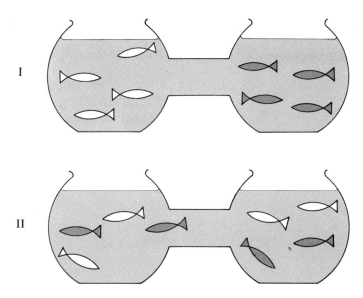

Figure B21.1
Fish-bowl analogy. (OHP 54)

As an additional example, consider the formation of a simple ester from an alcohol and an organic acid:

$$C_2H_5OH(l) + CH_3CO_2H(l) \longrightarrow CH_3CO_2C_2H_5(l) + H_2O(l)$$

*See the CHEMStudy film 'Equilibrium'.

or the hydrolysis of an ester to form an organic acid and an alcohol:

$$CH_3CO_2C_2H_5(l) + H_2O(l) \longrightarrow CH_3CO_2H(l) + C_2H_5OH(l)$$

If the oxygen atoms in, say, the alcohol were labelled in some way, we could 'see' what happened when the ester was formed or when it was hydrolysed. Chemists have been able to do this by using isotopic labelling: by using compounds containing oxygen-18 rather than oxygen-16. The analyses of reaction products formed from both of these reactions yield similar results and support the view that a chemical equilibrium is a dynamic rather than a static phenomenon.

Alternatively, radioactive tracer techniques can be used to show this effect with certain chemical systems.

Thus, by using such advanced techniques as these, we acquire evidence that both forward and back reactions do in fact occur when the system is in equilibrium. Indeed, many systems already studied in earlier Topics can be tested and shown to illustrate this view.

In addition, pupils should be made aware that physical equilibria, such as may exist between solids and liquids or liquids and vapour, are also dynamic in nature. This view can be established by using the film loops 'Solid–liquid equilibrium' and 'Liquid/gas equilibrium'. The first film loop shows what happens when a solid is melted into a closed vessel and an equilibrium condition is established. The second film loop illustrates the measurement of saturated vapour pressures of a liquid. It may prove practicable for the teacher to demonstrate one or other of such systems to pupils.

Suggestion for homework

Summarize the characteristics of chemical systems in equilibrium.

Summary

By the end of this Topic, pupils should understand the concept of a dynamic equilibrium. They should know how, in general, it is possible to shift an equilibrium by changing the concentrations of the chemicals taking part in that equilibrium.

Note. Some teachers may wish to introduce radiochemical tracer techniques so that direct evidence of the dynamic character of the equilibrium process can be presented in this Topic, whereas other teachers may prefer to come back to this discussion at an appropriate point in Topic B22. Relevant experimental work will be found in section B22.4.

Rate processes in chemistry

Timing

About four weeks are required.

Introduction to the Topic

In the first section pupils are introduced to radioactivity through a short account of its discovery. They see the effects of radioactive materials on photographic film and scintillation plates. Cloud chambers may be discussed and perhaps demonstrated to show differences in the range and charge of α-, β-, and γ-radiation. The ionization of a gas by radiation is shown to be the principle underlying Geiger-Müller (GM) tubes. A GM tube is used to investigate the activities of potassium and uranyl ions.

In B22.2 protactinium chloride is studied as an example of a radioactive isotope. Growth and decay curves can be plotted and the concept of half-life introduced. The separation of the products of the decay is by solvent extraction, and pupils do an experiment using non-radioactive substances to illustrate solvent extraction.

A short account of the discovery of thorium-234 and of radioactive decay series and isotopes follows in the next section. Uses of radioactive isotopes are reviewed in the following section, and the teacher demonstrates the investigation of an equilibrium system using lead-212 as a tracer.

Pupils should understand the significance of atomic number (proton content, Z) as a means of identifying an element; and the relation between the atomic number and the mass number (number of protons + neutrons, A), which they may have met in their first study of periodicity through non-radioactive isotopes such as chlorine-35, $^{35}_{17}Cl$, and chlorine-37, $^{37}_{17}Cl$.

Finally rates of chemical reactions are investigated, and the concepts of growth, decay, and half-life are applied to these reactions. Differences between chemical and nuclear reactions are stressed.

Topic B21 (chemical equilibria) referred to the use of radioactive tracers to investigate a system in equilibrium, and this is followed up in section B22.4.

Topic A24 contains approximately the same content as this Topic.

Handbook for teachers, Chapter 24 gives further information on radiochemistry and Chapters 20 and 21 on rates of reaction.
Nuffield Physics: the fifth Year includes a study of the properties of radiation. A wide range of experiments based on a standard kit and closed sources is described (see *Teachers' guide 5*).
Nuffield Secondary Science *Theme 7 Using materials*, Field of study 7.5, provides a comprehensive treatment of all aspects of radioactivity, including health hazards.
Nuffield Advanced Chemistry: brief references to radiochemical techniques appear in Topics 12 and 14 *(Teachers' guides I and II)*. An advanced treatment of reaction rates occurs in Topic 14.
Nuffield Physical Science: radiochemical techniques are used in Sections 3, 12, and Option G2. The treatment of reaction rates is divided between Section 8 and Option G2. (See *Teachers' guides I, II,* and *III*.)
Nuffield Advanced Physics Unit 5 *Atomic structure*.

Modern Physical Science Report No. 6 (ed. T. A. H. Peacock) (1971)
Radiochemistry (Association for Science Education/John Murray), provides a further range of experiments. It should be noted that most of these involve sources for which approval from the Department of Education and Science must be obtained.
Faires, R. A. (1970) *Experiments in radioactivity*. Methuen. A useful book for teachers who require more information and suggestions for practical work.
Hornsey, D. J. (1974) *Radioactivity and the life sciences*. Methuen. This book is written for the non-specialist and provides up-to-date information on current uses of radioactive isotopes in the life sciences.
Lewis, J. L. and Wenham, E. J. (1970) *Radioactivity*. Longman Physics Topics. This book is intended as a reader for pupils following a physics course and contains useful illustrations of cloud chamber tracks.

Film loops
2–13 'Catalysis in industry'
2–15 'Radioactive materials – uses'

Films
'Catalysis' ICI Film Library
Several films and film strips are featured in the current catalogue issued by the Education Division of The Chemical Society.

Overhead projection original
28 The thorium decay series *(figure A24.1)*

Handbook for pupils, Chapter 8 'Radioactivity'.
Chemists in the world, Chapter 1 'Finding out about the atom' and Chapter 2 'Using ideas about the atom'.
Curie, E. (1938) *Madame Curie*. Heinemann

Safety precautions to be followed in this Topic are outlined below. More information is given in Department of Education and Science (1976) *Safety in science laboratories* (HMSO); and in Department of Education and Science Administrative Memorandum 2/76 (17 May 1976) 'The use of ionizing radiations in schools, establishments of further education, and teacher training colleges' and the associated explanatory booklet, issued by the Establishment and Organization Branch, Department of Education and Science, Elizabeth House, York Road, London SE1 7PH.

In this Topic the teacher uses a Geiger-Müller liquid counter and a scaler, which are the most suitable instruments for use with low-activity sources. It should be remembered that a GM liquid counter will not normally respond to α-radiation.

Scintillation plates are available from Panax Equipment Limited, but since their radioactive paint contains a phosphor they should be stored in the dark for at least twenty-four hours before use.

All radioactive solutions and apparatus for use with these solutions should be kept on a plastic tray lined with absorbent paper. The tray should be placed on a bench which is protected by paper.

Liquids should be transferred by using a teat pipette set aside for this purpose and solids separated from liquids in a centrifuge.

Protective gloves must be worn, and test-tube holders (for example, wire loops or special forceps) used to hold test-tubes.

Solutions should be kept in stoppered vessels. If these bottles are fitted with droppers, the droppers must be long enough to allow for their easy withdrawal.

Centrifuge tubes should never be more than two-thirds full of liquid, to allow for the insertion of a dropper (with a depressed bulb) for withdrawal or mixing, so as to avoid spilling the contents of the tube.

Non-radioactive solutions and reagents should be kept apart from radioactive materials. The 'radioactive tray' should contain a stoppered vessel for radioactive waste.

The hazards of radiation are best explained to pupils when they see the precautions adopted. Exposure to radiations above certain levels, and especially to γ-radiations, causes internal damage. Protective clothing must be worn and mouth-controlled operations (such as with bulb pipettes) *must be avoided*, so that long-term contact with radiation sources is eliminated.

B22.1
Introduction to radiochemistry

In this section the teacher demonstrates various phenomena associated with radiochemistry. A variety of ways of detecting 'invisible rays' are used.

Pupils should be familiar with the idea that there are many different kinds of invisible radiation which can be detected in a variety of ways. Mention may be made of radiant heat, ultra-violet light, and radiowaves. They may also be aware that radiations from radioactive materials can be detected in different ways.

1. Awareness of the role of chance in scientific discovery and the work of Becquerel in radioactivity
2. Knowledge of the types of radiation emitted by radioactive substances
3. Awareness of the techniques needed for measuring radioactive radiations

Becquerel (in 1896) was interested in Roentgen's work on the discovery of X-rays. He found quite by accident that certain substances gave out a radiation which fogged photographic plates in the complete absence of light. Subsequently it was found that this new form of radiation was highly penetrating and also ionized air. In 1898, one of Becquerel's students – later to become Madame Curie – named this phenomenon radioactivity.

Experiment B22.1a

Becquerel's chance discovery

Apparatus

The teacher will need:

Box of photographic plates or X-ray film (e.g. Kodirex general purpose X-ray film from Kodak)

Access to a dark room

Coin or key

Uranium nitrate

Thorium nitrate

Developing solution and trays

Procedure

The X-ray film is wrapped inside some black paper. Place a key or coin on the paper which is also sprinkled with some powdered uranium nitrate or thorium nitrate. About 2 or 3 spatula measures will be needed. Leave the package in a cupboard for several days. When the film is developed it should be sufficiently 'fogged' to show up the shadow of the key or coin.

Next pupils examine a scintillation plate – which contains a radioactive material – and go on to compare it with some luminous paint or with the dial of a wrist-watch which has luminous figures.

Experiment B22.1b

An examination of scintillations

Apparatus

Each pair of pupils will need:

Experiment sheet 101

Scintillation plate* (taken from the packet and kept in the dark for *at least* 24 hours)

Magnifying lens

Dark room

Specimen of non-radioactive luminous paint

Procedure

It will be necessary for the teacher to explain the procedure to be followed: part of the lesson will be in artificial light and part in the dark. Non-radioactive luminous paint gives off light in the dark because of delayed emission of absorbed light. Note that scintillation plates must be stored, *and* issued for use, in the dark. *Experiment sheet* 101 gives further information.

Alternatively the experiment may be demonstrated. The teacher may use an optical microscope with slide platform and base inside a dark box covered with a piece of black cloth. A scintillation plate and slide treated with a non-radioactive luminous paint should be placed on the covered slide platform. After focusing onto specimens using a low magnification, allow the pupils to view the specimen.

> **Experiment sheet 101**
> *Note.* Most of this experiment must be done in the dark when you will not be able to read these instructions, so you must try to remember what you have to do.
>
> When the lights are on, put a specimen of non-radioactive luminous paint in front of you and observe it through a hand lens. Continue your observation when the lights have been turned off.
>
> Does the light from the paint fade as time goes on?

*Available from Panax Equipment Ltd, Willow Lane, Mitcham, Surrey CR4 4UX

Is the light given off all over the surface or from some parts only?

Next, you will be given, *in the dark*, a scintillation plate, which contains radioactive material. Put it on the bench next to the non-radioactive paint and look at both through the lens.

Is light given off from the scintillation plate?

Is the light given off all over the surface or from some parts only?

If you have a watch with luminous figures, look at them through the hand lens. Does the paint on them contain radioactive or non-radioactive material?

If your watch dial behaves like a scintillation plate, let your teacher know: it may be useful in a later experiment.

When the pupils have completed the experiment, restore normal laboratory illumination and collect the apparatus. During the discussion of the results, refer to the everyday experience of the production of light by the effect of radiation on the screen of a television set. Next explain the function of the radiation source on the scintillation plate and the role played by zinc sulphide or barium cyanoplatinate phosphor. Make the distinction between fluorescence and phosphorescence: with phosphorescence but not with fluorescence, the emission of radiation persists for some time after the illuminating radiation has been cut off.

Describe (and perhaps demonstrate) a simple diffusion cloud chamber, such as that used in Nuffield Physics (years 1 and 5). If pupils are unfamiliar with cloud chambers, it will be necessary to explain how they work. The 'seeding' of saturated vapour to form liquid droplets is caused by charged particles, and 'vapour trails' can be seen. The effect of electric and magnetic fields on α- and β-particles may also be shown, and used to distinguish them.

Charged particles, such as the β-particles, can be used to make gases conduct. This is in fact used in the GM tube (see Experiment B22.1c). It should be noted that α-particles cannot usually be detected using a GM liquid counter tube, due to their short range.

Experiment B22.1c

Apparatus
The teacher will need:

Scaler and GM liquid counter

Distilled water

2 measuring cylinders, 25 cm³

Beaker, 100 cm³

Stop-clock

'Radioactivity tray' (see page 587)

Polythene gloves

(*Continued*)

Detecting radiation produced by solutions
The experiment MUST be done by the teacher.

Procedure
Standard radiochemical precautions must be observed (see page 586).

The purpose of this experiment is to give some experience in the use of a scaler and GM tube (GM liquid counter), and to find out something about the radioactivity of potassium salts and uranium salts. (The limitations of the GM tube should be described beforehand.) The scaler is first switched on and adjusted to the recommended voltage for the GM tube.

Count in turn over a period of 100 seconds the activity from each of the following solutions: 4M hydrochloric acid; a mixture of equal volumes of 4M hydrochloric acid and 4M potassium hydroxide which has been cooled to room temperature (i.e. a prepared 2M

Bottle for waste liquid

6 test-tubes, 100 × 16 mm

Test-tube rack

4M potassium hydroxide

4M potassium carbonate

Saturated potassium hydroxide solution

Saturated potassium chloride solution

0.1M uranyl nitrate

4M hydrochloric acid

0.1M lead nitrate

potassium chloride solution); 4M potassium hydroxide; saturated potassium chloride; saturated potassium hydroxide. Wash out the GM liquid counter several times with distilled water between each measurement to remove the highly caustic liquid.

It should be clear from the results that the activity is due to the potassium ions, and that the more potassium ions there are, the greater the activity. There is only a very low count from the hydrochloric acid, due to background activity. The origin of background activity may be mentioned and used to lead to the need to compare results from radioactive sources with those from non-radioactive material.

Ask the class to predict the activity of 4M potassium carbonate solution. Check their prediction by experiment. From the data obtained using saturated solutions, derive some idea about the solubility of potassium chloride and potassium hydroxide.

From 0.1M uranyl nitrate solution prepare 0.05M, 0.25M, and 0.01M solutions. Take a background count using 0.1M lead nitrate solution. This illustrates the need to use similar non-radioactive systems for comparisons. Count the most dilute uranyl nitrate solution first and follow with the others in an *ascending* order of concentration. Rinse out the counter with the next solution in the series before taking measurements on a given solution.

The results will confirm that the count rate is proportional to the concentration of the radioactive ions. 'Twice as many uranium or potassium ions give twice as much radiation.'

By the end of this section, pupils should understand the principle of the GM liquid counter, and be aware of the effects of radiation on a photographic plate, and on a supersaturated vapour. They should be able to explain the effect seen in a scintillation plate. In addition, pupils should know that potassium and uranium ions give off radiation, that there are different types of radiation, and that the count rate is proportional to the number of radioactive atoms or ions in the counter tube.

Suggestion for homework

Find out what you can about the following discoverers of important aspects of radioactivity: Becquerel, Curie, Rutherford, Geiger, Wilson, Joliot-Curie. (See *Chemists in the world*, Chapter 1, 'Finding out about the atom'.)

B22.2
The study of a radioactive element

A short history of the discoveries of Madame Curie and her husband leads up to the idea that a radioactive element, while emitting radiation, is slowly transformed into another element. Sometimes the new element is itself radioactive. In this section, pupils study protactinium chloride and isolate protactinium from uranium and thorium by solvent extraction. Growth and decay curves may be plotted and the concept of half-life introduced.

Objectives for pupils

1. Awareness of the contribution of the Curies to the early history of radioactivity
2. Knowledge of the use of solvent extraction to separate a mixture of similar compounds
3. Awareness of the role of a comparative experiment in a scientific investigation
4. Knowledge of the growth and decay of activity in radioactive materials
5. Knowledge of the concept of half-life

After Becquerel had discovered that invisible rays are given off from chemicals containing uranium atoms (or ions), the Curies noted that pitchblende, one of the chief ores of uranium, had a greater activity than was expected from the uranium it contained. This result showed that the ore contained an element of even greater radio-activity than uranium. By using chemical means, two such elements were isolated; polonium and radium. About that time, Marie Curie and G. C. Schmidt independently showed that thorium compounds were radioactive. These discoveries led to a systematic study which showed that some 40 elements emit characteristic radiations quite spontaneously.

Pupils who want to know more about the earlier discoveries could read the book *Madame Curie* by Eve Curie (see Reading material for the pupil).

These investigations led to the idea that when an atom gives off radiation, the atom is changed into another atom of a different element. The product of this change may or may not be radioactive. In the next experiment, solvent extraction is used to isolate protactinium from uranium and thorium. This method of separation depends on using a difference in physical properties of chemically similar substances.

Having given some emphasis to this problem of isolating one substance from a mixture of substances with very similar properties, pupils can be asked to apply the idea of solvent extraction to a mixture of two very similar chlorides; barium chloride and calcium chloride. Butan-1-ol may be used to dissolve calcium chloride from an aqueous solution containing both barium and calcium chlorides.

Experiment B22.2a may be carried out by the teacher while pupils attempt Experiment B22.2b. Alternatively some teachers may prefer to deal with these experiments sequentially.

Apparatus

The teacher will need:

Scaler and GM liquid counter

Stop-clock

'Radioactivity tray' (see page 587)

Separating funnel, 50 cm³

Measuring cylinder, 25 cm³

Beaker, 100 cm³

(Continued)

Isolation of protactinium chloride, containing $^{234}_{91}$Pa, and measurement of its change in activity over a period of time
This experiment MUST be done by the teacher.

Procedure
Standard radiochemical precautions must be observed (see page 586).

Fill the liquid counter tube with amyl acetate (pentyl ethanoate) and use this to obtain the background count. Dissolve 1 g of uranyl nitrate in 3 cm³ distilled water. Add 7 cm³ concentrated hydrochloric acid. (This changes the molarity of the solution to 7M with respect to hydrochloric acid.) Transfer this solution to a separating funnel and shake it with 10 cm³ of amyl acetate for half a minute. Remove the lower aqueous layer. Transfer 5 cm³ of the amyl acetate layer to the liquid counter. This contains most of the pro-

Uranyl nitrate

Amyl acetate (pentyl ethanoate)

Concentrated hydrochloric acid

Distilled water

tactinium. Start counting immediately. Because it takes a finite time to make and record an observation, two people are needed for this experiment. A pupil may be asked to record the count at 15, 30, 45, etc., seconds. Thus, the first count occurs at 15 seconds and is used to calculate the value for the count-rate half-way through that period (i.e. at 7.5 seconds). The second count is at 30 seconds and is used to calculate the count-rate at 15 seconds, and so on.

The two layers can be shaken together and the experiment repeated using either one of the layers. The aqueous layer leads to a growth curve for the formation of protactinium-234.

A discussion of the results of this experiment is deferred until after Experiment B22.2b.

Experiment B22.2b

Solvent extraction, using a non-radioactive mixture

Apparatus

Each pair of pupils will need:

Experiment sheet 120

Test-tube, 150 × 16 mm, fitted with a cork

2 test-tubes, 150 × 16 mm

Test-tube rack

Teat pipette, to reach three quarters of the way down a test-tube (e.g. approximately 75 mm)

Beaker, 100 cm³

Watch-glass, to rest on top of beaker

Tripod and gauze

Bunsen burner and heat resistant mat

Platinum or nichrome wire for flame tests

Concentrated hydrochloric acid

Butan-1-ol, 10 cm³

Potassium chromate solution (10% *w/v*), 6 cm³

1M acetic acid (ethanoic acid), 6 cm³

0.2M barium chloride solution, 8 cm³

0.2M calcium chloride solution, 8 cm³

(Continued)

The preferential extraction of calcium chloride from a mixture containing barium chloride and calcium chloride in aqueous solution using butan-1-ol demonstrates the role of solvent extraction in separating mixtures of chemicals of a similar nature.

Procedure
See *Experiment sheet* 120.

Experiment sheet 120
In this experiment you will use a solvent (butan-1-ol) to extract calcium chloride preferentially from a solution containing calcium chloride and barium chloride.

Try out a test for barium ions in solution by putting about 2 cm³ barium chloride solution into a test-tube. Add 2 cm³ dilute acetic acid (ethanoic acid) and 2 cm³ potassium chromate solution. Mix the solutions by shaking gently. What happens?

Repeat this same test using calcium chloride solution in place of barium chloride solution. What happens?

Next, try out a test for calcium ions in solution. Transfer 2 cm³ calcium chloride solution to a *clean* test-tube, add 0.5 cm³ ammonium sulphate solution and 2 cm³ ammonium oxalate solution. Mix the solutions by shaking gently. What happens?

The solvent extraction experiment results in the transfer of calcium chloride from the water to the butan-1-ol. Put 10 cm³ of the mixed solution into a test-tube and add an equal volume of butan-1-ol. Cork the test-tube and shake it gently for *about 30 seconds*. Set the test-tube aside for several minutes until two layers separate. Remove the cork. Use a teat pipette to transfer as much of the upper-layer (butan-1-ol) as possible to a *clean* test-tube, *and* to a watch glass. Use the lower (water) layer for other tests.

Tests on the butan-1-ol layer
1. Evaporate about 2 cm³ of the butan-1-ol layer to dryness on a watch glass heated by steam from boiling water in a beaker. Record your findings.

2. Test 2 cm³ of the butan-1-ol layer with acetic acid (ethanoic acid) and potassium chromate solution. What happens?

3. Test 2 cm³ of the butan-1-ol layer with ammonium sulphate solution and ammonium oxalate solution. What happens?

Mixture of equal volumes of barium chloride and calcium chloride solutions, 10 cm^3

Saturated ammonium sulphate solution, 1 cm^3

Saturated ammonium oxalate solution, 2 cm^3

Tests on the water layer

1. Test 2 cm^3 portion of the lower layer with acetic acid (ethanoic acid) and potassium chromate solution. What happens?

2. Test 2 cm^3 portion of the lower layer with ammonium sulphate solution and ammonium oxalate solution. What happens?

Which layer contains barium ions?

What does this tell you?

What happens to the calcium ions? Are they completely or only partly transferred to the butan-1-ol layer?

When pupils have completed Experiment B22.2b, discuss their results before discussing the results of Experiment B22.2a. Provide pupils with the necessary data from B22.2a and ask them to present the data graphically, plotting counts per second against time. Sample results are shown in figure A24.3.

Relate the decay of protactinium-234 to the decrease in the number of source atoms with time. Use the graph to show how half-life is calculated. It is important to stress that the total number of atoms (active and successor atoms) in the liquid remains constant unless some are lost through conversion to the gaseous state or to the solid state.

It will be necessary to point out that the count rate gives no clue as to the type of radiation emitted or to its penetrating power. Remind the pupils about the limited range of α-particles (and, in this case, of the low energy of γ-radiation), and of the fact that GM tubes cannot be used to detect α-particles.

(*Note.* If the teacher wishes to carry out or discuss a second example of isotope separation and half-life determination, Experiment A24.2b may be used at this point.)

By the end of this section, pupils should know that physical and chemical methods may be used to isolate the products of radioactive decay of uranium or thorium salts. They should know that the radiation from radioactive materials will change with time. They should also be able to calculate the half-life of a given radioactive element.

Suggestion for homework   Read *Handbook for pupils*, Chapter 8 'Radioactivity'

B22.3
Relating products and radiations to the source of activity

An account is given of the work of Crookes, Rutherford, and Soddy, and mention is made of the thorium decay series.

Objectives for pupils

1. Knowledge of the differences between α, β, and γ-radiation
2. Knowledge of the contributions of Crookes, Rutherford, and Soddy to this branch of science
3. Knowledge of the decay pattern for thorium

E. Rutherford (1899) studied the penetrating power of radiation originating from radioactive materials. He identified two types of radiation: α-rays which were the more readily absorbed and β-rays which were not and which also produced stronger ionization. The influence of magnetic fields on these rays was also studied and led to the view that β-rays were negatively charged and α-rays were positively charged. A little later, P. Curie (1900) showed that part of the radiation from radioactive materials was not influenced by magnetic fields (i.e. that some radiation carried no charge), and later that year P. Villard showed that these rays had exceptionally high penetrating power. This third form of radiation was called γ-radiation.

During the period 1900–09, a series of independent investigations revealed that α-rays consisted of helium atoms carrying two units of positive charge, and that β-rays were fast-moving electrons. Rutherford and C. da C. Andrade in 1914 showed that γ-rays were electromagnetic radiations analogous to X-rays.

Sir William Crookes made many notable contributions. He discovered thallium (1861); investigated the discharge through gases and somewhat later the properties of cathode rays; contributed to our knowledge of the chemistry of the rare earths; and discovered thorium-234 or uranium-X (1900). This last discovery was made by adding ammonium carbonate solution to a solution of uranium salt until the precipitate first formed almost completely redissolved. The residue was uranium-X. Becquerel was able to show in 1901 that uranium-X contained almost the whole of the radioactivity of a sample of uranium whereas the uranium salt taken up in solution was almost inactive. His studies lasted some weeks and he found that the uranium in solution gradually regained its initial activity.

This unexpected result led Rutherford and F. Soddy (1902 onwards) to the discovery that uranium broke down to form a large number of products. Their problem was then to identify these substances. In 1903, they put forward a *theory of radioactive disintegration* to account for their observations. This suggested that a radioactive atom disintegrated spontaneously (unlike the atoms of other elements) with the emission of α- or β-particles and the formation of atoms of other elements which differed both physically and chemically from the parent atom. They suggested that this disintegration process could occur several times until a stable atom was produced.

This theory enabled their chemical separation experiments to be explained and led in 1913 to the wide generalization (by Fajan, Russell, and Soddy) known as the *group displacement law*: when a radioactive change occurs in which an α-particle is emitted, then the product atom is displaced two places to the left of the parent atom in the Periodic Table; but if a β-particle is emitted, then the product atom is displaced one place to the right of the parent atom. Soddy went on to coin the term *isotope* for those elements which occupied the same position in the Periodic Table (i.e. with the same atomic number) and with the same chemical properties but with different atomic masses.

It is convenient to indicate the significance of atomic number (proton content Z) as the means of identifying an element; and to mention the relation between atomic number and mass number (number of protons + neutrons, A) which they may have met in their first study of periodicity through non-radioactive isotopes such as chlorine-35 and chlorine-37.

To return to the earlier theme, other work revealed that radioactive isotopes possess specific half-lives. The thorium decay series (figure A24.1) illustrates how a variety of separately determined facts can be accommodated by the periodic law.

Assuming the theory of the nuclear atom, the group displacement law is consistent with the suggestion that α- and β-particles emitted during radioactive decay originate from the atomic nucleus. This is in harmony with the independent observation that the rate of emission of these particles is independent of the state of combination of the atom or of the physical conditions used. The thorium decay series may be used to illustrate these ideas:

$$^{232}_{90}\text{Th} \longrightarrow {}^{228}_{88}\text{Ra} + {}^{4}_{2}\text{He} \, (t_{\frac{1}{2}} = 1.41 \times 10^{10} \, \text{years})$$

$$^{228}_{88}\text{Ra} \longrightarrow {}^{228}_{89}\text{Ac} + \text{e}^{-} \quad (t_{\frac{1}{2}} = 6.7 \, \text{years})$$

$$^{228}_{89}\text{Ac} \longrightarrow {}^{228}_{90}\text{Th} + \text{e}^{-} \quad (t_{\frac{1}{2}} = 6.13 \, \text{hours})$$

and so on. Half-lives for the various nuclides may be mentioned to illustrate the big differences in time needed for the changes to occur.

B22.4
Uses of radioactive materials

Some discussion of uses of radioactive materials is followed by an experiment to illustrate the use of radioactive 'tracers'.

Objectives for pupils

1. Knowledge of the use of radioactive isotopes in chemistry and biochemistry
2. Application of radioactive tracer techniques to solve chemical problems, such as the study of chemical equilibrium

Some of the work in this section could be left for pupils to do as a library project but considerable help must be given to direct their studies before starting.

Alternatively, we might start with the use of these materials as sources of energy.

Natural radioactivity leads to the formation of a more stable nucleus and to the release of energy, some in the form of fast-moving particles and some as electromagnetic radiation. This occurs over a long period of time. A fast – almost instantaneous – conversion of matter into energy can be achieved by the breakdown of very heavy atoms. This 'fission' process was originally used in the so-called atomic bomb but is now being exploited for the generation of electricity. Another type of nuclear reaction is 'fusion' in which larger nuclei are formed from smaller ones with the release of energy.

This has been used in the 'hydrogen bomb' and is being investigated as a commercial source of energy.

Radioactive materials can also be used as 'tracers' – in much the same way as the addition of a dye might be used to trace the course of a particular supply of water – and as sources of radioactive particles. Some uses are mentioned in Alternative IIA Topic A24. Others are given below.

1. In the control of dredging in the Thames. The use of labelled sand to follow sand bank shifts in the river has been used to make dredging more efficient (see film loop 2–15 'Radioactive materials – uses').

2. Studying engine wear. A car engine component can be made radioactive by neutron bombardment in a nuclear reactor. If it is now fitted into an engine, the amount of wear can be continuously monitored by measuring the radioactivity in the oil.

3. To detect pipeline leaks. A radioactive solution is passed down the pipe followed by inactive materials. A survey of the outside of the pipe with a GM counter will soon show where the radioactive material has leaked out. If the isotope used decays to give harmless products and has a short half-life, there will be no permanent contamination of the surrounding ground.

4. To help medical diagnosis. A patient with suspected thyroid trouble can be given a small amount of radioactive iodine. How this iodine becomes distributed in the body allows the doctor to see if the thyroid gland is functioning properly.

Some of the most important uses of radiation from radioactive material occur in the field of medicine. γ-rays in particular can harm living cells, especially those that divide very quickly. Hence the use of isotopes such as cobalt-60 and radium-226 for tumour therapy. γ-radiation is also used for the sterilization of surgical dressings and instruments.

In industry, γ-radiation can also be useful. Just as an X-ray photograph will show up the faults and cracks in a bone, so a γ-ray photograph will show up faults or cracks in a weld or metal cast. β-radiation can be useful for measuring thickness, the amount of radiation absorbed being related to the amount of material through which it has passed. Thus, a small β-source and detector can be used to monitor the thickness of paper, or to check the level of the liquid in a cylinder, or to make sure that cartons of soap powder are properly filled.

Not all the uses mentioned above need to be discussed. It is better to deal only with those which can be illustrated, perhaps by use of the Nuffield film loop 2–15 'Radioactive materials – uses'. These suggestions may give some idea of the variety of different uses to which radioactive materials may be put. Pupils should be able to find many more examples.

In the last part of this section, a radioactive tracer is used to study the equilibrium between lead chloride and its saturated solution. The subject of dynamic equilibrium was raised in the previous Topic.

We want to find out if there is any evidence for both forward and backward reactions occurring when the system is in equilibrium. We attempt to find an answer to this question by labelling some of the reacting particles. The equilibrium between a solid and its saturated solution enables us to do this simply. Lead chloride solution in contact with precipitated lead chloride is chosen because the lead can be labelled with a radioactive isotope, lead-212.

Experiment B22.4

Apparatus

The teacher will need:

Beaker, 100 cm^3

Beaker, 500 cm^3

Stand, boss, and clamp

2 test-tubes, 100 × 16 mm

Teat pipette

Mechanical stirrer or shaker

Bunsen burner and heat resistant mat

Tripod and gauze

Scaler

GM liquid counter

Centrifuge and test-tubes, 100 × 16 mm

Tray lined with absorbent paper

Polythene gloves

Talcum powder

Thorium nitrate solution (5%)

Lead nitrate solution (5%)

2M hydrochloric acid

Acetone (propanone)

Distilled water

Investigating the equilibrium between solid lead chloride and its saturated solution using a radioactive tracer technique
This experiment MUST be done by the teacher.

Procedure
First, make a count of the background radiation. Make a saturated solution of lead chloride by adding just enough 2M hydrochloric acid to about 5 cm^3 of 0.2M lead nitrate solution to precipitate most of the lead. Shake well, centrifuge the mixture, and transfer 5 cm^3 of the solution to the GM tube. Take a count over a period of a few minutes.

To about 10 cm^3 of 0.1M thorium nitrate solution in a beaker, add 5 cm^3 of lead nitrate solution, and then precipitate lead chloride by adding hydrochloric acid. This precipitate will contain the radioactive isotope lead-212. Centrifuge the mixture and add the precipitate to some saturated lead chloride solution. Shake or stir mechanically for about twenty minutes. Centrifuge the mixture and transfer the clear liquid to the GM tube. Take a count over a period of a few minutes. The liquid will now be active: this is due not to the dissolution of pure lead chloride (the solution was saturated) but to the exchange of lead ions between the solution and the solid. The count rate will be found to be several times greater than that obtained with the original saturated lead chloride solution.

Miscellaneous experiments

The equilibrium between the element iodine, iodide ions, and triodide ions may be investigated using iodine-131. For this purpose, it is convenient to use a 1.5 μC potassium iodide tablet as a source of iodine-131. It will be recalled that iodine dissolves in tetrachloromethane but that iodide ions are insoluble in this solvent. So, if an aqueous solution of an iodide is shaken with iodine in tetrachloromethane solution some of the iodine will pass to the aqueous phase and react to form a triodide. The equilibrium:

$$I_2(s) + I^-(aq) \rightleftharpoons I_3^-(aq)$$

may then be investigated.

Rather similar to the lead chloride experiment (B22.4) is that using solid silver chloride and a solution of silver nitrate. The silver chloride is labelled with silver-110.

These and other experiments require a teacher to possess a license from the Department of Education and Science to use radioactive isotopes. The two experiments mentioned come from Faires, R. A. (1970) *Experiments on radioactivity*. Methuen.

Write a summary of the work done in this section.
(*Note. Chemists in the world,* Chapter 2 'Using ideas about the atom', provides a basis for this work.)

B22.5
Studies of rates of reaction

The second of two aspects of rate processes in chemistry, reaction rates, is discussed in this section and earlier work on measuring rates of reaction is revised. Provision is made for supplementary studies.

A suggested approach

Objectives for pupils

1. Awareness of two sorts of rate processes in chemistry (i.e. reaction rates and radioactive growth and decay)
2. Ability to investigate the rate of a reaction

Remind pupils of the two types of rate processes which they have studied so far: radioactive growth and decay, in earlier parts of this Topic, and rates of reaction, in Topic B13. To find conditions for an equilibrium to be set up in a chemical system it is necessary to use the idea of dynamic equilibrium and hence the matching of reaction rates. Thus, when hydrogen reacts with iodine to form hydrogen iodide, we may write:

$$H_2(g) + I_2(g) \longrightarrow 2HI(g) \qquad \text{(reaction A)}$$

and when hydrogen iodide decomposes to form hydrogen and iodine, we write:

$$2HI(g) \longrightarrow H_2(g) + I_2(g) \qquad \text{(reaction B)}$$

In due course and if the systems are closed, the concentration of the reactants in both cases will change. Indeed, the rates of these reactions at the same temperature and under comparable conditions will balance and we will be left with two systems in equilibrium:

$$H_2(g) + I_2(g) \rightleftharpoons 2HI(g)$$

The removal of any one of these reactants and/or products will upset such an equilibrium and the system will adjust itself to accept the change in conditions. This is obviously worth further exploration at a later stage and is fully discussed in the *Handbook for teachers*, Chapter 22.

This introduction is intended to confirm the impression that a system in equilibrium arises from opposing but equal reaction rates. Remind pupils that the rate of a chemical reaction is dependent upon a variety of factors. During Topic B13 the factors which were shown to influence the rate of reaction included:
1. the concentration of the reactants;
2. the presence or absence of catalysts;
3. the temperature.

It may be useful to revise these generalizations briefly by referring back to the various systems studied in Topic B13 before going on to further investigations. Remind pupils that such factors do *not* influence radioactive changes, only chemical changes. Indeed it may be necessary to emphasize the basic distinction between these two types of rate process by referring to the fact that radioactive changes are derived from changes in the nucleus whereas chemical processes relate only to changes in the electronic structure of the atom.

Experiment B22.5a

Investigations on chemical systems

Apparatus

Each pair of pupils will need:

Experiment sheet 121

4 test-tubes, 150×16 mm

Test-tube rack

Test-tube holder

3 teat pipettes

Bunsen burner and heat resistant mat

Hydrogen peroxide solution (2 volume), 6 cm³

Potassium iodide solution (5%), 5 cm³

Dilute sulphuric acid, 2 cm³

Acetone (propanone), 1 cm³

Iodine solution, 5 cm³

2M sodium hydroxide, 2 cm³

Procedure

The object of these investigations is to enable pupils to explore new systems and to relate their findings to earlier work on reaction rates. Details are given in *Experiment sheet* 121.

Experiment sheet 121

In separate test-tubes make two mixtures of equal volumes (about 2 cm³) of hydrogen peroxide solution and potassium iodide solution. What happens?

To one test-tube add 10 drops sodium hydroxide solution. Shake the mixture. What happens?

To the other test-tube add 10 drops dilute sulphuric acid. Shake the mixture. What happens?

Repeat the above procedure using a mixture of iodine solution (about 4 cm³) and acetone (propanone) (5 drops). When the alkali or acid has been added, warm each mixture gently.

What happens to the alkaline mixture?

What happens to the acidic mixture?

After the results of these experiments have been discussed in class, write notes below of any points of interest that they indicate.

After the experiment, the findings should be summarized. The first system is relatively simple to explain. The reaction between iodide ions and hydrogen peroxide under acid conditions may be expressed as:

$$H_2O_2(aq) + 2I^-(aq) + 2H^+(aq) = I_2(s) + 2H_2O(l)$$

Pupils find that iodide ions and hydrogen peroxide react somewhat slowly in the absence of hydrogen ions. The addition of acid affects the rate of reaction. The teacher should demonstrate the effect of adding starch to the system. This reaction is investigated in greater detail in the next experiment.

The formation of iodoform from acetone (propanone) under alkaline conditions should have been detected by the pupils and no further explanation of this reaction is required. The odour of iodoform is characteristic.

An investigation of the reaction between iodide ions and hydrogen peroxide in solution

Apparatus

Each pair of pupils will need:

Experiment sheet 122

Burette, $50 \times 0.1 \text{ cm}^3$

Burette stand

Stop-clock

750 cm^3 flask containing 50 cm^3 0.1M potassium iodide and 10 cm^3 1% starch solution diluted to 250 cm^3 – *solution A*

500 cm^3 flask containing 100 cm^3 3.5M sulphuric acid and 25 cm^3 0.05M hydrogen peroxide diluted to 250 cm^3 – *solution B*

200 cm^3 flask containing 100 cm^3 of 0.1M sodium thiosulphate solution – *solution C*

Procedure

See *Experiment sheet* 122. It is essential to label the three solutions clearly.

Experiment sheet 122

You will be provided with the following solutions in separate flasks:
Solution A containing 50 cm^3 0.1M potassium iodide solution and some starch solution in a total volume of 250 cm^3 solution.
Solution B containing 100 cm^3 3.5M sulphuric acid and 25 cm^3 0.05M hydrogen peroxide solution in a total volume of 250 cm^3 solution.
Solution C which is 0.1M sodium thiosulphate solution.

You will know by now that iodide ions react with hydrogen peroxide and hydrogen ions to produce iodine:

$$2I^-(aq) + H_2O_2(aq) + 2H^+(aq) \longrightarrow I_2(aq) + 2H_2O(l) \qquad (I)$$

and that iodine solution gives a blue colour with starch solution. But if we carry out the reaction in the presence of thiosulphate ions, these prevent the appearance of the blue colour by reacting with the iodine formed:

$$2S_2O_3{}^{2-}(aq) + I_2(aq) \longrightarrow S_4O_6{}^{2-}(aq) + 2I^-(aq) \qquad (II)$$

so that the appearance of the blue colour is delayed until all the thiosulphate ions have been used up. We can therefore estimate the rate at which reaction (I) proceeds by finding the time needed for the blue colour to appear in the presence of different amounts of sodium thiosulphate. This can be done as follows.

Fill a clean burette with solution C and adjust the level of the liquid to the 0 cm^3 mark, making sure that the jet is filled with solution. Add 2 cm^3 sodium thiosulphate solution from the burette to the flask containing solution A and shake the mixture. Now add the whole of solution B to this mixture as quickly as possible. Take the time (or start a stop-clock) when about half of solution B has been added. Shake the largest flask while you are pouring in solution B. Note the time when a blue colour appears in the mixture (it happens suddenly so you will have to watch all the time) and enter it in the table below. Keep the stop-clock going and add a further 2 cm^3 of solution C and shake well. The blue colour will disappear. Note the time when it returns again, and add a further 2 cm^3 solution C with shaking. Continue in this way until a total of 10 cm^3 sodium thiosulphate solution have been added.

Total volume of sodium thiosulphate sodium/cm^3	Time from start /sec
0	0
2	
4	

Plot the results on a graph with time on the horizontal axis and volume of sodium thiosulphate solution on the vertical axis.

What is the value for the half-life of the reaction studied?

Experiment B22.5c

The decomposition of hydrogen peroxide: a further investigation

This study is intended to complement Experiment B18.4c or Experiment B13.3c. It is optional and is more likely to interest schools where computer-based materials are in general use.

After reminding the pupils of the practical details of earlier experiments, summarize the general findings. This particular exercise (B22.5c) requires the use of computer-based material which is available to simulate the decomposition of hydrogen peroxide.

Computer simulation is in no way intended to replace practical work but rather to supplement it. Moreover it is important that before using the computer-based material pupils should have carried out the experiments which have been mentioned above, and maybe others. The rationale for the simulation lies in its potential for increasing the flexibility of the ways in which pupils can vary the reaction conditions – such as initial concentration, temperature, and catalyst. Such a wide range of investigations in the laboratory would require a prohibitively large allocation of time. More specifically, the simulation gives pupils a greater chance to design their own investigations, an important aspect of the way in which scientists work.

The computer-based material consists of students' notes, a teachers' guide, and a computer programme, and is one of the units produced by the Schools Council Project 'Computers in the curriculum'. Further information is available from the Centre for Science Education, Chelsea College, Bridges Place, London SW6 4HR.

At the end of this Topic, gather the class together for a discussion of the main findings. By considering simple models of an atom, pupils should be able to account for the two types of rate processes considered. Pupils should be aware of the distinction between radioactive changes and chemical changes.

Suggestion for homework

The rates of chemical reactions are affected by changes of temperature and by the concentrations of reactants, but the rates of radioactive decay processes are not. From your knowledge of atoms, suggest an explanation for the differences between the two types of processes.

Summary

By the end of this Topic, pupils should be aware of the phenomena associated with radioactivity and of the use of radiochemical techniques in chemistry. They should also know how to measure rates of reaction and understand the likely effect of concentration of reactants, temperature, and catalysts on reactions in general terms. By considering simple models of an atom, pupils should be able to suggest an explanation for the differences between the two types of rate processes considered in this Topic. They should be aware of the distinction between a radioactive change (which involves the nucleus of the atom) and a chemical change (which involves the electronic structure and is influenced by such variables as temperature, pressure, and the presence or absence of a catalyst).

Chemicals and energy

1. To use the kinetic theory to explain the three states of matter.
2. To provide some simple techniques for measuring energy change in a system.
3. To measure the energy needed to vaporize liquids.
4. To use heats of vaporization as a means of identifying the structures of materials.
5. To demonstrate energy transformation and the general applicability of the principle of conservation of energy.

B23.1 Energy changes and changes of state
B23.2 Energy changes and chemical reactions
B23.3 Chemicals as fuels
B23.4 Energy transformations and related subjects

About four weeks are needed to cover the main points of this Topic. About six weeks will be needed for an extended treatment.

Energy changes associated with changes of state are measured experimentally and interpreted using kinetic theory. The values of heats of vaporization for various substances are shown to be related to the structure of these substances in many instances.

In the second section, the energy changes associated with chemical reactions are measured, the terms exothermic and endothermic are introduced, and the ΔH notation is also used for the first time. Energy-level diagrams are used to summarize experimental determinations.

Next, the combustion of fuels is considered and in simple cases shown to be related to the composition and structure of the fuel. An elementary account is given of the origins of the heat of a reaction.

Finally, the problem of energy transformation is considered simply and some mention is made of the general applicability of the principle of conservation of energy before considering voltaic and fuel cells as examples of chemical systems undergoing change.

This is given in Topics A15 and A23 where the physical and chemical aspects are treated separately.

Handbook for teachers, Chapters 15, 16, 17, 18, and 19 provide an extensive commentary on this Topic.
Collected experiments, Chapters 16 and 17 offer additional material.
Nuffield Physics *Teachers' guide 1* and *Guide to experiments I*.
Nuffield Advanced Chemistry (Topics 7, 8, 11, 17 in *Teachers' guides I* and *II*) and

Nuffield Physical Science (Sections 4, 6, 7, and Option G1 in *Teachers' guides I* and *III*) provide advance treatments of the themes considered in this Topic.
Dawson, B. E. (1971) *Energy in Chemistry*. Methuen.

Film loops
2–4 'Heating water'
2–5 'Liquid–gas equilibrium'
2–6 'Solid/liquid equilibrium'
2–14 'Energy changes in HCl formation'
2–17 'Solids, liquids, and gases'

Films
'Energy in chemistry (1) Energy levels, latent heats, and heats of reaction' ESSO Films for Science Teachers No. 21
'Energy in chemistry (2) Chemical energy, electrical energy, and work' ESSO Films for Science Teachers No. 22
'Change of state' ESSO
'Molecular theory of matter' ESSO

Reading material

for the pupil

Chemists in the world, Chapter 5 'Energy and chemicals'.

B23.1
Energy changes and changes of state

In this section, we look at the energy changes associated with the conversion of a liquid to a gas, or vice versa. A simple kinetic theory model is used to interpret such changes. Finally, values of heat of vaporization are shown to be related to the structures of elements and compounds.

A suggested approach

Objectives for pupils

1. Knowledge of the energy needed to change the state of a substance
2. Application of kinetic theory to interpret changes of state
3. Relating energy changes and the structures of substances

The teacher will need

Film loops
2–5 'Liquid–gas equilibrium'
2–6 'Solid liquid equilibrium'
2–17 'Solids, liquids, and gases'
Film-loop projector

In Topic B21 we assumed there was a dynamic equilibrium between reactants and products in a reaction. Can we use the concept of dynamic equilibria when studying phase changes? Use film loops 2–5 and 2–6 to show what happens at the atomic level in liquid–gas and solid–liquid equilibria.

From the film loops it appears that when a solid turns into a liquid there is no great increase in the distance between particles, whereas there is a very large increase when a liquid turns into a gas. Compare the volume occupied by a known mass of water and the same mass of steam. At 100 °C, 1 mole of water occupies 18.8 cm^3 whereas 1 mole of steam occupies 30 600 cm^3. We would therefore predict that a large amount of energy will be required to convert a liquid into its vapour at the boiling point.

Simple heat-transfer experiments might now be demonstrated in order to show the need for a system of units with which to measure heat energy changes. One possibility is presented in Alternative IIA, in Experiment A23.1a in which the calorie is introduced. The SI heat unit is the joule which will be used in this Topic. If pupils have used heat units in their physics course, teachers will need to remind them of the meaning of the term joule before starting any practical work.

In Experiment B23.1a an electric kettle is used to convert a known quantity of liquid into its vapour. Quantities of energy are easily measured by means of the energy rating of the kettle. This is usually given in kilowatts and one kilowatt is equivalent to one kilojoule of energy per second. Thus, a 2.5 kilowatt element in an electric kettle supplies energy at the rate of 2.5 kJ per second. By using an electric kettle, we can establish the relationship between the energy supplied per second to the system and its subsequent rise in temperature. This relationship is used to relate the energy supplied to water at its boiling point and the quantity of water converted into steam.

Examining the boiling of water
This experiment should be done by the teacher.

Apparatus

The teacher will need:

Electric kettle (known rating)

Measuring cylinder, 1 dm³

Thermometer, −10 to +110 × 1 °C

Stop-clock

Access to balance

Procedure
Record the volume of water required to cover the electric element inside the kettle. Note the initial temperature of the water. Switch on the power to the kettle and record the temperature of the water at 10 second intervals until it boils. When there is no further change in temperature, remove the thermometer and replace the lid on the kettle to reduce losses through splashing. Allow the kettle to boil for about 5 minutes. Switch off the power and note the total time taken. Allow the kettle to cool.

Meanwhile, ask pupils to display the variation of temperature with time in the form of a graph. Draw attention to the fact that once the water boils, all of the thermal energy is being used to convert water into steam – as indicated by the constancy of this temperature.

Pour the cooled water from the kettle into the measuring cylinder and record the volume of water which remains. Figure B23.1 shows a typical set of results for the experiment.

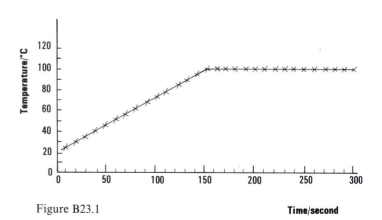

Figure B23.1

Topic B23 Chemicals and energy

Specimen results
Initial volume of water $1\,200\,cm^3$
Final volume of water $800\,cm^3$
Volume of water converted into steam $400\,cm^3$

Since all measurements of volume were carried out at room temperature, and since the density of water is $1\,g\,cm^{-3}$, we can write:

mass of water converted into steam $= 400\,g$
rating of kettle used $\qquad\qquad = 2.75\,kW$
$\qquad\qquad\qquad\qquad\qquad\quad = 2.75\,kJ$ per second
time of boiling $\qquad\qquad\qquad = 300$ seconds
Therefore energy requirement $\quad = 300 \times 2.75\,kJ$
$\qquad\qquad\qquad\qquad\qquad\quad = 825\,kJ$

Thus, $400\,g$ water requires $825\,kJ$ to form steam at $100\,°C$.

How much energy is needed to turn 1 mole of water into steam? One mole of water has a mass of $18\,g$.

$18\,g$ water therefore requires $\dfrac{825 \times 18}{400}\,kJ = 37\,kJ$

The energy required to turn one mole of water into steam (from this determination) is $37\,kJ$. This quantity of energy is the molar heat of vaporization of water and the value obtained is in fair agreement with the published value.

We might now enquire whether this value for the molar heat of vaporization of water differs from that for other compounds (Experiment B23.1b).

With some classes, the data obtained from this simple demonstration can be used to determine the specific heat capacity for water, that is, the energy needed to raise $1\,kg$ water through $1\,°C$. This constant is required in section B23.2. Alternatively reference back to this calculation may be made later.

Thus, from the specimen results quoted:
Energy needed to raise $1.2\,kg$ water from $20\,°C$ to $100\,°C$
$\qquad = $ (time taken to heat water from $20\,°$ to $100\,°$) $\times 2.75\,kJ$
$\qquad = 150 \times 2.75\,kJ$

Specific heat capacity of water $=$ energy needed to raise $1\,kg$ water
through $1\,°C$

$$= \frac{150 \times 2.75}{1.2 \times 80} \approx 4.3\,kJ$$

| Experiment B23.1b | **Measuring the energy required to vaporize a given liquid** |

Note. If enough immersion heaters and joule meters are available, this experiment can be a class investigation using a variety of organic liquids with boiling points of less than $120\,°C$. Alternatively, the teacher may prefer to use the experiment as a teacher demonstration with pupils helping.

Apparatus

The teacher or each pair of pupils will need:

Experiment sheet 123

Handbook for pupils

Joulemeter

Immersion heater, 12 V
(Nuffield Physics; item 75)

Power supply, 12 V

Test-tube with side arm,
150×25 mm

Block of expanded polystyrene,
$9 \text{ cm} \times 9 \text{ cm} \times 21 \text{ cm}$ (see
figure B23.2)

Delivery tube

Liebig condenser

2 lengths of connecting tubing

Measuring cylinder, 25 cm^3

Beaker, 100 cm^3

2 clamps, bosses, and stands

Connecting wire

One of the following: methanol,
ethanol, benzene, cyclohexane,
carbon tetrachloride
(tetrachloromethane)

Procedure

See *Experiment sheet* 123.

Figure B23.2

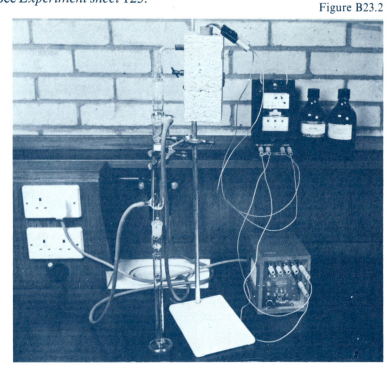

Experiment sheet 123
The apparatus for this experiment is shown below.

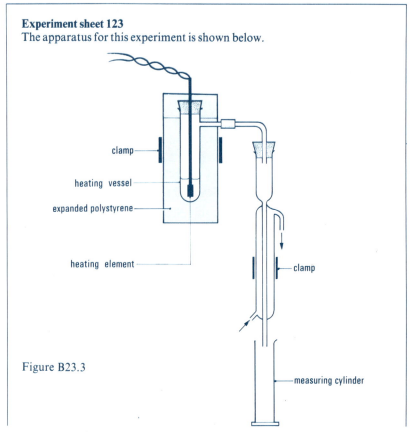

Figure B23.3

Into the heating vessel pour enough of the liquid to cover the heating element. Make sure that the stopper carrying the element is firmly inserted. Replace the measuring cylinder by a small beaker and switch on the heater. Allow the liquid to boil and to distil into the beaker until no drops of liquid form in the glass tube connecting the heating vessel to the condenser. Now replace the beaker by the measuring cylinder and read the joulemeter. When $10 \, cm^3$ of liquid have distilled read the joulemeter again and switch off the heater. Record your readings in the table below.

Liquid used:

First joulemeter reading .... J

Second joulemeter reading .... J

Energy used in vaporizing liquid .... J

Volume of liquid collected .... cm^3

Density of liquid (tables) .... $g \, cm^{-3}$

Mass of liquid collected:

.... × .... = ... g

Formula of liquid used: ....

Molar mass of liquid .... g

.... g liquid required .... J for vaporization

1 mole of liquid requires

$$\frac{.................................... \times ....}{.................................... \times 1000} \, kJ \qquad \text{for vaporization}$$

Molar heat of vaporization of

.... is .... $kJ \, mol^{-1}$

Typical results obtained using this apparatus are:

methanol, 40 kJ; ethanol, 48 kJ; benzene, 36 kJ; cyclohexane, 36 kJ; carbon tetrachloride (tetrachloromethane), 34 kJ; water, 41 kJ.

After discussing the results with pupils, refer them to the *Handbook for pupils*. The Reference section lists a number of heats of vaporization. It will soon be apparent that those substances which the pupils know to have a giant structure possess a higher heat of vaporization per mole than those which have a molecular structure.

Ask whether pupils think boiling point is a reliable guide to structure. Refer them again to the Reference section of the *Handbook for pupils*. For homework they could plot a graph of boiling point against heat of vaporization per mole for a range of representative chemicals, and offer an interpretation of their findings.

Or if time allows, complete this work in class and use the question set for A15.3 for homework.

If time is available, this section can be extended by adapting section A15.2 to cover energy changes needed to melt solids.

Finally, recapitulate the work done in this section by using the film loop 2–17 'Solids, liquids, and gases'.

B23.2
Energy changes and chemical reactions

In this section, heats of reaction are calculated from experimental data and energy-level diagrams are introduced as reaction summaries.

Objectives for pupils

1. Familiarity with examples of chemical reactions in which measurable energy changes occur
2. Understanding energy-level diagrams and the ΔH convention
3. Knowledge of the terms exothermic and endothermic
4. Familiarity with simple calorimetric procedures for measuring energy changes
5. Ability to calculate the heat of reaction from experimental data

When a chemical reaction takes place, heat energy can be released to or absorbed from the surroundings. This has been mentioned from time to time and it is the purpose of this section to explore and interpret such energy changes.

It is convenient to provide a few qualitative demonstrations of the energy changes which accompany chemical reactions. Some reactions which release energy to the surroundings could be demonstrated, such as:
1. The reaction of an acid with a base – such as the reaction of hydrochloric acid with sodium hydroxide, or of acetic (ethanoic) acid with calcium hydroxide.
2. The addition of concentrated sulphuric acid to water. (Use a beaker of water, add the acid slowly, and stir the mixture continuously.)
3. The addition of a limited quantity of water to a sample of freshly prepared anhydrous copper sulphate.

Systems which absorb energy from the surroundings include the following:
4. The solution of ammonium nitrate crystals in water. (Add the salt to a beaker of water and stir continuously.)
5. The addition of calcium chloride solution to sodium carbonate solution.
6. The mixing of molar proportions of barium hydroxide octahydrate with ammonium thiocyanate.

We need a convention to show whether heat is given out or absorbed. '$-$heat' or '$-q$' is an obvious convention and can be added to an equation on the products side when the energy change leads to a lowering of the total energy content of the reactants and a corresponding release of energy to the surroundings until the products of reaction lose energy and return to room temperature. The convention '$+$heat' or '$+q$' as an addition to the products is less simple to follow unless the teacher makes the point that to restore the products to room temperature heat needs to be added to the system from the surroundings. The overall energy content of the products is higher than that of the reactants.

A simple form of energy-level diagram assists the presentation of these points. Explain that for each system shown, heat is either given out or taken in. Ask whether the products of a specific reaction possess more or less energy than the reactants before displaying a diagram of the type shown in figure A23.2.

Next introduce the need for a calorimeter so that heats of reaction can be measured. The following points should be made.

1. The underlying assumption behind these studies is the principle of conservation of energy.

2. Heat losses should be minimal. This can be achieved by using small reaction vessels made from polythene or polystyrene. Both of these materials conduct heat badly.

3. Pupils already know that 1 kg of water requires 4.2 kJ to warm water 1 °C (see calculation based on Experiment B23.1a). Most solutions of not more than 1.0M concentration possess approximately the same specific heat capacity. Therefore if 1 mole of material is allowed to react in 1 dm^3 solution and the temperature rise is t °C, then the energy release is 4.2t kJ.

4. The energy change associated with a reaction can be written into the equation. We use the equation to represent mole quantities of reactants or products. The symbol H is used to represent the energy of a chemical and ΔH to represent the change in energy that has taken place, where:

$$\Delta H = H_{\text{final materials}} - H_{\text{initial materials}}$$

We regard the energy change from the viewpoint of the reactants so if, in changing, they warm up the calorimeter and its contents then energy is being lost and ΔH is negative.

It is important for pupils to appreciate that such a system does not transfer energy until it starts to cool down. An energy-level diagram is useful in conveying such an idea. The dilution of sulphuric acid, for example, may be represented by the equation:

$$H_2SO_4(l) + aq \longrightarrow H_2SO_4(aq, 1.0M): \ \Delta H = -71.4 \, kJ$$

or by the energy-level diagram shown in figure B23.4.

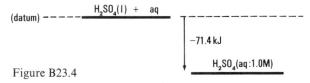

Figure B23.4

The concentrated sulphuric acid and water contain more energy than does the 'cold' diluted acid. The products of reaction cool down by releasing energy to the surroundings.

In the case of the solution of ammonium nitrate crystals in water, there is a drop in temperature. To restore to the original temperature, energy has to be absorbed from the surroundings, and so the ammonium nitrate solution at the end of the process has *more* energy than had the crystals and water from which it was formed at the same temperature (see figure B23.5).

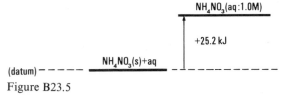

Figure B23.5

To measure the energy change which accompanies a chemical reaction: three examples:
To measure the heat of reaction of a neutralization reaction
To measure the heat of reaction of an ion combination reaction
To measure the heat of reaction for a displacement reaction

Apparatus

Each pair of pupils will need:

Experiment sheets 94, 95, and 96

Polythene bottle, 70 cm³, fitted with a bung and thermometer, -10 to $+110 \times 1\,°C$

Measuring cylinder, 25 cm³

2 beakers, 100 cm³

Access to stock solutions which have been kept overnight in the laboratory:

Experiment sheet 94 requires:

2M sodium hydroxide, 25 cm³

2M potassium hydroxide, 25 cm³

2M hydrochloric acid, 25 cm³

2M nitric acid, 25 cm³

Experiment sheet 95 requires:
0.5M sodium chloride, 25 cm³

or 0.5M potassium chloride, 25 cm³

or 0.5M ammonium chloride, 25 cm³

0.5M silver nitrate, 25 cm³

Experiment sheet 96 requires:

0.2M copper(II) sulphate

Zinc powder, 0.5 g in a specimen tube

or iron powder, 0.5 g in a specimen tube

Procedure

When using polythene bottles, the thermometer should be fitted into the bung such that its bulb just protrudes from the inside of the bung. Temperature readings are made with the bottle and thermometer inverted. Details are given in *Experiment sheets* 94, 95, and 96. Variations in the chemicals shown in these sheets may be tried (see list of chemicals).

Experiment sheet 94

In this experiment you are going to mix equal volumes of 2.0M solutions of sodium hydroxide and hydrochloric acid and find the temperature change during the reaction. Your teacher will tell you the actual volume of each that you are to use.

Measure the appropriate volumes of acid and alkali into separate beakers, and take the temperature of each solution.

Pour the alkali into the calorimeter (a polythene bottle) and add the acid, stir the mixture well, and record the highest temperature reached.

Results

Volume of each solution used .... cm³

Temperature of acid .... °C

Temperature of alkali .... °C

Average temperature .... °C

Highest temperature of mixture .... °C

Calculate the heat of reaction per mole of reactants as follows.

Since equal volumes of solutions are used, and since each solution is 2M, the concentration of each reactant in the mixture will be .... M.

Hence if the heat of reaction (ΔH) is measured in kJ,

$\Delta H = -(\text{temperature change}) \times 4.2 \text{ kJ}$

$\quad = -(\text{...............................} - \text{...............................}) \times 4.2 \text{ kJ}$

$\quad = \text{...............................} \text{ kJ}$

Write the equation for the reaction, and include the ΔH value that you have calculated.

Experiment sheet 95

The reaction whose heat of reaction you are going to measure is

$Ag^+(aq) + Cl^-(aq) \longrightarrow AgCl(s)$

You will be provided with 0.5M solutions of silver nitrate (source of $Ag^+(aq)$ ions) and of a soluble chloride (source of $Cl^-(aq)$ ions).

Write the name of the chloride solution you use here:

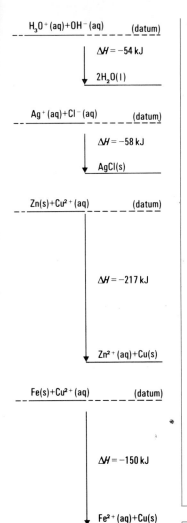

Figure B23.6

Put 25 cm³ (measuring cylinder) of the silver nitrate solution into a polythene bottle (the calorimeter). Insert the rubber bung and the thermometer and take the temperature of the solution; you may have to invert the bottle to make sure that the thermometer bulb is immersed in the solution. Wash the measuring cylinder with distilled water and use it to add 25 cm³ of the chloride solution to the silver nitrate solution. Stopper the bottle and shake it well. Watch the thermometer and record the highest temperature reached.

Results

Volume of 0.5M silver nitrate used   25 cm³

Volume of 0.5M .... used   25 cm³

Initial temperature   .... °C

Final temperature   .... °C

Temperature rise   .... °C

Total volume of solution used   50 cm³

Assuming that this has the same specific heat capacity as water, 4.2 joule $g^{-1}\,°C^{-1}$, and a mass of 50 g:

Heat evolved during reaction $50 \times 4.2 \times$ .... joules

Number of moles of silver chloride precipitated

$$\frac{25 \times 0.5}{1000} = 0.0125 \text{ mole}$$

Heat evolved if 1 mole of silver chloride was precipitated:

$$= \frac{\text{................................} \times \text{................................}}{\text{................................}} \text{joules}$$

$$= \text{................................ kilojoules}$$

Therefore,

$$Ag^+(aq) + Cl^-(aq) \longrightarrow AgCl(s): \quad \Delta H = \text{................................ kJ}$$

Experiment sheet 96

You will measure the heat of reaction for either:

$$Cu^{2+}(aq) + Fe(s) \longrightarrow Cu(s) + Fe^{2+}(aq)$$

$$\text{or } Cu^{2+}(aq) + Zn(s) \longrightarrow Cu(s) + Zn^{2+}(aq)$$

The procedure is, in general, the same as that outlined in Experiment 95, except that the second reagent to be added is a solid. Use 25 cm³ 0.2M copper(II) sulphate solution, measure its temperature, add about 0.5 g of the metal used (iron or zinc powder), shake the mixture, and record the highest temperature reached.

Make up your own table of results and calculate the heat of reaction. Assume that the specific heat capacity of the copper(II) sulphate solution is 4.2 joule $g^{-1}\,°C^{-1}$.

After the experiments, discuss the results obtained. Ask the pupils to construct energy-level diagrams for the reactions studied. Figure B23.6 shows energy-level diagrams to scale (for comparative purposes).

1. The heat of a coke or coal fire comes *chiefly* from the reaction:

$$C(s) + O_2(g) \longrightarrow CO_2(g)$$

Draw an energy-level diagram showing this change. You are told that on burning 1 g of pure carbon, 1 dm^3 of water is warmed though 7.8 °C. Your diagram should show the energy change for the reaction.
a. Cross out the unsuitable word.
The burning of carbon is an endothermic/exothermic reaction.
b. Complete the following:
A mole of pure carbon weighs ... g.
c. ΔH for the reaction is ... kJ (to the nearest kJ).
d. How much coal (at least) would be needed to heat water for a bath? You may assume that 22 gallons (equivalent to 100 dm^3) of water are needed and that water must be heated from 15 °C to 35 °C.
e. In (*d*) above it states 'at least'. In fact, much more than the minimum quantity is required. Suggest a reason for this statement.
2. Glucose is sometimes called 'bottled sunshine'. The equation for the photosynthesis of dissolved glucose and gaseous oxygen from carbon dioxide and water is:

$$6CO_2(aq) + 6H_2O(l) \longrightarrow C_6H_{12}O_6(aq) + 6O_2(g) : \Delta H = 2830.8 \text{ kJ}$$

a. What is the equation for the oxidation of glucose in living cells?
b. What is the ΔH for this oxidation?
c. What happens to the energy liberated?
d. Draw an energy-level diagram to illustrate the way living things depend for their energy on the energy from the Sun.

B23.3
Chemicals as fuels

After a brief discussion of the energy changes which take place during reactions, attention is focused on combustion processes. A method for measuring the heat evolved when different substances burn in air is worked out and the results are compared.

Objectives for pupils

1. Familiarity with examples of energy changes which accompany chemical reactions
2. Skill in determining heats of combustion
3. Ability to analyse an experimental procedure
4. Ability to interpret heats of combustion in simple terms

The lesson may be opened by reviewing examples of chemical reactions which are accompanied by a noticeable transfer of energy. Some examples (such as combustion of fuels or explosives) can be discussed and displayed on the blackboard or on an overhead projection transparency. It may be useful to compare such examples with the systems studied in the previous section.

The most common way of producing energy or heat is to burn a fuel in air (for example, oil or coke or coal). Point out that oxygen has to be supplied if the reaction is to take place and that boilers, furnaces, ovens, grates, and so on are designed to do this.

Ask the class to list some common fuels, to state their uses and why they are used in particular circumstances.

Cost, the amount of heat liberated per unit of fuel, handling problems, and effects on the environment are some of the more important factors to be mentioned. Matters of topical or local interest can add a further dimension to the discussion. The need for oxygen in the combustion of fuel can be brought out in a discussion of spacecraft fuels – for example, the supply of liquid oxygen, which has to be transported as well as the fuel itself.

How much energy is given out when fuels are burnt? In Experiment B23.3, we are going to measure the heat produced by a fuel by allowing the energy released to heat up a known mass of water and measuring the temperature rise. What fuel should be used? In this case the best fuels are those that burn easily and 'cleanly'. Four alcohols are chosen.

Since the fuels are liquids, the mass burned can be found by weighing a suitable 'spirit-lamp' burner before and after the experiment. Tests can be devised to identify the products of combustion. For comparative purposes, the same type of container must be used by different groups of pupils.

To measure the heat of combustion of a liquid fuel

Procedure
See *Experiment sheet* 97.

Experiment sheet 97
You can get an approximate value for the heat evolved when a known mass of a liquid burns in air by burning it in a spirit lamp which is used to heat a known amount of water. The class will investigate a number of different liquids in this way; your teacher will tell you which you are to use.

Write its name here:

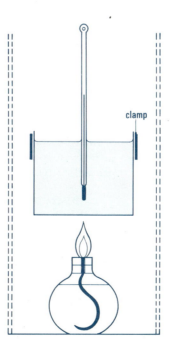

clamp

Figure B23.7

Support a metal can containing 300 cm^3 water in a clamp on a retort stand and place a thermometer in the water. Put some of the liquid in a spirit lamp, place it under the metal can, and adjust the height of the can so that there is a space of 3 cm between the top of the spirit lamp and the bottom of the can. Arrange to shield the flame of the lamp from draughts by using asbestos sheets.

Find the mass of the spirit lamp and liquid as accurately as you can. Replace it below the metal can. Take the temperature of the water in the can. Light the spirit lamp. Stir the water with the thermometer and watch the temperature; when it has risen about 30 °C, extinguish the lamp and record the highest temperature reached. Find the mass of the spirit lamp and remaining liquid as quickly as possible. Why?

Results
Volume of water used   300 cm^3

Mass of water used .... g

Mass of lamp plus .... before burning .... g

Mass of lamp plus .... after burning .... g

Mass of .... burnt   .... g

Initial temperature of water .... °C

Final temperature of water .... °C

Rise in temperature .... °C

Calculation
Number of joules of heat given to the water

$$300 \times 4.2 \times \text{..................} = \text{....................... joules}$$

Formula for .... is ....

Mass of 1 mole of .... is .... g

.... g .... gives out .... joules of heat when burnt

1 mole of .... gives out

$$\frac{\text{........................} \times \text{........................}}{\text{........................}} \text{........................ joules}$$

$$= \text{........................ joules}$$

Heat of combustion of .... is

.... joule mol^{-1}

$= $ .... kilojoule mol^{-1}

State any sources of error which you think might be present in this experiment.

This procedure yields consistent results which are usually about two-thirds of the accepted value. It will be found helpful to compare values for the various heats of combustion with the accepted values (see Table 23.1) and to ask why pupils' results are different. Does all of the liberated heat warm the water? What effect does this have on their results? A more efficient procedure may then be devised and the apparatus for Experiment A23.3b shown to the class. Solids, such as carbon, sulphur, food, and coal, as well as liquids, can be burned in this apparatus. Several versions of it are available from suppliers, although home-made versions made of metal and fitted with some form of heat insulation shield can be more effective.

If time allows, Experiment A23.3b can be demonstrated.

Compound	Molecular formula	ΔH /kJ mol^{-1}	Increment /kJ
methanol	CH_3OH	-726	—
ethanol	CH_3CH_2OH	-1367	641
propan-1-ol	$CH_3CH_2CH_2OH$	-2017	650
butan-1-ol	$CH_3CH_2CH_2CH_2OH$	-2675	658
pentan-1-ol	$CH_3CH_2CH_2CH_2CH_2OH$	-3323	648
hexan-1-ol	$CH_3CH_2CH_2CH_2CH_2CH_2OH$	-3976	653

Table B23.1 Heats of combustion of alcohols

Table B23.1 supports the view that increments in the heats of combustion of a series of related compounds – such as alcohols – are closely related to increments in composition. It seems that the methylene group, $-CH_2-$, contributes a definite quantity of energy to each ΔH value. Indeed, the teacher may go so far as to suggest that each C—H, C—C, C—O, and O—H (or, in other cases, other bonds present) must also have a specific quantity of energy associated with it. The presentation of these ideas can be assisted by the use of molecular models for each of the compounds listed.

A simple interpretation of the origins of energy change in both endothermic and exothermic reactions may now be put forward.

Figure B23.8
See figure A23.6 for a simple interpretation of the origins of energy changes for
(a) exothermic reactions and
(b) endothermic reactions.

Figure B23.8 suggests that these changes could be calculated from the knowledge of the energies required to make or break the various bonds. The determination of bond energies need not concern pupils at this stage. (The significant point is *not* in discussing detail but rather the idea conveyed by figure B23.8.) It needs to be shown to be no more than an application of the principle of conservation of energy.

If we concentrate part of the discussion on a simple chemical system – such as the burning of hydrogen and chlorine to form hydrogen chloride – then a detailed analysis of the process will show that:

The energy of a chemical	=	energy of bonding, holding the atoms together in the particles	+	energy of movement of the particles, giving them kinetic energy (which is proportional to temperature)

The teacher will need

Film loop 2–14: 'Energy changes in HCl formation'

Film-loop projector

and that when the reaction occurs, hydrogen and chlorine molecules must be broken apart into atoms – a process which requires energy – and that atoms must recombine – a process which gives out energy. The film loop 2–14 'Energy changes in HCl formation' may be used to summarize the discussion. However, pupils should understand the limitations of these diagrams. They do *not* imply that first all the hydrogen atoms have to be dissociated, that then all the chlorine atoms have to be dissociated, and lastly that the atoms recombine. A study of the rate of reaction suggests that a few chlorine molecules are dissociated into atoms and then they react with a few hydrogen molecules and the energy liberated enables a few more . . . etc. Nevertheless, however the reaction happens, the book-keeping of energy changes as done above, and also the generalization shown in figure B23.8, must hold.

Write an account of the origins of the energy liberated in a chemical change such as the synthesis of hydrogen chloride.

B23.4
Energy transformations and related subjects

In this section, the transformation of energy from one form to another is considered, together with some related subjects.

Objectives for pupils

1. Familiarity with examples of energy transformation
2. Awareness of the general applicability of the principle of conservation of energy
3. Knowledge of voltaic and fuel cells

Investigations so far in this Topic have been concerned with the transfer of thermal energy, which we have observed either as a heating or a cooling of a mass of water. Energy can be transferred in forms other than heat and we should ask pupils whether chemical change can be used to produce mechanical energy, or radiation energy (for example, light), or electrical energy. Thermal energy from a chemical reaction, such as combustion, can be used to produce steam which in turn can be used to drive a steam engine: or the steam can be used to drive a turbine which is connected to a dynamo to produce electrical energy. This can be used to do mechanical work through an electric motor, or to produce heat from an electric fire, or radiant energy in the form of light from a fluorescent tube. The question which remains to be investigated is 'Can these other forms of energy be derived directly from chemical changes?'

The combustion of a piece of magnesium draws attention to the fact that not all chemical changes produce thermal energy alone. In this case, an intensely bright light accompanies the reaction as well.

It is worth while pointing out that some highly specialized living organisms can produce light from chemical reactions without the evolution of heat – for example, the firefly and certain deep water fish. 'Can the reverse process of light energy promoting chemical reaction take place?' Photosynthesis in plants to convert carbon dioxide and water into sugars, and the photographic process, may be mentioned.

Can chemical reactions be used to provide mechanical energy? Pupils may suggest rockets. Remind them that they have seen a chemical reaction doing work in a much less spectacular way when they saw a gas being generated and used to push the piston of a gas syringe against friction and atmospheric pressure.

Can chemical reactions be used to provide electrical energy? Electrical energy requires the flow of electrons and so we need to consider a process where there is an electron transfer. The reaction used in Experiment B23.4a involves the transfer of electrons when zinc is added to copper sulphate solution.

Experiment B23.4a

Transfer of electrons

Apparatus

The teacher will need:

Porous pot

Beaker, 100 cm³

2 lengths of connecting wire

Milliammeter, range 0–500 mA

Copper foil electrode

Zinc foil electrode

1.0M copper(II) sulphate solution

1.0M zinc sulphate solution

Procedure
Connect the electrodes to the milliammeter, the zinc electrode to the negative terminal and the copper electrode to the positive terminal. Transfer copper(II) sulphate solution to the porous pot and zinc sulphate solution to the beaker. Stand the porous pot in the beaker. Insert the copper electrode into the copper(II) sulphate solution and the zinc electrode into the zinc sulphate solution. The milliammeter will register a current. If the cell is left assembled for some time, the zinc will be seen to be used up.

This experiment shows that a chemical reaction can be used to produce electrical energy. Both the zinc foil and the copper sulphate solution are being used up and only a small flow of electricity is being produced. We can interpret the reaction: ·

$$Zn(s) + Cu^{2+}(aq) + SO_4^{2-}(aq) \longrightarrow Zn^{2+}(aq) + SO_4^{2-}(aq) + Cu(s)$$

as two half reactions:

$$Zn(s) \longrightarrow Zn^{2+}(aq) + 2e^-$$
and
$$Cu^{2+}(aq) + 2e^- \longrightarrow Cu(s)$$

By arranging for the electrons to be transferred outside the reaction vessel we can detect this electrical effect.

Tell pupils that this arrangement is known as a Daniell cell.

In the Daniell cell, the zinc electrode is immersed in a solution of zinc sulphate contained in a porous pot, and this half of the cell is dipped into copper(II) sulphate solution, contained in a copper can which is usually used as the other electrode. The porous pot prevents the zinc sulphate solution from mixing with the copper(II) sulphate solution.

Next, pupils examine a series of simple energy transformations and construct some voltaic cells.

Experiment B23.4b

Using chemical sources of electrical energy

Apparatus

Each pair of pupils will need:

4 SP2 batteries – preferably mounted as a 'battery board'

1.25 V bulb

Bulb holder

(*Continued*)

Procedure
The object of this experiment is to show that electrical energy from a dry battery – such as an SP2 cell – can be used to light an electric bulb or drive an electric motor which can then raise a weight. Details are provided below.

(*Note.* Commercially available SP2 cells are more effective than Daniell cells for this type of experiment. Some explanation of the choice of dry cells for this demonstration will be needed.)

Electric motor and associated
solar cell

Length of cotton and small
weight

Electric motor

Connecting wire

Apparatus

Each pair of pupils will need:

Experiment sheet 98

2 beakers, 100 cm³

Simple voltaic cell (see figure
B23.9)

Filter paper

Glass rod

Emery paper

Access to high resistance
voltmeter (0–3 V)

Metal foil electrodes – zinc,
copper, silver, lead, and
magnesium

1.0M copper(II) sulphate
solution, 25 cm³

1.0M zinc sulphate solution,
25 cm³

1.0M silver nitrate solution,
25 cm³

1.0M magnesium sulphate
solution, 25 cm³

1.0M lead nitrate solution,
25 cm³

Saturated potassium nitrate
solution, 10 cm³

1. Use an SP2 cell to light the pea bulb. Use the light from this bulb
to activate a solar cell and then use this cell to drive a small electric
motor.

2. A cotton thread should be tied around the driving shaft of an
electric motor and a small weight attached to the other end of the
thread. Connect SP2 cells to drive the motor which will then lift the
small weight.

To measure the e.m.f.s of some voltaic cells

Procedure
Figure B23.9 shows the construction of a simple voltaic cell. Pupils
should construct one or more such cells and record their e.m.f.s
using a high resistance voltmeter. Full details are given in *Experiment sheet* 98. However, the teacher may need to introduce the convention for drawing cells (for details see A23.4) before the experiment is started.

Experiment sheet 98

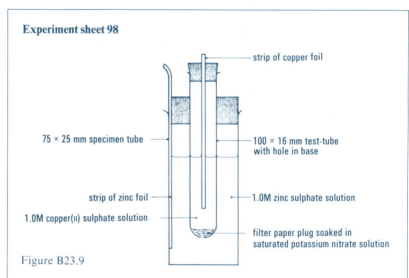

75 × 25 mm specimen tube

strip of copper foil

100 × 16 mm test-tube
with hole in base

strip of zinc foil

1.0M zinc sulphate solution

1.0M copper(II) sulphate solution

filter paper plug soaked in
saturated potassium nitrate solution

Figure B23.9

The method of preparing the cell shown in the diagram is described here.
You may be asked to prepare one containing different metals and solutions,
but the procedure is the same in all cases. Clean the surface of the metal strips
with emery paper.

The solutions which form part of the two electrode systems in the cell must
be kept from mixing and for this purpose a plug of filter paper soaked in
saturated potassium nitrate solution is used. Tear a filter paper into small
pieces and immerse them in saturated potassium nitrate solution. Remove
some of the pulp thus obtained, squeeze it with your fingers, and place it in
the bottom of the inner vessel of the cell (remove this inner from the outer
vessel to do this). Compress the filter paper into a layer about 1 cm thick
with a stout glass rod.

Two-thirds fill the outer vessel with 1.0M zinc sulphate solution, add a strip
of cleaned zinc foil (about 10 cm long) and insert the cork holding the inner
tube, so that the foil is held between the cork and the wall of the outer tube.
Pour 1.0M copper(II) sulphate solution into the inner tube, so that the levels
of liquid in both tubes are the same. Add a strip of cleaned copper foil (about
10 cm long) to the inner tube and hold it between the two halves of a split
cork.

To measure the e.m.f. of the cell connect the two pieces of foil to a high resistance voltmeter using wire and crocodile clips. You will have to find by trial and error which foil has to be connected to the positive terminal of the meter and which to the negative terminal.

Repeat the measurements using cells prepared by other members of your class. Record the results in the table below.

Negative electrode system	Positive electrode system	e.m.f. of cell /V
$Zn(s)\mid 1.0M\ Zn^{2+}(aq)$	$1.0M\ Cu^{2+}(aq)\mid Cu(s)$	

After the pupils have completed their measurements, discuss their findings and relate them to earlier work on the reactivity series (see Topics A6 or B10 in Stage I).

Pupils are next asked to construct a fuel cell which illustrates another way in which chemicals may be converted into sources of energy.

Experiment B23.4d

Apparatus

Each pair of pupils will need:

Experiment sheet 100

Electrolysis cell, with carbon electrodes (see figure B23.10)

2 test-tubes, 75×10 mm

d.c. supply, 4–6 V

2 leads fitted with crocodile clips on one end

Access to a high resistance voltmeter (0–3 V)

1.0M sodium hydroxide, 50 cm³

A very simple hydrogen/oxygen fuel cell

Procedure
Details are given in *Experiment sheet* 100. The gases absorbed on the carbon electrodes provide the reactants for this cell reaction. Some carbon dioxide may be produced during the electrolysis but this is absorbed by the alkali.

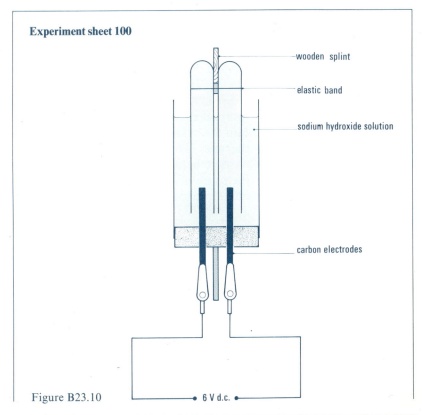

Experiment sheet 100

wooden splint

elastic band

sodium hydroxide solution

carbon electrodes

Figure B23.10   6 V d.c.

We are now going to try to 'burn' hydrogen in a cell, that is to make hydrogen and oxygen combine in an arrangement that will produce a voltage, or a source of electrical energy which will do work. This is known as a 'fuel cell'.

Set up the apparatus as shown in the diagram. Separate the two tubes with a wooden splint resting on the top of the wide glass tube, as shown above. Electrolyse the solution until *both* tubes are filled with gas. Now disconnect the d.c. supply and put a high-resistance voltmeter across the carbon electrodes.

What does it read? .... volt

Write equations:
1. For the reactions at the electrodes during electrolysis:

anode ....

cathode ....

2. For the reactions when the system is acting as a fuel cell:

positive electrode ....

negative electrode ....

The experiment shows one way in which scientists have been trying to increase the proportion of energy available as useful work from fuels. The more usual sequence such as fuel – steam boiler – steam engine or turbine – dynamo – electric motor, wastes a lot of energy. By converting energy from the oxidation of a fuel directly into electricity, a more efficient use of resources is possible.

Suggestion for homework

Write a summary of the principal ideas you have encountered in this Topic.

Summary

By the end of this Topic, pupils should be aware of the fact that when a chemical change occurs, there is a net energy change in the system. If the net energy change results in the liberation of energy to the surroundings, the process is said to be exothermic, whereas if thermal energy has to be absorbed from the surroundings to maintain the system at room temperature the process is said to be endothermic. These energy changes can be interpreted in terms of particle behaviour. Pupils should be aware of the value and use of fuels and of the various ways of transferring the energy of reaction into electrical energy which, in turn, is capable of doing work.

A second look at the Periodic Table

Timing

About 4 or 5 weeks are needed.

Introduction to the Topic

The main aim of this Topic is to bring Alternative IIB to a close with a piece of work in which various ideas met earlier in different parts of the course are collected together and seen to relate to each other.

In the first section the four elements carbon, silicon, tin, and lead are compared. The intention is to draw out the change in properties from those of a non-metal to those of a metal which occur in a group of the Periodic Table, rather than to learn a great deal of formal chemistry of these elements. (Option 7 'Periodicity, atomic structure, and bonding' of Stage III provides greater detail.)

The transition elements are studied in the second section. The final section rounds off the work of the Topic in a way which is especially useful if the Periodic Table is not studied further in the Stage III Options. It deals with the relationship between the properties of an element and its position in the Periodic Table.

Alternative approach

Topic A13 introduces the Periodic Table in a different way.

Background knowledge

This Topic continues work done in Topic B16.

Subsequent development

Option 7 'Periodicity, atomic structure, and bonding' takes the theme further.

Further references

for the teacher

Handbook for teachers contains many references to periodicity.
Collected experiments, Chapter 6 offers more experiments.
Topic A13 in Alternative IIA provides alternative experimental work.
Nuffield Advanced Chemistry (Topics 2, 4, 5, 6, 16, 19 in *Teachers' guides I* and *II*) and Nuffield Physical Science (Sections 2, 5, 10, and 11 in *Teachers' guides I* and *II*) provide advanced treatments of the Topic.
Greenwood, N. N. (1971) Nuffield Chemistry Special *Periodicity and atomic structure* offers another approach to this Topic.

Supplementary materials

Films
'The Periodic Table Part I: ionic and covalent compounds Part II' Open University. Available from Guild Sound and Vision Ltd.
'The mass spectrometer' (second half) Rank Film Library

Reading material
for the pupil

Handbook for pupils, Chapter 1 'Periodicity'.
Chemists in the world, Chapter 8 'Ceramics and glass'.
Greenwood, N. N. (1971) Nuffield Chemistry Special *Periodicity and atomic structure*.

B24.1
The Group IV elements

How are the elements in a vertical group related to one another? This question was posed in Topic B16 when pupils studied Groups I and VII and the approach used in this section is similar to that used earlier. The properties of carbon, silicon, tin, and lead, together with their oxides, hydroxides, and chlorides, are investigated by the pupils. The properties of germanium are considered at the end of the section so as to allow pupils an opportunity to test predictions based on their knowledge of the other elements in the group.

A suggested approach

Objectives for pupils

1. To appreciate that metallic character increases on going down the central group of the Periodic Table
2. To appreciate the physical and chemical changes that occur as the metallic character of elements increases in going down a group
3. To gain knowledge of the properties of the oxides, hydroxides, and chlorides of the Group IV elements
4. To test pupils' predictions about some of the properties of germanium
5. To appreciate the differences in the structure of Group IV elements

This section follows on naturally from Topic B16. The most obvious way to begin is to recall the conclusions that were reached by the end of that Topic. Pupils should have a picture of a group of metals (Group I, the alkali metals) and a group of non-metals (Group VII, the halogens) in their minds which will help them to appreciate the main point of this section. Group IV is selected because it illustrates the change from non-metallic to metallic properties in going down a group. Remind pupils of the work of Topic B11 to help them to recall the basic (sometimes alkaline) character of metallic oxides and the acidic character of non-metallic oxides. Point out that metallic elements (magnesium or zinc) tend to react with acids whereas non-metallic elements (the halogens) tend to react with alkalis.

Pupils carry out their own experiments on Group IV elements. They should be encouraged to make their own deductions from their results. Pupils should compare the chemistry of individual Group IV elements with elements in Group I and Group VII, which they have already studied. Where the experimental work is relatively simple, pupils could devise their own approach. The three experiments suggested below may be carried out either as separate investigations or as a continuous study.

Experiment B24.1a

Apparatus

Each pair of pupils will need:

Experiment sheet 124
(*Continued*)

The Group IV elements

Procedure
See *Experiment sheet* 124. Before pupils start their investigations, remind them about the precautions needed to prevent damage to ammeters during conductivity tests (use momentary contacts); and the need for caution when using hot concentrated acid or hot alkali.

6 V d.c. supply

6 V bulb and holder or ammeter, 0–5 A

Connecting wire and crocodile clips

6 hard-glass test-tubes, 100 × 16 mm

Test-tube rack

Test-tube holder

Bunsen burner and heat resistant mat

Access to graphite rod, lump of silicon, piece of tin, and piece of lead

Silicon powder, 0.5 g

Granulated tin, 0.5 g

2M sodium hydroxide, 10 cm³

Concentrated hydrochloric acid, 10 cm³

Discuss the experimental results before asking pupils to investigate the properties of some compounds of these elements. The conductivity test does not show a trend because carbon in the form of graphite is used. Otherwise conductivity is seen to increase with increasing metallic character: silicon, (germanium), tin, and lead.

Silicon does not react with water or concentrated hydrochloric acid when used under the conditions indicated. Silicon does however react with aqueous alkali forming hydrogen and sodium silicate.

Tin does not react readily with water or sodium hydroxide solution but does react with hot concentrated hydrochloric acid to form tin(II) chloride and hydrogen.

Experiment B24.1b

The oxides and hydroxides of Group IV elements

Apparatus

Each pair of pupils will need:

Experiment sheet 125

7 hard-glass test-tubes, 100 × 16 mm

Bunsen burner and heat resistant mat

Test-tube rack

Source of carbon dioxide

(Continued)

Procedure
Teachers are advised that pupils may find themselves in difficulty if they begin an experiment with too much of any one of the reactants. It will be seen from *Experiment sheet* 125 that they are asked to precipitate the hydroxide of a Group IV element and to see how it reacts with acid and with alkali. Teachers may wish to demonstrate the technique of encouraging the formation of certain precipitates by scratching the side of a reaction vessel with a glass rod. Other details are given in *Experiment sheet* 125.

0.1M sodium silicate solution, 2 cm^3

0.1M tin(II) chloride, 2 cm^3

0.1M lead(II) nitrate, 2 cm^3

2M nitric acid, 25 cm^3

2M sodium hydroxide, 25 cm^3

Full-range Indicator solution, few drops

Access to centrifuge

Estimate the pH of the final solution.

2. To about 2 cm^3 of sodium silicate solution add an equal volume of dilute nitric acid. Shake the mixture and allow it to stand for 5 minutes. Centrifuge to compact the precipitate of 'silicon hydroxide' at the bottom of the tube, and remove the solution above the precipitate, using a teat pipette. Add about 4 cm^3 distilled water, centrifuge, and remove the water. Transfer half the precipitate to another test-tube.

To about 2 cm^3 of separate portions of tin(II) chloride and lead(II) nitrate solution add sodium hydroxide solution, a drop at a time with shaking, until no more precipitate forms. Centrifuge each precipitate and wash with water, as above. Divide each precipitate into 2 parts occupying separate test-tubes.

Use the samples of 'silicon hydroxide', tin(II) hydroxide, and lead(II) hydroxide that you have prepared to examine the effects of (a) sodium hydroxide solution and (b) dilute nitric acid on them. Record the results in the table below.

Substance used	Action of sodium hydroxide solution	Action of dilute nitric acid
'silicon hydroxide'		
tin(II) hydroxide		
lead(II) hydroxide		

Write a brief account of the relative acidic and alkaline character of the four compounds you have examined.

After this set of experiments, discuss the results obtained. 'Silicon hydroxide' reacts readily with sodium hydroxide, and the other hydroxides are soluble in both sodium hydroxide and dilute nitric acid. Teachers may use the occasion to introduce the term amphoteric.

The next set of experiments requires close supervision or demonstration by the teacher. Each reaction requires discussion and interpretation.

Experiment B24.1c

Apparatus

The teacher or each pair of pupils will need:

3 hard-glass test-tubes

Test-tube rack

Bunsen burner and heat resistant mat

Carbon tetrachloride (tetrachloromethane)

Silicon tetrachloride*

Tin(II) chloride

The chlorides of the Group IV elements

These experiments should only be carried out by pupils under the strict supervision of the teacher. The reactions of silicon tetrachloride *must* be carried out in a fume cupboard and teachers are reminded of the violence of the reaction of silicon tetrachloride with water.

Procedure

Mix equal volumes of carbon tetrachloride (tetrachloromethane) and water and warm the mixture. The two liquids are immiscible and no reaction occurs even when the carbon tetrachloride is seen

*Silicon tetrachloride reacts violently with water forming hydrochloric acid and silica. After use, the stock bottle must be carefully re-sealed to exclude traces of moisture.

to reflux in the test-tube above the mixture. It will be recalled that the low boiling point of carbon tetrachloride is an indication of its molecular nature.

Add a few drops of silicon tetrachloride to some water. Note that the two liquids are immiscible. (This result is not unexpected and also indicates the molecular nature of silicon tetrachloride.) *Now carefully* shake the test-tube: a violent reaction occurs and hydrogen chloride fumes are evolved. Hydrated silicon oxide remains in the test-tube. *It is advisable to carry out this reaction in a fume cupboard!*

Finally, add a crystal of tin(II) chloride to water. In this case, a faintly cloudy solution is formed with very little temperature change.

When these experiments have been completed, and the experimental findings discussed, pupils should be asked to recall the experiment on the electrolysis of lead bromide. They should guess that the halides of tin and lead are ionic. Ask the pupils whether they would have predicted such changes in structure for the halides of the Group IV elements.

The character of the elements will be seen to become progressively more metallic going from carbon to lead. This is supported by the findings on conductivity. Pupils may be reminded of the property of allotropy shown by carbon at this point and some account given for the moderate conductivity shown by graphite. (There is little to be gained by referring to the allotropy of tin.) The structures of the elements could then be revised and the low number of nearest neighbouring atoms of the non-metals (4 in the case of carbon and silicon) contrasted with the high number of nearest neighbouring atoms of the metals (12 in lead).

Further evidence for the change in metallic character is obtained by showing that silicon (like the halogens) reacts with alkalis but not with acids, whereas tin (like a number of other metals) reacts with acids. With an able group of pupils, it could be pointed out that tin will react very slowly with *boiling* sodium hydroxide solution, thus showing that it is not 'fully' metallic. Tin only reacts very slowly with dilute acid, in contrast to silicon which does not react with acid but which reacts with sodium hydroxide solution in the cold.

Pupils met the relationship between metallic nature and the acid–base character of oxides in Stage I and in Topic B11. The experiments on the oxides and hydroxides of the Group IV elements show the change in metallic character in going down a group. Carbon dioxide is obviously an acidic oxide, and silicon 'hydroxide' can be shown to react with alkalis but not with acids, showing its acidic character. The hydroxides of tin and lead react with both acids and alkalis and thus behave more like metals than either carbon or silicon. Remind pupils that such oxides or hydroxides are said to be amphoteric.

Experiment B24.1c dealt with the chlorides of the Group IV elements. Both carbon and silicon tetrachlorides are typical molecular liquids immiscible with water, although silicon tetrachloride subsequently reacts with water to give hydrogen chloride and hydrated silica. The formation of molecular chlorides is typical of non-metals and carbon tetrachloride (tetrachloromethane) was taken as an example of a molecular substance in Experiment B23.1b (heat of vaporization determination). Point out that silicon tetrachloride is not exceptional in its reaction with water. Tin chloride dissolves in water and behaves like sodium chloride. This suggests that it is ionic. In Topic B19, lead bromide was shown to conduct electricity when molten. Thus, the metallic elements of Group IV (tin and lead) seem to form halides that are essentially ionic, whereas the non-metals (carbon and silicon) form molecular halides.

Pupils who have studied Supplement S1.3 could relate the change in metallic character on going down the group to increase in atomic size.

Now ask pupils to try to predict some of the physical and chemical properties of germanium. This enables them to revise one of the problems raised in B16.3. Point out how successful Mendeleev was in predicting the properties of 'eka-silicon'. Compare their predictions with the actual properties of germanium, making use of the *Handbook for pupils*.

By the end of this section, pupils should know that metallic character increases down a central group of the Periodic Table. They should be aware of how this shows itself in the structure of the elements, in the acid–base character of their oxides and hydroxides, and the nature of their halides. Pupils should also know that tin and lead are much less reactive than Group I metals. They should be able to predict the properties of any missing element in a group, given the properties of other elements in that group.

Suggestions for homework

1. Read *Chemists in the world*, Chapter 8 'Ceramics and glass'.

2. *(Library-based projects)*
a. Predict how the elements in Group V (nitrogen, phosphorus, arsenic, antimony, and bismuth) change on going down the group. A data book from your school library will help you to answer this question.
b. Silicon is the second most common element in the Earth's crust: try to find out what silicon-containing compounds there are in the Earth's crust.
c. Germanium is used as a semiconductor. What is a semiconductor? Why is germanium a semiconductor?
d. Herodotus (430 B.C.) called the British Isles the Cassiterides (tin isles). Find out how metallic tin was obtained from the ore first mined in Cornwall.

3. 'Lead pollution is the scourge of the twentieth century.' Find out how lead comes to pollute our atmosphere and our drinking water. What remedies can you suggest?

B24.2
The transition metals

Transition metal chemistry is complicated and it would be inappropriate for pupils to go further than appreciating the relative chemical inertness of the metals, the fact that the compounds and solutions are coloured, and that this colour varies with the environment. Pupils are NOT required to possess a detailed knowledge of the formation of complex ions or the different oxidation states formed by transition metals. (Words such as complex, coordination, and ligand do NOT form part of the course.)

When pupils use the Nuffield Periodic Table, they will notice that transition metals do not occur in main groups. Pupils will be familiar with Group I metals and Group IV elements and so the idea of a 'transition' from one type of reactivity to another can be introduced at the start of this section.

Pupils will know (from Topic B19) that copper forms more than one kind of ion. They will be aware of the same kind of behaviour in the chemistry of iron (see B20.4). They should be told that this behaviour is typical of all transition metals and contrasts with the behaviour of Group I metals. Therefore, the formulae of transition metal compounds cannot be predicted in quite the same way as those for Group I compounds. (Reasons for such behaviour could be discussed with those pupils who have met the ideas of atomic structure.)

Pupils should spend most of this section doing their own experiments. They should be encouraged to compare their results with those already obtained with Group I and Group IV elements.

The transition metals

Procedure

In this experiment, pupils are expected to devise their own approach to answer the questions 'How easily do transition metals react with (a) air and (b) water?' Most of them will probably try to burn the elements on pieces of asbestos paper. The surface of the copper will oxidize fairly easily but surfaces of iron and nickel will only oxidize slowly, in marked contrast to Group I metals. They might try the effect of boiling water on samples of the elements. No reaction will be observed, which again contrasts sharply with Group I metals. Iron filings react with steam (Topic A7) but copper does not.

Objectives for pupils

1. Awareness of the physical properties of transition metals such as density, melting point, and so on
2. Knowledge that transition metal compounds are coloured in solution and that the colour depends on the chemical environment
3. Knowledge that transition metal hydroxides are usually basic and sometimes amphoteric in character
4. Awareness that transition metals often form ions with several different charges

Experiment B24.2a

Apparatus

Each pair of pupils will need:

3 hard-glass test-tubes

Test-tube holder and rack

Bunsen burner and heat resistant mat

Tripod and gauze

Asbestos paper or equivalent material

Copper turnings, 1 g

Copper powder, 0.5 g

Iron nail, one

Iron filings, 1 g

Small pieces of nickel foil

Apparatus

Each pair of pupils will need:

Experiment sheet 126

Handbook for pupils

Test-tube rack

13 hard-glass test-tubes,
100×16 mm

Measuring cylinder, 25 cm^3

Conical flask, 100 cm^3, with cork

Buchner funnel

Filter pump

Filter paper

Bunsen burner and heat resistant mat

Test-tube holder

Beaker, 400 cm^3 (for use as an ice bath)

1.0M cobalt(II) nitrate, 10 cm^3

1.0M nickel (II) sulphate, 10 cm^3

1.0M copper(II) sulphate, 20 cm^3

0.67M chromium(III) sulphate, 10 cm^3

Note: The concentrations of these solutions, with the exception of copper sulphate, are not at all critical.

2M sodium hydroxide, 25 cm^3

2M ammonia solution, 20 cm^3

'0.880' ammonia, 10 cm^3

2M nitric acid, 20 cm^3

Concentrated hydrochloric acid, 10 cm^3

Ethanol, 15 cm^3

Ice (optional)

Procedure

See *Experiment sheet* 126. Pupils should also be encouraged to refer to the *Handbook for pupils*.

Experiment sheet 126

1. To separate portions (about 2 cm^3) of solutions of cobalt(II) nitrate, nickel(II) sulphate, copper(II) sulphate, and chromium(III) sulphate, add sodium hydroxide solution, one drop at a time with shaking, until no more precipitate forms. Centrifuge and wash the precipitates as described in Experiment 125, divide each into two parts, and examine the action of sodium hydroxide solution and dilute nitric acid on them. Record the results in the table below.

Substance used	Action of sodium hydroxide solution	Action of dilute nitric acid

Which of the substances are:
Acidic?
Basic?
Amphoteric?

2. Put 2 cm^3 portions (3 of each solution) of cobalt(II) nitrate solution, nickel(II) sulphate solution, and copper(II) sulphate solution into 9 separate test-tubes. Keep one test-tube of each solution as a control for comparison. To a second set of 3 test-tubes (one of each solution) add concentrated ammonia solution, a few drops at a time with shaking, until no further change is observed. To the third set of 3 test-tubes add concentrated hydrochloric acid, a few drops at a time with shaking, until no further change is observed. Note the colours of all the final solutions in the table below.

Solution used	Colour of solution	Colour after addition of excess ammonia	Colour after addition of excess hydrochloric acid
cobalt(II) nitrate	1	2	3
nickel(II) sulphate	4	5	6
copper(II) sulphate	7	8	9

Now fill each test-tube to within about 1 cm of the top with distilled water and shake. What do you observe?

Account for all your observations.

3. *To prepare crystals of tetrammine copper(II) sulphate*
Measure 10 cm^3 copper(II) sulphate solution (measuring cylinder) into a 100 cm^3 conical flask. Add 2M ammonia solution, a little at a time, with shaking, until the precipitate formed at first dissolves completely to form a deep blue solution. To this solution add an approximately equal volume of ethanol, carefully so that it forms a separate, clear, upper layer. Cork the flask securely and allow it to stand undisturbed for at least 24 hours. Filter off the deep blue crystals which separate, using a Buchner funnel (your

teacher will show you how to do this), wash them with ethanol containing a drop of concentrated ammonia solution, and allow them to dry on filter or blotting paper. Describe the appearance of the crystals obtained.

Heat a small crystal in a dry test-tube. What gas is given off (smell with caution)?

Write the formula of the compound that you have prepared.

When these experiments have been completed, gather the class together and discuss their findings. Pupils should have discovered that transition metals are much less reactive than Group I metals and, in certain respects, are of a similar reactivity to tin and lead. Their densities are also similar to those of tin and lead. However, their melting points are very much higher. Transition metal hydroxides are usually basic and insoluble in water (in contrast to the soluble hydroxides formed by the Group I metals) but there are exceptions to this, such as chromium(III) hydroxide which is amphoteric. The relationship of the acid–base character to the charge on the metal ion (oxidation number) is NOT required at this level. Pupils should also have observed that transition metal compounds are coloured in solution and that the addition of suitable compounds causes the colour to change. The colour changes indicate reactions which are examples of equilibria, and the reversibility of the reactions should be pointed out to the pupils if they are unable to see this clearly.

Finally, pupils should be able to contrast the tetrammine copper(II) sulphate they prepared with hydrated copper sulphate. The different colours of the two solids should convince pupils that it is the ammonia that is responsible for the colour change.

Suggestions for homework

1. Use the Reference section in the *Handbook for pupils* to tabulate the physical properties of any four transition metals, any three Group I metals, and tin and lead. From your table, make generalizations about the physical properties of (a) transition metals; (b) Group I metals; (c) metals like tin and lead.

2. Compare and contrast the chemical behaviour of (a) transition metals; (b) Group I metals; and (c) metals like tin and lead. Suitable properties for including in this study might be the colour of solutions of metal salts, the colour and nature of the oxides and hydroxides formed, the nature and formulae of the halides formed by the elements, and the chemical reactivity of the elements towards air and water.

B24.3
The position of an element in the Periodic Table

This section leads on naturally from the previous two sections and pupils can start by summarizing the properties of elements that they have met in this Topic or in Topic B16.

1. Ability to forecast the general behaviour of an element from the knowledge of its position in the Periodic Table
2. Ability to place an element in an appropriate position in the Periodic Table, given some of its physical and chemical properties

Ask pupils to classify the elements they have met so far. For example, they might use some of the following groupings:
1. non-metals
2. metalloids (that is, elements with properties which are mid-way between metals and non-metals)
3. reactive metals
4. heavy metals
5. transition metals.

Pupils could compare and contrast the properties of these different categories of elements, either in class or as homework. They should confine themselves to properties that they have met in this Topic and in Topic B16. Table B24.1 shows the type of information that can be tabulated in this way.

Property of element	Non-metal	Metalloid	Reactive metal	Heavy metal	Transition metal
Melting point	variable	variable	low	low	high
Density	low	low	low	high	high
Electrical conductivity	poor	variable	good	good	good
Chemical reactivity of the element	variable	variable	very reactive	generally unreactive	generally unreactive
Type of oxide	solid, liquid or gas acidic oxide	solid amphoteric oxide	solid alkaline (basic) oxide	solid often amphoteric oxide	solid basic or amphoteric oxide
Nature of chloride	molecular, often reacts with water	variable	solid ionic	solid ionic	solid ionic
Colour of aqueous solutions of its salts	colourless – if soluble in water	colourless – if soluble in water	colourless	colourless, but solid sometimes coloured	coloured, solids also coloured
Formulae of compounds	ionic predictable: molecular unpredictable	unpredictable	predictable	partly predictable	unpredictable

Table B24.1 A sample table of properties which pupils might produce in this section.

If pupils have studied atomic structure in Supplement S1–S3 or in Option 7, then details of structure should be added to the table.

Pupils should appreciate the difficulties of deciding which criteria are most suitable for a chemical classification of this kind. The section should succeed in tying up loose ends left over from earlier parts of the course and give the pupils a general view of the Periodic Table.

1. Summarize typical properties of (a) non-metals, (b) reactive metals, (c) heavy metals, such as tin and lead, and (d) transition metals. Use the *Handbook for pupils* and also refer to the results of practical work that you have carried out when studying Topics B16 and B24.

2. Predict some of the physical and chemical properties of the following elements and their compounds. Confine your answers to the type of properties you have met when studying Topics B16 and B24.

a. phosphorus, P;
b. rubidium, Rb;
c. thallium, Tl;
d. vanadium, V.

Before you attempt to answer the question, you should consult your Periodic Table and discover the positions of these elements.

3. Use the data given below to predict in which groups of the Periodic Table the approximate position of elements A to F in column 1 might be found.

Element	Melting point /°C	Density /g cm^{-3}	Properties of oxide	Formula of chloride
A	-189	gas	no oxide known	no chloride known
B	710	3.5	alkaline	MCl_2
C	1080	9.0	2 basic oxides	MCl, MCl_2
D	30	5.93	amphoteric	MCl_2, MCl_3
E	180	0.95	alkaline	MCl
F	119	2.1	2 acidic oxides	MCl, MCl_2, MCl_4

Which of the elements listed above would form compounds to give coloured aqueous solutions?

Summary

By the end of this Topic, pupils should have revised work done on the Periodic Table, and have learnt something of the chemistry of Group IV elements and of transition metals. They should know that the chemistry of an element is related to its position in the Periodic Table.

Stage II
Supplement
Atomic Structure and Chemical Bonding

Timing

S1 requires approximately 7 periods, S2 approximately 3 periods, and S3 approximately 5 periods. Total time needed is up to 15 periods if all three supplements are studied.

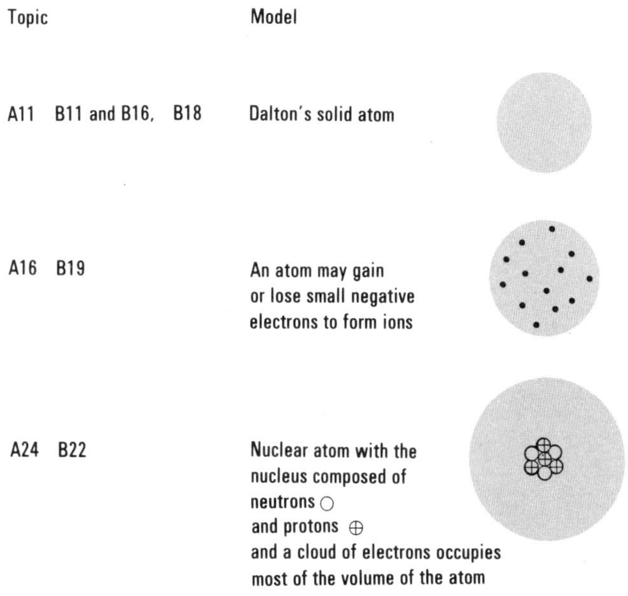

Topic			Model	
A11	B11 and B16,	B18	Dalton's solid atom	
A16	B19		An atom may gain or lose small negative electrons to form ions	
A24	B22		Nuclear atom with the nucleus composed of neutrons ○ and protons ⊕ and a cloud of electrons occupies most of the volume of the atom	

Figure S1

Note to teachers

These supplements are optional and non-examinable at Stage II. They relate and develop a number of ideas used in the basic course. Pupils find it helpful to go beyond the simple Daltonian atom and to acquire some basic ideas about ionic and covalent bonding. Alternatives IIA and IIB depend on an evolving model of the atom. This dependence is summarized in figure S1.

In addition, the proton is introduced in the discussion of acidity in Topics A20 and B20, and the symbol for the covalent bond is used in drawings of molecules in Topics A21 and B15.

Repeating patterns in the behaviour of the elements are explored in Topics A13, B16, and B24.

It would be possible to cover all three supplements at the end of Topics A16 or B19 but the work might be better interleaved throughout the course. Thus, in Alternative IIA, S1.1 and S1.2 might go between A15 and A16; S1.3 and S1.4 between A18 and A19; and S2 and S3 between A20 and A21. The corresponding pattern for Alternative IIB might be S1.1 and S1.2 between B18 and B19; S1.3 and S1.4 between B21 and B22; S2 and S3 between B23 and B24.

Further references
for the teacher

Bernal, J. D. (1969) *Science in history*. Penguin. Volume 3. Chapter 10 'The physical sciences in the twentieth century'.
CHEMStudy (1963) *Chemistry: an experimental science*. W. H. Freeman. Chapters 6, 14, 15, and 16.
Kneen, W. R., Rogers, M. J. W., and Simpson, P. (1972) *Chemistry: facts, patterns, and principles*. Addison-Wesley. Chapter 2 (atomic structure) and Chapter 4 (bonding).
Spiers, A. and Stebbens, D. (1973) *Chemistry by concept*. Heinemann Educational. Chapter 8 (bonding).
Nuffield Advanced Chemistry (1970) *Students' book I*, Chapter 4 'Background reading' (background to Rutherford's experiment with gold foil).

Supplementary materials

Films
'Conquest of the atom' EFVA
'Ionization energy' CHEMStudy. Distributed by Guild Sound and Vision Ltd.
Rank Film Library Atomic Physics Series:
'Part 2 Rays from atoms'
'Part 3 The nuclear structure of the atom'

Overhead projection originals
55 Interpreting the results of Rutherford's experiment *(figure S7)*
56 Two isotopes of carbon, $^{12}_{6}C$ and $^{14}_{6}C$ *(figure S9)*
57 Mass numbers of atoms in chlorine *(figure S10)*

Reading material
for the pupil

Handbook for pupils, Chapter 3 'Atomic structure and bonding'.
Greenwood, N. N. (1970) Nuffield Chemistry Special *Periodicity and atomic structure*.
CHEMStudy (1963) *Chemistry, an experimental science*. W. H. Freeman. Chapters 6, 14, 15, and 16.

S1
The Structure of the Atom

Supplement S1 traces the historical development of our ideas about the atom, from the simple model put forward by Dalton to the nuclear atom and isotopes.

S1.1
Introduction: Dalton's atomic theory

Objectives for pupils

1. Understanding of Dalton's model of the atom
2. Recognition that the atomic number of an element gives its position in the Periodic Table

Figure S2
Dalton's model of the atom.

The work may be introduced by a discussion of the ideas put forward in Topic A11 or B11, B16, and B18. A statement describing Dalton's model of the atoms is a useful starting point. He suggested that:

1. Matter consists of small particles called atoms. These atoms are indivisible and can be neither created nor destroyed.
2. Atoms of a given element are identical and are different in mass from the atoms of other elements.
3. Chemical changes are the result of the regrouping of atoms to form new substances.

At each stage in the development of our ideas about the atom, a diagram can be drawn and contrasted with Dalton's model. This device is helpful to pupils and enables them to recapitulate essential steps.

Plot a graph of relative atomic volume against atomic number and comment on the pattern, and on the relationship of relative atomic volume to the size of individual atoms.

S1.2
Can atoms be divided?

Objectives for pupils

1. Awareness that the electron is a constituent of all atoms
2. Familiarity with the small mass of the electron relative to the mass of the whole atom
3. Knowledge of the nuclear theory of the atom

The theme falls naturally into two parts. That dealing with the electron is considered first.

1. The electron
Understanding of the electron as a constituent of all atoms is required in Topics A16 and B19. Two alternative approaches are suggested:

Approach A
Some teachers may like to use the experiments in Topic A16 or B19 which investigate the quantity of electricity required to deposit a mole of atoms of various elements during electrolysis. The discussion might proceed as follows:

1. The charge required to deposit a mole of atoms of an element in electrolysis is always a whole number of Faradays.

2. A mole contains a definite number (the Avogadro number) of specified particles.

3. This suggests that the ions being discharged each carry a whole number of some small fundamental charge. The charge can be calculated by dividing the Faraday by the Avogadro number:

$$\text{the size of this charge} = \frac{96\,500}{6 \times 10^{23}} = 1.6 \times 10^{-19} \text{ coulombs}$$

The work done by Thomson and Millikan on the electron can be discussed, mentioning the small mass of the electron (1/1840 the mass of a hydrogen atom) and its charge (-1.6×10^{-19} coulombs). The formation of ions by the loss or gain of a whole number of electrons can then be explained.

Approach B
After a brief revision of the terms encountered in electrolysis, an experiment to produce cathode rays is demonstrated. Electrodes at high voltage are separated by a gas at very low pressure.

Experiment S1.2

Apparatus

The teacher will need:

Teltron Deflection Tube
TEL 525

Teltron Universal Stand
TEL 501

Power unit supplying 6 V and 5 kV

Connecting wires

Access to a room with black-out facilities

The production of cathode rays and their deflection in an electric field
This experiment **must** be done by the teacher.

Note. As pupils may have seen this experiment performed in physics, coordination with the physics department is desirable.

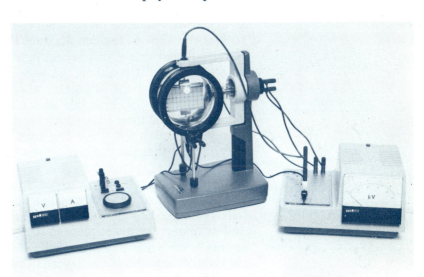

Figure S3
TEL 525 Deflection Tube (Teltron Limited).

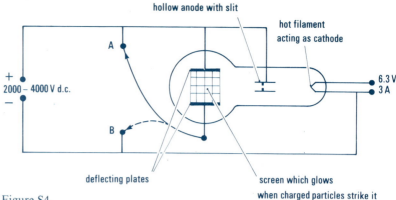

hollow anode with slit

hot filament
acting as cathode

A

+
2000 – 4000 V d.c.
−

6.3 V
3 A

B

deflecting plates

screen which glows
when charged particles strike it

Figure S4

Procedure
Warning. Care is needed with the high voltages required for this experiment. Note the use of 'hot cathode' tubes to minimize the risk of producing harmful X-rays.

The Teltron tube contains a screen which glows when the cathode rays strike it showing the path of the rays. The experiment demonstrates how the path of the rays is bent when an electric field is put at right angles to their initial path.

Lock the deflection tube into the stand, and connect up the circuit as shown with lower deflection plate connected at 'A'. In this way both the deflection plates are at the same voltage (the anode voltage) and there is no deflecting electric field. Leaving the H.T. voltage switched off, switch on the filament voltage V_f, and darken the room so that the glow of the hot filament is clearly seen on the screen. Point out that one is just seeing light rays from the hot filament passing through the slit in the anode. Now switch on the H.T. supply and turn up the voltage till, at about 3000 V, the cathode rays give rise to a visible blue line across the screen, and explain that Thomson called these rays which have come from the cathode and travelled through the slit in the anode 'cathode rays'. The pupils may suggest that the positive anode has drawn the rays from the cathode, so the rays may be a stream of negative particles. Ask them how they might test such a suggestion. Then show them that one can easily change the circuit so that the plate just below the screen becomes negatively charged. To do this, switch off the H.T. voltage and reconnect the lower plate to the negative terminal of the H.T. supply as shown as B in the circuit. Switch on the H.T. supply, and the path of the cathode rays is seen to bend away from the negatively charged deflection plate. This confirms that cathode rays consist of negatively charged particles, which Thomson named 'electrons'. (It may be noticed that the cathode glow is not deflected and this helps to draw a distinction between cathode rays, which are a stream of particles, and light rays, which are not.) The *Handbook for pupils* contains a simplified diagram of this experiment.

small, negative electrons
in a sphere of diffuse,
positive charge

Figure S5
Thomson's 'plum-pudding'
model of the atom.

Suggestions for homework

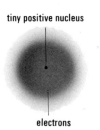

tiny positive nucleus

electrons

Figure S6
The nuclear atom.

It is suggested that neither the deflection of the cathode rays by a magnetic field nor the derivation of e/m is relevant at this stage. The pupils can be told briefly that the work of Thomson and of Millikan indicated that an electron has :

a negative charge size 1.6×10^{-19}

a mass equal to $1/1840$ mass of a hydrogen atom

By identifying a particle much lighter than the lightest atom, Thomson had discovered that the atom could be divided. Moreover he found cathode rays with identical properties whatever metal he chose as cathode, or whatever gas was present in the tube, and therefore proposed that the atoms of *all* elements contain electrons, embedded in a sphere of positive charge.

1. Calculate the fraction of the mass of a particular atom lost on losing an electron, remembering that relative atomic mass gives the number of times the atom is heavier than an hydrogen atom. (*This calculation reinforces the idea that the electron has negligible mass.*)

2. Calculate the charge on a mole of electrons in coulombs using the charge on an electron and the Avogadro constant.
(*They will find this equal to the Faraday, a constant charge of 96 500 coulombs required to discharge one mole of singly charged ions, as found in Topics A16 and B19.*)

2. Rutherford's nuclear theory

This section may be presented as a discussion illustrated by over-head transparencies and by the film 'Conquest of the atom'. The discussion should aim at making the pupils aware that new ideas about the atom were needed to interpret the experiments. Pupils should not learn details of the experiments.

Pose the question, 'What is the positive part of the atom like?' and remind them of Thomson's picture of the atom as a 'plum-pudding' (figure S5).

Rutherford's experiment can then be discussed. In this α-particles, regarded as minute positively charged 'cannon balls', were fired at a very thin gold foil. If Thomson's theory were true, one would expect only minor deflections of the α-particles. The surprising results of this experiment (illustrated possibly by the film 'Conquest of the atom') that most of the α-particles passed through the foil un-deviated while a few were deflected and even reflected through large angles, led Rutherford to suggest that the atom contains a very small positively charged nucleus surrounded by electrons. Furthermore, the nucleus is found to be roughly 5×10^{-15} m in diameter. Comparison with the size of an atom (2×10^{-10} m) shows how small the nucleus is (figure S6).

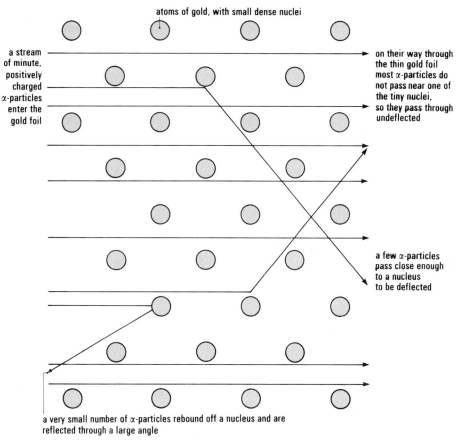

atoms of gold, with small dense nuclei

a stream of minute, positively charged α-particles enter the gold foil

on their way through the thin gold foil most α-particles do not pass near one of the tiny nuclei, so they pass through undeflected

a few α-particles pass close enough to a nucleus to be deflected

a very small number of α-particles rebound off a nucleus and are reflected through a large angle

Figure S7
Interpreting the results of Rutherford's experiment. (OHP 55)

Ask the pupils what positive charge the nucleus carries. Discussion leads to the answer that it would be equal in magnitude to the number of electrons around it, in units of 1.6×10^{-19} coulombs. Tell them that Rutherford's experiments with a few elements suggested that this charge is equal to the atomic number of the element, Z. Till then this defined the position of the element in the Periodic Table. Later work by Moseley confirmed this suggestion. (It is not suggested that the basis of Moseley's work on X-ray spectra can be usefully discussed at this stage.)

Suggestions for homework

The diameter of an atom is roughly 10^{-10} m, but the diameter of a typical nucleus is roughly 5×10^{-15} m.
1. How many times larger is the size of the whole atom than the size of the nucleus, if we compare (a) diameters, (b) volumes?
2. Estimate very roughly the size of your school assembly hall in metres. If we picture an atom as just filling the hall, estimate what size the nucleus would be on this scale. Suggest an everyday object of the correct size to represent the nucleus.
3. If the density of a solid is 1 g cm^{-3}, estimate from your answer in question 1 part (b) how dense the nuclei of its atoms are.

Summary

Pupils should now appreciate that the atoms of all elements contain light, negatively charged particles called electrons, which are identical in all substances. They need not learn details of Thomson's experiments but should appreciate that it is not very difficult to remove an electron from an atom. They should be able to picture the nucleus of the atom as a minute central particle in which all the positive charge and almost all the mass of atom is concentrated.

S1.3
What does the nucleus consist of?

A suggested approach

Objectives for pupils
Knowledge of protons and neutrons as constituents of the nucleus

From the previous section the pupils know that the size of the positive charge on the nucleus is equal to the atomic number Z in units of 1.6×10^{-19} coulombs. Tell the pupils that Rutherford suggested that the charge is carried by Z particles of unit atomic mass and a positive charge each of 1.6×10^{-19} coulombs, called protons. (This is seen to be a reaonable suggestion if the charge on the nucleus is always a multiple of this basic unit.) This means that the nucleus of a hydrogen atom consists of one proton (Greek *protos*, first), and this proton together with the almost negligible mass of one electron accounts for the mass of the hydrogen atom.

Ask the pupils to comment on the properties of a helium nucleus, perhaps referring to a Periodic Table with relative atomic masses given on it. It will be noticed that two protons ($Z = 2$) do not account for the relative atomic mass of 4. They may then be told that Rutherford accounted for this discrepancy by suggesting the existence of the neutron, a particle of unit mass like the proton, but with no electric charge. The presence of two neutrons in the helium nucleus in addition to the two protons accounts for the relative atomic mass of 4. The neutron was first detected a decade later by Chadwick in 1932.

proton ⊕
neutron ○
electron e⁻

Figure S8
An atom of helium.

A diagram representing helium (figure S8) may be used to summarize this discussion of the proton and the neutron.

In preparation for a subsequent discussion on the nature of acids (sections A20.3, B20.6), attention can be drawn to the size of an H^+ ion, which is a bare nucleus consisting of one proton. All other ions met in the course, such as Pb^{2+}, Na^+, Cl^-, retain a number of electrons round the nucleus and are of the order of 100 000 times larger than the proton.

Suggestion for homework

Read the *Handbook for pupils,* Chapter 3, up to the section 'Sub-atomic particles found in atoms'.

S1.4
Isotopes and relative atomic mass

Objectives for pupils

1. Understanding of the concept of isotopes
2. Understanding of the concept of relative atomic mass

Carbon—12 atom
Consists of 6 orbital electrons and a nucleus of 6 neutrons and 6 protons

$^{12}_{6}$C

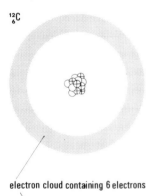

electron cloud containing 6 electrons

$^{14}_{6}$C

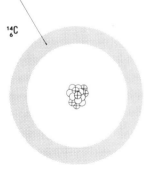

○ neutron
⊕ proton

Carbon—14 atom
Consists of 6 orbital electrons and a nucleus of 8 neutrons and 6 protons

Figure S9
Two isotopes of carbon, $^{12}_{6}$C and $^{14}_{6}$C. (OHP 56)

The teacher may provoke discussion by quoting the results of work with Aston's mass spectrograph. Aston measured the relative masses of individual atoms, and found that atoms of a pure element often differed from each other in mass by a few units. Class discussion may then be used to emphasize that all the atoms of an element have the same number of protons and electrons as each other, as this number (which is the atomic number) characterizes the chemical nature of the atom, but that number of neutrons may vary. They should be told that two atoms of the same element with different numbers of neutrons are different *isotopes* of that element (Greek *'isos topos'* meaning 'same place' – in the Periodic Table). They should be shown the standard notation for isotopes:

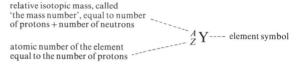

relative isotopic mass, called 'the mass number', equal to number of protons + number of neutrons

atomic number of the element equal to the number of protons

$^{A}_{Z}$Y ---- element symbol

and asked to draw diagrams of some isotopes similar to the helium atom (figure S8). Two examples of suitable complexity are $^{12}_{6}$C and $^{14}_{6}$C (figure S9).

The following points may also be discussed:
1. Most elements have several isotopes.
2. Not all isotopes are radioactive (to counter a common misconception that they are).
3. Carbon-12 is now the basis for our definition of atomic mass: the unit of the scale is one-twelfth of the mass of an atom of carbon containing 6 protons and 6 neutrons.
4. Carbon-14 is radioactive and is the basis of dating archaeological finds containing carbon.

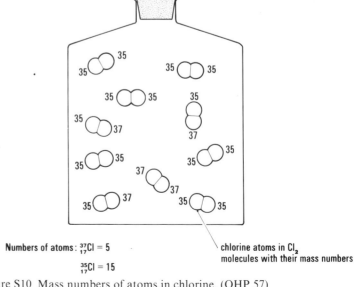

Numbers of atoms: $^{37}_{17}$Cl = 5

$^{35}_{17}$Cl = 15

chlorine atoms in Cl_2 molecules with their mass numbers

Figure S10 Mass numbers of atoms in chlorine. (OHP 57)

Ask, 'If the mass numbers of all isotopes are all (very nearly) whole numbers, why is the relative atomic mass of an element so often not a whole number?' In discussion the teacher should make clear that the relative atomic mass is the weighted mean of the masses of the naturally occurring isotopes, taking into account their abundance. This could be illustrated by figure S10, which illustrates the random distribution of isotopes of chlorine and provides data for a simple calculation of the relative atomic mass.

$$\text{mean mass of Cl atom} = \frac{(35 \times 15) + (37 \times 5)}{15 + 5}$$

$$= 35.5 \quad \text{the relative atomic mass}$$

1. Draw diagrams of selected isotopes.
2. Collect together the important properties of the subatomic particles discussed in Chapter 3 of the *Handbook for pupils*.
3. Take each point in Dalton's atomic theory and discuss how they have been superseded in our developing concept of the atom.

Summary

At the end of this supplement pupils should know that Dalton's simple model of the atom has had to be considerably revised. They should understand the nuclear nature of the atom and know that the nucleus contains a number of protons equal to the atomic number of the element, and may contain different numbers of neutrons in different isotopes. They should be able to interpret an isotope symbol in terms of the subatomic particles present.

S2
How are the Electrons Arranged in Atoms?

The main aim here is for the pupils to gain a clear and simple idea of the arrangement of the electrons in the atoms of the first twenty elements in the Periodic Table. This concept can then be used to gain an understanding of simple ionic and covalent bonding in S3. Further evidence of the periodicity in the properties of elements as arranged in the Periodic Table is discovered and pupils can gain a clearer idea of the chemical differences between metals and non-metals.

S2.1
First ionization energies of elements and the concept of 'shells'

A suggested approach

Start by explaining the exact meaning of the term 'first ionization energy'. It can be defined as 'the energy required to remove a mole of electrons from a mole of gaseous atoms of an element to form ions'. The teacher may possibly like to outline briefly the technique of measuring ionization energies by electron bombardment and with an able class he might even demonstrate the measurement of the first ionization energy of argon using the equipment recommended in Nuffield Advanced Chemistry (see Nuffield Advanced Chemistry *Teachers' guide I*, Experiment 4.4, *Students' book I*, Section 4.4) or that for xenon (see Mullard Educational Service 'Critical potentials in gases' *Educational electronic experiments*, Leaflet No. 7). The measurement of ionization energies is also illustrated in the CHEMStudy film 'Ionization energy'.

Tell pupils to use the data supplied below to plot a graph of the first ionization energy (the energy to remove the first electron from the gaseous atom) of the first twenty elements in the Periodic Table against atomic number. In addition they should write the correct symbol for each element beside the appropriate point on the graph.

Atomic number	Symbol	1st ionization energy/kJ mol^{-1}	Atomic number	Symbol	1st ionization energy/kJ mol^{-1}
1	H	1310	11	Na	500
2	He	2370	12	Mg	740
3	Li	520	13	Al	580
4	Be	900	14	Si	790
5	B	800	15	P	1010
6	C	1090	16	S	1000
7	N	1400	17	Cl	1260
8	O	1310	18	Ar	1520
9	F	1680	19	K	420
10	Ne	2080	20	Ca	590

Questions
1. (*i*) Which elements occur at the three *main* peaks in the graph?
   (*ii*) Will these elements readily lose electrons?
2. (*i*) Which elements occur at the three *main* troughs in the graph?
   (*ii*) What is the nature of these elements?
   (*iii*) Is it easy to remove an electron from each atom of these elements?
3. What is the *general trend* of the first ionization energies as one crosses a period in the Periodic Table, for example, Na, Mg, Al, Si, P, S, Cl, Ar?
4. Is it easier to remove an electron from the atoms of metals or non-metals?
5. What is the trend in the first ionization energies as one descends Group I, Li, Na, K, etc.?

Next we consider the concept of 'energy levels' or 'shells' in arrangements of electrons. Pupils can be asked to comment on the relative values of the first ionization energies of H, He, and Li (remembering, of course, that the nuclear charge is increasing). Provided that the pupils know beforehand that the forces between charges depend on the distance between the charges, they should be able to suggest that the low value of the first ionization energy of lithium indicates that the electron which is removed in this case must be further away from the nucleus. The 'energy levels' can be represented diagrammatically (figure S11).

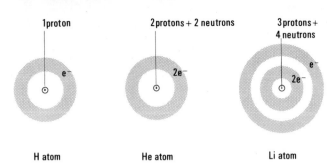

H atom   He atom   Li atom

Examination of the graph of first ionization energies indicates that this second 'shell' may hold up to eight electrons at which stage a noble (inert) gas is reached. The pupils should quickly see that the low value of the first ionization energy of sodium indicates that a third 'shell' has been started (figure S12).

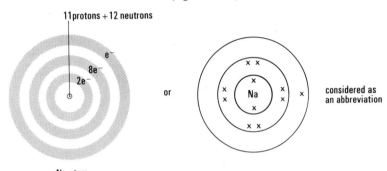

Figure S12

Na atom

Practice can then be gained in predicting the 'cloud diagrams' for other elements using this idea that the electrons in atoms exist in definite 'energy levels'.

In discussion of the graph of the first ionization energies, alert pupils may well point out the minor peaks and troughs which occur in the graph showing that the first ionization energies do not rise in a completely steady manner across a period. The teacher can mention that the theory which the pupils have developed on the electronic configurations of atoms is a relatively simple one and that a more elaborate one exists which will explain these minor fluctuations in the general trend.

Suggestion for homework

Use the graph of first ionization energies to predict 'cloud diagrams' to represent the arrangement of electrons in selected atoms.

S2.2
The successive ionization energies of sodium

A suggested approach

To test the hypothesis for the electronic configuration of sodium by studying the data for the successive ionization energies of that element

Can the pupils' hypothesis for the arrangement of electrons in a sodium atom be tested by examining how difficult it is to move all the electrons from the atom? Explain that after one electron has been removed from an atom of an element it becomes more and more difficult to remove successive electrons from the atom and that the energies required to do this are known as the successive ionization energies.

From the last section, pupils have been able to develop a rough idea about how the electrons in the atoms of the first twenty elements may be arranged. Now they should examine data for the *successive* ionization energies of sodium (atomic number 11), that is, the energies required to remove successively each of the eleven electrons of the sodium atom from the gaseous atom (or ion). Ask pupils to plot a graph of the successive ionization energies of sodium against the number of the electrons removed from each atom. If they cannot do this, what is the difficulty and can they suggest a means of overcoming it?

Number of the electron removed from each atom (or ion)	Ionization energy /kJ mol^{-1}
1	500
2	4600
3	6900
4	9500
5	13 400
6	16 600
7	20 100
8	25 500
9	28 900
10	141 000
11	158 700

Questions

1. How much more difficult is it to remove the eleventh electron from a sodium atom than it is to remove the first?

2. Comment briefly on the shape of the alternative graph you have drawn. For example, does it support or conflict with your earlier ideas about how the electrons may be arranged in a sodium atom? Justify your opinion.

3. Sketch a rough but clear diagram of the shape of the analogous graph that you would predict for potassium (atomic number 19).

A problem arises if the pupils try to represent the relation between successive ionization energies and the number of the electrons removed from each atom, as they will discover that simply plotting a graph of the data supplied will be unsatisfactory due to the wide variation in the values of the ionization energies. What mathematical method can they suggest to 'compress' the values of the ionization energies? The usual method of solving such a problem is to plot the *logarithm* of the ionization energies against the number of the electron removed from each atom and the pupils can do this. An equally good method would be to plot the *square root* of the ionization energies against the number of the electron removed from each atom and this is the basis of the method used in figure S13.

Figure S13
The area of each square is proportional to the successive ionization energy.

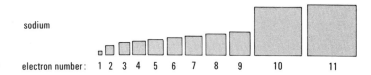

sodium

electron number: 1 2 3 4 5 6 7 8 9 10 11

Supplement S2 How are the electrons arranged in atoms?

Such a diagram, although unconventional, may perhaps give the pupils a better 'feel' for the energy which needs to be invested to remove each successive electron from an atom, than does a graphical method.

This study will confirm their earlier predictions on the arrangement of electrons in sodium and the section can be completed by a discussion of other atoms in the Periodic Table.

Note
1. It is suggested that *the term 'orbit' should be avoided* as this falsely indicates that we can locate precisely the position of an electron in an atom. *Drawings of electronic configurations in terms of circles should be regarded as abbreviations.*
2. No more than the first twenty elements in the Periodic Table need be covered in this simplified treatment and it is appropriate to warn pupils of the existence of a more sophisticated theory on the arrangement of electrons in atoms.

Suggestions for homework

1. Plot the logarithm *or* the square root of the successive ionization energies of sodium against the number of the electron removed from each atom.
2. Draw sketches to represent the variation of the successive ionization energies of the atoms of other elements such as Mg, K, F.
3. Read *Handbook for pupils*, Chapter 3, on the arrangement of electrons in atoms.

Summary

The pupils should now appreciate that the electrons in atoms exist in a series of 'energy levels' which we can represent diagramatically as 'shells' and they should be able to describe the arrangement of electrons in the first twenty elements in the Periodic Table.

S3
Chemical Bonding

The aim here is for the pupils to gain an understanding of the concepts involved in the simplest examples of ionic and covalent bonding. This in turn will improve their existing understanding of stoichiometry, structure, and electrolytes as well as introducing them to the symbol for the covalent bond required in Topics A21, B17, and B18.

S3.1
Ionic bonding

1. Understanding how ionic compounds are formed by the transfer of electrons between atoms
2. Appreciation that on reacting together the atoms achieve relatively stable arrangements of electrons which are frequently, although not always, the same as those of the atoms of the inert gases
3. Ability to explain the formation and stoichiometry of simple metal/non-metal compounds formed from the first twenty elements in the Periodic Table
4. Knowledge of the typical properties of an ionic compound

Start by discussing the nature of a simple compound such as lithium fluoride using ideas that the pupils will have met earlier in the course (the diagrams for LiF are slightly simpler than those of the more familiar example of NaCl). Valuable points to bring out include:

1. Lithium and fluorine react together violently (exothermically).
2. It can be shown that 1 mole of lithium atoms combines with 1 mole of fluorine atoms. Thus the formula of lithium fluoride is LiF (*cf.* ideas gained in Topic A11; B18).
3. The substance is an electrolyte; it must therefore consist of ions (*cf.* ideas gained in Topic A16; B19).
4. The charges on these ions can be shown to be Li^+ and F^- (*cf.* ideas gained in Topic A16; B19).

The formation of Li^+F^- from the atoms can then be discussed with the aid of figure S14, and the fundamental idea of *the transfer of electrons between atoms to form ions* established.

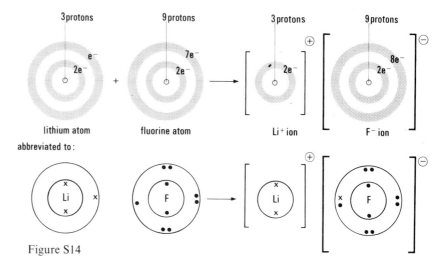

Figure S14

The pupils can then be asked to comment on the electronic configurations of the resulting ions, which are the relatively stable ones of inert gases. The teacher can tell the pupils that this is the type of bonding which occurs when a simple compound is formed from a metal and a non-metal and can then assist the class to predict electron-transfer diagrams to describe the formation and stoichiometry of other compounds such as NaF, NaCl, MgF_2, Na_2O, and MgO. The pupils will notice that in these examples the resulting ions all have the electronic configurations of inert gases, and the teacher should point out that this is by no means always the case by briefly discussing substances such as $PbBr_2$, CuO, and $FeCl_3$.

It is worth while discussing the crystalline giant ionic structure to be expected from these compounds and revising knowledge of the properties of electrolytes. With a very able class the teacher might stimulate discussion of the various energy changes involved in the formation of an ionic compound.

Note
1. It is not necessary to cover many examples nor is it desirable that the pupils should gain rigid ideas on the inert gas electronic configuration of ions.
2. It is however worth asking pupils to consider how many electrons can be removed relatively easily from Group I, Group II and Group III elements respectively (*cf.* ionization energies).
3. It is valuable to explain the reasons why it is unlikely that any single atom will lose or gain more than three electrons under normal circumstances.

Suggestions for homework

1. Read *Handbook for Pupils*, Chapter 3, section on 'ionic bonding'.
2. Predict electron-transfer diagrams to describe ionic bonding in sodium chloride.

S3.2
Covalent bonding

A suggested approach

Objectives for pupils

1. Understanding of how atoms can combine by the sharing of electrons to form a molecule in which each atom has achieved a stable arrangement of electrons
2. Recognition of the symbol for a covalent bond
3. Knowledge of the number of covalent bonds which atoms of hydrogen, oxygen, nitrogen, and carbon will form
4. Revision of the properties expected in substances consisting of small and of giant molecules

In S3.1 the class should have discussed the reasons why a carbon atom, for example, is unlikely to form either C^{4+} or C^{4-} ions under normal circumstances. They can also be reminded that most simple substances formed solely from non-metal atoms are non-electrolytes, for example, H_2, O_2, Br_2, CO_2, CH_4, and CCl_4. Then with the aid of suitable diagrams, explain the bonding in the hydrogen molecule and introduce the pupils to the symbol for the covalent bond (see figure S15).

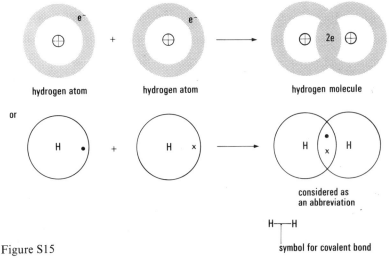

Figure S15

This simple example establishes the basic idea of *the sharing of electrons between atoms to form molecules*. Then reinforce pupils' understanding by helping them to predict the bonding in simple substances such as Cl_2, HCl, H_2O, CH_4, and CCl_4. Discussion of the bonding in CO_2 and N_2 will introduce the idea of double and triple bonds between atoms. Only a few examples of covalent bonding need be studied; an understanding of the symbol used for the covalent bond is the main requirement. If the class know the number of covalent bonds which atoms of hydrogen, oxygen, nitrogen, and carbon will form, this will prove to be of great value when, for example, Topic A21 is reached.

The section should be completed by a discussion of small and large molecules and the properties associated with these, thus reinforcing ideas gained previously in Topics A14 and A15; B17.

Suggestions for homework

1. Read *Handbook for pupils*, Chapter 3, section on 'covalent bonding'.
2. Draw dot-cross electron diagrams of simple molecules.
3. Draw structural formulae of simple molecules using the symbol for the covalent bond.
(*In preparation for the work they will meet in Topic A21 or B17 and B18.*)

Part 3 Appendices

Appendix 1

Apparatus, chemicals, and experiments for Stage II

Access to a laboratory is crucial for almost every lesson in the scheme. Without it, an investigational approach is almost impossible. During Stage II, the investigations are more complex than during the two introductory years of Stage I, and may require demonstrations, and class practical work. Some additional items of apparatus are necessary and these include electrical apparatus, molecular models, and a number of small items that can be most readily made in the school workshop. Film loops, slides, and overhead projector transparencies play an important part in the development of this stage of the scheme and to some extent there is an overlap with the requirements of Nuffield Advanced Science courses in Chemistry and Physical Science.

The range of chemicals and apparatus does not differ markedly from that already used in many schools, although there are several chemicals which are not commonly used at this level and a few pieces of apparatus which will be found helpful. Much will depend on the choice of pathway through the Stage II materials.

Matters of general concern are considered at length in the Appendices of *Teachers' guide I*, and are not repeated here.

Safety precautions

Both teachers and pupils need to be aware of potential dangers in the laboratory. Essential requirements for teachers and laboratory assistants are given in publications from the Department of Education and Science, in particular *Safety in science laboratories* (which has been revised extensively and replaces an earlier edition and *Safety at schools*, Educational Pamphlet No. 53) and in Administrative Memoranda on the use of apparatus and chemicals in schools, colleges of education, and major establishments of further education. At the time of going to press, the following memoranda were in force:

A.M.3/70   Avoidance of carcinogenic aromatic amines in schools and other educational establishments.

A.M. 7/70   Use of lasers in schools and other educational establishments.

A.M. 2/76   The use of ionizing radiations in educational establishments: and the booklet to be read in association with A.M. 2/76 'Notes for the guidance of schools, establishments of further education, and colleges of education on the use of radioactive substances and equipment for producing X-rays'.

A.M. 6/76   The laboratory use of dangerous pathogens.

A.M. 7/76   The use of asbestos in educational establishments.

(These and other memoranda may be obtained from D.E.S., York Road, London, SE1 7PH, *or* S.E.D., 43 Jeffrey Street, Edinburgh 1, *or* Ministry of Education, Dundonald House, Belfast 4.)

Teachers are in a position to warn pupils about potential dangers in the laboratory. Many local education authorities encourage teachers to display a short list of safety rules in each laboratory. The Association for Science Education publishes such a list, together with much else on this subject including *Safety in the lab*, a leaflet for distribution to pupils. A film, 'In the movies it doesn't hurt', made as a companion to *Safety in science laboratories*, is available on free loan from the Central Film Library, Government Building, Bromyard Avenue, Action, London W3 7JB.

The form and purpose of Experiment sheets II

During Stage II, the pupils' approach to Chemistry will be more mature and the experiment sheets become more demanding on the pupil. The requirements of Alternatives IIA and IIB differ and appropriate modifications are made in the series of sheets.

Experiments for Stage II

The lists given below contains all the experiments described in this book. The second list starting on p.661 contains the sheet numbers and titles in the pupils' book, *Experiment sheets II* and their correlation with the *Teachers' guide* experiments.

Key
Column 1: number of experiments in *Teachers' guide II*.
Column 2: ES41, ES42, etc., refer to the pupils' book, *Experiments sheets II*. P indicates an experiment for pupils for which no Experiment sheet is supplied. T indicates that the experiment is to be carried out by the teacher.
Column 3: title of experiment.

Alternative A

A11.1a	T	Diffusion in solutions
A11.1b	T & ES41	Diffusion of gases
A11.1c	P	Demonstration of Brownian motion by smoke particles
A11.2a	ES42	Into roughly how many particles might a crystal of potassium permanganate be divided?
A11.2b	T	How small are the particles of matter?
A11.2c	ES43	How small are the particles of oil?
A11.3	P	Exercises in obtaining moles of atoms (gram-atoms) of various elements
A11.4a	ES44	To find the formula of magnesium oxide
A11.4b	ES45	To find the formula of black copper oxide
A11.4c	T	What is the formula of mercury chloride?
A11.4a	ES117	Finding the formula of magnesium oxide by experiment
A11.4b	ES45	To find the formula of black copper oxide
A11.4c	T	What is the formula of mercury chloride?
A12.1a	ES46	Investigating some properties of 'salt gas'
A12.1b	T	The fountain experiment

A12.1c	T	Preparing a solution of 'salt gas'
A12.1d	ES47	Finding out what causes the steamy fumes when 'salt gas' is exposed to air
A12.2a	T	Finding one of the elements in 'salt gas'
A12.2b	ES48	To find out what other substance is present in 'salt gas'
A12.3	T	Burning hydrogen in chlorine
A12.4	T	What is the formula of hydrogen chloride?
A13.2a	T	The action of (1) air; (2) chlorine on sodium and potassium
A13.2b	T	The action of sodium and potassium on water
A13.2c	T & ES49	What are the reactions of lithium?
A13.3a	T	Some properties of the halogens
A13.3b	ES50	Investigating some of the properties of iodine
A13.4a	ES51	Looking at the properties of some transition metals
A13.4b	ES52	Looking at the properties of some compounds of certain transition metals
A13.5a	ES53	To find out if carbon is formed when substances from plants and animals are heated
A13.5b	ES54	To confirm the presence of carbon in various substances
A13.6	T	The reaction between magnesium and silica
A14.1	P	Alternative route: exercises in using moles of atoms (gram-atoms) of various elements
A14.3a	ES55	Watching crystals grow
A14.3b	T	Watching growth in a model of a crystal
A14.4a	T	Crystal structure analogy using a ripple tank and a 'model' crystal
A14.4b	T	An optical analogue for X-ray diffraction patterns
A14.5a	ES56	Growing crystals of metals
A14.5b	T or P	How do crystals form from molten lead?
A14.5c	T or P	How do crystals form from molten tin?
A14.5d	P	How might atoms be arranged in metals?
A14.6a	T	How strong are metals?
A14.6b	T	Comparing the ductility of metals
A14.6c	T	Investigations on the hardness of metals
A14.6d	T	The effect of heat treatment on steel
A14.6e	T	The preparation of a metallic surface to show slip lines
A14.6f	P	Making a bubble raft

A14.7a	T & ES57	Making various allotropes of sulphur
A14.7b	ES58	Observing what happens when sulphur is heated: making plastic sulphur
A14.8	P	Making models of graphite and diamond structures
A14.9	ES59	What volumes are occupied by one mole of molecules (gram-molecule) of various gases?
A14.10a	T	Models of molecular substances in the liquid and solid states
A14.10b	T or P	How easy is it to pull atoms apart in silicon or graphite (giant structures) and molecules in iodine (a molecular structure)?
A14.11a	T	Can the atoms in sodium chloride and magnesium oxide be pulled apart readily?
A14.11b	P	Making models of the structure of magnesium oxide, sodium chloride, and caesium chloride
A15.1	P	Evidence for the movement of molecules in the liquid state
A15.2a	ES60	Is energy needed to melt a solid?
A15.2b	ES61	How much energy is required to vaporize 1 mole of water?
A15.2c	T	How much energy is required to vaporize 1 mole of heptane molecules?
A16.1a	T	Simple demonstrations of electrostatic phenomena
A16.1b	T	Experiments with an electrometer
A16.1c	T	The effect of an electric field on a charged sphere
A16.1d	T	Finding out which substances conduct electricity
A16.2	T	Investigating the electrolysis of lead bromide
A16.3a	T	Looking at the migration of ions
A16.3b	ES62	The migration of ions
A16.4a	T	A quantitative study of the electrolysis of lead bromide
A16.4b	ES63	How much electricity is needed to deposit 1 mole of copper atoms and 1 mole of silver atoms from solution?
A16.5	P	Investigating the electrolysis of solutions of electrolytes
A16.6	ES64	Is the same quantity of electricity always required to liberate 1 mole of copper atoms?
A17.1	T	Preparation of 1.0M solution of sodium chloride
A17.2a	ES65	To verify the composition of the precipitate formed during the reaction between lead nitrate solution and potassium iodide solution
A17.2b	ES66	To verify the composition of the precipitate formed during the reaction between barium chloride and sodium carbonate solutions

A17.2c	ES67	To verify the composition of the precipitate formed during the reaction between barium chloride and potassium chromate solutions
A17.3a	ES68	To determine the volume of 1.0M hydrochloric acid which reacts completely with $10.0\ cm^3$ 1.0M sodium carbonate
A17.3b	ES69	To measure the volume of carbon dioxide formed when a known amount of sodium carbonate reacts with dilute hydrochloric acid
A17.3c	ES70	To find whether hydrochloric acid, sodium carbonate, and sodium chloride solutions are electrolytes
A17.3d	ES71	To measure the volume of hydrogen produced when 0.024 g magnesium reacts with dilute hydrochloric acid
A17.3e	T	To verify the equation for the reaction between gaseous ammonia and gaseous hydrogen chloride
A17.4	ES72	An examination of the reaction between iron and copper(II) sulphate solution
A18.1a	T	The reaction between marble chips and hydrochloric acid
A18.1b	ES73	What is the effect of the concentration of the reactants on the rate of reaction between sodium thiosulphate and hydrochloric acid in solution?
A18.2	ES74	What is the effect of temperature on the rate of reaction?
A18.3	T	What is the effect of particle size on the rate of a reaction?
A18.4a	T	A study of the effect of metal oxides on the thermal decomposition of potassium chlorate
A18.4b	T	The reaction between hydrogen and oxygen in the presence of a platinum catalyst
A18.4c	P	Investigation of the effects of a catalyst on the rate of decomposition of hydrogen peroxide
A19.1	T	A study of the reaction between iodine and chlorine
A19.2a	ES75	The distribution of iodine between two immiscible solvents
A19.2b	ES76	Two examples of reversible reactions
A19.2c	ES77	Two other examples of systems in chemical equilibrium
A19.3	T	Investigating the equilibrium between solid lead chloride and its saturated solution using a radioactive tracer technique
A20.1a	T	Investigating the acidic properties of some compounds
A20.1b	ES78	Investigating the acidic properties of some compounds
A20.2a	T	Properties of acids and alkalis
A20.2b	ES79	To examine the reaction between dilute sulphuric acid and copper (II) oxide

A20.3a	T	A comparison of the properties of hydrogen chloride in water and in toluene solution
A20.3b	T	The preparation of solutions of hydrogen chloride in toluene and in water
A20.4a	ES80	To find the ratio in which barium and sulphate ions combine
A20.4b	T	To find the ratio in which barium and sulphate ions combine
A20.5a	ES81	Preparation of copper sulphate from (1) copper(II) oxide; (2) copper(II) carbonate
A20.5b	ES82	Preparation of magnesium sulphate from (1) magnesium; (2) magnesium carbonate
A20.5c	ES83	Preparation of lead chloride from lead nitrate solution
A21.1	ES84	The detection of carbon and hydrogen in organic chemicals
A21.2	ES85	Identification of the breakdown products of starch using a chromatographic method of analysis
A21.3	ES86	To break down the glucose molecule by fermentation
A21.4	ES87	Can ethanol be broken down further?
A21.5	ES88	Cracking petroleum products
A21.6	ES89	The reaction of castor oil with (1) sodium hydroxide; (2) sulphuric acid
A21.7	T	Depolymerizing perspex and polymerizing the monomer
A21.8	ES90	To prepare some nylon
A22.1a	P	Investigating food stuffs (1)
A22.1b	ES91	Investigating food stuffs (2)
A22.1c	T	The detection of ammonia as a product of the decomposition of a protein
A22.2	ES92	To remove hydrogen from ammonia
A22.3	T	The volume composition of ammonia
A22.4	T	The synthesis of ammonia
A22.5	ES93	Preparation of ammonium sulphate
A23.1a	T	Heat transfer experiments
A23.1b	ES94	To measure the heat of reaction of a neutralization reaction
A23.1c	ES95	To measure the heat of reaction of an ion combination reaction
A23.1d	ES96	To measure the heat of reaction for a displacement reaction
A23.3a	ES97	To measure the heat of combustion of a liquid fuel
A23.3b	T	To measure the heat of combustion of a fuel
A23.4a	T	The interconversion of different forms of energy

A23.4b	ES98	To measure the e.m.f.s of some voltaic cells
A23.4c	ES99	To determine the maximum amount of energy available as work from a voltaic cell
A23.5	ES100	A very simple hydrogen-oxygen fuel cell
A24.1a	T	Becquerel's chance discovery
A24.1b	ES101	An examination of scintillations
A24.1c	T	A first use of a Geiger-Müller tube and scaler
A24.2a	T	The decay and growth curves of lead-212
A24.2b	T	The decay of bismuth-212
A24.2c	T	The decay of protactinium-234
A24.3	T	Evidence for the dynamic nature of equilibria

Alternative B

B11.2	T	The formation of oxides
B11.3a	ES46	Investigating some properties of 'salt gas'
B11.3b	T	The fountain experiment
B11.3c	T	Preparing a solution of 'salt gas'
B11.3d	ES47	Finding out what causes the steamy fumes when 'salt gas' is exposed to air
B11.4a	T	Finding one of the elements in 'salt gas'
B11.4b	ES48	To find out what other substance is present in 'salt gas'
B11.4c	T	Burning hydrogen in chlorine
B12.1a	T	Investigating the acidic properties of some compounds
B12.1b	ES78	Investigating the acidic properties of some compounds
B12.2	ES102	The preparation and properties of some bases
B12.3a	ES81	Preparation of copper sulphate from (1) copper(II) oxide; (2) copper(II) carbonate
B12.3b	ES82	Preparation of magnesium sulphate from (1) magnesium; (2) magnesium carbonate
B12.3c	ES83	Preparation of lead chloride from lead nitrate solution
B13.1a	ES103	What is the effect of the concentration of hydrochloric acid on the rate of reaction between magnesium ribbon and dilute hydrochloric acid?
B13.1b	T	What is the effect of concentration of hydrochloric acid on the rate of reaction between marble chips and hydrochloric acid?
B13.2	ES74	What is the effect of temperature on the rate of reaction?
B13.3a	T	The reaction of zinc with dilute sulphuric acid

B13.3b	T	The reaction between hydrogen and oxygen in the presence of some platinum
B13.3c	P	Investigating the effect of metal oxides on the rate of decomposition of hydrogen peroxide
B14.1a	P	Investigating foodstuffs (1)
B14.1b	ES91	Investigating foodstuffs (2)
B14.1c	T	The detection of ammonia as a product of the decomposition of a protein
B14.1d	ES104	Investigating the properties of ammonia
B14.1e	T	To find out if ammonia will burn in oxygen
B14.2a	T	Can ammonia be broken down by passing it over heated iron?
B14.2b	ES92	To remove hydrogen from ammonia
B14.3	T	Can ammonia be made from nitrogen and hydrogen?
B14.4a	ES93	Preparation of ammonium sulphate
B14.4b	ES105	Preparation of ammonium nitrate
B15.1	ES106	Food tests
B15.2a	T	Detecting the presence of carbon and hydrogen in starch
B15.2b	ES85	Identification of the breakdown products of starch using a chromatographic method of analysis
B15.2c	ES107	To attempt a synthesis of starch from glucose
B15.3	ES86	To break down the glucose molecule by fermentation
B15.4	ES89	The reaction of castor oil with (1) sodium hydroxide; (2) sulphuric acid
B15.6	ES88	Cracking petroleum products
B15.7a	ES108	To compare the properties of polythene with those of wax
B15.7b	T or P	To make a sample of perspex
B15.7c	ES90	To prepare some nylon
B15.7d	ES109	To make a sample of urea-formaldehyde resin
B15.7e	ES110	To compare the melting and burning behaviour of some common polymers
B15.7f	ES111	To compare the properties of some plastics with those of some metals and wood
B15.8a	T	How strong are metals?
B15.8b	T	Comparing the ductility of metals
B15.8c	T	Investigations on the hardness of metals
B15.8d	P	Using polystyrene spheres to investigate how atoms may be arranged in a metal

B15.8e	P	Making a bubble raft
B15.8f	T or P	How do crystals form from molten tin?
B15.8g	T or P	How do crystals form from molten lead?
B15.8h	T	The preparation of a metallic surface to show slip lines
B15.8i	ES56	Growing crystals of metals
B16.1	T	Getting some idea of the size of atoms
B16.2	T or P ES112	Getting an idea of the relative masses of some 'model' atoms
B16.3	ES113	How can elements be classified?
B16.4a	T	To investigate the action of (1) air; (2) chlorine, on heated samples of sodium and potassium
B16.4b	T	To investigate the action of water on samples of sodium and potassium
B16.4c	ES49	What are the reactions of lithium?
B16.5a	T	Properties of the halogens
B16.5b	ES50	Investigating some of the properties of iodine
B16.6a	T	To examine the reaction of magnesium with silicon dioxide (silica)
B16.6b	P	Making a 'crystal garden'
B17.1a	T	Demonstration of molecular diffusion and motion
B17.1b	T	Demonstration of high speed molecular motion
B17.2	ES55	Watching crystals grow
B17.3a	T	Molecular models of liquid bromine and of solid iodine
B17.3b	T or P	Comparing how easily the atoms may be pulled apart in sodium chloride (a giant structure) with how easily the molecules may be pulled apart in iodine (a molecular substance)
B17.4a	T	Comparing how easily the atoms may be pulled apart in graphite (a giant structure) with how easily the molecules may be pulled apart in iodine (a molecular substance)
B17.4b	T T or P ES57	Making various allotropes of sulphur 1. Preparation of rhombic sulphur 2. Preparation of monoclinic sulphur from liquid sulphur
B17.4c	ES58	Observing what happens when sulphur is heated: making plastic sulphur
B18.1a	ES114	Counting the spheres in a model of magnesium oxide
B18.1b	ES115	The principle of counting by weighing
B18.1c	ES116	Using relative atomic masses
B18.2	ES117	Finding the formula of magnesium oxide by experiment

B18.3	ES118	The mole of atoms
B18.4a	ES45	To find the formula of black copper oxide
B18.4b	ES119	The formula of lead bromide
B18.4c	T	Determination of the formula of mercury chloride
B18.4d	T	Determination of the formula of water
B18.4e	T	Determination of the formula of sodium oxide
B18.7a	T	Finding the formula of hydrogen chloride
B18.7b	T	The volume composition of ammonia
B19.1a	T	Some properties of a simple electrical current
B19.1b	P	Some investigations of the properties of an electrolytic cell
B19.1c	T or P	Investigating electrolysis of a molten salt
B19.2a	T	Simple demonstrations of electrostatic phenomena
B19.2b	T	Experiments with an electrometer
B19.2c	T	The effect of placing a charged sphere in an electric field
B19.3a	T	Looking at the migration of ions
B19.3b	ES62	The migration of ions
B19.5a	ES63	How much electricity is needed to deposit 1 mole of copper atoms and 1 mole of silver atoms from solution?
B19.5b	T	A quantitative study of the electrolysis of molten lead bromide
B19.5c	ES64	Is the same quantity of electricity always required to liberate 1 mole of copper atoms?
B19.5d	P	Investigating the electrolysis of solutions of electrolytes
B20.1	T	Preparation of 1.0M solution of sodium chloride
B20.2a	ES65	To verify the composition of the precipitate formed during the reaction between lead nitrate solution and potassium iodide solution
B20.2b	ES66	To verify the composition of the precipitate formed during the reaction between barium chloride and sodium carbonate solutions
B20.2c	ES67	To verify the composition of the precipitate formed during the reaction between barium chloride and potassium chromate solutions
B20.3a	ES68	To determine the volume of 1.0M hydrochloric acid which reacts completely with $10 \, cm^3$ 1.0M sodium carbonate
B20.3b	ES69	To measure the volume of carbon dioxide formed when a known amount of sodium carbonate reacts with dilute hydrochloric acid
B20.3c	ES70	To find whether hydrochloric acid, sodium carbonate, and sodium chloride solutions are electrolytes

B20.3d	ES71	To measure the volume of hydrogen produced when 0.024 g magnesium reacts with dilute hydrochloric acid
B20.3e	T	To verify the equation for the reaction between gaseous ammonia and gaseous hydrogen chloride
B20.4	ES72	An examination of the reaction between iron and copper(II) sulphate solution
B20.5a	T	A study of the reaction between chlorine and iodine
B20.5b	ES76	Two examples of reversible reactions
B20.6a	T	A comparison of the properties of hydrogen chloride in water and in toluene solution
B20.6b	ES80	To find the ratio in which barium and sulphate ions combine
B20.6b	T	Alternative method of finding the ratio in which barium and sulphate ions combine
B21.2a	T or P ES75	The distribution of iodine between two immiscible solvents
B21.2b	ES77	Two other examples of systems in chemical equilibrium
B22.1a	T	Becquerel's chance discovery
B22.1b	ES101	An examination of scintillations
B22.1c	T	Detecting radiation produced by solutions
B22.2a	T	Isolation of protactinium chloride, containing $^{234}_{91}Pa$, and measurement of its change in activity over a period of time
B22.2b	ES120	Solvent extraction, using a non-radioactive mixture
B22.4	T	Investigating the equilibrium between solid lead chloride and its saturated solution using a radioactive tracer technique
B22.5a	ES121	Investigations on chemical systems
B22.5b	ES122	An investigation of the reaction between iodide ions and hydrogen peroxide in solution
B22.5c	P	The decomposition of hydrogen peroxide: a further investigation
B23.1a	T	Examining the boiling of water
B23.1b	T or P ES123	Measuring the energy required to vaporize a given liquid
B23.2		To meaure the energy change which accompanies a chemical reaction; three examples
B23.2a	ES94	To measure the heat of reaction of a neutralization process
B23.2b	ES95	To measure the heat of reaction of an ion combination reaction
B23.2c	ES96	To measure the heat of reaction for a displacement reaction
B23.3	ES97	To measure the heat of combustion of a liquid fuel
B23.4a	T	Transfer of electrons

B23.4b	P	Using chemical sources of electrical energy
B23.4c	ES98	To measure the e.m.f.s of some voltaic cells
B23.4d	ES100	A very simple hydrogen/oxygen fuel cell
B24.1a	ES124	The Group IV elements
B24.1b	ES125	The oxides and hydroxides of Group IV elements
B24.1c	T or P	The chlorides of the Group IV elements
B24.2a	P	The transition metals
B24.2b	ES126	Transition metal compounds

Supplement

S1.2	T	The production of cathode rays and their deflection in an electric field

Experiment sheets

This list gives the number of the *Experiment sheet* in the first column, and the number of the experiment in the *Teachers' guide* to which it corresponds in the second column, followed by the title of the experiment.

ES	Teachers' guide II		
41	A11.1b.3		Are the particles of a gas in motion?
42	A11.2a		Into roughly how many particles might a crystal of potassium permanganate be divided?
43	A11.2c		How small are the particles of oil?
44	A11.4a		To find the formula of magnesium oxide
45	A11.4b	B18.4a	To find the formula of black copper oxide
46	A12.1a	B11.3a	Investigating some properties of 'salt gas'
47	A12.1d	B11.3d	Finding out what causes the steamy fumes when 'salt gas' is exposed to air
48	A12.2b	B11.4b	To find out what other substance is present in 'salt gas'
49	A13.2c	B16.4c	What are the reactions of lithium?
50	A13.3b	B16.5b	Investigating some of the properties of iodine
51	A13.4a		Looking at the properties of some transition metals
52	A13.4b		Looking at the properties of some compounds of certain transition metals
53	A13.5a		To find out if carbon is formed when substances from plants and animals are heated
54	A13.5 b		To confirm the presence of carbon in various substances
55	A14.3a	B17.2	Watching crystals grow
56	A14.5a	B15.8i	Growing crystals of metals
57	A14.7a.3	B17.4b.2	Preparation of monoclinic sulphur from liquid sulphur

58	A14.7	B17.4c	Observing what happens when sulphur is heated: making plastic sulphur
59	A14.9		What volumes are occupied by one mole of molecules (gram-molecule) of various gases?
60	A15.2a		Is energy needed to melt a solid?
61	A15.2b		How much energy is required to vaporize 1 mole of water?
62	A16.3 b	B19.3 b	The migration of ions
63	A16.4b	B19.5a	How much electricity is needed to deposit 1 mole of copper atoms and 1 mole of silver atoms from solution?
64	A16.6	B19.5c	Is the same quantity of electricity always required to liberate 1 mole of copper atoms?
65	A17.2a	B20.2a	To verify the composition of the precipitate formed during the reaction between lead nitrate solution and potassium iodide solution
66	A17.2b	B20.2b	To verify the composition of the precipitate formed during the reaction between barium chloride and sodium carbonate solutions
67	A17.2c.	B20.2c	To verify the composition of the precipitate formed during the reaction between barium chloride and potassium chromate solutions
68	A17.3a	B20.3a	To determine the volume of 1.0M hydrochloric acid which reacts completely with 10.0 cm^3 1.0M sodium carbonate
69	A17.3b	B20.3b	To measure the volume of carbon dioxide formed when a known amount of sodium carbonate reacts with dilute hydrochloric acid
70	A17.3c	B20.3c	To find whether hydrochloric acid, sodium carbonate, and sodium chloride solutions are electrolytes
71	A17.3d	B20.3d	To measure the volume of hydrogen produced when 0.024 g magnesium reacts with dilute hydrochloric acid
72	A17.4	B20.4	An examination of the reaction between iron and copper(II) sulphate solution
73	A18.1b		What is the effect of the concentration of the reactants on the rate of reaction between sodium thiosulphate and hydrochloric acid in solution?
74	A18.2	B13.2	What is the effect of temperature on the rate of reaction?
75	A19.2a	B21.2a	The distribution of iodine between two immiscible solvents
76	A19.2b	B20.5b	Two examples of reversible reactions
77	A19.2c	B21.2b	Some other examples of systems in chemical equilibrium
78	A20.1b	B12.1b	Investigating the acidic properties of some compounds
79	A20.2b		To examine the reaction between dilute sulphuric acid and copper(II) oxide
80	A20.4a	B20.6b	To find the ratio in which barium and sulphate ions combine

81	A20.5a	B12.3a	Preparation of copper sulphate from (1) copper(II) oxide; (2) copper(II) carbonate
82	A20.5b	B12.3b	Preparation of magnesium sulphate from (1) magnesium; (2) magnesium carbonate
83	A20.5c	B12.3c	Preparation of lead chloride from lead nitrate solution
84	A21.1		The detection of carbon and hydrogen in organic chemicals
85	A21.2	B15.2b	Identification of the breakdown products of starch using a chromatographic method of analysis
86	A21.3	B15.3	To break down the glucose molecule by fermentation
87	A21.4		Can ethanol be broken down further?
88	A21.5	B15.6	Cracking petroleum products
89	A21.6	B15.4	The reaction of castor oil with (1) sodium hydroxide; (2) sulphuric acid
90	A21.8	B15.7c	To prepare some nylon
91	A22.1b	B14.1b	Investigating food stuffs
92	A22.2	B14.2b	To remove hydrogen from ammonia
93	A22.5	B14.4a	Preparation of ammonium sulphate
94	A23.1b	B23.2a	To measure the heat of reaction of a neutralization reaction
95	A23.1c	B23.2b	To measure the heat of reaction of an ion combination reaction
96	A23.1d	B23.2c	To measure the heat of reaction for a displacement reaction
97	A23.3a	B23.3	To measure the heat of combustion of a liquid fuel
98	A23.4b	B23.4c	To measure the e.m.f.s. of some voltaic cells
99	A23.4c		To determine the maximum amount of energy available as work from a voltaic cell
100	A23.5	B23.4d	A very simple hydrogen-oxygen fuel cell
101	A24.1b	B22.1b	An examination of scintillations
102		B12.2	The preparation and properties of some bases
103		B13.1a	What is the effect of the concentration of hydrochloric acid on the rate of reaction between magnesium ribbon and dilute hydrochloric acid?
104		B14.1d	Investigating the properties of ammonia
105		B14.4b	Preparation of ammonium nitrate
106		B15.1	Food tests
107		B15.2c	To attempt a synthesis of starch from glucose
108		B15.7a	To compare the properties of polythene with those of wax
109		B15.7d	To make a sample of urea-formaldehyde resin

Sources of supply of 16 mm films

The use of films in the planned presentation of the course requires considerable forethought. Often such films will need to be ordered several months in advance and, where appropriate, the video-recording of television programmes must, of course, be done at the time of transmission. Details of the supply of films relating to Stage I appear in *Teachers' guide I* and some of this information could apply to Stage II.

Complete lists of films for hire or loan are relatively easy to obtain from the suppliers listed below. In addition, the Chemical Society publishes an index of films and filmstrips on chemistry and related topics which is revised at regular intervals.

16 mm films. Catalogues and distributors
Alcan Film Library, 303 Finchley Road, London NW3
BBC Enterprises Film Hire, 25 The Burroughs, Hendon, London NW4 4AT
British Gas Film Library, 16 Paxton Place, London SE27 9SS
British Oxygen Company Ltd, Film Library, 42 Upper Richmond Road West, London SW14 8DD
British Petroleum Film Library, 15 Beaconsfield Road, London NW10 2LE
British Steel Corporation, Films Section, 151 Gower Street, London WC1E 6BB (catalogue only)
British Transport Films, Melbury House, Melbury Terrace, Central Film Library, Government Building, Bromyard Avenue, London W3 7JB
The Chemical Society, The Publications Sales Officer, Blackhorse Road, Letchworth, Hertfordshire SG6 1HN (catalogue only)
Concord Films Council Ltd, Nacton, Ipswich IP10 0JZ
EFVA, the National Audio-Visual Aids Library, Paxton Place, Gipsy Road, London SE27 9SR
Encyclopaedia Britannica Films, Fergus Davidson Associates Ltd, 22 South Audley Street, London W1Y 6ES
Eothen Films Ltd, EMI Film Studios, Shenley Road, Boreham Wood, Hertfordshire WD6 1JG
Esso Films, Golden Films Ltd, Stewart House, 23 Francis Road, Windsor, Berkshire
Gateway Educational Films, St. Lawrence House, 29/31 Broad Street, Bristol BS1 2HF
Guild Sound and Vision Ltd, Woodston House, Oundle Road, Peterborough PE2 9PZ (distributors for the Open University Film Library, and other sponsored films)
ICI Film Library, Thames House North, Millbank, London SW1P 4QG

The Institution of Metallurgists, Northway House (8th Floor), High Road, Whetstone, London N20 9LW (catalogue only)

Longman Group Ltd, Longman House, Burnt Mill, Harlow, Essex CM20 2JE (all Nuffield visual aid materials)

National Coal Board, Film Library, Hobart House, London SW1X 7AE

Rank Film Library, PO Box 20, Great West Road, Brentford, Middlesex TW8 9HR

Shell Film Library, 25 The Burroughs, Hendon, London NW4 4AT

Unilever Film Library available from EFVA (see above)

United Kingdom Atomic Energy Authority, Film Library, 11 Charles II Street, London SW1Y 4QP

Viscom Audio-Visual Library, 16 Paxton Place, London SE27 9SS

Note. The Petroleum Films Bureau no longer exists. Films formerly distributed by the Bureau can be obtained from the film libraries of the various oil companies.

Appendix 3

School and public examinations

The present public examination arrangements

A discussion of the impact of examinations on learning and teaching, and on matters of direct concern to Stage I of the Nuffield Chemistry proposals, appears in *Teachers' guide I*. This appendix relates primarily to Stage II of the proposals and to public examinations designed to encourage and assess the response made by pupils in schools and colleges to the course materials.

The examinations are at GCE Ordinary level and are conducted by the University of London in association with the other GCE examining boards. Schools wishing to enter candidates for Nuffield Chemistry (full subject), or for Nuffield Chemistry of Physics-with-Chemistry, at GCE Ordinary level for the first time should inform the GCE examining board to which the school (college) is attached of this intention at least one year before the actual examination.

The Nuffield Chemistry examinations are designed to encourage and assess those abilities expressly associated with the Nuffield proposals. These were first published in the *Introduction and Guide* (1966) and are reproduced below for the convenience of teachers.

1. Facility in recalling information and experience.
2. Skill in handling materials, manipulating apparatus, carrying out instructions for experiments, and making accurate observations.
3. Skill in devising an appropriate scheme and apparatus for solving a practical problem.
4. Skill in handling and classifying given information (including graphical representations and quantitative results).
5. Ability to interpret information with evidence of judgement and assessment.
6. Ability to apply previous understanding to new situations and to show creative thought.
7. Competence in reporting, commenting on, and discussing matters of simple chemical interest.
8. Awareness of the place of chemistry amongst other school subjects and in the world at large.

The manner in which these abilities are assessed has evolved gradually to meet the needs of both teachers and pupils. Initially, the Nuffield Chemistry examinations were used by only a small sample of candidates from a few schools. During the initial trials of course materials, the progress of these candidates could be monitored very closely. The techniques used for these trials were developed further to suit the assessment of much larger groups of candidates from a wider range of schools. Table A3.1 records the development of Nuffield Chemistry examinations in schools and colleges entering candidates for Chemistry (full subject) at GCE O-level. Table A3.2 relates to the use of Nuffield Chemistry examinations by candidates entering for the Physics-with-Chemistry examination.

Year	1965	1966	1967	1968	1969	1970
Entries	167	677	1050	1860	3769	6572
Centres	6	20	32	55	105	202

Year	1971	1972	1973	1974	1975	1976
Entries	10 179	14 218	15 705	17 142	18 130	19 332
Centres	297	416	452	474	497	506

Table A3.1
Centres and candidates for GCE O-level Nuffield Chemistry: 1965–1976

Year	1965	1966	1967	1968	1969	1970
Entries	58	280	343	317	438	705
Centres	3	12	16	12	20	32

Year	1971	1972	1973	1974	1975	1976
Entries	895	1245	1393	1286	1203	982
Centres	41	54	62	52	50	41

Table A3.2
Centres and candidates for GCE O-level Nuffield Chemistry of Physics-with-Chemistry: 1965–1976

The schemes of examination

As indicated by the tables, two schemes of examination make use of the course materials. In the discussion which follows, the schemes of examination apply to those examinations up to and including June 1975. More recent information is provided in the current issue of the leaflet issued by the University of London in association with the other GCE examining boards.

The *Nuffield Chemistry* (full subject) examination at GCE O-level consists of two written papers: Chemistry 1 ($1\frac{1}{4}$ hours) and Chemistry 2 (2 hours), both of which must be taken. Both papers are based on Stages I and II of the published materials for the course and, in addition, Chemistry 2 contains questions based upon Stage III option materials.

Paper 1 consists entirely of multiple-choice questions. Each question has five alternative responses and the candidate is required to choose one response for each question. It has been usual to set 70 questions although a reduction to 60 questions is planned for the examination in and after June 1979.

Paper 2 consists of two sections, A and B. Section A consists of 5 questions and candidates are required to attempt 3. For examinations in and after June 1979, it is planned to require candidates to answer all questions in this section (about 5) and it should be noted that such questions will be similar to, but shorter than, those set at present. Each of these questions is divided into several parts and each part requires a brief answer. A reduced version of the list of abilities given at the beginning of this Appendix is used to characterize these structured questions. An inspection of recent examination papers shows that the examiners required candidates to show abilities under five main headings:

1. Ability to recall knowledge.
2. Ability to plan experiments and design apparatus.
3. Ability to handle and classify information:
(*a*) the qualitative use of data;
(*b*) in calculations;
(*c*) in graphical work;
(*d*) in writing chemical equations.
4. Ability to suggest explanations or to make predictions about a familiar situation.
5. Ability to suggest explanations or to make predictions about an unfamiliar situation.

Questions on Stage III of the course materials appear in section B of paper 2. This form of examination was used for the first time in 1968. Candidates are required to offer two Options, such that *either* both Options can be examined by this paper *or* one may be so examined and one is examined by a school-based assessment procedure determined by the Examining Board. It should be noted that in and after June 1979, the questions set will be based *only* on the revised materials.

The essay of section B is designed to encourage candidates to demonstrate their ability to report on and discuss matters of chemical interest in a manner which cannot readily be tested by any other means in a limited time. There can be no written questions for 'school assessed' options and so candidates who offer a school assessed option are required to leave the examination room 30 minutes before the end of the examination.

Paper 1 carries 40%, Paper 2 (section A) carries 40%, and Paper 2 (section B) carries 20%, of the total mark for the examination.

The Nuffield Chemistry component of the Physics-with-Chemistry examination at GCE O-level also consists of two written papers: Chemistry 1 ($\frac{3}{4}$ hour) and Chemistry 2 (1 hour), both of which must be taken. Both papers are based on Stages I and II of the published course materials.

Paper 1 consists entirely of multiple-choice questions. Each question has five alternative responses and the candidate is required to choose one response for each question. It is usual to expect candidates to attempt 40 questions.

Paper 2 consists of five questions comparable to section A of Paper 2 of Nuffield Chemistry (full subject) examination. Candidates are normally required to attempt 3 questions, although in and after June 1979, it is planned to require candidates to attempt all questions and to set 4 questions of the projected newer type mentioned earlier.

In this scheme Papers 1 and 2 carry equal marks. The mark which a candidate obtains for the Nuffield Chemistry component of the examination is sent to the GCE examining board through which the candidate has made his entry and is then combined with the marks awarded for the 'normal' or 'Nuffield' physics component. The result is used to award a grade in Physics-with-Chemistry.

Multiple-choice questions in Nuffield Chemistry examinations

The fixed-response questions used in Paper 1 for either examination are intended to complement all other measures of ability used in the Nuffield Chemistry examinations. Multiple-choice questions are a simple means of ensuring an even assessment of the entire course. Thus, for some years, Paper 1 has been constructed according to a pre-determined specification. A two-dimensional grid is used, setting abilities against activities encountered during the course. For this purpose, the abilities and activities considered are:

Abilities
1. Knowledge
2. Comprehension
3. Application
4. Analysis/evaluation

Activities
1. Composition and changes in materials
2. Practical techniques
3. Patterns in the behaviour of materials
4. Essential measurements
5. Concepts concerning the particulate and electrical nature of matter.

In addition, another two-dimensional grid is used which sets Topics against Abilities.

Figure A3.1 shows the Ability/Activity specification for the multiple-choice paper for Nuffield Chemistry. The figures show the desired weighting for a 70-item paper: the actual weighting is not shown.

Activities

$d = 18$
$a =$
$d = 28$
$a =$
$d = 14$
$a =$
$d = 10$
$a =$

$d = 21$ $d = 7$ $d = 12$ $d = 19$ $d = 11$
$a =$ $\quad a =$ $\quad a =$ $\quad a =$ $\quad a =$

$d =$ desired weighting
$a =$ actual weighting

Figure A3.1
Ability/Activity specification for the multiple-choice paper

Appendix 3 School and public examinations

The term 'activities' is used to cover broad groupings of the subject matter of the scheme, as indicated above, and the term 'abilities' is used in a technical sense, based on Bloom, B. S. (ed.) (1965) *Taxonomy of educational objectives*. Longman.

1. *Knowledge.* The ability to recall facts, nomenclature, classification, practical techniques, laws and theories.
2. *Comprehension.* The ability to calculate, to translate data from one form to another (verbal into mathematical or graphical), to interpret and deduce the significance of data, and to solve problems in which the mode of solution of the problem should be familiar.
3. *Application.* The ability to apply knowledge, experience, and skill to new situations presented in a novel manner. The method of solution will not be implied in the question.
4. *Analysis/evaluation.* The ability to analyse given information into its various parts and, as a result, to make a judgement as to its value.

The decision to use multiple-choice objective questions as part of the scheme of examination was based on several considerations. Thus:
1. It is possible for the candidate to answer a large number of questions in a relatively short time; this permits a wide and systematic coverage of both subject matter and abilities.
2. Multiple-choice questions may be written to emphasize one particular ability, for example, comprehension or analysis. It is possible for the candidate to answer these questions without depending too much on other skills, such as fluency in writing. The inclusion of both multiple-choice questions and others seemed to the examiners to give a more balanced examination; it ensures both a wide coverage of the scheme and examining in depth.
3. Multiple-choice questions (as well as other forms of question) can be pre-tested on a group of students similar to the one which will take the examination. The questions may then be amended to ensure that they are as free from ambiguity as possible, that they are of appropriate levels of difficulty for the candidates, and that the questions discriminate well between candidates.
4. Multiple-choice questions can reflect a situation in class. After studying something in the classroom or laboratory, the pupils may be asked to provide an explanation. Several possible explanations may be suggested, each reflecting a different degree of understanding. Similarly, one can devise a multiple-choice question which offers a set of possible answers in which only one is a complete explanation. These written responses enable a candidate to consider a question which deals with a complex or unfamiliar situation more carefully than in a normal teaching situation. The careful use of multiple-choice questions, both in teaching and in examinations, can guide pupils towards a degree of understanding which might not have been possible otherwise.
5. A multiple-choice paper can be constructed to a pre-determined specification in which the abilities and activities receive an agreed weighting. This technique can be applied to structured questions

where there is no choice. However, the technique is not so readily or precisely applied to these other types of question.

From this it follows that some assurance can be given that Paper 1 is measuring systematically a wide sample of the educational objectives of the Nuffield Scheme.

It should be noted from past examination papers that many questions contain material which is not in the published Nuffield course materials. It is an important objective of the scheme that candidates should be able to demonstrate their ability to transfer their expertise and experiences to situations which are similar, but not identical, to those used in the classroom. The freedom of the examiners to use additional material in this way, with proper safeguards, must be retained. Such additional material should *not* become the focus of classroom activities and lead to general adoption and consequently to overburdened courses. Material should not find its way into common use simply because it formed part of a previous examination!

Setting examination questions

The art of teaching is closely related to the art of asking questions. If a new approach to the teaching of chemistry is to be widely adopted, teachers must devise questions which are in tune with the new approach. This is not easy: each examination question requires much thought, discussion, and trial before it can be said to test specific abilities which its author claims it can test. Teachers will find that devising such questions can be useful and stimulating. Setting questions to test specific abilities nearly always takes longer than setting the conventional essay question, although the required answer to such a question will be shorter. Also they require more time to read and comprehend. It follows that if there must be a time limit for examinations using questions of this kind, it should be a generous one. Many such questions will need to be broken down into several parts. If a pupil is asked to make a judgement or express an opinion, the question must be so devised that he is doing so on one or two specific matters unclouded by side issues or irrelevant information.

Above all, questions should be set because they reflect the spirit of the course and not because they are easy to set and easy to mark!

Publications relating to Nuffield Chemistry examinations

During the period of development of the Nuffield Chemistry examinations, the University of London was undertaking a special review of the techniques used in the GCE examinations administered by its University Entrance and School Examinations Council. It seemed appropriate for the University to provide the Chemistry papers for the Nuffield Chemistry project while other examining bodies collaborated with other Nuffield pilot schemes in Physics and Biology. In many ways, these Chemistry papers foreshadowed the use of fresh techniques in many other subjects. In common with other aspects of the Nuffield Science Teaching Project, the examinations were arranged across all of the GCE examination boards.

Initially, those boards associated with the University of London in this chemistry project were the Welsh Joint Education Committee and the Northern Ireland Examination Committee.

Copies of multiple choice question papers used in either of the Nuffield Chemistry examinations may now be obtained from the University of London Publications Office, 52 Gordon Square, London, WCIH OPJ. Copies of other examination papers which have been used in this examination can also be purchased from this same office.

Changes in the examination following the revision

The structure and design of the examination and any changes in the examination arrangements will be implemented by the examining board following consultation with teachers on its committees.

The revision of the Sample Scheme has not changed the aims of the course, but greater prominence has been given to the applied, social, and historical aspects of chemistry. In the future, examiners will expect candidates to give evidence of having read—with understanding—about these aspects of the subject. Suitable reading material is given in Part 2 of the *Handbook for pupils* and in *Chemists in the world*. However, pupils will not be expected to recall detailed factual information from these books.

Teachers should note that the aims and objectives of Stage II are defined by the *Teachers' guide II*, and that the compulsory parts of the public examinations will be based on the content common to the teaching schemes. Part I of the *Handbook for pupils* contains ideas and information required for Stage III and more advanced work. This extension material is not examinable as part of Stage II.

INDEX